MW01235496

VOLUME I

VADOSE

ZONE

SCIENCE AND TECHNOLOGY SOLUTIONS

EDITED BY

Brian B. Looney, Ph.D.

AND

Ronald W. Falta, Ph.D.

Columbus • Richland

Library of Congress Cataloging-in-Publication Data

Vadose zone science and technology solutions / edited by Brian B. Looney and Ronald W. Falta

p. cm.

Includes bibliographical references and index.

ISBN 1-57477-085-3 (alk. paper)

1. Soil pollution. 2. Zone of aeration. 3. Soil remediation. I. Looney, Brian B., 1956– . II. Falta, Ronald W., 1960–

TD878.V76 2000
658.5'5–dc21

00-023080

Printed in the United States of America

Battelle Press
505 King Avenue
Columbus, OH 43201-2693, USA
614-424-6393 or 1-800-451-3543
Fax: 614-424-3819
E-mail: press@battelle.org
Website: www.battelle.org/bookstore

Contents

VOLUME I

VOLUME II

Preface

A dominant theme in this book is that practical, science-based solutions to vadose zone problems can be simultaneously efficient, clever, and—importantly—technically justifiable. We have attempted to provide information to support people with diverse backgrounds who serve various roles in environmental programs. Our goal is to help each individual develop the best possible conceptual model and technical understanding of the scientific concepts so that no one will have to view the vadose zone in "black box" form. Everyone can contribute to solving the remaining scientific challenges. We admit, therefore, to some unevenness in the level of this book. As discussed below, some portions of the book provide a high-level understanding of vadose zone processes and complexities; some portions provide specific mathematical, geological, and biochemical information; and other portions address practical engineering issues. We encourage readers to sample the diversity represented in the book—we have tried to take apart the black box and put it back together for your scrutiny. In the process, we have identified what we believe are the most significant scientific challenges and data gaps. These are highlighted to stimulate future scientific progress.

This book is unique in several ways. For each topic considered, we offer a certain amount of general information, a certain amount of theory and mathematics, and a significant number of "real-world" examples. The contributors provided brief, compelling case studies that exemplified important vadose zone concepts or challenges. The content of the book was the result of three national workshops, supplemented by countless hours of post-workshop writing and coordination by dozens of contributors. Importantly, the book is not a collated proceedings, but rather an eclectic and comprehensive presentation that will be useful to advanced undergraduate and graduate students in science and engineering, working professionals, researchers, managers, and others. Our challenge was to find the right balance among the various ingredients to generate a product that would be useful to this broad audience.

Vadose Zone Science and Technology Solutions is composed of 10 chapters grouped into five general categories. The overall structure correlates roughly to workshop topics, and the chapter titles include "standard" vadose zone topics as well as a few nonstandard topics.

- *Introducing Vadose Concepts*

  Chapter 1, The Vadose Zone: What It Is, Why It Matters, and How It Works

  Chapter 2, Managing an Effective Vadose Zone Program

- *Characterizing and Monitoring Vadose Systems*

  Chapter 3, Vadose Zone Characterization and Monitoring—Current Technologies, Applications, and Future Developments

  Chapter 4, Performance Monitoring

- *Understanding and Modeling Vadose Systems*

  Chapter 5, Flow and Transport Modeling of the Vadose Zone

  Chapter 6, Biogeochemical Considerations and Complexities

- *Solving the Problem of Vadose Contamination*

  Chapter 7, Remediation of Organic Chemicals in the Vadose Zone

  Chapter 8, Remediation of Inorganic Contamination in the Vadose Zone

  Chapter 9, Barriers and Containment Methods

- *Identifying Scientific Challenges and Opportunities*

  Chapter 10, Future Science and Technology Focus

In the end, the presentation is more detailed than the available introductory text and less detailed than books that are dedicated to only a single aspect of vadose zone science (for example, characterization). We employed summary tables and similar approaches to make the book as user-friendly as possible. The practical experiences and observations within the case studies are particularly valuable to illuminate the ideas in the text. The case studies provide a fascinating means to sample the depth and breadth of vadose zone science. We solicited case studies that show both failure and success and that suggest approaches to overcome scientific uncertainties. The entire book and all of the case studies are provided in electronic format on the enclosed CD-ROM.

We are pleased with the effort and the final product. It is, first and foremost, the result of the work and contributions of the scientists who participated in the workshops and the individuals who served as lead authors for various chapters. We thank these individuals for their contributions. It was exciting and humbling to work with individuals who embody the finest work on this topic. Workshop attendees and book contributors included Paul Witherspoon, Lorne Everett, Rien Van Genuchten, Glendon Gee, and many others too numerous to list here. We would also like to express our appreciation to agencies and institutions who participated in the effort: U.S. Department of Agriculture, U.S. Environmental Protection Agency, U.S. Geological Survey, various universities, industry, and others. We offer special thanks to the EPA Environmental Measurement Systems Laboratory and the Lawrence Berkeley Laboratory for hosting workshops. Primary credit for *Vadose Zone Science and Technology Solutions* goes to the U.S. Department of Energy and to a key group of dedicated individuals. Ernie Moniz and the Office of Science and Technology conceived of and supported the project. Tom French, of the Savannah River Technology Center, was the driver who kept the project on course. The effort would have been impossible without the support of Kim Sharpe, Joann Hafera and her colleagues at InnerLink, Michelle Silbernagel and her colleagues at EnviroIssues, Roy Gephart, and others.

As editors, we wanted to produce a useful book. We recognized at the outset that there are many quality vadose zone references on the market and we wanted to extend this work rather than repeat it. We believe that the unique presentation will supplement the existing vadose zone literature and provide a helpful source of information and perspective to those interested in this topic.

BRIAN B. LOONEY, *Savannah River Technology Center*
RONALD W. FALTA, *Clemson University*

Forewords

– I –

O.E. Meinser, the father of hydrology, referred to the vadose zone as *no-man's land*. As the terms "zone of aeration" and "unsaturated zone" have given way to the more logical American Society for Testing and Materials (ASTM) use of the term "vadose zone," scientists of all disciplines have begun to consider in great detail the last unexplored frontier of hydrology.

The vadose zone is that part of the lithology which extends from the earth's surface to the water table and includes the soil zone, the intermediate vadose zone, and the capillary fringe. The name vadose is derived from the Latin noun *vadosus* which translates to "shallow." The study of the vadose zone, with a few noted exceptions, has been ignored in most university programs throughout America.

This state-of-the-knowledge vadose zone book was developed after the nation became aware of the failure of existing approaches to adequately characterize, monitor, remediate, and predict solute fate and transport behavior in the vadose zone. It is perhaps even more poignant to recognize that the Hanford Reservation, with its leaked high-level radioactive waste, is the setting for the largest perceived failure of vadose zone modeling techniques to predict contaminant transport. Recent events have acutely identified weaknesses relative to vadose characterization and have brought to light national regulatory weaknesses that exist within the Comprehensive Environmental Response

Compensation and Liability Act and the Resource Conservation and Recovery Act (CERCLA and RCRA). In effect, since vadose zone regulatory requirements generally do not exist within federal mandates, government and industrial entities neglect or place a very low priority on characterization and monitoring in the vadose zone. This illogical national regulatory climate, which allows contamination to move through the vadose zone and only begins regulation at the water table, is a flawed philosophical approach. This philosophy is akin to monitoring a patient to indicate when the patient died!

The misleading concepts of field capacity and specific retention, with its attendant term, "specific yield," have masked the behavior of water transport in soils as we believe it occurs. The notions that water can move in conditions drier than field capacity, and that water can move through soils without changing their water content, mean that a change in capillary pressure at a particular soil depth is not sufficient to determine the direction and magnitude of the rate at which water moves through the vadose zone.

As pointed out during a session of the Dr. Chester C. Memorial Lecture, which is held annually at the University of Arizona, the challenging issues in the vadose zone are related to the following: (1) the hysteretic nature of soil water content relative to its energy status; (2) the temperature dependencies of soil water properties; and (3) the transport of water in both the vapor and liquid phases under isothermal and non-isothermal conditions.

The major shortcoming of the science of vadose zone hydrology is our inability to accurately predict contaminant transport. As noted by the State of California, the modeling assumptions of instantaneous adsorption, steady-state flow, and hysteresis, the magnitude of anisotropy in hydraulic conductivity, and the use of adsorption isotherms must be fully appreciated. Serious questions exist about using deterministic coefficients and stochastic coefficients in the convection-dispersion equation. Advantages have been seen, however, with the use of transfer function models.

The fundamental problem originates with the input parameters, which are derived from site characterization. These input parameters appear to have enormous scale issues relative to their representative nature. In the late 1980s, scale issues related to three fundamental length scales were presented: (1) laboratory, local, and regional scales used to characterize flow and transport phenomena; (2) the correlation scale associated with heterogeneity; and (3) the scale of measurement and sampling. These issues have resulted in the statement by the California Water Resources Control Board that "because the reliability of models for contaminant transport has not been established even for site-specific conditions, it appears that direct monitoring of the constituents in the vadose zone remains a necessity into the foreseeable future."

As a national strategy, the phrase "into the foreseeable future" simply is not acceptable for protecting human health and the environment unless direct monitoring is implemented across America. This book therefore focuses on the weakness, failures, and lack of understanding within the vadose zone that have resulted in our inability to accurately predict contaminant transport. Logic says that we can only begin to fix the problems when we understand the weakness of our science and the gaps in our knowledge. Based upon these perceived gaps, we can begin to focus needed resources on moving the state of our knowledge forward. Since it is impossible to prove a negative, regardless of the sophistication of the science, it may turn out that direct monitoring in the vadose zone is our best defense in protecting the groundwater of the United States.

LORNE G. EVERETT, PH.D., D SC.
Chief Scientist and Senior Vice President, The IT Group
Director, Vadose Zone Monitoring Laboratory, UCSB
Member, Russian Academy of Sciences
Chancellor, Lakehead University, Thunder Bay, Ontario, Canada

– II –

The need to understand and manage the vadose zone has been growing for the past decade or more because the proper management of this zone is a key factor in improving and protecting groundwater supplies. Surface and groundwater are linked by the degree of communication that exists in the unsaturated zone. This communication has not always been recognized, especially as it relates to the disposal of liquid and solid wastes in near-surface repositories that were not carefully designed. The resulting contamination has involved the migration of fertilizers and pesticides from agricultural and domestic sites, solvents and toxic substances from industrial plants, and many other chemicals from the surface into the unsaturated zone. The subsequent pollution of the underlying groundwaters has led to legislative measures by state and federal agencies in an effort to control or regulate these releases and identify local environmental conditions that mitigate against transport.

These difficulties have prompted a substantial effort to investigate the technical problems involved in achieving effective measures of remediation. The important problem of characterizing and modeling flow and transport in the vadose zone has received much attention because of the need to simulate the behavior of rock systems under ambient conditions as well as when these systems are perturbed. One of the critical parts of this process is to predict the response of the contaminated system to some proposed program of remediation. The accuracy of this prediction depends on having reliable field characterization of the controlling parameters of the system. An important use of numerical simulators is to design and analyze laboratory and field tests that can produce reliable data for evaluating these parameters.

The unsaturated, or vadose, zone involves many aspects of hydrology: infiltration, evaporation, groundwater recharge, soil moisture storage, and soil erosion. The traditional approaches to these subjects have generally assumed that the soil layers under consideration could be treated like porous media that is reasonably homogeneous.

However, as a result of the various geologic processes that lead to the formation of soils and sediments, there are heterogeneities in these materials over a wide range of length scales. The invasion of the unsaturated zone by root systems and burrowing animals adds one level of micropores, and the construction activities of man add another level of micropores. Discontinuities in the rock systems, such as fractures and faults, introduce another source of heterogeneity in the flow systems. This is especially important for locations in arid climates, where the water table may be hundreds of meters below the surface. Rock systems in these locations can have a complicated geology with very heterogeneous conditions for flow and transport.

These heterogeneities in the flow paths of the vadose zone are a critical feature because they can lead to the development of preferential flow. Instead of a relatively uniform movement of a wetting front moving through the rock system, a preferential movement of fingers that are spatially non-uniform in length and width can occur. These are caused by isolated zones of higher liquid saturation moving vertically downward, under the influence of gravity, more rapidly than in adjacent zones of lower saturation. Laboratory experiments and field observations have revealed the reality of these non-uniform flow patterns.

In the case of undisturbed fractured rock, there will often be several sets of fractures, and usually at least one set has an orientation that reflects the tectonic history of the area and a dip that is near vertical. The other sets can have different orientations with dips that range down to sub-horizontal. Preferential flow in these fractures will be controlled by the heterogeneity in the aperture distributions within the fractures and the degree of verticality within each different set. The complex flow patterns that develop in such fractures have been demonstrated in modeling studies as well as in a number of laboratory and field investigations.

One of the fundamental complications when water is flowing through the vadose zone is that the capillary pressures and relative permeabilities are different depending on whether the porous medium is undergoing wetting or drainage. The variation of these parameters with a decrease in water saturation will not be the same as when the saturation increases. A hysteresis occurs because of the variability in pore geometry and differences in contact angles, depending on whether water is advancing or draining and on any trapped air that may be present. The situation may also be complicated by the point at which a change in saturation occurs. In other words, the history of how the saturations have either increased or decreased can impact the hysteretic effects.

The flow of gases in the vadose zone is another complicating factor that must be considered. The density of the gas may vary, depending on its composition and/or temperature, and the migration will be caused by the effects of gravity. For example, gases from volatile organic chemicals can be lighter or heavier than soil air. Another factor that leads to gas flow is the pneumatic effect from changes in barometric pressure. These effects cause the soil to "breathe" with atmospheric air migrating into the porous media or soil air migrating out. The infiltration of rainfall can also cause a pressure buildup that produces migration of soil air and leads to a certain amount of residual, or trapped, air.

The effects of heat in the vadose zone in shallow systems near the surface are usually minor because the temperature gradients, mostly due to seasonal changes and diurnal variations, do not extend to any significant distance below the surface. Where there are sources of heat in shallow systems, such as microbial activity in landfills or steam injection for NAPL remediation, the temperature gradients can be sufficient to produce significant effects from heat flow through the matrix and fluid phases.

The amount of heat released in the vadose zone by high-level radioactive waste can create a much more complicated situation. For example, at the Yucca Mountain Project in Nevada, the proposed waste repository will be located in an unsaturated, fractured tuff at a depth of about 300 m. This waste is spent fuel that releases significant amounts of heat over extended periods of time. As this heat is conducted into the rock mass, the cumulative effects will increase rock temperatures to the boiling point, and the subsequent generation of steam will produce a two-phase flow system whose migration and thermal effects are difficult to analyze.

Additional factors to consider in understanding the vadose zone are the geochemical and microbiological processes that occur in the unsaturated environment. In a landfill, the environment can involve geochemical processes (such as adsorption, cation exchange, precipitation) as well as microbiological activity (biodegradation, hydrolysis, transformation) on the mixtures of contaminants and assorted materials that are usually present. One needs to understand how these processes affect the partitioning of contaminants between the solid and liquid phases to describe the migration of contaminants through the system. Of course, an appropriate characterization of the environment is needed to determine the extent and rate of the controlling processes. The transport of the various fluids within the zones of activity in a landfill often controls the magnitude and extent of reactions.

An entirely different class of problems is encountered when the vadose zone involves an extensive mass of fractured rock, as is the case in the Yucca Mountain Project mentioned previously. The effects of the thermal field generated by the waste heat require the development of a thermo-hydrological-chemical model. This model is needed to simulate the complex coupling of rate-limited reactions of minerals, gases, and water that develops under the influence of the thermal perturbation and lasts for many thousands of years.

Modeling fluid migration through fractured rock in the vadose zone raises a perplexing problem—numerically representing the fracture/matrix interaction. When fluids are flowing in fractured rocks, whether they are saturated or unsaturated, one needs to know how the liquid component will be divided between fractures and matrix. This is a crucial problem in modeling transport because retardation depends on the magnitude of this separation of the liquids between fractures and matrix. Investigation of this problem has been limited due to the difficulty in obtaining direct measurements of the fracture and matrix components of flow *in situ*. The current understanding of fracture/matrix flow is based primarily on indirect evidence from field studies and numerical modeling investigations. The problem is complicated further by the considerable uncertainties that are associated with fracture and matrix properties, their spatial distributions, and the mechanisms governing interactions.

Paul A. Witherspoon
University of California, Berkeley and
Lawrence Berkeley National Laboratory, Berkeley, CA

Acknowledgements

GENERAL

The "Vadose Book Project Team" would like to acknowledge those who played a key role in the development and publication of this work. The Department of Energy (DOE) Office of Science and Technology sponsored this effort. We appreciate the guidance and support of C. L. Huntoon, E. J. Moniz, G. G. Boyd and G. Chamberlain, Jr. of DOE Headquarters and J. A. Wright, Jr. of the Subsurface Contaminants Focus Area. We thank all of the workshop participants for their time and thoughtful contributions. We acknowledge the participation and collaboration of other federal agencies, notably, the U.S. Department of Agriculture (USDA), the U.S. Geological Survey (USGS), the U.S. Department of Defense (DOD) and the U. S. Environmental Protection Agency (EPA). Special thanks to the EPA National Exposure Research Laboratory in Las Vegas, NV and to the Lawrence Berkeley National Laboratory (LBNL) for hosting workshops.

We would like to extend our specific acknowledgement to the following people and organizations.

- The Book Project Steering Committee:

Skip Chamberlain	DOE, EM-50
Mary Harmon	DOE, EM-40
Jim Wright	DOE, Subsurface Contaminants Focus Area
Lorne Everett	The IT Group
Clay Nichols	DOE, Idaho

Rich Holten	DOE, Richland
Jim Hanson	DOE, Richland
Kevin Leary	DOE, Nevada
Jack Corey	Savannah River Technology Center (SRTC)
Roy Gephart	Pacific Northwest National Lab (PNNL)
Bill Isherwood	Lawrence Livermore National Lab (LLNL)
John Koutsandreas	Florida State University
Joann Hafera	InnerLink Publication Services
Michelle Silbernagel	EnviroIssues
Joette Sonnenberg	SRTC

- The lead authors for the eight core chapters of the book:

Marilyn Quadrel	PNNL
Boris Faybishenko	LBNL
Eric Lindgren	Sandia National Lab (SNL)
Bo Bodvarsson	LBNL
Cal Ainsworth	PNNL
Larry Murdoch	Clemson University
Jim Phelan	SNL
David Daniel	University of Illinois-Urbana/Champaign

- All of the authors of the case studies.

- Lorne Everett, The IT Group; Dr. Fred Molz, Clemson University; and Dr. Tetsu Tokunaga, LBNL, for their expert and timely peer review of the book.

- Joann Hafera and the staff at InnerLink for the skilled assistance they have provided in making this book such a success.

- Louise Dressen and the staff at EnviroIssues for the careful planning that enabled the success of the workshop process.

- Joe Sheldrick of Battelle Press and Susan Vianna of Fishergate for their expert and patient support of the book publication process.

- Kim Sharpe, SRTC, for expertly coordinating all of the book activities.

CHAPTERS 1 AND 10

Peer Review of Chapters 1 and 10 were performed by all of the chapter lead authors, independent external reviewers, and several members of the steering committee. We appreciate the efforts of Roy Gephart in coordinating the peer review process. The writing of chapters was coordinated through SRTC. SRTC is a contractor of the U.S. Government that performs work under Contract No. DE-AC09-96SR18500.

CHAPTER 2

This chapter benefited from the contributions of a diverse set of colleagues with extensive personal and intellectual investment in the Department of Energy's cleanup program. Special thanks to Roy Gephart, Amorett Bunn, Gariann Gelston, Jean Shorett, Kristi Branch, and Judith Bradbury (PNNL), Dave Rice (LBNL), Clay Nichols (DOE-ID), Barbara Harper (Yakama Indian Nation), Stuart Harris (Umatilla), Moses Jarayssi (Bechtel Hanford Incorporated), Mike Sully, and Tom Woods. These individuals, as well as the workshop participants, provided substantive contributions and overall perspective that helped select the focus of this discussion. PNNL coordinated the writing of this chapter. PNNL is operated by Battelle for the U.S. Government and performs work under Contract No. DE-AC06-76RLO 1830.

CHAPTER 3

Appreciation is extended to Rien van Genuchten of "USDA George E. Brown, Jr. Salinity Laboratory" (USSL), Joe Wang and Terry Hazen of LBNL, Tom Wood of INEEL, and Eric Lindgren of Sandia Labs for their review of different sections of Chapter 3 and their comments and useful suggestions. Special thanks to Rien van Genuchten and the staff of USSL for providing numerous useful references used in preparing Chapter 3. We would like to recognize the co-authors for the main body of Chapter 3:

M. Bandurraga, M. Conrad, P. Cook, S. Hubbard, P. Jordan, E. L. Majer, A. Simmons, and T. Hazen of LBNL; L. Everett of The IT Group; C. Eddy-Dilek, B. Riha and J. Rossabi of SRTC; A. Hutter of DOE Environmental Measurements Lab; C. Keller of

Flute Inc.; F. J. Leij of USSL; N. Loaiciga and S. Renehan of the Univ. of California, Santa Barbara; L. Murdoch of Clemson University; Y. Rubin of the Univ. of California, Berkeley; S. Weeks and C. Williams of Sandia National Lab (SNL).

Sally Benson coordinated the LBNL support for this chapter—her assistance is gratefully acknowledged. LBNL is a contractor of the U.S. Government and work is performed under Contract No. DE-AC03-76SF00098.

CHAPTERS 4 AND 8

SNL contributed about $30K of program development funds. These funds were made available through George Allen. An additional $15K came from DOE, Albuquerque, with Missy Klem's assistance. SNL is a multiprogram laboratory operated by Sandia Corporation, a Lockheed Martin Company, for the United States Department of Energy under Contract DE-AC04-94AL85000.

CHAPTER 5

We thank the peer reviewers Q. Hu, J. Liu, G. Moridis, and Y. Tsang of LBNL, and M. O'Sullivan of the University of Auckland, New Zealand for their careful critique of the sections. We also appreciate the technical editing by D. Hawkes and J. McCullough, generation of illustrations by R. Hedegaard and D. Swantek, and text preparation by R.M. Bradley and M. Villavert, all of LBNL. Special thanks to L. Murdoch of Clemson University for his technical review of this chapter. Sally Benson coordinated the LBNL support for this chapter—her assistance is gratefully acknowledged. LBNL is a contractor of the U.S. Government and work is performed under Contract No. DE-AC03-76SF00098.

CHAPTER 6

We are grateful for those who provided valuable peer review, including Andy Thompson of LLNL, several PNNL scientists, and the book project peer review team. PNNL coordinated the writing of this chapter.

PNNL is operated by Battelle for the U.S. Government and performs work under Contract No. DE-AC06-76RLO 1830.

CHAPTER 7

Methods for remediating organic chemicals in the vadose zone have evolved rapidly over the past few years, and important advances in technologies have often outpaced descriptions in technical journals. In order to document the current status of each technology, we have solicited input from leading experts who are aware of the most recent activities and developments. The majority of part 1 of this chapter consists of their contributions, and many of them helped with the integration efforts in part 2. The contributing authors to this chapter are cited below.

John S. Gierke, Michigan Technological University; Joe Rossabi, SRTC; John Ree and, Denis Conley, TerraTherm; Jim Phelan, SNL; Ron Falta, Clemson University; William Heath, PNNL; Terry Hazen, LBNL; Robert L. Siegrist, Michael A. Urynowicz, Colorado School of Mines; Olivia R. West, Oak Ridge National Laboratory (ORNL); Wilson Clayton, IT Corporation; Bill Slack, FRx, Inc.; Paul Bishop, University of Cincinnati; Larry E. Erickson, L. C. Davis, and P. A. Kulakow, Kansas State University.

CHAPTER 9

The Civil Engineering Department at the University of Illinois coordinated the writing of this chapter. We gratefully acknowledge the assistance of the university staff in the preparation and revision of the manuscript. Also, we appreciate the valuable peer review provided by the book project team.

Thank you all,
Tom French, *Project Manager*
Brian Looney and Ron Falta, *Editors*

Acronyms

AEC	Atomic Energy Commission
AODC	acridine orange direct counts
ASTM	American Society for Testing and Materials
ASV	anodic stripping voltammetry
ATV	acoustic televiewer
BERT™	barometrically enhanced remediation technology
bgs	below ground surface
BTEX	benzene, toluene, ethylbenzene, and xylene
CAMU	Corrective Action Management Unit
CCL	compacted clay liner
CERCLA	Comprehensive Environmental Response Compensation and Liability Act
CFC	chloro-fluorocarbons
CHC	chlorinated hydrocarbon
CMB	chloride mass balance method
CML	Charles Machine Works, Inc.
CMP	common midpoint
COC	contaminant of concern
CPT	cone penetrometer tests
CQA	construction quality assurance

CRCIA	Columbia River Comprehensive Impact Assessment
CRDF	Civilian Research & Development Foundation
CSM	conceptual site model
CZT	cadmium zinc telluride
DC	direct current
DCE	dichloroethylene
DFA	direct fluorescent antibody
DGA:PLFA (ratio)	Diglyceride fatty acid to phospholipid fatty acid (ratio)
DLVO (theory)	Derjaguin, Landau, Verwey, Overbeek (theory)
DNA	deoxyribonucleic acid
DNAPL	dense nonaqueous phase liquid
DoD	Department of Defense
DOE	Department of Energy
DOE-GJO	DOE Grand Junction Office
DOE-RL	DOE Richland Operations Office
DQO	data quality objective
DST	double-shelled tank
DUS	dynamic underground stripping
ECD	electron capture detectors
ECFH	early-time constant head-falling head
ECM	effective continuum method
ECP	Eastern coastal plain
EDTA	ethylenediaminetetraacetic acid
EDTA	Ethylenediaminetetraacetate
Eh	Measurement of reductive potential
EM	electromagnetic
EM	environmental monitoring
EM-50	Office of Science and Technology (of the DOE Office of Environmental Management)
EML	Environmental Management Laboratory
EMWD (system)	environmental measurement-while-drilling (system)

EMWD-GRS	environmental measurement-while-drilling gamma ray spectrometer
EOR	enhanced oil recovery
EPA	Environmental Protection Agency
EPRI	Electric Power Research Institute
ER	electrical resistivity
ERT	electric resistivity tomography
ESC	Expedited Site Characterization
ET	evapotranspiration
FDA	Food and Drug Administration
FGZ	fine-grained zone
FID	flame ionization detectors
FITC	fluorescien isothiocyanate
FLUTe™	Flexible Liner Underground Technology
FOSM	first-order-second-moment
FPD	flame photometric detectors
FRC	Field Research Center
FTIR	Fourier transform infrared
GAC	granular activated carbon
GC	gas chromatography
GCL	geosynthetic clay liner
GEM	genetically engineered microorganism
GPR	ground-penetrating radar
GRS	gamma ray spectrometer
HA	Health Advisory
HDPE	high-density polyethylene
HELP	Hydrologic Evaluation of Landfill Performance
HPGe	high purity germanium
HPLC	high pressure liquid chromatography
HPO	hydrous pyrolysis/oxidation
ICP	inductivity-coupled plasma emissions spectroscopy

IDT	interdigital transducers
INEEL	Idaho National Engineering and Environmental Laboratory
ISTD	in situ thermal desorption
ISV	in situ vitrification
ITRD	innovative treatment remediation demonstration
KAFB	Kirtland Airforce Base
KOH	potassium hydroxide
KUT (logging)	potassium-uranium-thorium (logging)
LANL	Los Alamos National Laboratory
LBNL	Lawrence Berkeley National Laboratory
LIF	laser-induced fluorescence
LLNL	Lawrence Livermore National Laboratory
LNAPL	light nonaqueous phase liquid
LPM	low permeability media
LSIT	Large Scale Infiltration Test
LUST	leaking underground petroleum storage tank
MCA	multichannel analyzer
MCL	maximum containment level
MDL	method detection limit
MINC	multiple interacting continua
MIP	membrane interface probe
MIRFEWS	mid-infrared fiber-optic evanescent wave sensor
MLAAP	Milan Army Ammunition Plant
MNA	monitored natural attenuation
MPN	most probable number
MRCO	mixed region chemical oxidation
MRVS	mixed region vapor stripping
MSE	mean square error
MWLID	mixed waste landfill integrated demonstration
NAPL	nonaqueous phase liquid

NOM	natural organic matter
NPP	nuclear power plant
NPS	non-point source
NRC	National Research Council
NTA	nitrilotriacetic acid
NTS	Nevada test sites
NURE	National Uranium Resource Evaluation
O & M	operating and maintenance
OBG	open burning ground
ORNL	Oak Ridge National Laboratory
PAH	poly(allylamine) hydrochloride
PAH	polycyclic aromatic hydrogen
PAWS	portable acoustic wave sensor
PCB	polychlorinated biphenyl
PCE	perchloroethylene
PDE	partial differential equation
pE	-log (e-)
PID	photoionization detectors
PITT	partitioning interwell tracer test
PiX	precision injection/extraction
PLFA	phospholipid fatty acid
PMT	photomultiplier tube
PNNL	Pacific Northwest National Laboratory
POA	poly(o-anisidine)
POLO (system)	(subsurface) position locating (system)
POM	pariculate organic matter
PTF	pedotransfer function
PVC	polyvinyl-chlorine
PVET	pilot vapor extraction test
PVT	pressure-volume-temperature
QCM	quartz crystal microbalances

RCRA	Resource Conservation & Recovery Act
RDX	cyclotrimethylenetrinatramine
Redox	Oxidation-reduction
REV	representative elemental volume
RF	radio frequency
RFI	RECRA facility investigations
Rh	relative humidity
RI	remedial investigation
RLS	radionuclide logging system
RNA	ribonucleic acid
ROI	radius of influence
ROST™	Rapid Optical Screening Tool
RTD	resistance temperature device
RWMC	Radioactive Waste Management Complex
S/S	solidification/stabilization
SAC	systems assessment capability
SAW	surface acoustic wave
SCAPS	site characterization and analysis penetrometer system
SCFH	steady-state constant head-falling head
SEA	Science and Engineering Associates
SERDP	Strategic Environmental Research and Development Program
SFA	shape factor analysis
SGLS	spectral gamma logging system
SI	saturation index
SNL	Sandia National Laboratories
SOH	hydroxylated surface sites
SP	spontaneous potential
SPH	six phase heating
SRS	Savannah River Site
SRTC	Savannah River Technology Center

SST	single shell tank
SVE	soil vapor extraction
SWASV	square wave ASV
TCA	time compression analysis
TCA	l,l,l-trichloroethane
TCE	trichloroethylene
TCLP	toxicity characteristics leaching procedure
TDR	time domain reflectometry
TNRCC	Texas National Resources Conservation Commission
TOC	total organic carbon
TP	thermocouple psychrometer
TSM	thickness shear mode
TU	temporary unit
TVD	total variation diminishing
UFA	unsaturated/saturated flow apparatus
UMEA	ultramicroelectrode arrays
UMTRA	uranium mill tailings remedial action
UNSODA	unsaturated soil hydraulic
USGS	U.S. Geological Survey
UV-FEWS	ultraviolet fiber-optic evanescent wave sensors
VOC	volatile organic compound
WY	water year

VOLUME I

VADOSE ZONE

SCIENCE AND TECHNOLOGY SOLUTIONS

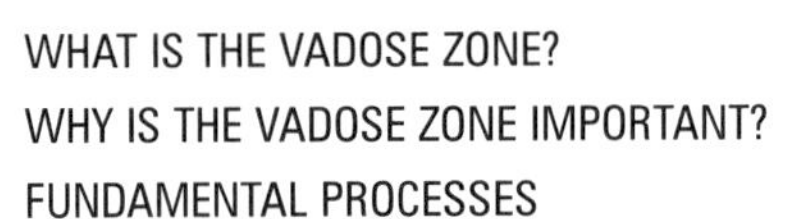

CHAPTER 1 CONTENTS

1

The Vadose Zone

What It Is, Why It Matters, and How It Works

Brian Looney and Ron Falta

WHAT IS THE VADOSE ZONE?

The official textbook definition of the vadose zone is quite simple: "*The geologic media between land surface and the regional water table*" (Stephens 1996). "Vadose" is derived from the Latin noun *vadosus*, which translates to "shallow." This definition and derivation are deceptively simple and yet elegant.

That scientists settled on such a simple term is, in fact, a testament to the difficult and complex nature of the processes occurring in this important subsurface environment. Where possible, scientists prefer terminology that is specific and descriptive of physical or chemical processes and/or status. With respect to the vadose zone, however, most of the available terms (such as unsaturated zone, zone of aeration, and soil zone) are inaccurate or are subject to technical exceptions or inconsistencies (Cullen *et al.* 1995). Scientists were left with the option of consolidating various terms into one that does not attempt to represent a specific technical environment and accepting this more general term. The broad term, "vadose zone," indicates the general compartment and its position within the subsurface hydrologic system (that is, being shallow or close to the land surface). Widespread use and acceptance of the term vadose zone has evolved over the past several years, although many other terms (unsaturated zone, zone of aeration) still appear in scientific literature.

In accordance with the simple definition, the vadose zone includes the soil zone, the intermediate vadose zone, and the capillary fringe (Figure 1-1). The soil zone often exhibits weathered soil horizons and may support a plant root zone. Throughout the vadose zone, the pores and fractures typically are partially filled with water. In a simple sense, two processes control water in the vadose zone. The first is gravity, which moves water downward. The second is a capillary process that, similar to water dripped slowly on a sponge, moves water in all directions and stores and releases it.

The capillary process is controlled by the nature of the sediment and rock and the pores and fractures within it. Throughout this book, these processes are discussed and mathematically described in detail with examples from various settings. It is important to develop a general

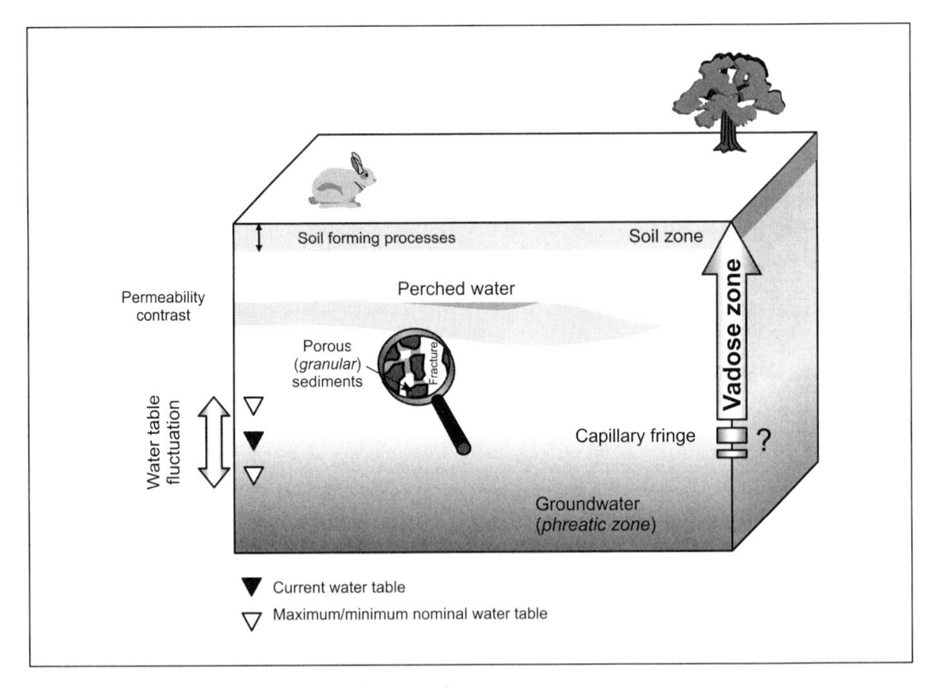

Figure 1-1. Schematic diagram of the vadose zone.

conceptual model of the relative importance of the various processes under different conditions. In most cases, capillary processes are dominant in fine-grained sediments (clay, silt, and the like), while gravity is dominant in course-grained sediment and large fractures. Pores and fractures can be completely filled with water in places within the vadose zone. These saturated portions of the vadose zone are typically referred to as "perched water" or the "capillary fringe," depending on where they occur.

As implied in Figure 1-1, just understanding and describing the distribution and movement of water in the vadose zone is a challenge. Concepts from meteorology, botany, geology, hydrology, physical chemistry, and other disciplines must be taken into account in meeting this challenge. At most sites, the boundary conditions are dynamic, and the water content is changing throughout the system over time. The behavior of chemicals in the vadose zone adds more layers of complexity as the chemicals interact with the soil and other constituents in the solution. For example, chemicals have been observed to interact with small, mobile particles in an apparently chaotic fashion. Despite many years of high-quality science, these challenges stretch the limits of current understanding and complicate the process of choosing practical technical responses at real vadose zone sites.

The general objective of our book is to document important concepts of vadose zone science, and to describe the state of practice, the state of the art, and future directions in this important area. We are focusing on this overall objective, along with the simplicity and elegance of the textbook definition, to overcome some of the questions of vadose zone definition that arise—questions that *actually* arose throughout our workshops and writing process—such as, what about perched water? capillary fringe? variations in water table? fractured rock?

In the simplest sense, the vadose zone consists of the subsurface region above the water table, including the capillary fringe. From a practical perspective, we are interested in portions of the subsurface where vadose zone challenges are significant. These challenges are seen, and are often acute, in areas near a fluctuating water table. This type of dynamic condition can move contaminants to unexpected locations and can trap them and make them difficult to access. As a result, we have adopted a practical approach that defines the bottom of the vadose zone, based on the technical challenge, as extending to the nominal depth of fluctuation of the regional water table (Figure 1-1). While this modification of the definition is operational, rather than formal and semantic, it establishes a rough boundary for the extent of vadose zone influences and helped us to define the scope of this book.

One of the unique features of this book is the use of case studies to illustrate important vadose zone science concepts: the case studies document the state of knowledge through specific and compelling examples. Respected scientists who represent a cross section of federal agencies, universities, and industries prepared the various case studies. A series of workshops provided the forum for frank discussion of possible case studies. The participants, through their writings, describe successes and failures in vadose zone science. The result is a diversity of approach and opinion that is both useful and entertaining.

Events and conditions in the vadose zone greatly influence the behavior of contaminated water being discharged into an underground system. As an example, we developed a short case study describing the migration of a tritium plume at Brookhaven National Laboratory on Long Island, New York. This particular example has several characteristics that make it appropriate for the first case study in a book on the state of vadose zone science. Some of these characteristics include the relatively straightforward geology and geochemistry of the site, a reasonable description of the discharge source, and the availability of detailed plume characterization data.

The migration of a tritium plume is described in the case study "Tritium Source Characterization at the High Flux Beam Reactor, Brookhaven National Laboratory" by Brian Looney and Douglas Paquette. *See page 50.*

Although it seemed simple to predict the behavior of tritium (a contaminant that behaves almost exactly like water) moving through permeable sands, it wasn't. Vadose zone behaviors controlled and shaped the contaminant plume, and influenced its location. Unfortunately, the importance of the vadose zone and vadose zone science was not clearly recognized when plume characterization and interpretation was initiated at Brookhaven. As a result, opportunities to improve and optimize characterization were missed, and public credibility was eroded. Statements about plume behavior were frequently revised. Expectations for particular characterization tools were oversold, then not met. Unexpected data, which were generated from vadose zone controlled behaviors, were treated as surprises and resulted in rapid shifts in the technical program and associated press releases. The net result was that the description of the nature and extent of the problem changed frequently. Each version was altered just enough to be consistent with the latest data set. Despite these problems, which we have purposely and selectively highlighted, the characterization work was outstanding. The resulting data are unique and compelling. The data provided a relatively clear idea of the behavior of the tritium plumc and, in fact, confirmed that the small quantity of tritium in the plume had remained onsite and that risk was near zero.

This particular case study illustrates a broad range of vadose zone science concepts and their importance. It provides several valuable lessons about the importance of incorporating vadose zone science when planning and implementing characterization and interpretation activities. In preparing this particular case study, a few additional ground rules were imposed that were not generally applied throughout the book. Specifically, we present and discuss vadose zone behaviors/science and their significance without using excessive jargon or complicated mathematical equations. The reason for this, of course, is that our goal so far has been to define the vadose zone. As the case study unfolds, various behaviors of water in vadose systems are described in everyday terms. The result is a list of broad assertions that we hope will introduce some of the interesting and, at first blush, unusual characteristics of vadose zone processes. We do not attempt to justify or prove these assertions within the case study.

A second guideline used in this case study is to emphasize simple geometric analysis rather than detailed flow and transport modeling as a scoping method for determining the impact of the vadose zone on plume behavior. This format clarifies the linkage between the listed broad behaviors and their specific impacts on the plume. In the following chapters, the conceptual model implicit in the broad, plain language assertions is supported by first principles and by detailed theory and mathematics.

As you work through the various chapters, we encourage you to periodically reevaluate the ideas in this case study and decide if the broad concepts are consistent with the detailed theory. Were the ideas over-simplified? If so, how? What types of complications are caused by contaminants that are more challenging than tritium? How much more characterization data are needed to improve understanding beyond the broad conceptual model described in the case study? What can scientists and engineers do to coax benefit from vadose zone science? What is the best way to implement environmental programs that include necessary vadose zone science for a reasonable cost? What are the most significant data/knowledge gaps in vadose zone science?

WHY IS THE VADOSE ZONE IMPORTANT?

Vadose zone processes play a pivotal role in the behavior of subsurface contaminants and determine the options and opportunities for environmental cleanup. The significance and control exerted by the vadose zone is often overlooked in characterization and monitoring. More importantly, the influence of the vadose zone is not routinely incorporated into the conceptual model of contaminant behavior at waste sites and industrial facilities. Throughout the 20th century, the vadose zone was often assumed to simply "hold up contaminants and protect the groundwater." This official policy was assumed to be "true" until conclusively disproven—typically without direct vadose zone monitoring.

Complicating matters further, environmental regulations (such as those imposed by CERCLA[1] and RCRA[2]) identify environmental impacts based on measurement of contamination already in the groundwater. As noted by Lorne Everett in his compelling charge to the scientists and engineers contributing to our book, "A philosophy that allows contamination to move through the vadose zone, and only begins regulation at the water table, is flawed. This approach is akin to monitoring a patient who has already died!"

The simple and optimistic concept that the vadose zone is a shield, and the broad philosophy that understanding and monitoring the vadose zone aren't necessary to protect the environment are a convenient, but treacherous, combination. The growing body of environmental data and scientific study documents a more interesting reality—a reality in which

[1]Comprehensive Environmental Response, Compensation, and Liability Act
[2]Resource Conservation & Recovery Act

vadose zone processes play an important, complex, and often controlling role in contaminant transport.

Since waste sites are not normally built in swamps, virtually all subsurface contamination, except for injection well fluids, must pass though the vadose zone.

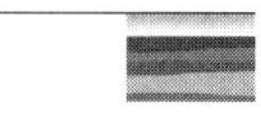

The vadose zone is typically the first subsurface environment encountered by contaminants. As a result, all subsequent groundwater and surface water concentrations, and any resulting environmental impact, are influenced by the complexities in this dynamic system. Measurable contaminant penetration in the vadose zone is the rule rather than the exception, even in cases where most of the contaminant is "held up." Scientific studies have helped us understand transport mechanisms (for example, colloids, extreme chemistry near the waste discharge, and microbiology) and have clarified the remaining challenges and data gaps. The ultimate goals of vadose zone science are to understand and to predict behaviors so that we can prudently manage facilities in a safe and protective manner, and to identify creative and cost-effective options for cleaning up contaminants before they enter the groundwater. Scientific studies indicate that state-of-the-practice modeling tools must be modified in a stepwise fashion so that realistic predictions are generated and observation of contaminant penetration is not treated as a surprise.

Understanding and incorporating the relevant scientific and technical concepts is a critical step in developing rational, protective, and credible environmental policy.

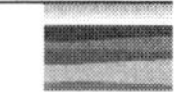

The importance of vadose zone complexities has been explicitly recognized at a few high-visibility sites—primarily nuclear waste handling and disposal sites—but the important role that the vadose zone plays in contaminant transport and risk is universal. Vadose zone controls are not limited to arid climates. With few exceptions, contaminant leaks and disposal into subsurface systems begin in a vadose zone in every region of the United States and throughout the world. The seriousness and complexity of this fact are evident at the United States Department of Energy (DOE) Hanford Site in Washington State. Vadose zone contamination from leaking high-level radioactive waste storage tanks and other Hanford facilities has confounded characterization and cleanup efforts. The complicated scientific and technical problems that emerged have shaken public confidence and are a significant barrier to future progress. Over the past several years, Hanford has worked with regulators and its neighbors to address this environmental challenge. Through the Columbia River Comprehensive Impact Assessment, CRCIA, and the subsequent Systems Assessment Capability (SAC) and Groundwater Vadose Zone Integration Project, the parties have steadily worked toward an integrated solution to the challenge—a solution that addresses contaminants from the contaminant source through long-term impacts (DOE-RL 1998, 1999).

The CRCIA/SAC approach is summarized in Figure 1-2. The vadose zone is an important component in this illustration, but it is not the only component, and Hanford has appropriately emphasized integration as a way to prioritize needs and resources and to identify the most important science and technology data gaps (DOE 1999). In general, this prioritization process has identified 1) contaminant migration in the vadose

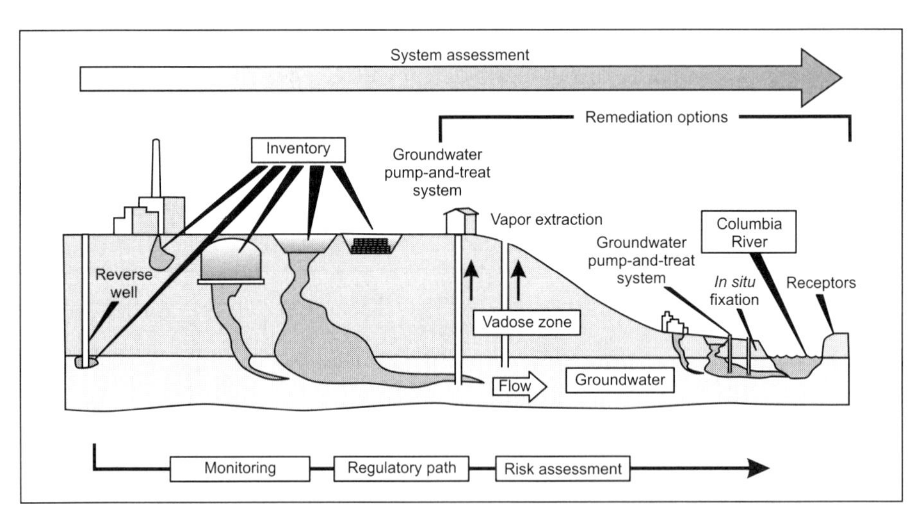

Figure 1-2. Example of an integrated approach to environmental management that shows the role of the vadose zone.

zone and 2) understanding exposure and risk to human health and the environment (the Columbia River, for example) as the two most critical study topics. Reducing uncertainty in these areas was determined to be most beneficial in understanding and improving environmental efforts (DOE 1999). The science and technology integration process provides a framework for efficiently identifying and balancing stakeholder and regulator concerns and overcoming institutional barriers.

In a comprehensive view, improved understanding of the complex behavior of water and contamination in the vadose zone is only one of many science and technology needs. It is, however, one of the most critical. Vadose zone behaviors impact all subsequent measurements and predictions. Monitoring and understanding the vadose zone has historically received less attention than groundwater and surface water.

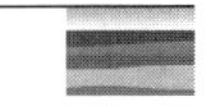

A simplified analysis of the relationship of climate to the vadose zone provides an example of the type of information presented and its use. Figure 1-3 is a diagram of precipitation minus potential evapotranspiration across the continental United States (Geraghty *et al.* 1973). *Precipitation* is a measure of water that falls on the land surface as rain, snow, and sleet in an average year. *Evapotranspiration*, the reverse of precipitation, represents the transport of water from the earth back to the atmosphere through evaporation and plant transpiration. *Potential evapotranspiration* is the loss of water that would occur if there were sufficient water in the soil at all times for maximum use by vegetation.

Dark shading on Figure 1-3 (positive numbers) indicates regions where more precipitation is available than can be returned to the atmosphere by evaporation and plants. Such regions tend to have more runoff, higher vadose zone water content, and more perched water in the vadose zone. There is a substantial driving force (downward water flow) for moving contaminants through the vadose zone in these regions, and the relative importance of other sources of water (such as water line leaks) is reduced.

The light shading (negative values) on Figure 1-3 indicates regions where the gross climactic balance is reversed (less rain and/or more evaporation and transpiration). These regions generally have less runoff and a drier vadose zone. Capillary forces are more important, and slower

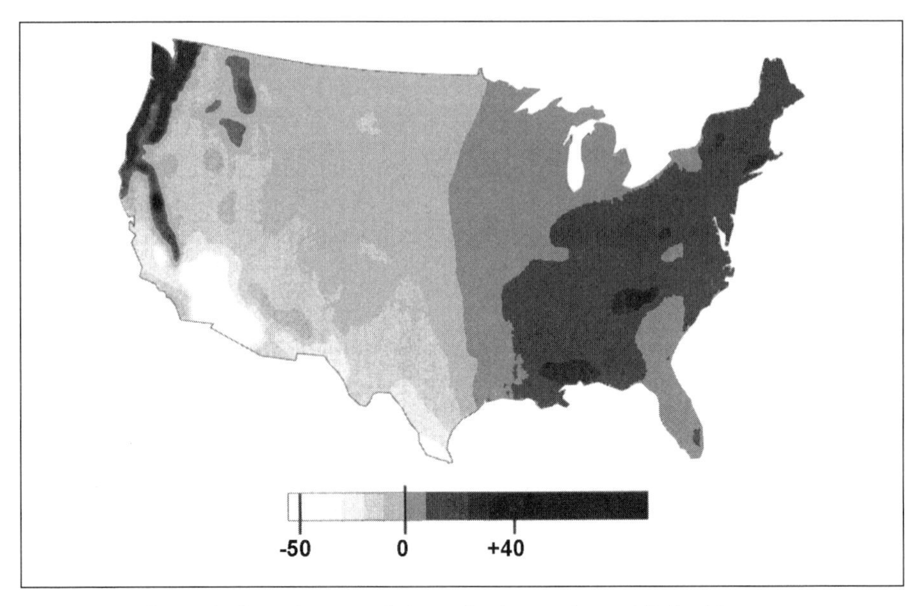

Figure 1-3. General climactic water balance in the continental United States. Data are precipitation minus potential evapotranspiration in inches (1 inch = 2.54 cm) (Geraghty *et al.* 1973).

natural downward migration is observed; however, water line leaks and other anthropogenic sources of water are more dominant. As a result, these sources of water are more important in moving contamination downward. Characterizing and stopping these sources of water is significant in reducing environmental impacts at such sites.

The simple map shown in Figure 1-3 does not directly predict vadose zone thickness. In "arid" eastern Washington, the vadose zone ranges from 0 to approximately 100 m thick; in "humid" South Carolina, the vadose zone is similar in thickness (0 to 60 m thick). The map does not provide a complete water balance because it does not account for runoff, excursions in climate, periods where potential evapotranspiration is not possible due to moisture limitations, or highly localized factors. Negative values on the map do not indicate areas where there is no recharge (water is still present in the vadose zone and moving in response to gravity and capillary forces). Cullen *et al.* (1995) and Stephens (1996) provide more detail on these limitations and on interpretation of the relationship between precipitation and potential evapotranspiration.

Even in this highly simplified form, however, the map (Figure 1-3) does provide interesting vadose zone information. This information includes the general nature of the vadose zone in various regions, the relative importance of possible water sources at contaminated sites, and the relative applicability of alternate environmental strategies. This map also indicates why vadose zone science is important to federal agencies like the DOE and the Department of Defense (DoD), large industries, and others. These organizations have facilities located throughout the climactic regime and their environmental policies and procedures must be appropriate for the entire range of conditions.

The trends in the map are also useful in making engineering judgments about possible barrier technologies for isolating contaminants. A variety of innovative capping systems that rely on capillary barriers and plants have been developed or proposed. Since these technologies are fundamentally based on the large-scale climactic driving forces, the map provides a good idea where they will be appropriate and provide robust performance (light shading) and where they will be less appropriate and more subject to failure (dark shading). An initial evaluation of this type can be supplemented by a more detailed evaluation of climate variability/dynamics and specific design features. Such analysis generates more credible and quantitative assessments of expected performance and probability of failure, but the initial screening facilitates inexpensive identification of areas where this type of technology is appropriate.

To fully understand the conceptual model of the vadose zone used in this book—and the importance and influence of the vadose zone in environmental problems—the reader should become familiar with the terminology and the approaches used in the field. These are introduced below and expanded upon in the following chapters.

FUNDAMENTAL PROCESSES

The processes governing fluid, heat, and chemical flow in the vadose zone are both complex and interrelated. These processes depend on the physical and chemical nature of the geologic materials that make up the vadose zone as well as on the types, amounts, and compositions of the fluids that occupy the pore spaces. This section provides a brief introduction to many of these properties and processes. Most of these topics are covered in depth in the following chapters of the book. To supplement this information, the reader is referred to a number of excellent texts on the physics, chemistry, and mathematics of flow and transport in the vadose zone (Campbell 1985, Hillel 1998, Jury *et al.* 1991, Stephens 1996, Selker *et al.* 1999, Tindall and Kunkel 1999, and Taylor 1972).

Geologic Media and Fluid Properties

Natural vadose zone materials include both fractured rock and granular substances such as sand and silt. Various fluid properties are associated with these materials and with combinations of them. The interrelationships of elements and properties must be known and understood in order to study the vadose zone. The following paragraphs discuss the basics of geologic media and their associated fluid properties.

Consolidated and Unconsolidated Porous, Fractured, and Porous Fractured Materials

The earth materials of the vadose zone depend on the geological history of the specific site, and many distinctly different types of vadose systems are found in nature. For example, the vadose zone may be composed of unconsolidated porous materials such as sands, silts, and clays, or of nonporous but fractured crystalline rocks. With regard to fluid, heat, and chemical transport, a critical feature of vadose zone materials is the type and distribution of the void spaces.

Porous materials contain widely distributed void spaces that typically are interconnected. These voids are usually individually very small (on the order of fractions of a millimeter), but numerous, and it is common for earth materials to contain as much as 40 percent void space. These porous media may be loose unconsolidated soils and sediments such as gravel, sand, silt, and clay, or they may be consolidated rocks such as sandstone, limestone, siltstone, volcanic tuff, or basalt.

Some dense crystalline rocks—such as unweathered granites—have very small amounts of internal void space. These rocks, however, are often naturally fractured, which gives rise to a certain amount of interconnected void space. Porous consolidated rocks, and some unconsolidated soils, may also contain fractures, and these are known as fractured porous media.

Grain Size Distributions, Bulk, and Grain Density

The distribution of soil grain size and the bulk densities of soil and rock are commonly measured during geotechnical investigations, but they are usually of limited use in vadose zone studies. Grain-size distributions are measured by passing soils through mesh screens of different sizes and noting the retained fraction in each screen. Soils are characterized by average grain size and by the distribution of grain sizes. Soil particles larger than about 2 mm are considered to be gravels, while sands range from about 0.05 mm (fine sand) to 2 mm (coarse sand). Fine-grained soils such as silts and clays have grain diameters of 0.05 to 0.005 mm, and < 0.005 mm, respectively. Soils having a wide range of grain sizes are called "poorly sorted," or "well-graded;" soils having a narrow range of grain sizes are called "well-sorted" or "poorly graded."

The dry bulk density of a soil or rock, ρ_b, is the dry field density of the material based on the total volume of the sample (including the void space). This quantity can often be related to the fraction of void space in the sample, and is important for certain geotechnical and geophysical applications. The dry bulk density is used to compute certain types of contaminant concentrations, and it plays a role in the storage of thermal energy in a rock mass. The dry bulk density of rocks ranges from about 2000 kg/m^3 to about 3000 kg/m^3 (Rahn 1996), while unconsolidated soils typically have dry bulk densities of about 1200 to 2000 kg/m^3.

The grain density of a soil or rock, ρ_s, is simply the density of the rock grains without the void spaces. The grain density is thus related to the bulk density through the fraction of void space in the sample. Civil engineers sometimes refer to the specific gravity of solids, G_s, which is defined as the density of soil grains divided by the density of water. Most soils have a grain density in the range of 2.6 to 2.8 times the density of water.

Porosity

The ratio of the volume of voids to the total volume of a rock or soil sample is known as the porosity (ϕ). The interconnected or effective porosity largely determines the volume of fluid that a given volume of rock or soil can contain. Fractured porous rocks contain two types of porosity, a primary porosity associated with the interconnected fracture system, and a secondary porosity, associated with the unfractured "matrix" rock blocks.

Fluid-Phase Volumetric Saturation and Volumetric Fluid Content

The rock or soil void spaces are filled with one or more fluids. In the vadose zone, a gas phase is almost always present in the pores, along with a liquid water (aqueous) phase. At contaminated sites, a nonaqueous phase liquid (NAPL), such as a chlorinated solvent or hydrocarbon fuel, may form an immiscible third fluid phase which is present in the pore space. The volumetric fractions of the porosity occupied by the various fluid phases are known as the fluid phase saturations. Therefore, a given soil or rock sample in the vadose zone will have a gas phase saturation, S_g, a liquid water or aqueous phase saturation, S_w, and possibly a NAPL saturation, S_n. These quantities always sum to 1:

$$S_g + S_w + S_n = 1 \tag{1.1}$$

When more than one phase is present in the pore space, the fluid system is referred to as a multiphase system to distinguish it from a single-phase system. Vadose zone problems are inherently multiphase, compared to groundwater problems, which are typically single phase except when NAPLs are present.

A related quantity known as the volumetric content is commonly used in soil physics to describe the fluid content of a rock or soil. The volumetric water content, or moisture content, θ_w, is equal to the product of the water saturation and the porosity, $\theta_w = \phi S_w$. Water content may also be defined on a mass fraction or weight fraction basis (mass or weight of water divided by mass or weight of oven-dried soil), in which case the water content is called the gravimetric water content. Gas phase and NAPL volumetric contents are defined similarly, and the sum of all of the volumetric contents is equal to the porosity. The volumetric gas content is sometimes referred to as the "air-filled porosity."

Each fluid phase is composed of several components or compounds. The gas phase contains a mixture of gases such as nitrogen, oxygen, carbon dioxide, water vapor, and occasionally, volatile organic chemical (VOC) vapors. The aqueous phase is mostly liquid water, but it contains dissolved organic and inorganic solutes. The NAPL phase may be a pure chemical (for example, pure trichloroethylene), but it is more commonly a mixture of several organic compounds.

Fluid Densities and Viscosities

The aqueous, gas, and NAPL phase densities, ρ_w, ρ_g, and ρ_n, are functions of phase composition, temperature, and pressure. The density of multicomponent liquids (NAPL and aqueous phases) may be computed to a good approximation by assuming volume additivity. That is, the volume of a mixture of components is assumed to be equal to the sum of the individual component volumes. Given the pure component densities, calculation of the phase density as a function of composition is straightforward. The aqueous and NAPL phases are only slightly compressible, and in the vadose zone, liquid compression effects are expected to be small. The liquid phase densities are weak functions of temperature. For example, pure water has a density of 999.7 kg/m^3 at 10°C, and this drops to a value of 958.4 kg/m^3 at a temperature of 100°C (Bejan 1984). NAPL phase densities have a similar dependence on temperature.

The gas phase density is very sensitive to variations in composition, temperature, and pressure. The density of the gas phase is calculated using the real gas law:

$$\rho_g = \frac{P_g M_{wt}}{zRT} \tag{1.2}$$

where P_g is the total gas phase pressure, M_{wt} is the average molecular weight of the gas phase, z is the gas compressibility factor, R is the universal gas constant (8314.4 Joules/mol K; using gram-moles), and T is the absolute temperature (Kelvin). The assumption of ideal gas behavior is often appropriate for vadose zone studies, due to the relatively small variations in gas temperature and pressure. In this case, the gas compressibility factor is equal to 1. Following Dalton's Law (see, for example, Sonntag and Van Wylen 1982), the ideal gas law can be written as

$$\rho_g = \frac{\sum_{i=1}^{n} P_g^i M_{wt}^i}{RT} \tag{1.3}$$

where P_g^i is the gas partial pressure of component i, M_{wt}^i is the molecular weight of component i, and n is the total number of gas phase components. The strong influence of composition, temperature, and pressure on gas phase density often gives rise to buoyancy-driven gas flows in the vadose zone. Comparing the density of a gas computed using equation (1.2) with that of air ($M_{wt}^{air} \approx$ (29 g/mol), it is apparent that if a gas has an average molecular weight greater than that of air, that the gas will be denser than air. The reverse also true, and low molecular weight gases like methane are lighter than air.

Fluid dynamic viscosities are also functions of fluid composition, temperature, and, for gases, pressure. Liquid viscosities are fairly strong functions of temperature, and the viscosity decreases with increasing temperature. Pure liquid water has a dynamic viscosity, μ_w, of 1.304×10^{-3} kg/ms at 10°C, and this drops to a value of 2.83×10^{-4} kg/ms at a temperature of 100°C (Bejan 1984). NAPL-forming organic liquids show a similar behavior with increasing temperature, but there is a wide range of viscosity depending on the specific chemical. Some common NAPLs have a viscosity (μ_n) less than that of water (trichloroethylene, benzene, and toluene, for example), while others are much more viscous than water (diesel fuel, crude oils, and lubricating oils, for example). Estimating the viscosity of liquid mixtures is complicated by molecular interactions, but calculation methods are available (Reid *et al.*1987).

The gas phase dynamic viscosity (μ_g) is much smaller than liquid viscosities. Dry air at 10°C has a viscosity of 1.76×10^{-5} kg/ms (Bejan 1984), which is about 75 times smaller than the liquid water value at the same temperature. For this reason, the gas phase can flow in response to very small pressure and density variations. Unlike liquids, gas viscosity tends to increase slightly with increasing temperature, and it is a function of pressure as well. Estimation of gas mixture viscosity is also complicated by molecular interactions, but several techniques are available for calculating gas mixture viscosity (Reid *et al.* 1987).

Vadose Zone Fluid Statics

The multiphase nature of the vadose zone gives rise to strong capillary effects which cause each fluid phase to have a different local fluid pressure. Combined with gravitational effects, these capillary effects largely determine the static fluid distributions in the vadose zone.

Atmospheric and Gage Pressure

The top of the vadose zone directly interacts with the atmosphere unless the ground surface is sealed. Rainfall, and atmospheric pressure, temperature, and gas composition, strongly influence fluid flow and chemical transport in the vadose zone. The average atmospheric pressure at a location is mainly determined by its elevation. A standard atmosphere (1 atm) is defined as the average gas pressure at sea level and is equal to 101,325 N/m^2 (or Pascals, Pa). At higher altitudes, the atmospheric pressure is substantially lower. At an elevation of 6,000 feet (1,829 m), for example, the atmospheric pressure is only 83.6 percent of a standard atmosphere (Hemond and Fechner 1994).

Atmospheric pressure is a dynamic quantity that fluctuates at several time scales due to wind, solar heating, and weather systems. Typical ranges of pressure fluctuation are roughly ±1 kPa or about ±1 percent. These fluctuations can give rise to significant barometric gas pumping effects in the vadose zone. Gage pressure is the pressure of a fluid relative to the local atmospheric pressure. Some vadose zone and groundwater measurements are commonly recorded as gage pressures (for example, water manometers), and these must be corrected for the fluctuating barometric pressure.

Surface and Interfacial Tension, Fluid Wettability

Multiphase systems are characterized by fluid-fluid interfaces. Unequal molecular forces at these phase interfaces give them a membrane-like elastic quality. The amount of energy required to expand a fluid-fluid interface area by a unit amount is known as the surface, or interfacial, tension. Surface tension refers to a liquid-gas interface such as gas-aqueous (σ_{gw}) or a gas-NAPL system (σ_{gn}). Interfacial tension (σ_{nw}) refers to a liquid-liquid aqueous-NAPL interface. Surface and interfacial tensions can also be thought of as a force required to extend the interface by a unit length, so the standard units can be either J/m^2 or N/m (Adamson 1982). Surface and interfacial tensions are more commonly reported in units of either erg/cm^2 or dyne/cm. Pure water at room temperature has a surface tension of 72.8 dyne/cm, while many NAPLs have surface tensions between 20 and 40 dyne/cm (Reid *et al.* 1987).

The interfacial tension between NAPL and water is a very strong function of composition, but typical values for pure, hydrophobic organic liquids are in the range of about 40 dyne/cm. The interfacial tension, and to a much lesser extent, the surface tension, can be reduced through the presence of surfactants or cosolvents. In many cases, the interfacial tension can be eliminated through the addition of a surfactant or cosolvent, making the aqueous/NAPL fluid pair fully miscible with one another.

Fluid wettability refers to a fluids affinity for a particular solid surface. This affinity can be measured by the contact angle that a drop of the fluid makes with the solid surface. Figure 1-4 shows the behavior of a wetting/nonwetting fluid pair. The wetting fluid drop on the left has a small contact angle (γ), and it preferentially wets the solid surface. The nonwetting phase drop on the right does not wet the solid surface, and the contact angle (measured through the fluid) is large. A perfectly wetting fluid would have a contact angle of 0.

In almost all porous and fractured geologic media, and for almost all liquid-gas pairs, the liquid preferentially wets the solid surfaces.

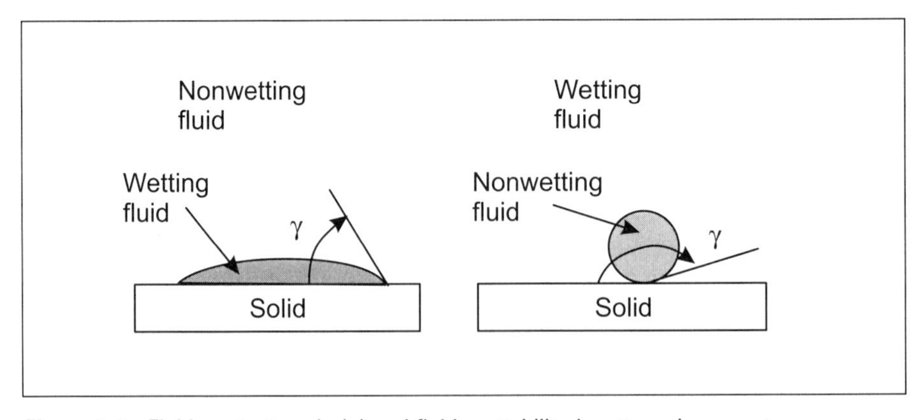

Figure 1-4. Fluid contact angle (γ) and fluid wettability in a two-phase system.

Therefore, in a water/gas system, the water phase is virtually always the wetting phase. Likewise, in a NAPL/gas system (with little or no water), the NAPL phase normally wets the solid surface. There are exceptions to this rule, including gas/mercury systems, and some hydrophobic soils, but these are isolated cases. In liquid-liquid NAPL/water systems, the aqueous phase is usually, but not always, the wetting phase. The usual order of phase wettability in three-phase gas/aqueous/NAPL systems is aqueous>NAPL>gas. The aqueous phase is considered to be the ultimate wetting phase, while the gas is considered to be the ultimate nonwetting phase. The NAPL has an intermediate wettability, depending on the amounts of gas and water present.

At low water contents, water adsorption to the solid surfaces is important, and virtually all vadoze zone materials are at least coated with a film of water several molecules thick. The surface chemical forces that cause this water adsorption are so strong that it may take a pressure difference of tens of atmospheres to displace the water from the solid surface.

Capillary Action

The affinity of the wetting phase for a solid surface causes the wetting phase to displace the nonwetting phase from small diameter tubes (capillary tubes). Figure 1-5 shows a small diameter capillary tube placed in a large container filled with a wetting fluid such as water. The nonwetting fluid could be a gas such as air. The capillary tube has a radius of r, so the circumference of the wetted surface is $2\pi r$. Given a surface tension of σ_{gw}, and a contact angle of 0, the water/solid interface exerts an upward capillary force of $2\pi r\sigma_{gw}$. At equilibrium, the upward capillary force is balanced by the weight associated with the density difference between wetting phase and the nonwetting phase. If the nonwetting phase in Figure 1-5 is a gas, its density is very small compared to the liquid wetting phase, and the weight of the water in the zone of capillary rise is $\pi r^2 h\rho_w g$, where g is the magnitude of gravitational acceleration. Equating these forces, the height of capillary rise, or capillary head, h is equal to

$$h = \frac{2\sigma_{gw}}{\rho_w g r} \qquad (1.4)$$

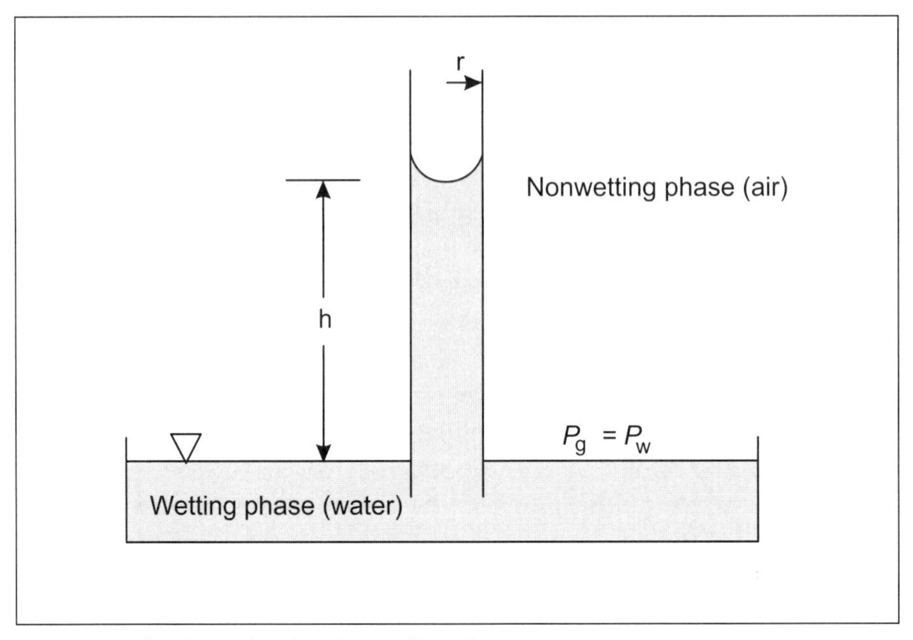

Figure 1-5. Capillary tube showing capillary rise.

The height of capillary rise is therefore inversely proportional to the capillary tube radius, and it is proportional to the water surface tension. At the flat water-air interface in the large container, there is no pressure drop across the phase interface, and $P_g = P_w$. Under static conditions, the water pressure in the tube is hydrostatic, so at an elevation of h above the free water surface, the water pressure in the capillary tube is $P_g = P_w - \rho_w gh$, and the water (wetting phase) pressure is less than the gas (non-wetting phase) pressure. This pressure drop occurs over the curved phase interface in the capillary tube, and it is called the capillary pressure, P_{cgw}. Here, the g and w subscripts indicate a pressure difference between adjacent gas and aqueous phases. Using equation (1.4), the capillary pressure across the phase interface in a capillary tube is found to be

$$P_{cgw} = P_g - P_w = \frac{2\sigma_{gw}}{r} \tag{1.5}$$

A similar derivation for the capillary rise between two parallel plates gives

$$P_{cgw} = \frac{2\sigma_{gw}}{b} \tag{1.6}$$

where b is equal to the aperture between the parallel plates.

It is conceptually helpful to visualize the porous soil or rock as a bundle of different size capillary tubes. The lefthand side of Figure 1-6 shows a bundle of capillary tubes placed in a container of water, open to the atmosphere at the top. As in the single capillary tube case (Figure 1-5), there is no pressure drop across the phase interface in the container, and the water is pulled to different heights in the different radius capillary tubes due to the pressure drop across the curved phase interfaces. If the volume average water saturation in the bundle of tubes is plotted as a function of elevation above the free water surface, a capillary pressure curve such as the one shown in the right hand side of Figure 1-6 can be drawn. This curve then gives the average water saturation associated with a particular capillary pressure. While real soils and rocks are certainly more complex than this simple model, they tend to exhibit similar characteristics when tested experimentally. As one would expect from the capillary tube model, finer grained soils tend to have higher capillary pressures than coarser grained soils for a given water satura-

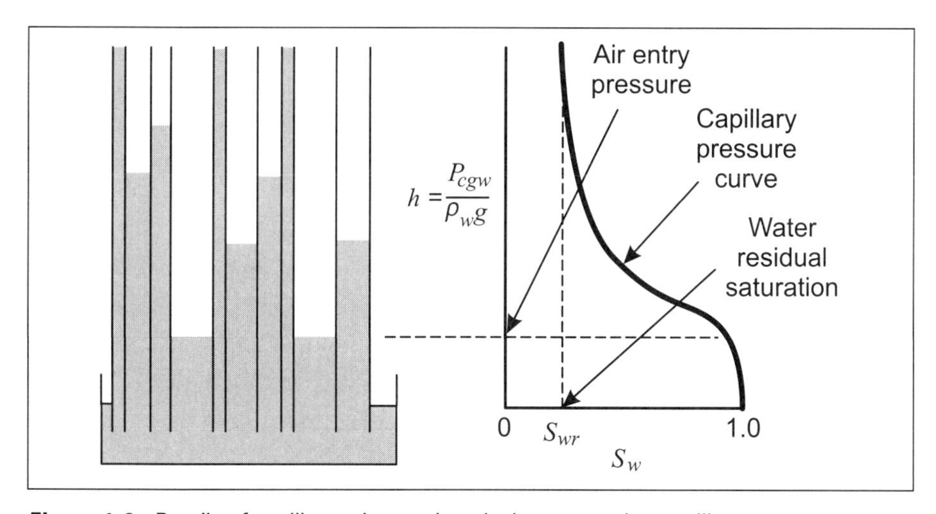

Figure 1-6. Bundle of capillary tubes and equivalent saturation-capillary pressure curve.

tion. If it is assumed that pore sizes are linearly proportional to grain sizes, then the capillary pressure would scale with the inverse of the grain size, $P_c \propto 1/d$, where d is the grain diameter.

Two of the significant landmarks in the capillary pressure curve are shown in Figure 1-6. The water residual saturation is the water saturation at which further increases in capillary pressure fail to displace (much) more water. In reality, the residual water saturation depends on the method of measurement and on the magnitude of displacing pressures used, because if the capillary pressure is increased enough, more water can always be removed from the system. The residual water saturation is usually associated with the smallest pores in the system and with adsorbed water films. The capillary tube analogy is the small diameter tubes shown in the illustration. At high capillary pressures, above 1 atm, the capillary tube analogy breaks down because the water in the capillary tube would boil; however, other phenomena such as water adsorption and capillary condensation on the solid surfaces take effect. Thus at low capillary pressures, (less than 1 atm), the water is retained in the porous media mainly by the pressure drop across the curved fluid phase interface, while at high capillary pressures, (above 1 atm), the water is held by mainly by surface chemical forces. These mechanisms are experimentally impossible to separate, and are usually considered together.

The other landmark in the capillary pressure curve shown in the illustration is the air (nonwetting phase) entry pressure. This is the pressure required to displace the wetting phase from the largest pores in the material when it is initially saturated with the wetting fluid. This entry pressure must be overcome, for example, to inject air into a formation below the water table. Many soils and rocks, particularly those with a wide range of pore sizes, do not have an air entry pressure, and water drainage begins as soon as the capillary pressure is increased.

Although we have been referring to the capillary pressure versus water saturation curve in Figure 1-6 as a capillary pressure curve, many other names are commonly used to describe this curve. Soil physicists commonly refer to the water matric potential. The matric potential is the actual water pressure (or head), and it is typically measured as a gage pressure. Thus a plot of matric potential versus volumetric water content is essentially the same as a capillary pressure-water saturation curve. The following names all refer to the same types of curves (or functions used to fit the curves): capillary pressure curves, matric potential curves, soil moisture curves, water retention curves, water characteristic functions, soil water characteristics, characteristic curves, matric suction curves, and constitutive relationships.

The capillary pressure curve can be measured either under wetting phase drainage conditions or under wetting phase imbibition conditions. The capillary pressure curves often exhibit hysteresis, with the drainage curve giving a higher capillary pressure for a given saturation than the imbibition curve. Chapter 3 details the methods for measuring these curves and contains a more in-depth discussion of vadose zone water.

Fluid Pressure and Saturation Distributions in the Vadose Zone

Under static conditions, there is no fluid flow, water pressure is hydrostatic, and gas pressure is gas static. At the ground surface, the gas pressure is atmospheric, so if gas density is nearly constant with depth, gas pressure is

$$P_g = P_{atm} + \rho_g g z \tag{1.7}$$

where z is the depth from the ground surface. At the water table, by definition, gas and water pressures are equal, and capillary pressure is 0. Proceeding upwards from the water table, the absolute water pressure is

$$P_w = P_g - \rho_w g h \tag{1.8}$$

where h is the height above the water table, and gas density has been neglected in the gravity term. From equation (1.8), the static capillary pressure at any elevation can be determined, and this pressure is used with the capillary pressure-saturation curve for the media to determine the static water saturation. Again, the simplified model shown in Figure 1-6 is helpful for visualizing this phenomenon.

The static fluid distribution (also known as gravity capillary equilibrium) in heterogeneous layered soils or rocks is easily computed using the capillary pressure and capillary pressure-saturation curves for the different materials. Figure 1-7 shows a cross-section of the vadose zone from the water table to the ground surface. The dashed capillary pressure curves on the right hand side of the figure correspond to the three

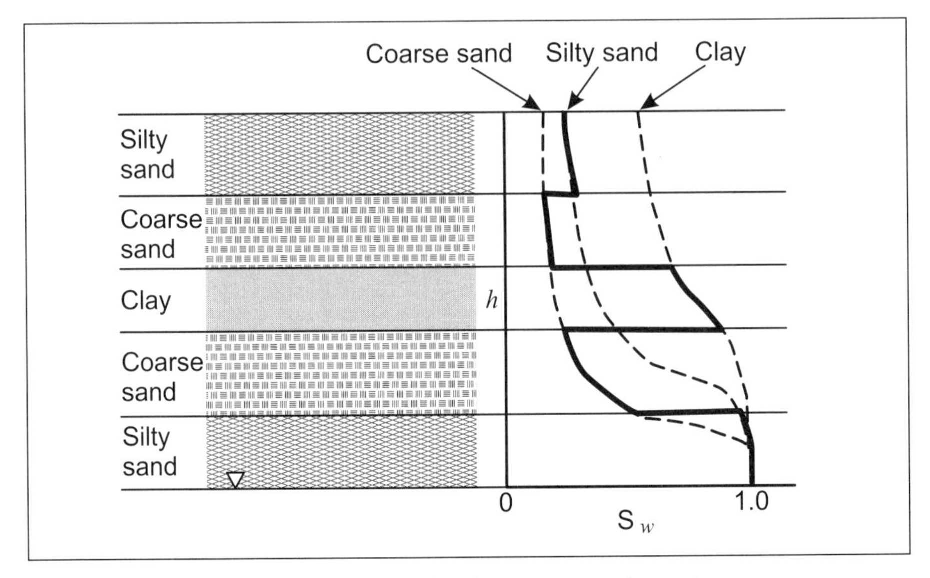

Figure 1-7. Gravity capillary equilibrium in a heterogeneous layered system.

media which make up this vadose zone: coarse sand, silty sand, and clay. At gravity capillary equilibrium, the water saturation at any elevation is found from the capillary pressure curve for the appropriate media. In the perfectly layered system shown here, the capillary pressure is continuous and linear with depth; however, there are sharp contrasts in water saturation at the media boundaries. These types of sharp saturation contrasts occur in natural and engineered systems, and they can result in various capillary barrier effects. Engineered capillary barriers are discussed in Chapter 9.

The example shown in Figure 1-7 applies only to vadose systems in which the amount of infiltration is small or zero. Rainfall events would temporarily modify the saturation profile, especially near the ground surface, and periodic infiltration can have a great effect on the water saturation profile. Calculation of these profiles is more complicated, and generally requires the use of a numerical model.

Three-Phase Capillary Pressures

When NAPLs are released into the vadose zone, it becomes a three-phase gas/aqueous/NAPL system. Unfortunately, three-phase capillary pressures are difficult, if not impossible, to measure. The usual approach to estimating the three-phase behavior is to use the three two-phase capillary pressure-saturation curves, along with an assumption about the fluid wettability order. The three capillary pressures are the gas-water capillary pressure (P_{cnw}) defined before, the NAPL-water capillary pressure,

$$P_{cnw} = P_n - P_w \tag{1.9}$$

and the NAPL-gas capillary pressure

$$P_{cgn} = P_g - P_n \tag{1.10}$$

By definition, each of the three two-phase capillary pressures is equal to the nonwetting phase pressure minus the wetting phase pressure. Recalling the capillary tube development, these capillary pressures are expected to scale with the interfacial or surface tension,

$$\begin{aligned} P_{cgw} &\propto \sigma_{gw} \\ P_{cnw} &\propto \sigma_{nw} \\ P_{cgn} &\propto \sigma_{gn} \end{aligned} \tag{1.11}$$

These relationships can be used to estimate the NAPL-water and the NAPL-gas capillary pressure curves from gas-water capillary pressure data.

Most modeling approaches force the three-phase capillary pressures to smoothly revert to the appropriate two-phase curves as one of the phases disappears. For example, as the NAPL phase disappears,

$$P_{cgn} + P_{cnw} \rightarrow P_{cgw} \tag{1.12}$$

It was mentioned earlier that the NAPL phase is usually the intermediate wetting phase with respect to the gas and water. NAPL may behave either as a wetting phase (with respect to gas), or as a nonwetting phase (with respect to water), so the capillary behavior of the NAPL in the vadose zone depends on the amounts of the other phases present. If a given vadose zone material has a very high water saturation, such as a wet clay, then the NAPL will act as a nonwetting phase, and the nonwetting phase entry pressure may be large enough to exclude the NAPL from the material. On the other hand, if there is a continuous gas phase present in the clay, the NAPL phase may displace the gas, acting as a wetting phase. These types of three-phase interactions with the NAPL in the vadose zone may be important for vadose zone remediation efforts, but they are poorly understood.

FLUID FLOW IN THE VADOSE ZONE

The gas, aqueous, and NAPL phases each flow in response to phase pressure and gravitational forces. The phases interact with one another in such a way that the flow of each fluid is coupled with flow of the other fluids. The driving forces for the fluid flows arise from natural phenomena such as rainfall infiltration and barometric pressure fluctuation, as well as from man-made influences.

Single-Phase Flow

The essential issue for the quantification of fluid flows in porous materials is the relationship between the volumetric fluid flux through the media and the gradient of the driving force causing the flow. This relationship was first established for the single-phase flow of water through sand-packed columns by Henry Darcy in 1856. He found that the water volumetric flow rate, Q_w, divided by the column cross-sectional area, A, was a linear function of the hydraulic head (h) drop across the sand column:

$$\mathbf{V}_w = \frac{Q_w}{A} = K\frac{\Delta h}{L} \tag{1.13}$$

where $\mathbf{V}_w$ is known as the Darcy velocity or flux (m^3/m^{2}s), L is the length of the column, and K is called the hydraulic conductivity of the media (m/s). The hydraulic head contains a pressure head and an elevation head term, but the kinetic energy term is neglected:

$$h = \frac{P_w}{\rho_w g} + z \tag{1.14}$$

and z is the elevation relative to some reference elevation. This formulation for fluid flow works with any orientation, and it is known as Darcy's Law. Darcy's Law can be written in terms of the hydraulic head gradient

$$\mathbf{V}_w = -K\nabla h \tag{1.15}$$

and this form is commonly used in groundwater modeling studies.

The hydraulic conductivity depends on both the water and the porous media properties. It can be shown experimentally that fluid and porous media properties may be isolated

$$K = \frac{k\rho_w g}{\mu_w} \tag{1.16}$$

where k is called the intrinsic permeability of the media. With the exception of some clay soils, the intrinsic permeability is only a function of the porous media, and it is proportional to the square of the grain diameter, $k \propto d^2$. Recalling that the capillary pressure is proportional to the inverse of the grain diameter, $P_{cgw} \propto 1/d$, it is expected that the capillary pressure should scale as the inverse square root of the permeability,

$$P_{cgw} \propto \frac{1}{\sqrt{k}} \tag{1.17}$$

This relationship is commonly used in numerical modeling studies to scale the magnitude of the capillary pressure curves in the absence of direct measurements of capillary pressure.

Darcy's Law is also used to describe fluid flow in fractured rocks. It can be shown theoretically that for laminar flow between two smooth parallel plates, the equivalent permeability is

$$k = \frac{b^2}{12} \tag{1.18}$$

where b is the aperture between the plates. Equation (1.18) is used to estimate the intrinsic permeability of a single fracture, and is known as the "cubic law" relationship for fracture flow (the flow rate is proportional to the cube of b). Numerical models of fractured rocks can model the fractures discretely, using individual fracture permeabilities, or as a continuum, using an equivalent permeability for the fractured rock mass. If an impermeable rock has N parallel fractures per unit length perpendicular to the flow, then the equivalent permeability of the fractured rock mass can be estimated from the cubic law (Snow 1968) as

$$k = \frac{Nb^3}{12} \tag{1.19}$$

with an equivalent porosity of *Nb*. The continuum approach is only valid when the fracture network is dense compared to the scale of numerical discretization.

Fractured porous media such as fractured sandstones and volcanic tuffs require a special mathematical treatment because both the fractures and the porous matrix are active parts of the flow and transport regime. These dual media are discussed in Chapter 5.

Multiphase Flow

Equation (1.15) can be extended for modeling multiphase flows if the conductivity is made to be a function of fluid saturation. Using equations (1.14) and (1.16), Darcy's Law can be written as

$$\mathbf{V}_\beta = -\frac{kk_{r\beta}}{\mu_\beta}(\nabla P_\beta + \rho_\beta g \nabla z) \qquad (1.20)$$

where β indicates either the gas, water, or NAPL phase, and $k_{r\beta}$ is called the β-phase relative permeability. The relative permeability is a phase-saturation-dependent scaling factor used to account for pore blockage by other phases, and it is equal to 1 for single phase flow. This form of Darcy's Law is convenient for multiphase flow modeling because 1) the fluid properties are separated from the rock properties, and 2) because the pressure and buoyancy terms are separated, which allows the capillary pressure gradient to easily be incorporated in the overall phase pressure gradient.

The dependency of the phase relative permeabilities and capillary pressures on the phase saturations means that the different fluid flows are strongly coupled. These dependencies also lead to strong mathematical nonlinearities that require special mathematical solution techniques. These issues are fully discussed in Chapter 5.

Figure 1-8 shows a typical set of two-phase relative permeability curves for a wetting/nonwetting phase pair (gas and water). The phase relative permeabilities vary from 0, at the phase residual saturation ($S_{\beta r}$), to 1 at a phase saturation of 1. Between these limits, the wetting phase relative permeability is often a power function of the wetting phase saturation. The nonwetting phase relative permeability is also sometimes a power function of its phase saturation, but in some materials, it may have a shape that is closer to $(1 - k_{rw})$. In any case, the sum of the phase

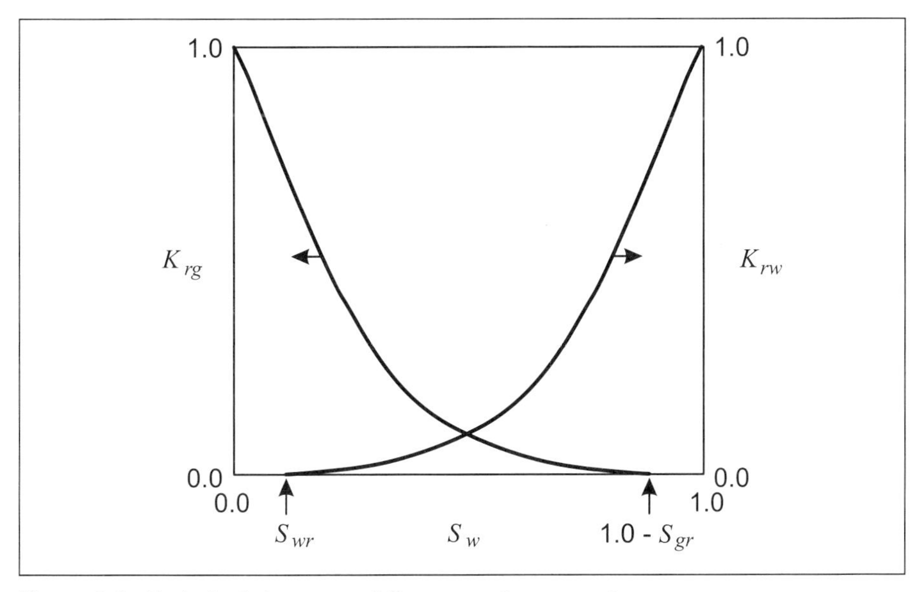

Figure 1-8. Typical relative permeability curves for a two-phase gas-water system.

relative permeabilities must always be less than or equal to 1, and the sum is often much less than 1. While methods are available to estimate the relative permeability curves from capillary pressure data, these methods may not always give accurate relative permeability curves. The two-phase relative permeability curves have a critical influence on numerical simulations of two-phase flows, and reliable measurement of these curves, along with the capillary pressure curves, are central to any vadose zone transport study. Methods for measuring two-phase relative permeability curves are discussed at length in Chapter 3.

Although we have called the relationship between the effective phase permeability and the fluid saturation the relative permeability curve, other names are used. Soil physicists commonly refer to the unsaturated hydraulic conductivity-water content curve, and sometimes measure the unsaturated hydraulic conductivity as a function of matric potential (in other words the relative permeability as a function of capillary pressure). The gas phase relative permeability is not commonly measured in soil physics studies.

When a NAPL is present in the vadose zone, three-phase relative permeabilities come into play. Unfortunately, it is difficult or impossible to measure the phase relative permeabilities over the range of three-phase conditions, so they are typically estimated from the three sets of two-phase curves—(k_{rg}, k_{rw}); (k_{rw}, k_{rn}); (k_{rg}, k_{rn}). This approach is commonly used in petroleum reservoir engineering studies and is similar to the method used to estimate the three-phase capillary pressures from two-phase data. Assuming that water is the ultimate wetting phase, and that gas is the ultimate nonwetting phase, k_{rw} and k_{rg} are usually assumed to be only functions of their respective phase saturations. The NAPL has an intermediate wettability, and k_{rn} is taken to be a function of both S_w and S_g. The three-phase k_{rn} function reverts to the appropriate two-phase functions as the third phase (gas or water) disappears.

It is not known how well this modeling approach works for NAPLs in the vadose zone. The situation is further complicated by the fact that NAPL flow in the vadose zone is almost always unstable. Unstable NAPL flow can lead to the development of small high-phase saturation fingers, which conduct most of the NAPL flow. These fingers occur at a smaller scale than that of the grid blocks in a numerical simulation, and these effects are not accounted for with the usual k_{rn} functions.

Vadose Zone Chemical Transport

Chemicals move through the vadose zone by a variety of mechanisms. These mechanisms include advection with the bulk flow of the gas, aqueous, and NAPL phases; diffusion and dispersion within the fluid phases; and mass transfer between the phases. Many compounds interact with the solid (soil or rock) matrix physically or chemically, and chemicals undergo a multitude of chemical and biological reactions in the subsurface.

Chemical Concentrations

The fact that a given chemical or compound may be present in the gas, aqueous, or NAPL phases, combined with the multidisciplinary nature of vadose zone research, has resulted in common usage of several different, but related, measures of chemical concentration.

The mass concentration of a chemical, C_β^i, is the mass of chemical i per unit volume of the phase β, where β can be the gas, aqueous, NAPL, or solid (for chemical adsorption) phase. The mass concentration has units of kg/m^3 or g/l, and is often reported as mg/l or μg/l. Aqueous phase samples are typically reported as mass concentrations, and when the density of the aqueous phase is ~1000 g/l, 1 mg/l is approximately 1 part per million (ppm, a mass fraction). The sum of all of the chemical concentrations in a phase is equal to the phase density for the fluid phases:

$$\rho_\beta = \sum_{i=1}^{n} C_\beta^i \tag{1.21}$$

A total mass concentration, C_T^i, can also be defined on a total volume basis. The total concentration of a chemical includes the amounts of the chemical in each of the fluid phases, as well as the adsorbed concentration

$$C_T^i = \phi S_g C_g^i + \phi S_w C_w^i + \phi S_n C_n^i + C_s^i \tag{1.22}$$

where C_s^i is the adsorbed mass of chemical i per unit bulk volume of the solid media.

The mass fraction of a chemical, X_β^i, is the mass of the chemical i per unit mass of the phase β. The sum of the component mass fractions in a phase is equal to 1, and the chemical mass fraction in a phase is equal to the mass concentration divided by the phase density

$$X_\beta^i = \frac{C_\beta^i}{\rho_\beta} \tag{1.23}$$

Mass fractions are dimensionless and are often reported as weight-percent or parts per million. The total mass fraction of a chemical describes the mass of chemical per unit mass of dry (clean) soil or rock:

$$X_T^i = \frac{C_T^i}{\rho_b} \tag{1.24}$$

where ρ_b is the dry bulk density of the soil or rock. The total mass fraction is commonly used to describe chemical concentrations in soil borings and rock samples.

Molar concentrations (c_β^i) are similar to mass concentrations, and are defined as the moles of component i per unit volume of phase β. These are mainly used for aqueous concentrations, where units are mol/m^3 or mmol/l. The sum of all of the component molar concentrations in a phase gives the phase molar density:

$$c_\beta = \sum_{i=1}^{n} c_\beta^i \tag{1.25}$$

The mole fraction, χ_β^i, is used in many chemical calculations and is particularly convenient for problems involving ideal gases. The mole fraction of a chemical is defined as the moles of component i per mole of the β phase, so it is dimensionless like the mass fraction. The sum of the component mole fractions in a phase is equal to 1, and the mole fraction is equal to the molar concentration divided by the molar density. The mole fraction is related to the mass fraction through the component molecular weights

$$\chi_\beta^i = \frac{X_\beta^i / M_{wt}^i}{\sum_{i=1}^{n} X_\beta^i / M_{wt}^i} \tag{1.26}$$

and

$$X_\beta^i = \frac{\chi_\beta^i M_{wt}^i}{\sum_{i=1}^{n} \chi_\beta^i M_{wt}^i} \tag{1.27}$$

The concentration of radioactive compounds is usually reported in terms of the amount of radioactivity present per unit volume. The amount of radioactivity is determined by the rate of nuclear disintegrations that the substance undergoes during radioactive decay. The standard unit is the becquerel (Bq), which corresponds to one nuclear disintegration per second. Curies (Ci) are widely used as well, and 1 curie is equal to 3.7×10^{10} Bq. Aqueous concentrations of radioactive substances are typically reported as μCi/l or pCi/l.

Gas-Phase Measurements

Gas-phase concentrations are measured and reported in a variety of related units. Assuming ideal gas behavior, the gas mass concentration may be computed from the gas partial pressure

$$C_g^i = \frac{P_g^i M_{wt}^i}{RT} \tag{1.28}$$

where P_g^i is the partial pressure of i. The gas mole fraction is determined by the ratio of the partial pressure to the total pressure

$$\chi_g^i = \frac{P_g^i}{P_g} \tag{1.29}$$

where P_g is the total gas phase pressure. The gas volume fractions are also equal to gas mole fractions for ideal gases, and a common unit for the measurement of gas concentrations is the part per million by volume (ppmv). This unit is completely different from a mass fraction ppm used to describe aqueous mass fractions.

Equilibrium Phase Partitioning

Chemicals in the vadose zone undergo phase partitioning between the gas, aqueous, NAPL, and solid phases. This process is transient and is known as kinetic interphase mass transfer. The driving force for any type of interphase mass transfer is the degree of chemical potential disequilibrium, and the final state is one of chemical potential equilibrium. The following discussion is a basic description of equilibrium phase partitioning in multiphase vadose systems. A more comprehensive treatment of chemical equilibrium is given in Chapter 6.

Phase equilibrium between a gas phase and a NAPL depends on the chemical makeup of the NAPL. If the NAPL consists mainly of a single chemical, the equilibrium gas partial pressure (P_g^i) is equal to the chemical vapor pressure, P_{vap}^i. Then the gas concentration is given by equation (1.28) or (1.29). If the NAPL is composed of many compounds, the multicomponent chemical equilibrium is more complex and is, in general, a function of the molar composition (Prausnitz *et al.* 1986). To a first approximation, the equilibrium gas composition may be calculated using Raoult's Law, where the gas partial pressure is a linear function of the chemical mole fraction in the NAPL:

$$P_g^i = \chi_n^i P_{vap}^i \tag{1.30}$$

Thus, the equilibrium partial pressure of a chemical in equilibrium with a multicomponent NAPL is always less than the chemical's pure vapor pressure.

Phase equilibrium between the gas phase and the aqueous phase often involves relatively dilute aqueous concentrations. In the limit of dilute solution (a relative definition), equilibrium gas and aqueous concentrations have a linear relationship, known as Henry's Law. Henry's Coefficient, H^i, is generally defined as the ratio of the gas concentration to the aqueous concentration. Unfortunately, this is the source of much confusion, because a variety of different units are used for the gas and aqueous concentrations. Some common definitions of Henry's Constant include:

$$H^i = \frac{P_g^i}{\chi_w^i} \tag{1.31}$$

where H^i has units of pressure,

$$H^i = \frac{P_g^i}{c_w^i} \tag{1.32}$$

where H^i has units of (pressure x volume / mole), and

$$H^i = \frac{C_g^i}{C_w^i} \tag{1.33}$$

in which H^i is dimensionless. At higher aqueous concentrations, Henry's Law is no longer valid, and the solution begins to behave as a concentrated solution. At these higher aqueous concentrations, Raoult's Law may be used to estimate the gas-aqueous phase equilibrium using the mole fraction in the aqueous phase.

Equilibrium between the NAPL and the aqueous phases may be calculated by first computing the equilibrium gas phase concentration, then using Henry's Law. Then for single component NAPLs (with a limited aqueous solubility), equations (1.31) through (1.33) can be used with the chemical's vapor pressure to compute the solubility. Conversely, if the chemical's pure aqueous solubility is known, Henry's Constant may be estimated in the same way. If the NAPL is a multicomponent NAPL, Raoult's Law is used to get the gas equilibrium, followed by Henry's

Law to get the aqueous equilibrium. This method is equivalent to weighting the pure aqueous solubility of the individual compounds by their mole fraction in the NAPL:

$$C_w^i = \chi_n^i \bar{C}_w^i \tag{1.34}$$

Here, $\bar{C}_w^i$ is the aqueous solubility of a pure compound i.

Many organic, inorganic, and radioactive contaminants partition strongly to the solid soil or rock grains. This phenomenon is known as adsorption, and it results from a variety of physical, chemical, and biological mechanisms, which are discussed in Chapter 6. Because water is almost always the wetting phase in the vadose zone, the soil or rock grains are almost always covered with a layer of liquid water. Therefore, chemical adsorption in the vadose zone occurs primarily from the aqueous phase. The degree of adsorption at equilibrium can be plotted as a function of aqueous concentration to produce an adsorption isotherm. The adsorbed concentration is sometimes a linear function of the aqueous concentration, especially at low aqueous concentrations. Linear adsorption isotherms are characterized by K_d, the soil-water distribution coefficient for a chemical. This coefficient is the ratio of the adsorbed mass fraction to the aqueous mass concentration:

$$K_d = \frac{X_s^i}{C_w^i} \tag{1.35}$$

and it has units of inverse concentration, m^3/kg. The distribution coefficient is a function of both the chemical and the solid, and the value of K_d ranges over several orders of magnitude for various chemicals and solids. The solid adsorbed mass fraction is related to the adsorbed mass concentration through the dry bulk density of the soil or rock:

$$C_s^i = \rho_b X_s^i = \rho_b K_d C_w^i \tag{1.36}$$

The equilibrium partitioning of a chemical into different phases means that only a fraction of the total chemical mass will be present in any single phase. If chemical transport occurs in one phase, it is retarded by the overall partitioning into the other phases. An equilibrium multiphase retardation coefficient can be defined as the ratio of the total concentration of a chemical to its mass in a single fluid phase (per unit volume):

$$R_{\beta}^{i}=\frac{C_{T}^{i}}{\phi S_{\beta}C_{\beta}^{i}} \tag{1.37}$$

Then if the chemical is transported by the gas phase, and no NAPL is present, its movement is slowed by a factor of R_{g}^{i},

$$R_{g}^{i}=1+\frac{S_{w}}{S_{g}H^{i}}+\frac{\rho_{b}K_{d}}{\phi S_{g}H^{i}} \tag{1.38}$$

Similarly, if the chemical is transported by the aqueous phase, and no NAPL is present, its movement is slowed by a factor of R_{w}^{i},

$$R_{w}^{i}=1+\frac{S_{g}H^{i}}{S_{w}}+\frac{\rho_{b}K_{d}}{\phi S_{w}} \tag{1.39}$$

Equation (1.39) is identical to the standard groundwater retardation coefficient if $S_w = 1$. The concept of chemical retardation is only valid when there is a linear relationship between the total concentration (C_T^i), and the individual phase concentrations (C_g^i or C_w^i). When a single-component NAPL is present, the equilibrium gas and aqueous concentrations are fixed at their saturation values (vapor pressure and solubility), and the total concentration mainly depends on the NAPL saturation. The retardation coefficient cannot be used under these conditions because the total concentration is not a linear function of the phase concentrations.

However, if a chemical exists only as a "dilute" component of a NAPL, a partition coefficient similar to Henry's Constant may be defined:

$$K_{p}^{i}=\frac{C_{n}^{i}}{C_{w}^{i}} \tag{1.40}$$

Then the gas and aqueous phase retardation coefficients include a term for chemical partitioning in the NAPL:

$$R_{g}^{i}=1+\frac{S_{w}}{S_{g}H^{i}}+\frac{S_{n}K_{p}^{i}}{S_{g}H^{i}}+\frac{\rho_{b}K_{d}}{\phi S_{g}H^{i}} \tag{1.41}$$

and

$$R_{w}^{i}=1+\frac{S_{g}H^{i}}{S_{w}}+\frac{S_{n}K_{p}^{i}}{S_{w}}+\frac{\rho_{b}K_{d}}{\phi S_{w}} \tag{1.42}$$

Equation (1.41) may be used to analyze the results of NAPL partitioning interwell gas tracer tests conducted in the vadose zone. These types of tracer tests are discussed in Chapter 3.

Multiphase Transport Methods

Chemicals are transported through the vadose zone in the gas, aqueous, and NAPL phases by advection, molecular diffusion, and mechanical dispersion. Chemical advection is often the dominant transport process, and it is simply the transport of a chemical with the bulk movement of the phase. The chemical mass flux due to advection is the product of the chemical mass concentration and the Darcy velocity.

$$\mathbf{F}_\beta^i = C_\beta^i \mathbf{V}_\beta = -\frac{C_\beta^i k k_{r\beta}}{\mu_\beta}(\nabla P_\beta + \rho_\beta g \nabla z) \tag{1.43}$$

Molecular diffusion of chemicals in the gas phase is an important transport mechanism in the vadose zone. The standard formulation, valid for "dilute" concentrations of noncondensing vapors, uses Fick's Law of diffusion with a correction for the porous media influence:

$$\mathbf{J}_g^i = -\phi S_g \tau_g D_g^i \nabla C_g^i \tag{1.44}$$

Here, $\mathbf{J}_g^i$ is the gas phase diffusive mass flux, D_g^i is the chemical diffusion coefficient in free gas, and τ_g is called the tortuosity factor. The product $\phi S_g \tau_g$ is used to account for the reduced cross-sectional area available for diffusion in multiphase porous media, and for the longer, more tortuous diffusion paths in multiphase porous media. An equivalent formulation may be used to describe diffusion in the aqueous and NAPL phases:

$$\mathbf{J}_\beta^i = -\phi S_\beta \tau_\beta \mathrm{D}_\beta^i \nabla C_g^i \tag{1.45}$$

The liquid molecular diffusion coefficients are about 10,000 times smaller than the gas phase diffusion coefficient, so liquid diffusion processes are mainly important at smaller, local scales. Liquid diffusion plays an important role in the local kinetic interphase mass transfer of chemicals, where it often dominates the mass transfer process.

The gaseous diffusion of condensable vapors such as water vapor and some NAPL chemical vapors can be enhanced by local condensation and evaporation of the vapors across liquid films. This phenomenon is called "enhanced vapor diffusion," and it is important in some systems.

Mechanical dispersion is a diffusion-like spreading of chemical in a phase due to small-scale velocity variations. Dispersion theory is widely used in groundwater studies of dissolved plumes, where the dispersive flux is used to account for chemical spreading due to velocity variations at the subgrid-block scale. The standard formulation for mechanical dispersion assumes a Fickian diffusion process, using an anisotropic velocity-dependent dispersion coefficient. For water flow in the x-direction, the longitudinal (in the direction of flow), dispersive mass flux is calculated by

$$\mathbf{J}^i_{w,l} = -D_{w,l} \frac{\partial C^i_w}{\partial x} \tag{1.46}$$

where the longitudinal dispersion coefficient is defined as

$$D_{w,l} = \alpha_l \frac{\mathbf{V}_w}{\phi S_w} \tag{1.47}$$

The longitudinal dispersivity, α_l, depends on the scale of transport, as well as on the mathematical resolution of the velocity field. If the velocity field is accurately modeled, the value of α_l needed to fit experimental data is relatively small. The dispersive spreading in the direction transverse to the flow is calculated as the transverse dispersive mass flux:

$$\mathbf{J}^i_{w,t} = -D_{w,t} \frac{\partial C^i_w}{\partial y} \quad or \quad \mathbf{J}^i_{w,t} = -D_{w,t} \frac{\partial \mathbf{C}^i_w}{\partial z} \tag{1.48}$$

where the transverse dispersion coefficient is defined as

$$D_{w,t} = \alpha_t \frac{\mathbf{V}_w}{\phi S_w} \tag{1.49}$$

and α_t is called the transverse dispersivity. The transverse dispersivity is typically more than ten times smaller than the longitudinal dispersivity.

A similar mathematical formulation could be used to describe dispersion in the gas and NAPL phases, but this is rarely if ever done in vadose zone studies. Issues related to mechanical dispersion modeling are discussed further in Chapter 5.

Mass Balance Equations

The governing differential or integral equations for fluid flow and chemical transport are derived using the conservation of mass law. For a given volume, conservation of mass requires that for each component

$$\begin{aligned}&\textit{(rate of mass accumulation)} = \\ &\textit{(rate of mass flow in)} \\ &-\textit{(rate of mass flow out)} \\ &+\textit{(rate of internal mass generation)}\end{aligned} \qquad (1.50)$$

This equation is applied to the components of interest, yielding one equation for each mass component. The accumulation term includes the time derivative of the component total concentration. The mass flow terms include the advective, diffusive, and dispersive fluxes of the component in all of the fluid phases present. The rate of internal mass generation accounts for production or decay of the component due to chemical or biological reactions, or due to other sources. These chemical reactions are discussed at length in Chapter 6. The mathematical steps leading to differential and integral governing equations, and their solutions, are presented in Chapter 5.

MULTIPHASE THERMODYNAMICS AND HEAT TRANSFER

Subsurface heat transfer in the vadose zone is of interest because high-level radioactive wastes produce large amounts of heat. It is also of interest because several new methods for organic chemical remediation involve thermal energy (Chapter 7).

Heat, Energy, Work, and Power

Heat and work are both forms of energy, and have the standard metric unit of the Joule. A Joule is defined as the energy required to move 1 Newton 1 meter, J=Nm. Calories and BTUs are also commonly used energy units, and the conversions are: 1 calorie = 4.1868 J; 1 BTU = 1055.056 J. Power is defined as energy per time, and the standard unit is the Watt, which is equal to 1 Joule per second, W=J/s. A kilowatt-hour (kWhr) is a measure of energy equal to 3,600 kJ.

Internal Energy and Enthalpy

Specific internal energy, u, is defined as the internal heat energy of a unit mass of a substance (J/kg). The internal energy is a relative quantity which must be defined as 0 at some reference state. The calculation of specific internal energy depends on whether or not the substance undergoes phase changes (boiling or freezing) over the range of expected temperatures.

The specific internal energy of the solid parts of the rock or soil is calculated from

$$u_R = \int_{T_{ref}}^{T} C_R d\xi \tag{1.51}$$

where C_R is the solid specific heat capacity (J/kg°C), and T_{ref} is usually chosen to be 0°C. The heat capacity is usually a weak function of temperature, so the integral in equation (1.51) can be replaced by

$$u_R = C_R T \tag{1.52}$$

if the reference temperature is 0. The specific heat capacity of liquids is calculated in much the same way:

$$u_l = \int_{T_{ref}}^{T} C_{v,l} d\xi \cong C_{v,l} T \tag{1.53}$$

where $C_{v,l}$ is the liquid specific heat capacity at constant volume. If the liquid is expected to undergo freezing and melting, the liquid internal energy must also include terms for heating the solid form of the substance to the freezing point, along with the specific internal energy of fusion (melting).

The specific internal energy of condensable gases (defined as those which may condense under system conditions such as water vapor) includes both the substance liquid internal energy, and the internal energy of vaporization:

$$u_g = u_l + u_{g,vap} \tag{1.54}$$

where $u_{g,vap}$ is the energy required to evaporate a unit mass of the liquid substance (J/kg). The internal energy of vaporization is very large for liquids such as water, and it dominates the gas internal energy of most condensable vapors. The specific internal energy of noncondensable gases such as air and nitrogen, is computed by

$$u_g = \int_{T_{ref}}^{T} C_{v,g} d\xi \cong C_{v,g} T \tag{1.55}$$

where $C_{v,g}$ is the gas specific heat capacity at constant volume.

Specific enthalpy, h, is closely related to specific internal energy, but it includes a compression work term that is important for energy transport because of the flow of compressible gases. Specific enthalpy is related to specific internal energy by

$$h_\beta = u_\beta + \frac{P_\beta}{\rho_\beta} \tag{1.56}$$

and it may be calculated from integrals similar to equations (1.51), (1.53), and (1.55), using the specific heat capacity at constant pressure rather than the heat capacity at constant volume.

Internal Energy and Enthalpy of Mixtures

The previous definitions of internal energy and enthalpy assumed that the phases consisted of only a single component (that is, liquid water composed only of H_2O). The internal energy and enthalpy of multicomponent mixtures is calculated with mass fraction weighting using the individual pure component internal energies and enthalpies

$$u_\beta = \sum_{i=1}^{n} X_\beta^i u_\beta^i \tag{1.57}$$

and

$$h_\beta = \sum_{i=1}^{n} X_\beta^i h_\beta^i \tag{1.58}$$

The total energy content of a unit volume of the vadose zone is computed in a way that is exactly analogous to the total mass concentration presented earlier:

$$M^h = (1-\phi)\rho_R u_R + \phi S_g \rho_g u_g + \phi S_w \rho_w u_w + \phi S_n \rho_n u_n \tag{1.59}$$

It includes the energy contributions from the solid rock or soil and all of the fluid phases. Equation (1.59) can be used in a simple energy balance equation to estimate the amount of heat required for some types of thermal remediation, and it appears in the energy accumulation term used in the governing equation for numerical modeling studies.

HEAT TRANSFER BY CONVECTION AND CONDUCTION IN POROUS MEDIA

The flow of heat in porous and fractured rocks is dominated by thermal convection and thermal conduction. Thermal radiation is also important at the ground surface, but is not significant in the subsurface under normal conditions. The energy transport processes of convection and conduction are analogous to the mass transport processes of advection and diffusion, and they have similar mathematical forms.

Thermal convection is the movement of thermal energy with the bulk flow of fluid phases. The phase heat flux is calculated as the product of the Darcy velocity, the phase density, and the phase specific enthalpy:

$$F_\beta^h = \mathbf{V}_\beta \rho_\beta h_\beta = -\frac{h_\beta \rho_\beta k k_{r\beta}}{\mu_\beta} (\nabla P_\beta + \rho_\beta g \nabla z) \tag{1.60}$$

This formulation includes the transport of both latent and sensible heat, and accounts for compression work done on the fluid. The total convective heat flux is the sum of the phase convective heat fluxes and has units of W/m^2.

Thermal conduction in porous media follows Fourier's Law of conduction

$$F^h = -\lambda_T \nabla T \tag{1.61}$$

where λ_T is an overall effective thermal conductivity for the subsurface volume (W/m°C). The overall thermal conductivity is a function of fluid saturations, and is largest in fully water-saturated media. Thermal conduction is a relatively fast process (compared, for example, to gaseous diffusion), and heat can be transferred through impermeable barriers by this mechanism.

Energy Balance Equations

The governing equation for heat transfer is developed from the conservation of energy law:

$$\begin{aligned}&\textit{(rate of energy accumulation)} =\\&\textit{(rate of energy flow in)}\\&-\textit{(rate of energy flow out)}\\&+\textit{(rate of internal energy generation)}\end{aligned} \tag{1.62}$$

This law, and the governing equation for heat transfer, are completely analogous to the mass transport equations and can usually be solved using the same mathematical techniques.

The energy accumulation term includes the time derivative of the total energy, while the energy flow terms include the convective and conductive heat fluxes. The rate of internal energy generation accounts for heating due to internal sources such as radioactive decay, electrical resistance heating, and chemical reactions. Heat transfer in the vadose zone is strongly coupled with the fluid flow, and the mass and energy balance equations must be solved simultaneously (see Chapter 5).

SUMMARY

Vadose zone refers to a specific subsurface environment and its close-to-the-surface position within the subsurface hydrologic system. Acceptance of this term has evolved recently, although many other terms still appear in current scientific literature. In developing, understanding, and using vadose zone science, we must account for the historical body of scientific literature. This literature includes references that use alternative names (such as unsaturated zone or soil zone) or references to work by scientists from many countries. For example, terminology that translates as “zone of aeration” is still the most common usage in Russia.

The vadose zone includes the soil zone, the intermediate vadose zone, and the capillary fringe. A significant challenge of vadose zone science is understanding and describing the distribution and movement of water (and contaminants) within this subsurface environment. At most sites, the boundary conditions are dynamic, and water content changes throughout the system over time. The behavior of chemicals in the vadose zone adds more layers of complexity as the chemicals interact with the soil and other constituents in the solution. Despite many years of high-quality science, these challenges stretch the limits of current understanding and complicate development and implementation of technically based vadose zone actions. A recurring theme in the book is the formulation and use of conceptual models. There are many types of conceptual models (such as diagrams of expected behaviors, analytical models, and risk pathways) that span the disciplines and topics in the book. The various authors unanimously recommend early development of conceptual models and stepwise refinement of the model using scientific methods and based on prioritized-focused data collection.

One of the unique features of this book is the use of case studies to illustrate important vadose zone science concepts: the case studies document the state of knowledge through specific and compelling examples. Respected scientists who represent a cross section of experience and opinion prepared the various case studies and brought a diversity and immediacy to this book that illustrate the breadth and depth of our current knowledge of the vadose zone.

Vadose zone processes play a pivotal role in the behavior of subsurface contaminants and determine the options and opportunities for environmental cleanup. However, the influence of the vadose zone is not routinely incorporated into the conceptual model of contaminant behavior at waste sites and industrial facilities. In addition, environmental regulations identify environmental impacts based on measurement of contamination in groundwater, which is beyond vadose zone contamination.

The vadose zone is typically the first subsurface environment encountered by contaminants. As a result, all subsequent groundwater and surface water concentrations, and any resulting environmental impact, are influenced by the complexities in this dynamic system. Measurable contaminant penetration in the vadose zone is the rule rather than the exception. The ultimate goals of vadose zone science are to understand and to predict behaviors so that we can prudently manage facilities in a safe and protective manner and to identify creative and cost-effective options for cleaning up contaminants before they enter the groundwater.

The importance of vadose zone complexities has been explicitly recognized at a few high-visibility sites, but the important role that the vadose zone plays in contaminant transport and risk is universal. With few exceptions, contaminant leaks and disposal into subsurface systems begin in a vadose zone in every region of the United States and throughout the world.

The basic processes governing fluid, heat, and chemical flow in the vadose zone are complex and interrelated. These processes depend on the physical and chemical nature of the geologic materials that make up the vadose zone as well as on the types, amounts, and compositions of the fluids that occupy the pore spaces. Various fluid properties are associated with the naturally occurring materials of the vadose zone, and the interrelationships of elements and properties must be known and understood. The multiphase nature of the vadose zone gives rise to strong capillary effects that cause each fluid phase to have a different local fluid pressure. Combined with gravitational effects, these capillary effects largely determine the static fluid distributions in the vadose zone. Each of the gas, aqueous, and NAPL phases flow in response to phase pressure and gravitational forces. The driving forces for the fluid flows arise from natural phenomena such as rainfall infiltration and barometric pressure fluctuation, as well as from man-made influences. Chemicals move through the vadose zone by a variety of mechanisms, such as advection with the bulk flow of the gas, aqueous, and NAPL phases; diffusion and dispersion within the fluid phases; and mass transfer between the phases. Many compounds interact with the solid matrix physically or chemically, and chemicals undergo a multitude of chemical and biological reactions in the subsurface. Subsurface heat transfer in the vadose zone is also of interest because high-level radioactive wastes produce large amounts of heat. It is also of interest because several new methods for organic chemical remediation involve thermal energy.

A discussion of vadose zone science and technology is, by nature, interdisciplinary and wide ranging. It is easy to identify a list of knowledge gaps or data gaps, but it is more difficult to prioritize the issues. Prioritization is critical because of the scope of the problem, the realities of limited resources, and the appropriate need to responsibly balance expenditures and expected results. The complexities in the vadose zone are compelling, and could occupy scientists and engineers for a long time (ala global climate change). How much more understanding is enough for the objective? How do we determine that? An overview of these meta-science issues is provided in Chapter 2 and specific recommendations are provided in the various characterization, modeling, and remediation chapters.

The vadose zone is a fascinating and complex environmental setting. It controls the initial transport of contaminants released at most sites—industrial sites, military bases, landfills, waste repositories. It influences data and cleanup performance at large governmental and industrial sites and at small mom-and-pop operations like gas stations and dry cleaners. Understanding the vadose zone is critical for designing and siting new facilities. When we monitor the vadose zone we create opportunities for sensitive early-warning of potential environmental problems and we create opportunities for creative and cost-effective isolation and cleanup. When we incorporate existing vadose zone knowledge into environmental activities, and combine our knowledge with prioritized research to fill our knowledge gaps, we protect the environment in a prudent, responsible, and credible manner.

In the chapters that follow, the basics of vadose zone science are discussed, along with an illuminating look at both the state of the practice and the state of the art. Illustrative case studies that document vadose zone problems and solutions link the science, in a practical fashion, to the real world. All aspects of vadose zone science are addressed in turn, including characterization, monitoring, modeling, and remediation.

REFERENCES

Adamson, A.W. *Physical Chemistry of Surfaces*. John Wiley and Sons, New York, NY (1982).

Bejan, A. *Convection Heat Transfer*. John Wiley and Sons, New York, NY (1984).

Campbell, G.S. *Soil Physics with BASIC: Transport Models for Soil-Plant Systems*, Elsevier, New York, NY (1985).

Cullen, S.J., Kramer, J.H., Everett, L.G., and Eccles, L.A. "Is our Ground Water Monitoring Strategy Illogical?," in *Handbook of Vadose Zone Characterization and Monitoring*, Wilson, L.G., Everett, L.G., and Cullen, S.J. (Eds.), Lewis Publishers, Ann Arbor, MI (1995).

Geraghty, J.J., Miller, D.W., VanDerLeeden, F., and Troise, F.L. *Water Atlas of the United States*, Water Information Center Inc., Port Washington, NY (1973).

Hemond, H.F., and Fechner, E.J. *Chemical Fate and Transport in the Environment*. Academic Press, New York, NY (1994).

Hillel, D. *Environmental Soil Physics*. Academic Press, San Diego, CA (1998).

Jury, W.A., Gardner, W.R., and Gardner, W.H. *Soil Physics*. John Wiley and Sons, New York, NY (1991).

Prausnitz, J.M., Lichtenthaler, R.N., and Gomez de Azevedo, E. *Molecular Thermodynamics of Fluid-Phase Equilibria*. Prentice-Hall Inc., Englewood Cliffs, NJ (1986).

Rahn, P.H. *Engineering Geology*. Prentice Hall PTR, Upper Saddle River, NJ (1996).

Reid, R.C., Prausnitz, J.M., and Poling, B.E. *The Properties of Gases and Liquids*. McGraw-Hill Book Company, New York, NY (1987).

Selker, J.S., Keller, C.K., and McCord, J.T. *Vadose Zone Processes*. Lewis Publishers, Boca Raton, FL (1999).

Sontag, R.E. and Van Wylen, G. *Introduction to Thermodynamics*. John Wiley and Sons, New York, NY (1982).

Snow, D.T. "Rock Fracture Spacings, Openings, and Porosities," *Journal of Soil Mechanics*, Proc. Amer. Soc. Civil Engrs. V94 (1968): 73-91.

Stephens, D.B. *Vadose Zone Hydrology*. Lewis Publishers, NY (1996).

Taylor, S.A. *Physical Edaphology, the Physics of Irrigated and Nonirrigated Soils,* W.H. Freeman, San Francisco, CA (1972).

Tindall, J.A. and Kunkel, J.R. *Unsaturated Zone Hydrology for Scientists and Engineers*. Prentice Hall, Upper Saddle River, NJ (1999).

U.S. Department of Energy (DOE-RL). "Screening Assessment and Requirements for a Comprehensive Assessment, Columbia River Comprehensive Impact Assessment," DOE/RL-96-16, Rev. 1, U.S. Department of Energy, Richland Operations Office, Richland, Washington (1998).

U.S. Department of Energy (DOE-RL). "Groundwater/Vadose Zone Integration Project Management Plan," DOE/RL-98- 56, Rev. 0, U.S. Department of Energy, Richland Operations Office, Richland, Washington (1999).

CASE STUDY

TRITIUM SOURCE CHARACTERIZATION AT THE HIGH FLUX BEAM REACTOR, BROOKHAVEN NATIONAL LABORATORY

Brian B. Looney, *Savannah River Technology Center, Aiken, SC*
Douglas E. Paquette, *Brookhaven National Laboratory, Upton, NY*

BACKGROUND

Brookhaven National Laboratory is located on Long Island, New York, in a region of underlying fine to coarse sands and gravel that receives significant recharge from snow and rain. The vadose zone is relatively thin (approximately 15 m), and groundwater levels vary by approximately 1 meter throughout the year. Groundwater is an important resource on Long Island, and when high concentrations of tritium in groundwater were identified near the High Flux Beam Reactor (HFBR) in 1996, Brookhaven initiated an aggressive characterization effort and installed an interception/diversion system. The Brookhaven characterization clearly demonstrates important vadose zone behaviors and illustrates several important vadose zone lessons.

GENERAL BEHAVIORS OF WATER IN VADOSE ZONE AND SHALLOW GROUNDWATER SYSTEMS

Like dripping water on a dry sponge, moisture in a dry vadose zone tends to spread out by capillary forces, accumulating in the small pores and moving downward slowly. The large pores (holes) in the sponge are not filled with water. Conversely, pouring water onto the sponge fills most of the pores, and much of the water can flow downward by gravity through the large pores and leak rapidly from the bottom of the sponge.

Dry conditions in the vadose zone (analogous to a dry sponge) can result from arid climate, topography, surface barriers or caps, or structures such as roads, parking lots, and large buildings. Wet conditions can result from humid climates, topography, surface water bodies, engineered drainage, irrigation, or large volume water leaks.

Water from the vadose zone is not mixed into the aquifer; instead, vadose zone water/contaminants enter groundwater at the very top. As the groundwater flows away horizontally, additional vadose zone water is added to the top of the aquifer, and the original water appears to migrate downward into the aquifer along a discrete angled flow path. As water moves further from the original vadose zone source area, its position is controlled by geology and by location of the ultimate discharge (the flow path of water approaching a discharge region is controlled by, and moves toward, the collection point). Contamination from sources with low vadose zone seepage rates initially form thin plumes, while high discharge and seepage rates (or concentrated dense wastes) initially form thicker, block-shaped, deeper contaminant plumes.

CHARACTERISTICS OF THE LEAK FROM THE HFBR

The HFBR is an important and heavily used scientific resource. The reactor provides access to research stations in which scientists and engineers from universities, federal agencies, and industry can access particle beams for research. These beams have unique properties for studying properties of materials, complex fluid/flow processes, and many other topics. The inside of the reactor and the radioactive spent fuel storage canal are shown in Figure 1.

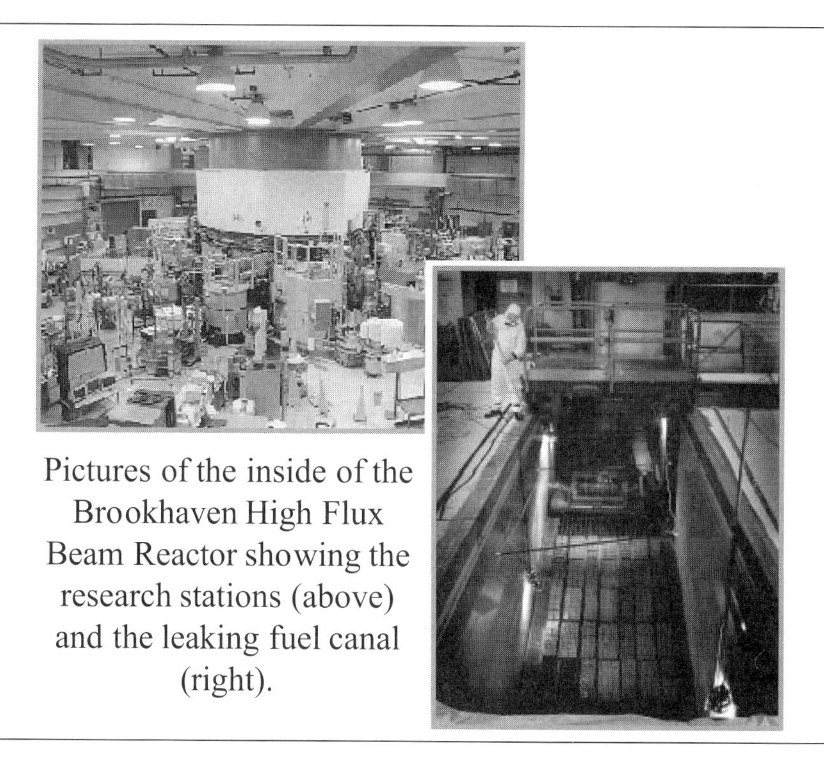

Pictures of the inside of the Brookhaven High Flux Beam Reactor showing the research stations (above) and the leaking fuel canal (right).

Figure 1. Photographs from the Brookhaven High Flux Beam Reactor.

In 1996, groundwater data from the site indicated that tritium was leaking from the HFBR and was entering the groundwater (although measured concentrations varied widely). Detailed study of the reactor history and engineering indicated that the most likely source of the leakage was the spent fuel storage canal. The canal is inside the large HFBR building and is constructed of concrete walls (installed in several lifts), covered by ceramic tile on the inside with an asphalt sealant on the outside. While several initial tests were not sensitive enough to identify a low-volume leak, a high-precision leak test, which was carefully controlled for temperature and evaporation, was ultimately performed. This test revealed a very small leak of about 6 to 9 gallons per day from the canal. Unfortunately, this leak contained high levels of tritium. Importantly for this simplified case study, tritium behaves exactly like water in this system because it is an integral part of some of the water molecules. Tritium (a form of hydrogen) replaces regular hydrogen in a small fraction of water molecules.

To address the tritium plume, Brookhaven scientists and engineers performed wide-ranging studies. These included the following efforts:

- Installing two horizontal wells under the reactor to confirm the canal as the source
- Installing additional monitoring wells
- Detailed profiling of geology and tritium to trace the plume as it moved away from the reactor

The completed work provided clear and unambiguous answers about the tritium from the HFBR. Along the way, however, the data sometimes appeared confusing: optimistic interim interpretations resulted in technical and public relations problems. These interim interpretations could have been significantly improved and refined by accounting for vadose zone processes.

Given the conditions at the HFBR and the concepts listed above, how would the leak behave in the vadose zone and how would this influence the observed plume? How can vadose processes and behaviors lead to confusing data in this case?

CONCEPTUAL TRITIUM PLUME BEHAVIORS AND FIELD OBSERVATIONS

The behavior of water and tritium in the vadose zone immediately below the HFBR reactor building is shown schematically in Figure 2. In this situation, the vadose zone is dry because it is capped by the large building. The small leak from the fuel storage canal spreads out laterally and makes its way slowly to the groundwater. The lateral spread in the vadose zone is enhanced as the water tends to accumulate in and move through fine grained zones (silt and sand) and around coarse grained (gravel) zones (that is, water does not fill the holes in the sponge under these conditions). These vadose zone behaviors caused problems in interpreting data from two horizontal wells that were installed to confirm the source of contamination. The two 125 meter long horizontal wells were installed upgradient and downgradient of the canal and between 0.6 to 1.6 meters below the water table, respectively (figure 2). Both wells were constructed with six separated screen zones running parallel to and within five meters of the canal footprint. The strongly stated goal of the wells was to rapidly and absolutely confirm that the canal was the source by showing that the upgradient well was clean and the downgradient well was contaminated. Unfortunately, the lateral spread of moisture and tritium resulted in tritium concentrations that were similar in the two wells. Also, the concentrations measured in the horizontal wells during their first sampling were <5,000 pCi/L, significantly lower than the 140 million pCi/L concentration of the canal water and ≈600,000 pCi/L concentrations detected in nearby downgradient monitoring wells. At the time, the project was viewed by some as a failure because it did not provide rapid and absolute confirmation. In fact, the project resulted in the successful and high-quality installation of two horizontal wells that would have provided useful confirmatory information when sampled over a period of several years. Importantly, vadose zone processes influenced the data in a manner consistent with theory.

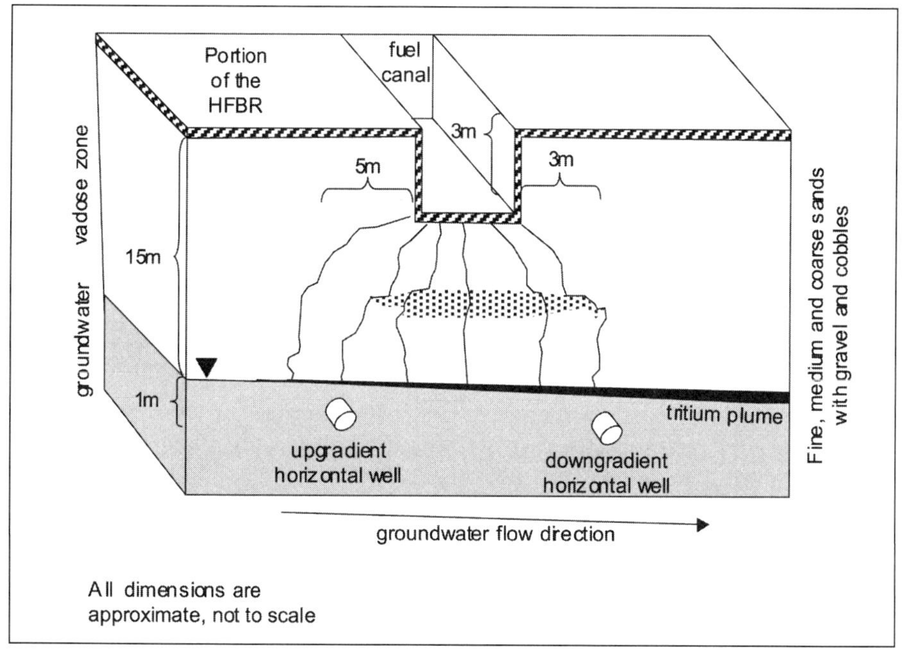

Figure 2. Detail schematic of vadose zone and shallow groundwater beneath the HFBR. Flow lines from a small continuous leak in a dry vadose zone spread out widely, especially when they encounter gravel and cobble zones.

The overall geometry of the contaminant plume beneath the reactor is a direct result of the slow leak rate and the lateral spread in the vadose zone. As the contaminated vadose zone moisture slowly enters the relatively fast moving aquifer (~0.3 m/d), the plume forms a thin plume at the top of the water table. The thickness of this contaminated layer can be estimated from a simple analysis of the relative flow rates and areas (Figure 3). Using leak rate data, measured lateral migration in the vadose zone, and typical hydrogeological values for this site, the tritium plume immediately beneath the reactor is expected to be very thin (<0.2 m). Even if this layer is smeared by seasonal water level fluctuations and other complexities, a simple evaluation of vadose zone delivery versus groundwater flow provides a clear understanding of why the initial sampling of the horizontal wells yielded low concentrations. Despite the wells being installed within a few feet of the water table, each well collects water beneath the main body of the plume throughout most of the year (Figure 2). Subsequent sampling of the horizontal wells during a seasonal

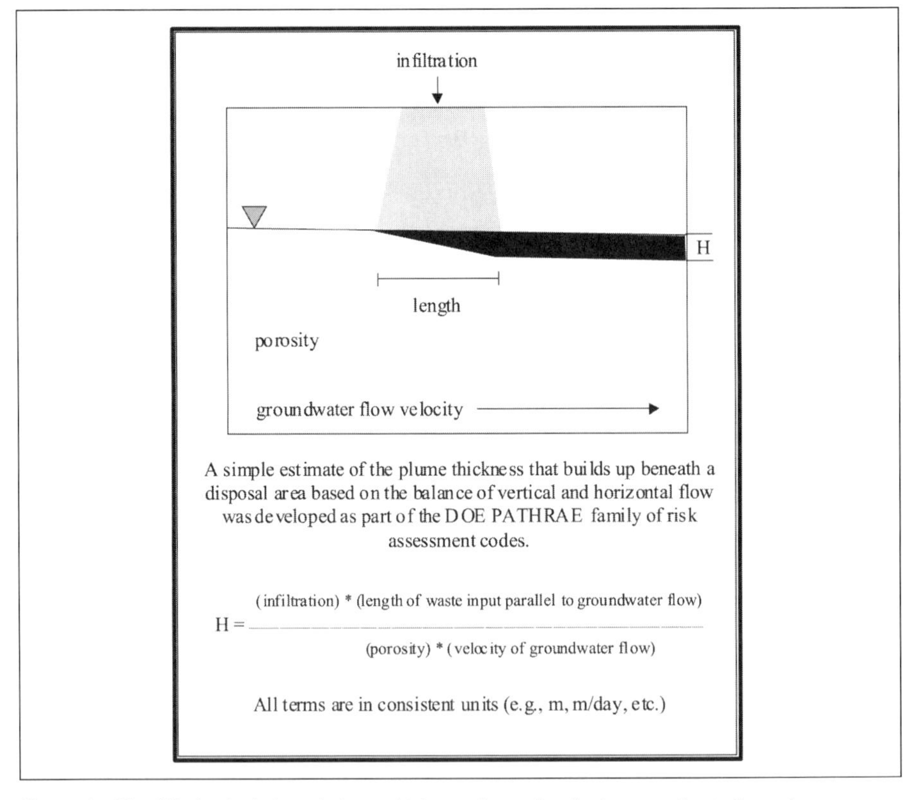

Figure 3. Simplified calculation of plume thickness immediately downgradient of a vadose zone source.

drop in the groundwater level confirmed that the tritium plume beneath the HFBR is primarily within a thin discrete zone at the water table surface. Tritium concentrations of >650,000 pCi/L were detected in the upgradient well when the groundwater level was within 0.3 meters of the well's screened zone. However, during the same period tritium concentrations continued to be low in the downgradient horizontal well (<2,000 pCi/L), because they remained approximately 1 meter below the water table. Because water levels have remained greater than 0.6 meter above the downgradient well since its installation, the originally expected high levels of tritium have not been observed.

When the plume exits the footprint of the HFBR, infiltration places clean water above the plume. Vertical migration of the plume is accelerated, and the plume is expected to exhibit a classic downward trajectory (Figure 4). Tritium that enters the water table at position A will migrate toward position A'. Once again, the actual monitoring data proved to be of high quality, and the large-scale measured Brookhaven plume behavior matches the expected pattern (Figure 5). This highly discrete vertical plume behavior resulted in additional complexity in the data interpretation from monitoring wells—highly variable measurements for samples collected at different times. Figure 6 documents the principal source of this variability for an example water table well located immediately downgradient of the source.

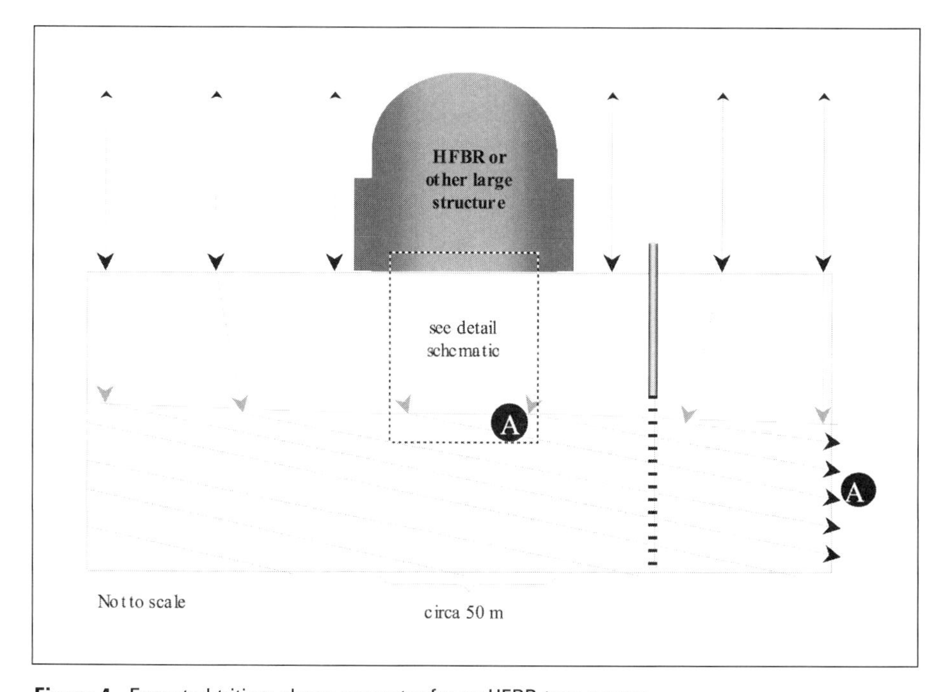

Figure 4. Expected tritium plume geometry for an HFBR type source.

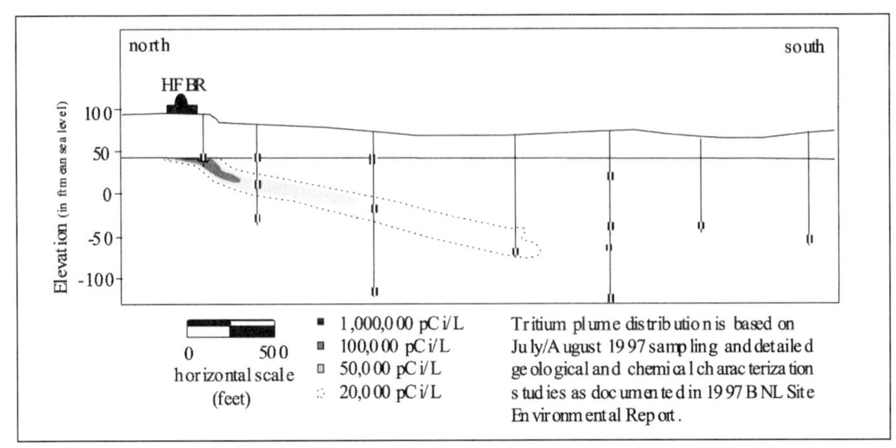

Figure 5. Observed tritium plume geometry at the HFBR.

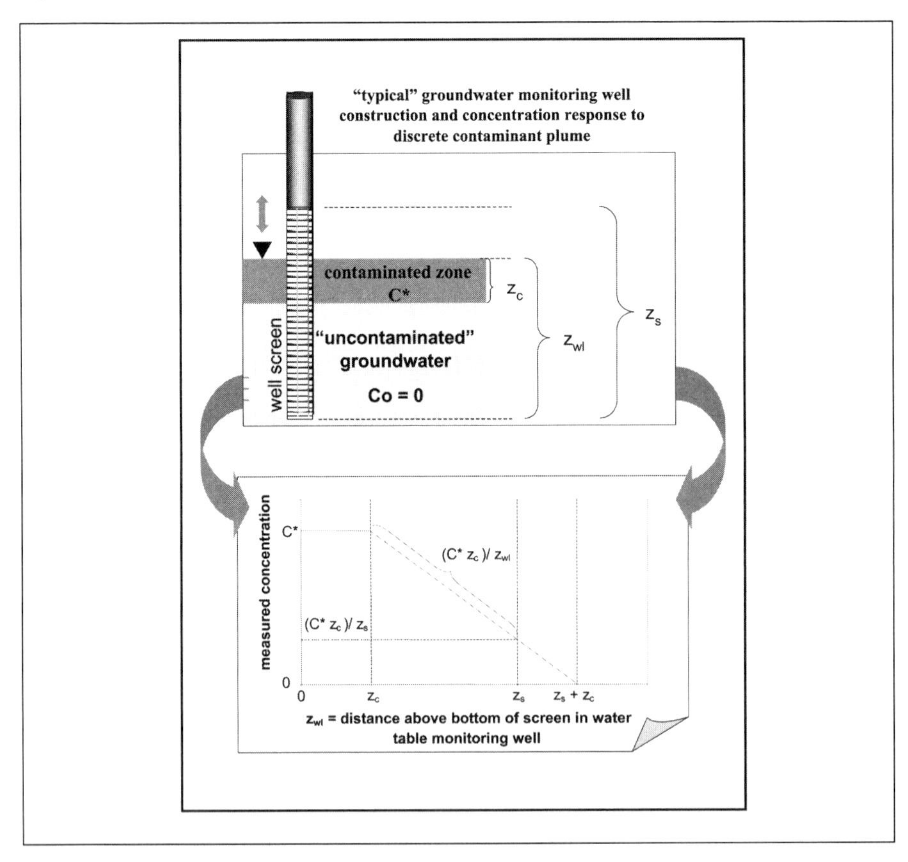

Figure 6. Expected measurement variation in a water table well intercepting a thin contaminant plume.

Based on simple geometry for the case of a thin plume (<0.3 m thick), the tritium measurement in a water table well is simply the plume concentration adjusted by the ratio of the plume thickness to the wetted screen thickness. As site operations and seasonal events impact groundwater levels, tritium levels in vertical monitoring wells will vary widely. While most pronounced for water table wells, the issue of plume/well geometry impacting concentration data is general and should be evaluated for all sites.

There are many lessons learned from the work at Brookhaven. This site was impacted by the simplest possible contaminant and a straightforward hydrogeology. Even under these conditions, complexities in the vadose zone strongly influenced and often controlled the environmental measurements. Sites with extreme chemistry, fractured rock, unstable flow and transport processes (such as colloids), and other scientific challenges add layer upon layer of complexity that needs to be addressed to support prudent environmental management.

Vadose zone programs must be planned and organized within the limitations of budget and time. As an example, we tabulated a series of options that were provided to the Brookhaven HFBR (Table 1). It is hoped that this type of evaluation, which incorporates vadose zone behaviors, will improve site characterization and cleanup efforts. For each tabulated technology, we identified the expected performance and results. Such information, if available early, would assist in stepwise refinement of the conceptual model of site processes and would help to reduce costs and to improve cleanup performance. Further, such evaluation would help maintain stakeholder credibility. Based on the HFBR experience, the technical staff at Brookhaven has modified sampling strategies and data interpretation for all of their facilities and waste sites.

TABLE 1 **Partial list of HFBR characterization technologies evaluated.**

Technology or Strategy	Description / Advantages and Disadvantages
Horizontal Wells	Collection of data using the horizontal wells available beneath the HFBR
Periodic sampling of horizontal wells during seasonal water level changes	Relies on natural process to confirm thin contaminated zone. Sample protocols can be developed based on available water level monitoring data. Generates information to help in 3D interpretation of plume and represents a technically defensible use of high-quality wells. Requires sampling over a long period to determine valid conclusions. Requires frequent monitoring of water levels to assure that samples are collected at the correct time. Water levels at installation were relatively high (based on historical record) and seasonal fluctuations might not bring water table to necessary levels for an extended period. This activity was performed by Brookhaven.
Vadose Zone Sampling	Several locations were identified and procedures were developed to access the vadose zone in case samples were needed. All these required core drilling through the concrete floor and deployment of various samplers into the vadose zone (including capillary fringe and shallow water table). Several of the listed activities are compatible and can be used together. These activities were not performed because of the high cost of drilling and potential breach of containment in the reactor building.
Depth discrete "soil" samples with analysis by either(a) simple suspension in distilled water combined with moisture content or (b) moisture samples collected by freeze drying	Moderate-cost technology that has a high probability of obtaining a valid sample from a thin contaminated zone due to the favorable geometry of the sample hole intersecting the plume and the capability to collect small-volume closely spaced samples.
Soil gas samples with cryotrap to collect HTO and T2O from humidity in soil gas	Inexpensive technology if hole is completed as a soil gas sampling well. Samples may require extended periods (days) to collect.
Suction lysimeter installation and direct collection of soil moisture from the vadose zone/ capillary fringe	Inexpensive (standard) method to collect vadose zone water for analysis. May be difficult to collect water from the driest portion of the vadose zone, but very good near the capillary fringe. Requires dedication of the hole and precludes later activities in the same hole.

continued

TABLE 1 **Partial list of HFBR characterization technologies evaluated.** *(continued)*

Technology or Strategy	Description / Advantages and Disadvantages
Soil gas measurements of 3He/4He ratios (3He is the breakdown product of tritium radioactive decay and is a marker of tritium activity in the vadose zone)	Interesting (but expensive) method. Only an indirect measure of tritium concentration in vadose zone moisture. Other sampling and analysis methods listed above are probably less expensive and more informative.
Groundwater Sampling	Sampling groundwater using monitoring wells, geoprobe, and the like
Detailed 3D plume delineation	Important activity that provides data to assist in environmental stewardship and to reconcile mass balance with engineering studies and models. Shape of plume provides information about vadose zone and groundwater boundary conditions and controlling processes.
Measure 3He/4He ratios in groundwater beneath the HFBR and in several downgradient wells	Ratios of helium isotopes in different portions of the plume indicate age and an provide an independent estimate of flow rates, dispersion process, and the like. Relatively expensive indirect test that does not provide information that is as accurate as 3D characterization.
Engineering Studies	Studies inside HFBR (in addition to engineering review teams, procedure review teams, inspections, and so on)
High-precision leak test	Good engineering method to provide lead information and to support mass balance calculations. Requires careful control of temperature, evaporation, and other variables to achieve required sensitivity. This activity was successfully performed.
Use tracers, such as halogenated gases dissolved in the fuel canal, to confirm leakage (through the detection of the tracer in the soil gas outside the reactor)	Test will confirm if canal is leaking or not. Indirect measure that will generate data that are less accurate and quantitative than leak test. May require extended time and require additional sampling infrastructure. Relatively expensive.

CHAPTER 2 CONTENTS

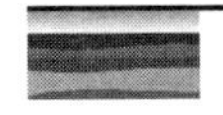

2

Managing an Effective Vadose Zone Project

Marilyn Quadrel and Regina Lundgren

INTRODUCTION

Chapters 1 and 3 through 10 address the technical complexities of the vadose zone. As we saw in Chapter 1, even its definition is subject to varying interpretations, theories, and concerns. This chapter focuses on how the state of technical understanding combines with a political landscape to make vadose zone projects especially challenging, and on what managers and others can do to facilitate an effective response. After delineating why vadose zone management is difficult, we describe three "first principles" of vadose zone management: *establish and manage to endpoints, develop a roadmap,* and *practice deliberate and disciplined engagement.* These principles focus on what is necessary to generate the understanding and consensus needed to make progress toward the ultimate objective of vadose zone management: long-term protection from subsurface contamination.

VADOSE ZONE MANAGEMENT: WHY IT'S HARD AND WHY WE CARE

A major objective of vadose zone projects is to generate useful information for making decisions, yet projects typically operate within a requirements vacuum in which political, organizational, programmatic, and technical uncertainties abound, while specific guidance regarding what information is needed for what purpose is absent. This situation derives fundamentally from two sources:

1. Absence of firm and predictable regulatory requirements directed specifically at anticipatory or corrective actions in the vadose zone.
2. Insufficient scientific and technical basis for promulgating such guidance or readily supporting case-by-case corrective actions.

These conditions put vadose zone managers and regulators in a difficult position. When responsible parties suspect a loss of some hazardous or radioactive inventory to the subsurface, they are required to respond with a corrective action. However, regulators are understandably reluctant to approve plans (especially very expensive plans) when there is little scientific or technical basis for determining their adequacy. When managers of vadose zone facilities proceed with such plans, they do so "at risk," sometimes expending very significant resources on actions that have no explicit support. The problem is only magnified when the data resulting from implementing those plans are inconclusive.

This "Catch-22" is largely the result of decades of relative inattention to the vadose zone. Historically, scientists, regulators, managers, and decision makers concerned with subsurface contamination have focused on groundwater and contaminant movement below the water table. This focus seemed warranted because groundwater is the principal system for moving contaminants away from a disposal site and toward the people and ecological systems that need protection. In contrast, the vadose zone has been looked upon as a natural contaminant buffer, and not as an important and dynamic part of the contaminant "delivery system." Today, the vadose zone is recognized as a key player in determining the long-term impacts of contamination, but this player is unpredictable. It may deliver some contaminants to the groundwater quickly and hold others for long periods of time. Responsible parties typically do not know how much contamination is on hold, what form it is in, and if or when the vadose zone will release it. Further, the mechanisms for answering these questions are not as well developed as mechanisms for studying other contaminant sources or transport pathways.

From a project perspective, the vadose zone is, therefore, mostly characterized by uncertainty: data are sparse, theory is limited, surprise is unexceptional, and cleanup interventions don't always produce predictable outcomes. This uncertainty is especially unsettling when viewing the long-term risks from subsurface contamination.

The situation is further complicated by additional programmatic challenges:

- Vadose zone monitoring and remediation decisions typically occur without clear agreements about the endpoints for the sites they serve.
- Because the vadose zone is an integrating physical system, interest in it typically cuts across program and regulatory boundaries, complicating the questions of what information needs to be collected and what level of intervention is required to adequately meet requirements.
- As part of the delivery system for contamination to the groundwater, the vadose zone is of great concern to local and regional government and to the public. This concern raises the stakes related to programmatic decisions and management considerably and introduces a host of additional variables into the management mix.
- Vadose zone projects often operate in a political "fish bowl," with every action scrutinized. Common criticisms of vadose zone and related projects within the national cleanup effort include: repeatedly attempting to gather information that turns out to be no more helpful than that already in hand; poorly coordinated activities; lack of agreement on measures of success; lack of progress; and misalignment of data collection, modeling, and decision (GAO 1994; NRC 1994; NRC 1995b; NRC 1996a; NRC 1997; GAO 1998; GAO 1999; NRC 1999a; NRC 1999b).
- Finally, the combined uncertainty about the physical system parameters (vadose zone properties and processes) and the regulatory and political environments is typically far greater than in traditional engineering programs, which frustrates the application of good engineering practice and the ability of traditional management practices to produce successful results.

As a result of these challenges, many managers charged with vadose zone responsibilities find themselves in the very unappealing situation of crafting a responsible characterization or remediation project with insufficient or conflicting requirements, minimal resources, and few means by which to change the equation, except to practice extraordinary management.

An enormous body of literature exists on defining and managing complex site-monitoring and remediation projects, from discipline-based theory to training materials synthesized from those disciplines, to the written reviews and advice of appointed and voluntary critics of the

Regulation of the Vadose Zone—A Unique Management Challenge

Consistent with its advocacy of early detection of contaminant releases, the U.S. Environmental Protection Agency proposed regulatory amendments to the Resource Conservation and Recovery Act (RCRA) in 1988. These amendments give RCRA permit writers discretion to require vadose zone monitoring at hazardous waste landfills, surface impoundments, and waste piles on a case-by-case basis. The rationale for this proposal was that such early monitoring would enable facility operators to take action before plumes became unmanageable or prohibitively expensive to remediate. However, to make an early warning system work, project managers must be able to accurately anticipate, acquire, make sense of, and respond to the data such a system could generate. This requires a level of understanding about how contaminants in the vadose zone behave that is not currently available. The U.S. Environmental Protection Agency's final action on vadose zone monitoring considered the practical aspects of an early warning system that detract from the benefits such systems would actually provide (for example, state of technology, predictive potential). The final amendment—elective or case-by-case action—puts the burden of protection in the vadose zone on the individual regulator, who is operating without the benefit of broadly prescribed scientific theory and technical data. The result is a broad range of vadose zone management requirements and actions, differing regulatory authories, regions, and facilities. Many believe that this situation is not providing the benefits of an early warning system that the agency had envisioned. In this context, projects that can effectively integrate actions in the vadose zone (characterization, monitoring, remediation) with the development of a broader understanding of contaminant transport and release (research and development) will help to build the technical base that regulators and policy makers need to more effectively protect our water assets.

nation's cleanup programs. Of course, not all elements of this literature are directly relevant, and many elements that are highly relevant in theory would be impractical, impossible, or impolitic to implement. Rather than present a specific management prescription derived from a single discipline (many of these are available) or a systematic review of the relevant literature, we offer three *first principles* of vadose zone management, designed to help managers overcome the regulatory, technical, and political challenges just described. These principles are:

1. *Establish and manage to endpoints*, specific and agreed-upon—to galvanize, focus, and sustain action.
2. *Develop a roadmap* for generating effective data, information and insight—to form a credible basis for decisions.
3. *Practice deliberate and disciplined engagement* of stakeholders—to form a broad constituency for project results.

By first principles, we mean those things to which a manager must pay full and focused attention. Without them, progress will be elusive. With them, the going may still be slow and difficult, but progress will be far more likely. These principles may seem obvious, but that doesn't mean they are easy. The fact that so few projects actually do these things successfully indicates that there are forces at work that detract from or subvert even the best-intentioned efforts.

What follows, then, is not breakthrough theory or a new and innovative solution to vadose zone management, but rather a discussion of these three fundamental first principles of effective management—what they are, why they are difficult to achieve, some tools for making them work, and traps and gaps that might hinder their success (Figure 2-1). The message is simply this: the challenge is great, the job is hard, and it will not be solved without smart, focused, painstaking attention to the basics.

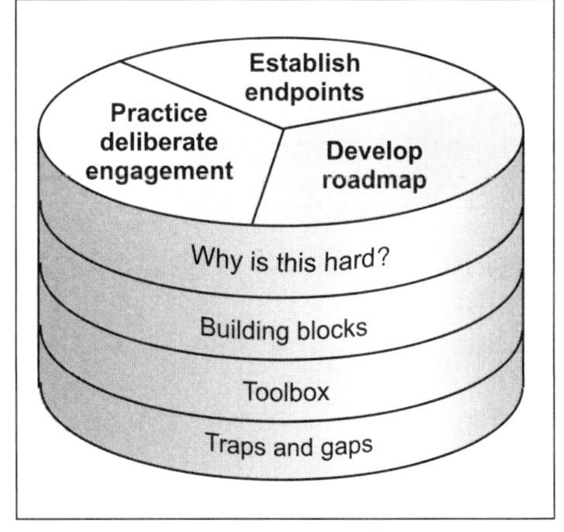

Figure 2-1. First principals of vadose zone management.

Vadose Zone Management: A Critical Role in the Cleanup and Stewardship of the Nation's Legacy Waste Sites

> *"One product clearly dominates other outputs as the raison d'être of [the U.S. Department of Energy's] cleanup: the management and reduction of risks that arise from wastes produced as by-products of weapons development and production during the Cold War."* (Bjornstad et al. 1998)

The primary goal of the U.S. Department of Energy's (DOE's) contaminated site cleanup is the reduction of risk, and the regulations governing that cleanup were designed to manage and reduce risks. However, actual risk reduction anticipated from the cleanup is uncertain. While managers can describe the endpoints of most cleanup projects, they cannot say whether those endpoints, in combination, comprise an endstate resulting in acceptable long-term risk. This problem is particularly disconcerting given that the DOE cleanup program is the largest civic works project in the country, projected by current estimates to cost American taxpayers more than 100 billion dollars. This "endstate question" is at the heart of why the DOE must care about the vadose zone. Effective strategies for protecting groundwater at large sites with multiple contaminated areas cannot be developed without understanding the transport and fate of contaminants in the vadose zone that lies beneath the *entire* site. Good management, therefore, dictates that an enhanced understanding of the vadose zone is necessary to make credible and enduring cleanup decisions.

The current uncertainty over endstates derives in part from how regulations driving the scope and schedule for cleanup have been implemented. Following a logical path in the context of environmental laws and regulations, the drafters of the consent agreements that govern the cleanup of large, complex DOE sites typically divided the sites into smaller cleanup units. Then, the drafters specified cleanup requirements to comply with the applicable regulations on an individual "unit basis." The schedule of milestones for the individual unit cleanups—each with its own priorities and limited budget—typically extends over many years. The resulting "compliance-driven" cleanup is defined by the collection of activities that emerge from this piecemeal approach. Taken together, however, these unit-by-unit activities do not generally account for the contamination existing in the vadose zone, which crosses all of the individual units, nor do they guarantee that the cumulative impact from the residual waste left across all of the units will fall within acceptable standards.

The endstate question is directly tied to and complicated by how contaminants in the subsurface behave and, in particular, how the vadose zone affects the long-term fate and transport of those

continued

contaminants. The vadose zone is a collection point for many contaminants from multiple source terms at large DOE waste sites. The vadose zone holds these contaminants, releasing them to the groundwater over time in unpredictable amounts and on an unpredictable schedule. Surface and subsurface conditions can combine to recontaminate an area that may have already been cleaned up, which further complicates the picture. The uncertainty over what may be released and when it will be released to the groundwater makes the vadose zone a critical actor in the assessment of long-term risks from these contaminants. As a result, decision makers, regulators, and stakeholders may legitimately question whether the collective level of cleanup resulting from project-by-project, unit-by-unit remediation (and compliance) will be adequate to protect public health and the environment. Will the cumulative residual still constitute a significant hazard?

Resolution of this question is hampered by a combination of limited budgets and a lack of specific regulatory policy and technical guidance for the vadose zone. So, while regulators and managers can always find a regulatory requirement that touches the vadose zone, they are not generally compelled to perform a specified level of vadose zone characterization or monitoring. Although central to any effort to bring cleanup endstates into focus, vadose zone activities therefore operate in something of a compliance and management netherland. That is, although releases to the environment are the basic subject matter of the regulations governing cleanup (Resource Conservation and Recovery Act; Comprehensive Environmental Response, Compensation, and Liability Act; and their state counterparts), compliance requirements do not specifically direct investments in or to the vadose zone. As a result, compliance-driven budgets (that is, budgets driven by consent agreement milestones, work unit definitions, and objectives) rarely support the effort needed to obtain the understanding required to describe and support an endstate for long-term cleanup.

Thus, the particular difficulties facing vadose zone projects result, in large part, from the greater challenge facing the nation's cleanup effort as a whole. That is, how will the government and its stakeholders develop a credible and enduring basis for cleanup and, subsequently, for long-term management of its sites?

PRINCIPLE ONE: ESTABLISH AND MANAGE TO ENDPOINTS

"The better we know where we are going, the more likely we are to get there."
R. Olson, *The Art of Creative Thinking* (Olson 1986)

The scope and focus of vadose zone management differ across sites and organizations. Some programs are concerned only with the delivery of information, such as characterization data or assessments; others are directly concerned with remediation or achievement of management goals. Whatever the program's focus, it is unlikely that its goals or endpoints will initially be obvious or that they will be shared. Where information generation is the target, data objectives and decision requirements are generally not readily accessible. Where remediation or management goals are the targets, regulatory and closure requirements needed to design the program are generally not established or agreed upon. This first principle emphasizes the importance of starting immediately to define and gain agreement on the project goals or endpoints in spite of these fundamental uncertainties.

WHY IS THIS HARD?

Establishing clear, shared endpoints is hard for three reasons. The first, and most fundamental reason, is that multiple, legitimately conflicting objectives and goals typically are placed on the program by influential or authoritative sources. It is not always apparent that these multiple objectives conflict, and the significance of the conflicts upon allocating resources and maintaining focus is usually underestimated. Because the objectives, and the endpoints that derive from them, are based on values, clarifying objective and goal conflicts may require values to be formalized. Resolving the conflicts may require analysis, even quantification, of values. Neither is a task for the faint-hearted.

The second reason why endpoint setting is hard involves a set of external factors over which projects have little control. Projects for which priorities are not compliance-driven (that is, "regulatorily challenged" projects) require the commitment of effective senior leaders and politicians for continued attention. For federally funded projects, that commitment is subject to budget and political cycles that are often far shorter than a program's lifespan. Project goals and, therefore, to some

Endpoint Technology

Constraint: A constraint is a specific, measurable guideline that must be adhered to in the pursuit of an objective.

Endpoint: An endpoint describes the products or environmental conditions expected when a project or program ends.

Endstate: An endstate describes the final or long-term condition of a site.

Goal: A goal is a specific and measurable accomplishment, ideally derived from and directly related to an objective.

Interim endpoint: An interim endpoint is the product or environmental condition expected at the close of a significant activity, at which time an anticipated endpoint may be re-evaluated, decided, or further defined.

Objective: An objective is the fundamental driver for programmatic activity; its achievement is the primary reason for funding a program.

extent, project endpoints depend on 1-year budget cycles and 2-year budget forecasts—so what is feasible this year may not be so next year. This leads to a slow deviation from original intentions. Underlying budget issues are 2- and 4-year political cycles, and the limited attention spans of a project's political champions or detractors.

The third, and perhaps most intractable reason, is the fact that vadose zone management objectives and goals will necessarily evolve over time. This evolution is in large part because the information needed to state unequivocally what the cleanup and management activity must protect or what the resulting cleanup standard must be does not currently exist. To know what must be protected and how to protect it requires greater social and political clarity and better science than projects usually have. Without it, managers and leaders must settle for an ongoing process of successive approximations, resolving the endpoint question on a rhythm established by budgetary and political drivers and within other technical, regulatory, and contractual constraints. They must define interim endpoints as evaluation points and understand that the program direction may change as a consequence of that evaluation. More rigor is needed in developing the activities that lead to defining interim endpoints, with an emphasis on maximizing information gathering and resolving conflicts while minimizing constraints on future options.

Having overcome these challenges and established agreed-upon endpoints, managers and leaders must manage these endpoints. This requires carefully and thoughtfully determining where the finish line is, and focusing and coordinating the people, activities, and resources of the project toward reaching it.

BUILDING BLOCKS

This section addresses aspects of establishing endpoints that managers may be able to affect. They are intended to provide additional insights and clarifications on the problem, and to suggest elements of successful solutions.

Clarify Fundamental Objectives

When the potential scope of work is not driven by compliance, yet addresses issues of significant interest to multiple organizations and stakeholders, it is critical to articulate the project's fundamental objectives. A project's fundamental objectives should describe why it is being funded.

There are strong political and institutional barriers to obtaining clarity and consensus around vadose zone management objectives. At the DOE's Hanford Site, for example, the DOE and Tribal governments have agreed to disagree about the interpretation of treaty rights. Tribal governments and the DOE agree that treaty rights relating to taking fish at all "usual and accustomed" places applies to the Hanford Reach of the Columbia River where it passes through the Hanford Site. Tribal governments and the DOE disagree, however, over the applicability of Tribal members' treaty-reserved rights to hunt, gather plants, and pasture livestock on the Hanford Site (DOE-RL 1999b). That disagreement affects efforts to articulate cleanup objectives, including selection of the exposure scenarios for assessments that would support cleanup. The legal and political liabilities associated with this ongoing dispute frustrate efforts to represent fundamental objectives when structuring program decisions and supporting assessments; yet *not* representing those objectives ensures that subsequent assessments or decision processes will be disputed by Tribal governments and the stakeholders who support those treaty rights. Many sites operating under treaty rights have similar issues. Sites that do not have treaty agreements are likely to be challenged by other political or institutional issues related to describing desired outcomes or objectives. In all cases, it is important to define and communicate clearly all critical objectives, even if it means explicitly communicating about the lack of consensus around objectives.

Clarify the "Demand Function"

Most vadose zone programs would benefit from being more driven by the demands of their constituents. This means involving the consumers of the project's ultimate product (that is, long-term protection from contamination) in its definition. For remediation programs affecting or affected by the vadose zone, mechanisms are needed that assess the value of cleanup outcomes (for example, with regard to resource management or recovery, land uses, cultural lifestyles, or economic activities). For projects collecting data, delivering information, or generating

Does Answering All Demands Give You A Better Program?

Will legal standards for cleanup protect Native American as well as dominant culture lifestyles tens and hundreds of years into the future? Will they protect regional economic interests? These and similar questions comprise the potential "demand function" for cleanup at a given site. It seems straightforward enough to determine cleanup requirements for a site by evaluating various scenarios for future site uses against the anticipated cleanup standards for that site. However, programs often face enormous sensitivities around the selection and development of such scenarios. The sensitivities derive from access to information required to pose a credible scenario, from the political and legal basis to select those scenarios, and from competition for resources to fund the work. A fundamental question for those responsible for cleanup—and thereby for understanding vadose zone contamination—is whether or not the cost of pursuing these questions is worth the potential benefit in terms of clarifying fundamental requirements for cleanup and subsequent management of complex waste sites.

new ways of looking at remediation issues, mechanisms are needed that assess the value of the information on which decisions will be built (such as data quality objectives and value of information analysis). Defining consumer demand requires programs to go beyond an objective statement to develop a level of detail that allows the benefits of planned activities to be assessed. Though often expressed in dollar terms, projects need not reduce demand to a dollar value (Bentkover *et al.* 1986). Through this definition process, participants and stakeholders can understand the ratio of program value to program cost and scope.

Bound the Playing Field

Once the underlying objectives and benefits of the program are better understood, it is possible to determine, generally, or formally (to quantify), the point at which additional effort produces little value (the point of diminishing returns). With that information, it is possible to begin bounding the reasonable range of effort for the project (Figure 2-2). This

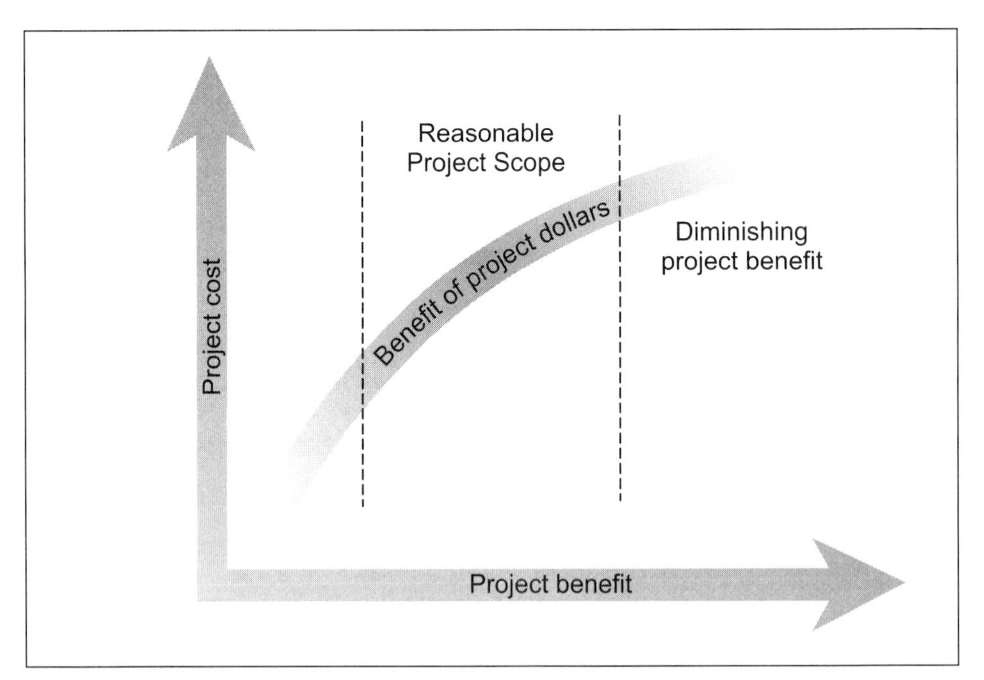

Figure 2-2. A Bounded Decision Field represents up front agreements on the acceptable cost-benefit envelope to keep focus on the effective use of resources.

exercise ensures that the project doesn't waste time and effort supporting or debating activities that produce very little value for the extra effort. Once the participants and ultimate beneficiaries of the product project's have agreed generally on a scope of work, management can focus its efforts on developing a strategy for continuing to more closely define the scope, either by conducting assessments to better resolve the cost-benefit question or by establishing a path forward that maximizes the experience base or data needed to eventually close on the endpoint without foregoing future options.

Stay the Course

One way to enhance the survivability of an agreement on program endpoints is to clearly communicate the accountability for those endpoints and formalize the commitment to their achievement. The process of narrowing and refining the strategy to the point of defining accountabilities and gaining commitments forces clarity and increases dedication and enthusiasm. Performance measures offer one way to formalize those commitments; well-designed and assigned performance measures can go a long way toward clarifying, legitimizing, and coordinating program goals and activities (Corbeil 1992; DOE 1996).

Learn from Mid-Course Corrections

The ability to recognize the need for, then make, mid-course corrections is probably as important as establishing initial endpoints and is the critical companion to staying the course. Surprise results, external events, new or changing values, new technologies, and new knowledge arc just some of the reasons to expect that initial endpoints will evolve or change with time. If program planners and managers don't expect this, they will not set up the internal review mechanisms to recognize when change is required or do the groundwork necessary to allow those changes to occur.

TOOLBOX

There are a number of tools, processes, and organizational constructs to help with the endpoint-setting task. These tools have broad application (many of them support two or all three of the principles discussed in this chapter) and are easily accessible. This list is by no means exhaustive, but may provide a good starting point for evolving an appropriate support system for this task.

Objectives Hierarchy

An objectives hierarchy is a construct for articulating and gaining agreement about why a project is being funded and what the critical tasks in the project are. An objectives hierarchy describes the basic objectives of the program, defines the primary concerns of the program manager, and provides guidance for program actions, including data gathering, modeling, and analysis or R&D to support the program (Keeney 1992). Objectives hierarchies deal with two kinds of objectives: *fundamental* objectives (why you care) and *means* objectives (what can be done) (Figure 2-3). Fundamental objectives describe what people care about and should describe why the project is being funded. Means

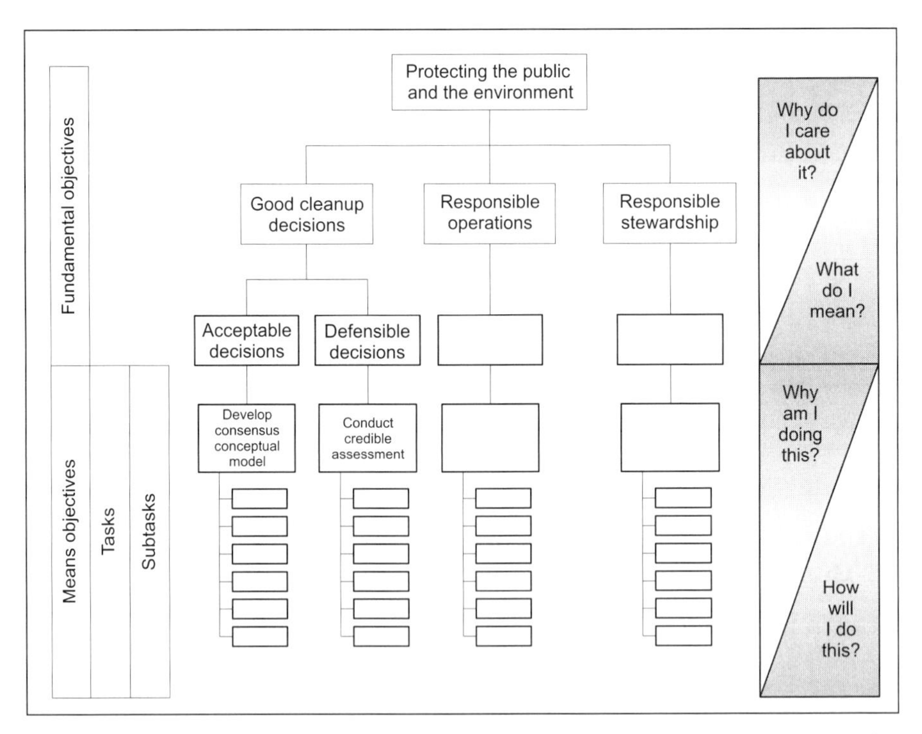

Figure 2-3. Objectives Hierarchy defines fundamental objectives to define values, and means objectives to support facts and activities.

objectives are the things a project can do to achieve a fundamental goal and should reflect (or encourage) agreement about how the program will proceed. While fundamental objectives are value-based, means objectives should be fact-based. That is, project actions should be guided by explicit hypotheses about how an action (such as further characterization or modeling) will influence another action (such as resolving an open issue precluding a decision) or a fundamental goal (such as improved decisions). Clear fundamental objectives are necessary for program stability; clear means objectives are necessary for efficient and effective progress. The objectives hierarchy pays special attention to keeping the two types of objectives distinct while clearly linking them. Fundamental objectives are successively defined by repeatedly asking "what do we mean?" Means objectives are elicited and linked together by repeatedly asking "why is this activity important?"

Theory of Constraints

The theory of constraints is a problem-solving process that uses logical tools to generate a series of pictures designed to clarify objectives and define a realistic path forward to achieve those objectives (Dettmer 1997). The theory addresses three basic questions that must be answered during the problem-solving process:

- *What* to change?
- What to change *to*?
- How to *effect* the change?

All three of these questions can be answered by using the five theory of constraints tools, either individually, or in concert as an integrated "thinking process." These tools are the Current Reality Tree, the Conflict Resolution Diagram, the Future Reality Tree, the Prerequisite Tree, and the Transition Tree.

"Nonquantifiable problems of broad scope and complexity are particularly prime candidates for a complete thinking process analysis." H.W. Dettmer, *Goldratt's Theory of Constraints: A Systems Approach to Continuous Improvement* (Dettmer 1997)

The most significant difference between the theory of constraints process and traditional problem analysis is that a series of rules governs the development of the diagrams produced by each of the five theory of constraints tools. These rules are designed to challenge individual perceptions of a situation and encourage consensus to generate representations that reflect reality as closely as possible. That consensus, combined with a hard and honest look at the entire problem and situation, allows a program to move forward more rapidly and with greater focus.

Balanced Scorecard

The balanced scorecard is a construct for measuring performance that clarifies the implications, responsibilities, and accountabilities associated with endpoints. The balanced scorecard emphasizes three aspects of performance assessment that may be of particular benefit to vadose zone programs. The first is its simultaneous focus on internal results and external validation—the process is designed to assess progress within the program while probing the external environment for signs that those results are having the intended effect. The second is its focus on the process for arriving at results (the efficiency and effectiveness of internal systems) as well as on the results themselves. The balanced scorecard process is particularly useful when results are ambiguous and may be initially best expressed in terms of the process for obtaining the information and the consensus necessary to clarify endpoints. Finally, the balanced scorecard method emphasizes a cross-functional view of success, resulting in a set of measures that are consistent and mutually reinforcing. This latter effect is critical when the program is operating under apparently or potentially conflicting objectives (Kaplan and Norton 1993).

Project Integrator

Most vadose zone programs exist in an environment where the responsibility for and interest in vadose zone conditions is distributed across programs. Complicating this is the need to integrate the management, project engineering and science, regulatory, and stakeholder communities into the program as active participants. Traditional project management models may not account for the disjointed activities and the diversity of cultures that need to be coordinated in an effective and efficient approach. The DOE's Hanford Site has established an entire project that serves as the integrator for site vadose zone activities (the Groundwater/Vadose Zone Integration Project). While other sites may not require the same level of integration, most vadose zone projects will require an integration function. At a minimum, the integration function ensures that activities affecting the vadose zone are not working at odds with one another (see the General Accounting Office report, *Nuclear Waste: Understanding of Waste Migration at Hanford is Inadequate for Key Decisions* ([GAO 1998]). Ideally, the function coordinates characterization, modeling, and monitoring activities across projects to optimize the amount and quality of information coming forth. Not only the project integrator, but also the management and science integrator and stakeholder manager (discussed later in this chapter), must communicate clearly and effectively with others, especially stakeholders. If they do not, they will be seen as roadblocks.

TRAPS AND GAPS

To use the tools, processes, and functions described in the preceding section, you need to be aware of some of the potential pitfalls. Some of these pitfalls are the result of obvious and typical programmatic distractions. Others are the result of technical or practical difficulties that arise when these tools are implemented.

Management Traps

Most of the management traps related to setting endpoints have to do with the illusion of clarity or agreement that encourages projects to cut short the process of formalizing outcomes and delineating responsibilities, accountabilities, and commitments. Specifically, projects need to watch for:

- *The illusion of endpoints.* Misunderstanding can easily occur around the definition of project objectives and endpoints because they are often described in insufficient detail to guide or measure subsequent management decisions. To avoid this situation, programs must grapple with what it means to define an endpoint (what must be described and in what detail), how that definition will be used to guide program decisions (its immediate and medium-term implications), and how the definition will be maintained in the program (its long-term implications).

- *The illusion of agreement*. Most people don't like conflict, and everybody eventually gets tired of managing it. Without the help of tools or processes that require precise descriptions and that force the program participants to work through their implications, programs are likely to carry disagreements forward unintentionally and unknowingly. In the end, this apparent harmony is unlikely to serve the project well.
- *The illusion of standing still*. Short political and budget cycles demand evidence of progress. Project managers will always be under pressure to move forward rapidly, which may conflict with the need to establish clear and agreed-upon endpoints. Establishing clear and agreed-upon endpoints can be a resource- and time-intensive process, filled with conflict and turmoil, that produces few visible results. Conflict is seldom interpreted as a sign of good management. In fact, most projects and supporting political agendas have little tolerance for conflict and few tools for dealing with it. Weathering this process is critical and will require sustained communication and cooperation with those in positions to redefine the project's scope.

Technical Gaps

Achievement of most fundamental objectives will not be measured in terms of "yes" or "no," but rather in terms of "how much." The greatest technical challenge associated with setting endpoints is probably the task of clarifying how many of the project's fundamental objectives will be achieved and to what degree. In cost-benefit terms, this clarification involves the very difficult task of qualifying or quantifying the overall benefit of the project and relating that to cost in a way that allows participants to locate the point of diminishing returns. This exercise may involve valuing benefits that do not have a market value (for example, the value of protecting ecological systems or native lifestyle practices) as well as valuing more direct benefits (such as avoiding socio-economic consequences of a "runaway waste" scare in the region). While there are many ways of measuring a program's achievement, none seems to be sufficiently practical to be properly implemented, sufficiently credible to claim ready adherence across the scientific and management communities, or sufficiently accepted to avoid challenges by Congress and other authorization or oversight bodies.

PRINCIPLE TWO: DEVELOP A ROADMAP

> "Mountains of data are collected in the management of hazardous, toxic and radioactive wastes. Unfortunately, the majority of these data have little or no value except to maintain a prosperous three-ring binder industry in our country. With shrinking resources for waste disposal and cleanup, we can no longer afford to collect data for the sake of collecting data." R.B. Gilbert, "Think Before Testing," *Practice Periodical of Hazardous, Toxic & Radioactive Waste Management* (Gilbert 1997)

Developing a roadmap is about learning. More than that, it's about infusing learning into the program through deliberate, often formalized and quantified, processes and practices. It is premised on the need to know more about the vadose zone to develop and reach agreement on remediation decisions and best management practices for existing and (eventually) remaining subsurface waste, especially where the contamination is long-lived. It is also premised on the equally (perhaps more) compelling fact that time and money are limited and that learning by design is more effective than learning by trial and error. We need to be reasonably clear about what is critical to know, what would be good to know, and what information, if presented, wouldn't matter. Every opportunity to obtain additional data about the system should be treated as an opportunity to build on the knowledge base through controlled and progressive hypothesis testing.

In this context, managers need to include a formal process for identifying and evaluating uncertainties, allocating resources to reduce them, reviewing the results of those investments and actions, and redirecting resources as necessary. It is critical that the program implement a mechanism for establishing a shared understanding of the physical, biological, and chemical systems, their behavior under different conditions, and their critical uncertainties. Further, that mechanism must be capable of evolving so that this understanding progresses as the program progresses. Whereas principle one, described in the preceding section, is about defining a program's destination, principle two is about drawing and updating a "roadmap" to get there. This roadmap describes the *current understanding* of the physical system and its uncertainties, the *information needed* to deal with critical uncertainties, and the *actions and investments* that will close the gaps.

As with endpoint setting, this principle seems obvious but is not easy to implement. The convergence of significant, technical, social, and institutional barriers impedes the effective pursuit of information.

WHY IS THIS HARD?

Determining what data, information, or insights are needed is difficult when the decisions they must support are not clearly defined or communicated. The decisions that depend upon vadose zone data and models are often the responsibility of several organizations. Processes to coordinate making decisions across organizations are seldom in place. Because systematic approaches to learning are more likely to succeed when the value of reducing identified uncertainties is explicit, decision makers, or decision-making bodies, need to be clear about the point at which they are willing—collectively or as individual groups—to move forward without additional information. Without this clarity, tools such as defining data quality objectives or performing a value of information analysis will be of little practical use.

Evaluating information requirements takes time and resources. The processes of defining, communicating, and evaluating information requirements demands a level of formality and quantification that may not be readily supported by the program, key participants, or stakeholders. Pressure on management to deliver products may result in shortcutting the process. Program scientists may be reluctant to have their discovery process judged by peers, management, and stakeholders. And, even when done openly and with the intent of communicating broadly, stakeholders may view significant activity around information planning as a distraction from commitments to progressing with the job of characterizing or remediating the problem.

Finally, consensus about methodologies is difficult to obtain. Opinions differ within the technical community about how to combine historical data and expert judgment with actual measurements to populate models; what level of modeling is necessary to reliably locate knowledge gaps in a decision model or risk assessment; how to best characterize those gaps; and, how to close the gaps. Notable examples of good, structured, and progressive methodologies that capture the value of new information are available, and significant capability for facilitating such processes exists at many sites. Although relevant experience does exist within other fields such as mining, petroleum, and water resource development, vadose zone managers often have little experience with and no standards for what information is needed to support decisions regarding the development of highly complex models or transparent (that is, containing no hidden assumptions or agendas) characterization plans.

BUILDING BLOCKS

Fundamental to many organizations is the task of directing organizational learning with respect to how, when, and for what purpose data are gathered and analyzed or R&D is conducted. To achieve successful solutions, this learning must be tailored to a program's particular technical, organizational, and political situation. As was true for principle one, this section describes some of the elements of good solutions. While not all

Roadmap Terminology

Projects may implement a roadmapping process intuitively, but rarely completely understand or manage it. With a few exceptions (for example, a conceptual model), the specific constructs are intended to outline a reasonable logic. This outlining process entails the following steps: 1) determine what you know, 2) figure out what you need to know, and 3) decide how you will obtain the required knowledge. How this is done—for example, whether a descriptive model is captured in pictures, spreadsheets or text—is less important than that it be done in a form useful for project planning.

Assessment model: An assessment model depicts the accurate and operational renderings of the conceptual model, given the constraints dictated by selected modeling architectures or codes. Assessment models answer the

continued

question "how does the system of interest behave under specific conditions?" Assessment models can help make predictions or answer questions about the condition and behavior of waste. To that end, they are critical to making credible cleanup or corrective decisions. Synonyms for "assessment model" include "predictive model" or "assessment."

Conceptual model: A conceptual model provides a detailed, technical description of the system. It answers the question "how do we believe the system actually operates?" A conceptual model is used as a blueprint to guide specific assessments, which in turn must conform to the rules and relationships (usually mathematical) defined by the conceptual model. A thorough conceptual model articulates both what is known and what is not known (uncertainties) about the system, and documents important assumptions.

Descriptive model: A descriptive model bounds the system of concern to the project, identifies important components, and sets forth the relationships of concern among those components. It answers the questions "what concerns us?" or "what do we need to understand?" The descriptive model is used to agree on the scope and areas of emphasis for project attention. To this end, it is not important that the resulting picture conform to accepted technical principles governing how the system actually functions, but rather that it communicates aspects of the physical, biological, chemical or social system that should be considered in project assessments. Synonyms for "descriptive model" include "scoping model," "public model," and "consensus model."

Planning model: A planning model presents a "reduced form" of the conceptual model suitable for quick and relatively inexpensive exploration of the system or its outputs. It answers the question "how variable is the system under different conditions?" If cautiously used, planning models can help distinguish important and unimportant variables in the system and, thereby, help gain agreement on the variables or outcomes of greatest interest. To that end, they are useful for project planning, including the prioritization of data collection, modeling enhancements or research and development intended to further enhance the system understanding. Synonyms for "planning model" include "analytic model," "reduced form model," "policy model," "rough-order-magnitude (ROM) model," or "minimal credible model."

Roadmap: A roadmap describes the project endpoints and the activities needed to achieve those endpoints. For vadose zone projects designed to gather information and close on corrective actions, a useful roadmap starts with the decision endpoints of concern (for example, descriptive model, conceptual model). It describes the information needed to make those decisions; lists the activities useful for obtaining that information; articulates the hypotheses and assumptions that guide those activities; describes the anticipated results; and documents conclusions resulting from those activities. A project's roadmap would include activity schedules, interfaces, and endpoints. Synonyms for "roadmap" include "science and technology roadmap."

of them will be relevant to a given program, together they can act as a checklist of things that may be part of a successful approach.

Describe the System of Interest—The Scoping Model

The vadose zone is part of a larger ecosystem. Indeed, that is the reason it is of such great interest to organizations concerned with the remediation and management of local, regional, or national resources. A sound beginning point for any vadose zone program is to describe the physical, chemical, biological, and social system of interest. This description should provide enough detail to adequately scope and communicate the project, and should include term definitions as well as potential impacts. In systems engineering language, this probably means a "level 2" or "level 3" description. The intent is to clarify the scope of concern, the system elements of interest, and the important relationships among elements. The description may not be sensitive to issues of scale or diversity, and is probably not directly useful for analysis. But, it captures the things the program and its stakeholders care about and expresses conceptually how those things are related. To this extent, it can be an important part of the project's goal-setting and planning process.

The case study "Hanford Groundwater/Vadose Zone Integration Project," by Barbara Harper and Stuart Harris, describes dependency webs. *See page 125.*

Describe the System to be Analyzed—The Conceptual Model

When broad acceptance of analyses is critical for success, one of the major vulnerabilities is failure to adequately describe and communicate the system that can or will actually be analyzed. This is the job of the conceptual model. The conceptual model transforms the system description into a real-world problem suitable for further analysis. It differs from the scoping model in that it conforms to rules, which, it is hoped, are broadly accepted, about how the real world operates. It recognizes and resolves scaling, diversity, and interface issues that the scoping model can ignore. Because there are significant gaps in our knowledge about the vadose zone, there is no single, clearly correct conceptual model, and vadose zone projects may elect to carry multiple conceptual models forward. A complete characterization should clarify *what we know* (or *what we don't know*), *how well we know it*, and *how we know it* for each major element of every model. In this sense, the conceptual model functions as the "baseline" for the program's understanding of the system, as well as the blueprint for numerical models that will guide further information gathering, R&D, program actions, and, ultimately, cleanup and management decisions.

Such a conceptual model—accessible to internal and external stakeholders and reflective of their interests—may be a valuable element of a project's planning portfolio. This is especially so when there are significant gaps in theory around the constructs being modeled and significant interest among stakeholders in the details of how these will be addressed.

"In Las Vegas, you can play roulette and many other games of chance. While you do not know if you will win or lose at a particular time, you can at least accurately characterize the probability of the outcome—you know (or should know) the odds, if you know the game. In hydrogeologic and geochemical systems, not only are we unable to accurately characterize the properties of a system, but there may also be significant uncertainty in the estimates of uncertainty. Therefore, we do not know the odds. In fact, we probably do not even know all the rules of the game (or perhaps even which game we are playing); that is, for these natural systems, there will be uncertainty in the conceptual models and in the complex nonlinear coupling between models." L.F. Konikow and R.C. Ewing, "Is a Probabilistic Performance Assessment Enough?" (Konikow and Ewing 1999)

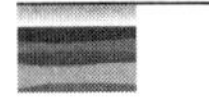

Establish a Formal Basis for Evaluating Knowledge Gaps—The Planning Model

The conceptual model will undoubtedly be characterized by significant uncertainty. However, not all those uncertainties will have implications for management. A critical task facing projects is to differentiate those uncertainties that affect current and future management decisions

"The concept of ecosystem management...has a logical requirement: that one be able to see the ecosystem as a whole in some fashion. This requires information, together with an analytic capability that can select from the information useful portraits of what is being managed." K.N. Lee, *Compass and Gyroscope: Integrating Science and Politics for the Environment* (Lee 1993)

from those that do not. Project activities should focus on reducing uncertainties that impact decision making.

The system of interest is typically too complicated, the stakes too high, and the resources too limited to delegate uncertainty identification, evaluation, and subsequent investments in new information to expert judgment alone. Explicit but relatively high-level modeling can:

- Clarify problems
- Enhance communication among scientists, managers, and stakeholders
- Screen policies to eliminate remediation or management options unlikely to add significant net value
- Identify knowledge gaps that make more sophisticated model predictions suspect and that undermine the ultimate acceptability of remediation or management decisions.

Such summary-level models have been referred to as "planning," "analytic," "reduced form," and "rough order of magnitude" models to distinguish them from predictive models designed to set standards that support cleanup decisions.

"The modeling step ... allows us, at least in principle, to replace management learning by trial and error (an evolutionary process) with learning by careful tests (a process of directed selection)." C. Walters, *Challenges in Adaptive Management of Riparian and Coastal Ecosystems* (Walters 1997)

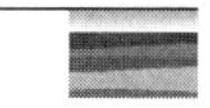

By definition and design, these planning models do not necessarily capture the complete source term or analyze the full range of impacts included in the system description (or scoping model). Instead, they use indicators that provide some level of confidence that the parameters and relationships of greatest interest are accurately reflected. They provide a potentially useful bridge between the conceptual and the more detailed and often less holistic predictive models that may be necessary for supporting final decisions. As such, policy models may be invaluable for limiting an otherwise potentially intractable prediction problem. They also have the benefit of being constructed and implemented in shorter

A Requirement for Identifying and Managing Dominant Factors

At the Hanford Site, regulators and stakeholders worked together with the U.S. Department of Energy (DOE) to establish requirements for the planning approach and methodology for conducting a site-wide impacts assessment. The following excerpt underscores the value of a planning model for identifying knowledge gaps.

In virtually all things a relatively small number of factors dominates the outcome. It is of the utmost importance that this assessment not leave out any factors which dominate the results. Yet the magnitude of work and cost of the analysis must be responsibly managed. Sensitivity analyses, parametric analyses, and related methods will be used to identify and rank order the factors which dominate the outcome of this assessment. These factors may be physical attributes of the Hanford Site or waste disposal, or they may be technical characteristics and challenges within the study itself. Assumptions framed through expert judgement (in lieu of repeatable analyses) will not be used to identify dominant factors or discard smaller contributors. The resulting understanding of relative importance will be used to focus technical emphasis, management oversight, and assessment planning…"

U.S. Department of Energy, *Screening Assessment and Requirements for a Comprehensive Assessment, Final; Columbia River Comprehensive Impact Assessment (CRCIA), Part II: Requirements for a Comprehensive Assessment* (DOE-RL 1998)

time and often at significantly less expense than more sophisticated models. This relative speed allows the program to demonstrate some progress and to "bound the playing field," relative to larger modeling efforts, in a reasonable amount of time.

Map Out the Activities Required to Fill Knowledge Gaps—The Roadmap

It is hard enough to identify the important uncertainties; it's yet another to fill these gaps. Knowledge gaps with which vadose zone projects will be presented will likely include conceptual gaps (requiring theory and methods), modeling gaps (requiring better predictive tools), sampling gaps (requiring better characterization methods), and measurement gaps (requiring better measures of accessible data). A plan, or map, for filling these gaps needs to address the full range of required

"A *roadmap* is an extended look at the future of a chosen field of inquiry composed from the collective knowledge and imagination of the brightest drivers of change in that field. Roadmaps can comprise statements of theories and trends, the formulation of models, identification of linkages among and within sciences, identification of discontinuities and knowledge voids, and interpretation of investigations and experiments. Roadmaps can also include the identification of instruments needed to solve problems, as well a graphs, charts, and showstoppers."
—R. Galvin, Chairman of the Executive Committee of Motorola,
Science, May 8, 1998. (Galvin 1998)

activities, show how these activities logically relate to one another, and explicitly specify the expected result and how it will be used by the program. Only then will the project have the basis for considering the relative value of possible activities, determining the best path forward, obtaining the best value in terms of better information, managing activities to maximize the total effect, and managing expectations around the final product.

Formalize Working Hypotheses and Document Results—Applying and Updating the Models and Roadmap

One of the most discouraging cost drivers in the often reactive or opportunistic approaches to managing vadose zones is the opportunity cost. Opportunities to learn are lost because of poor planning, poor execution, poor reporting, and poor communication, among other reasons. Natural resource management theory offers a basic principle for learning that has some relevance to vadose zone management (although it doesn't translate fully to the federal cleanup mission). The principle is "adaptive management" (Holling 1978; Lee 1993). The idea is to recognize that interactions with complex natural systems are, at best, experiments and to treat them as such. Similarly, programs operating in and around the vadose zone would do well to be explicit about the assumptions and hypotheses underlying selected approaches to characterization, monitoring, or remediation; design an approach consistent with that understanding; and, where possible and appropriate, consider control or analogous site designs that will help resolve alternative explanations for inevitable surprises. These assumptions and hypotheses can

"Adaptive management takes [uncertainty] seriously, treating human interventions in natural systems as experimental probes. Its practitioners take special care with information. First, they are explicit about what they expect, so that they can design methods and apparatus to make measurements. Second, they collect and analyze information so that expectations can be compared with actuality. Finally, they transform comparison into learning—they correct errors, improve their imperfect understanding, and change action and plans."—K.N. Lee, *Compass and Gyroscope: Integrating Science and Politics for the Environment* (Lee 1993)

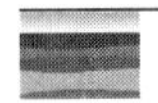

(and should) be recorded in the documentation for the conceptual model or roadmap to enhance institutional memory and learning.

TOOLBOX

The tools available to find and fill knowledge gaps are certainly numerous. We have described here several tools that we have found useful. This list is not exhaustive or even representative of the suite of tools available.

Structuring and Modeling Tools

There is significant technical debate over using "reduced form" models in program planning. Legitimate and significant concerns are related to the ability of these models to meaningfully represent the uncertainties of complex natural systems or to provide reliable predictions, even at a high level. However, simplified models can provide highly valuable adjuncts to more detailed modeling efforts (Merkhoffer 1987). Among the tools for providing that capability, influence diagrams offer valuable versatility.

Influence diagrams provide a logical framework for identifying and structuring the relationships among variables affecting a decision (such as setting a cleanup standard). They provide a relatively simple, visual representation of the system of concern, showing the critical actors and the relationships among those actors relative to a specific decision. The diagrams can be used to represent and quantify the relationships described in a conceptual model of a physical system (for example, dependency webs) or in a decision process. Applied to a complex problem such as the vadose zone, influence diagrams can be used to create, integrate, analyze, and communicate the results of detailed, but independent, quantitative models. Influence diagrams are generally more flexible and more visual; in many cases, they can require less computational power than detailed predictive models. Consequently, they function well as a bridge between science and policy, allowing "what if" analyses that help stimulate thinking about further assessment and management options. Combining software tools that use influence diagrams with tools such as hierarchical modeling (which embeds models within models to reduce complexity and enhance understanding), uncertainty and sensitivity analysis, and object-oriented environments, can result in a relatively low-cost, user-friendly vehicle to describe systems and conceptual models. Many such tools are available (for example, Decision Programming Language [DPL], Decision Analysis [DATA], HUGIN, Setica, Strategist, ithink, and STELLA). Most are credible and relatively easy to use, but practical differences will determine the best match to a specific project's needs.

Information Valuation Tools

Decision makers who face decisions with uncertain consequences have many choices about whether or not to gather information, what information to target, and how to get it. Decision makers who gather new information, through data collection, assessments, or R&D investments, do so because they believe it will help them make better decisions. However, even when new information addresses an acknowledged uncertainty or knowledge gap, its addition to the decision problem may not change how choices or outcomes are viewed. The data quality objectives process and value of information analyses are two of the better-known methods for ascertaining the value of gathering additional information to make better decisions.

The *data quality objectives process* is probably the most familiar method for formally incorporating information needs into program planning. This process is essentially a series of focused questions leading to an assessment (and documentation) of the type, quantity, and quality of data necessary to fulfil specific project objectives and ensure confidence in decisions. Used correctly, the data quality objectives process can assist in analyzing regulatory constraints; selecting the instruments, frequency, duration, sample locations, type, and extent of characterization or monitoring activities; establishing levels of acceptable uncertainty related to risk; identifying data limitations; and enhancing ultimate agreement and acceptance of final characterization or monitoring systems to be implemented (EPA 1993). The process explicitly addresses uncertainty in decision making caused by incomplete information and permits the examination of tradeoffs between better information, costs, and probability of incorrect decisions. Time and again, the data quality objectives approach has been demonstrated as an effective tool for reaching and documenting agreement among stakeholders on key decision criteria in the presence of uncertainty. "The Use of DQOs in Designing Vadose Zone Monitoring Systems," a case study by Kevin D. Leary contained on the accompanying CD, provides a good example of how data quality objectives are used in a planning problem related to the vadose zone.

Value of information analysis shares many of the qualities of the data quality objectives process, but ties the value of additional information more explicitly to the information's ability to change a decision maker's course of action. Beginning with a well-structured decision, a value of information analysis includes the specification of important decision variables, the relationship among those variables, the uncertainties around them, and some indication of the change in the anticipated value of a choice if uncertainty is reduced. That detail provides a formal basis for assigning value to new information specifically by testing whether or not reductions in uncertainty anticipated from new information will change the anticipated value of a decision. Like the data quality objectives process, value of information analyses are helpful for structuring decision problems as well as evaluating information needs (Wood *et al.* 1995; Clemen 1996). Fassbender *et al.* (1996) and Colson *et al.* (1997) provide good examples of how to apply value of information analysis to a cleanup problem.

Both processes require upfront agreement about the decisions and explicitly address decision uncertainties. Value of information analysis tends to be a more rigorous approach that dissects the uncertainties in more detail and integrates across multiple decisions simultaneously, whereas the data quality objectives process deals with the total decision error and addresses each decision independently. To that extent, value of information analysis returns a potentially more meaningful measure of value for decision making, if adequate information is available or if the added cost and complexity of structuring the decision or related decisions warrant the investment to obtain it.

Management and Science Partnerships

To generate usable vadose zone information, projects will likely need to address some of the basic knowledge and methods gaps that are highlighted in this book. To that end, science needs to be integrated into the project. This inclusion is not trivial, in part because of how differently scientists and managers behave and how the interaction between them is typically managed. Scientists tend to seek problems of intellectual difficulty and are therefore mostly in the business of raising questions. Managers are generally interested in solving known problems and are therefore in the business of closing questions. Scientists and managers also operate under reward systems that promote these differences. The challenge is to design a program interface that will promote a strong partnership between scientists and managers (Rogers 1998). To build this interface, scientists and managers need to recognize that the problem has not been solved until the desired management endpoint is reached, and that the desired endpoint won't be reached without specific contributions from the science community (as defined in a roadmap or other plan). The purpose of the partnership should be to develop consensus on goals, structure, and process. The underlying premise for this scientist-manager partnership is that the most enduring management solutions will be ones that are technically sound and scientifically defensible.

To establish this partnership, the project management team should create and staff a position that will act as the interpreter, integrator, advocate, and business manager for effectively infusing science into the project. This function may be filled by a chief scientist, a science advisory board, an expert panel, or some combination of these. Regardless of how the position is filled, the ability to operate effectively and communicate within both management and scientific circles, as well as with diverse stakeholders, is critical. Structuring scientific and technical tasks in a way that is sensitive to the reward systems of both communities will also help (for example, by specifying an activity that not only generates a product on schedule, but that also supports information sharing through conferences or publication). Most useful, though, will be ensuring that key participants in this partnership are clear about and agree with project objectives and endpoints, and understand how science and technology feed project priorities.

Applied Science and Technology Roadmaps

An applied science and technology roadmap lays out the vision, logic, schedule, and expected products of science and technology activities that will benefit the project. The roadmapping process uses structured tools to plan and manage the infusion of science and technology into a project to improve the project's chances of reaching a desired endpoint. In contrast to traditional research process plans, science and technology roadmaps emphasize consensus and building around goals and gaps. The product reflects the collective knowledge of a cross section of professionals from diverse scientific and management disciplines. Ideally, the process incorporates historical data as well as expert judgment and modeling results in its assumptions, logical relationships, and anticipated results. The roadmap provides the project manager with an organized way to update assumptions, uncertainties, hypotheses, and knowledge gaps.

Hanford's Groundwater/Vadose Zone Integration Project has developed a science and technology roadmap for filling knowledge gaps related to the development of a site-wide system assessment for establishing cleanup requirements (DOE-RL 1999a). That roadmap was created through a process in which scientists and engineers from national laboratories and universities worked with their customers (the DOE, Tribal Nations, regulators, stakeholders, and remediation contractors) to define important knowledge gaps and establish a path toward solutions. The main objective of the roadmap was to provide the new knowledge, data, tools, and understanding needed to enable site remediation and closure decisions. The roadmap links science and technology activities and outcomes, key decisions, and project execution to support those decisions, and then it presents budgets and schedules associated with those activities. Further, the roadmap establishes priorities based on the anticipated relative value of the science and technology outcomes and their likely success (Figure 2-4).

Roadmapping Logic		Timeline 1	2	3	4	5
Integration Logic	Identify project activities affecting vadose zone. Clarify relevant decisions, decision process, and schedules	▲ Key project milestone		▲ Key project milestone		▲ Key project milestone
Assessment Logic	When and what assessments are required to support project milestones?	▲ Rev. 1	▲ Rev. 2	▲ Rev. 3		
Inventory Modeling Improvements	What inventory improvements will reduce uncertainty the most?	▲	▲			
Vadose Zone/ Groundwater Improvements	What fate & transport improvements will reduce uncertainty the most?	▲	▲			
Risk Modeling Improvements	What exposure & impacts improvements will reduce uncertainty the most?			▲	▲	
Regulatory Activities	What programmatic resolutions will reduce uncertainty the most?	▲		▲	▲	
Science and eTchnology Requirements	What science and technology is required to support improvements?	Characterizing, modeling, monitoring, applied R&D requirements				
Science and eTchnology Investigations, Expertise, & Supporting Data	Critical path for generating data? Mechanisms for delivering expertise as needed?	Specific activities useful for meeting science and technology requirements with schedules and interfaces				

Figure 2-4. Example conceptual set roadmapping logic.

Prioritization

As distasteful is it often is, resources must be prioritized. Managers can prioritize either using implicit or intuitive rules (for example, squeaky wheel gets the grease) or using an explicit process. When prioritizing R&D or technology development investments, the basic questions are:

1. Where are the greatest uncertainties?
2. Which of these uncertainties involve the greatest stakes (such as budget, schedule, or other potential consequences)?
3. Which of these "high stakes" uncertainties can be most readily reduced?

These questions are the topic of significant research and debate, and the object of a plethora of decision analysis tools (NRC 1995a; NRC 1998; Matheson and Matheson 1998; Ulvila and Chinnis pending). In lieu of a complete overview of this topic, we offer the following practical principles for prioritization:

- Decide the purpose of the prioritization task early, and design the process to that task; re-examine the purpose as you go, and change the process or prioritization model as appropriate.
- Clarify the characteristics of the process most important for the problem at hand, and design to those characteristics. For example, if consensus is the most important attribute of the end product, the process should involve all important parties, and supporting models should support "what if" options that allow the group to explore alternatives. If technical defensibility is most important, the process should be formalized, documented and tied to broadly accepted methods. If efficiency is critical, the process should be simplified as much as possible.
- The approach, including any prioritization models, must be accessible and understandable to critical internal and external stakeholders.
- Significant benefits can be derived simply from clarifying the objectives. Articulate all of the important considerations among the evaluation criteria; do not withhold political or hard-to-measure considerations.
- Improve analytic techniques incrementally; take your best shot at doing it right, and refine the analyses only as useful.
- Use numerical formulas or algorithms for assigning weights (e.g., to criteria) or values (e.g., ratio of value to cost) cautiously; use sensitivity analyses liberally to investigate the implications of alternative schemes.
- Communicate results in major groupings (e.g., clear winners, contenders, and clear losers); do not attribute great meaning to small differences.
- Use a decision analysis facilitator to initiate and manage the process, especially where multiple stakeholders and potentially conflicting objectives are involved or where significant importance is placed on numerical models.

Structured Expert Judgment

Expert judgment is prevalent in solving any significant technical problem, and it will be a critical source of information in any vadose zone program. Judgment is involved in determining that a problem is worth attention. Judgment is needed to understand the dimensions of a problem, consider the best approaches to solving it, determine when and how to collect data, and interpret the results. Nevertheless, project managers, participants, and stakeholders are often resistant to and suspicious of expert judgment when it is used as a source of information in prioritization, modeling, assessment, and other technical activities. Explicit judgments are easier to communicate precisely, but require greater effort, than implicit judgments. As the use of expert judgment is unavoidable, the real management challenge is deciding when to augment it with formal analysis. The principle suggestion here is to be deliberate about these decisions and thorough in their implementation. At a minimum, this involves attention to when they are best used. Explicit structuring or elicitation techniques help to formalize judgments and obtain the greatest value from their use.

Structured expert judgments typically break an implicit thought process into smaller parts and logically integrate those parts to obtain the desired data (an example would be the level of uncertainty around specific inventories of contaminants). Judgments can be rendered either quantitatively or qualitatively, though quantification has more advantages. Four general principles of engineering apply to the process of assessing expert judgments:

1. Try a reasonable approach to interpret, explain, and reveal the implications of expert judgments. If this fails try a different reasonable approach.

2. Use successive approaches to challenge or budge initial judgments.

3. Use successively better approximations to converge to and bound elicited statements.

4. Use independent approaches to obtain reliable judgments (Keeney and von Winterfeldt 1989; Keeney and von Winterfeldt 1991).

The bottom line is 1) to recognize that experts are biased; 2) to consider reasonable approaches to removing bias; and 3) to validate the results.

Technical Peer Reviews

Handled appropriately, technical peer reviews can enhance the validity of the technical work and increase confidence in its use. Internal peer reviews improve the understanding and coordination of management activities that share a common vadose zone boundary. External peer reviews ensure that the program benefits from knowledge, methods, and lessons learned and developed elsewhere. Careful thought should be given to the scope and schedule for such reviews and attention paid to the charter and protocols governing expert review processes and panels. The following items need to be communicated clearly to panel members, program participants, and stakeholders:

- How reviewers are selected
- What is the scope of the review
- How specific review topics or meeting agendas are established
- How meetings will be run
- What resources are available to support reviews
- How panel members will interact with program participants and stakeholders
- What products are expected from review panels
- To whom the reviewers are accountable
- How the program will respond to expert advice.

The potential for conflicts of interest must be carefully addressed, as it can jeopardize the validity and credibility of the process. Many talented experts serve in multiple review capacities, and there is a danger that they may end up reviewing the results of panels on which they also served, or reviewing material related to decisions in which they have a commercial or personal stake. It is best to avoid conflicts of interest, but if that is not possible, it is very important that they be identified, disclosed, and managed. As with expert opinions, technical reviewers bring biases that, though not inherently bad, need to be carefully addressed.

Traps and Gaps

To use the tools and processes described in the previous section, you need to be aware of some of the following potential pitfalls.

Management Traps

One of the critical barriers to enhancing our understanding of the vadose zone is that *progress is predicated on management's demand for vadose zone information*. That demand for vadose zone information, in turn, requires that management be responsible for the nature and scope of decisions around closure and long-term management, which drives the need for vadose zone information. Significant political sensitivities and institutional complexities, as already discussed, frustrate even the best-intentioned efforts to do this.

Another trap is *letting passive management reinforce inaction by staff*. In this situation, managers and project participants realize that mistakes resulting from action are often more painful to the individual than mistakes resulting from inaction. The opposite may be true for the organization as a whole; that is, the system may benefit more from action than inaction.

Another management trap is *the belief that models or experts will eliminate bias* from the program. Biases are pervasive. Managers, scientists, stakeholders, regulators, and reviewers all have them. The management challenge is not to eliminate them but to manage them. The first step is to recognize the various biases. The second is to identify where the project may be most adversely affected by those biases. The third is to find ways to incorporate them into the project through processes that neutralize, or at least balance, those biases that would otherwise undermine progress.

A fourth trap involves *convictions, sometimes deeply rooted, about what constitutes a "real model."* Many projects are reluctant to invest in descriptive and policy-level models, preferring to direct all available resources to complex simulation or predictive models. This preference may reflect a presumption that sound predictions (and, hence, good decisions) will be found by looking more precisely, and in more detail, at more variables and factors (Walters 1977). While legitimate in some important respects, this line of thinking breaks down in other important respects. First, critical interactions or events can be highly concentrated in space and time at scales, locations, and/or times that may be ignored in even the most detailed models. Second, adding more detail adds more parameters to the model but doesn't necessarily guarantee that these will be supported by field data; the resulting "overparameterization" can degrade predictions. Third, some ecological interactions result in feedback from local events across local scales that produces highly irregular patterns at much larger scales (Holling 1992). When the measures of interest include local and regional effects, the value of very large, highly detailed models can be significantly undermined. Such concerns should be considered, along with the difficulty of maintaining the political will and program budget to develop, resolve, and eventually execute detailed and complex models. These concerns should be well-considered when defining any assessment effort.

A fifth trap lies in the mindset that *science and technology is a luxury rather than a necessity* for getting the job done. Related to this is the tendency to assign accountability, planning, and investment for science and technology to the periphery of the organization, distinct from other program activities. This practice almost guarantees that those investments will not deliver the anticipated benefit. Program managers need to be involved, as owners, in developing science and technology roadmaps and the resulting program and products. If budgets were unlimited and engineered solutions were available that eliminated concerns about endpoints being sufficiently protective, the need to improve the understanding of the current and future behavior of contaminants in the vadose zone would be significantly less. However, budgets are limited, and inexpensive engineering solutions have so far been elusive. Science and technology offer one important bridge between today's budgets and tomorrow's solutions.

Technical Gaps

One of the reasons we do not understand more about the vadose zone is that significant increments in our understanding will likely require improvements in our methods of characterizing, modeling, and investigating the vadose zone. Similarly, one of the reasons vadose zone management is hard is because we lack the methods to efficiently and credibly evaluate what projects to do, in what order, and how they should be done. Embracing the principle of finding and filling knowledge gaps means that project planners and managers will need to be determined and creative about pursuing methods to articulate, document, and assign value to what they care about, and then to allocate the resources to address important knowledge gaps.

There is also a lack of precedence or guidance for representing many significant uncertainties in policy-level models without sacrificing the ability of those models to meaningfully identify dominating parameters and sensitivities that will spotlight important gaps in our knowledge.

Related to this is a specific gap in the methodology for defining, selecting, and measuring nontraditional values or impacts of concern, especially when these have no clear market value (such as cultural lifestyle impacts) (Cummings *et al.* 1986; Peterson *et al.* 1988; Harris and Harper 1997). Even if it were currently possible to agree on and measure selected indicators for the values of interest, meaningfully integrating these measures into traditional modeling and planning tools, and interpreting and communicating the results, has eluded researchers to date.

PRINCIPLE THREE: PRACTICE DELIBERATE, DISCIPLINED ENGAGEMENT

Like determining project goals and developing an effective basis for understanding the system, engaging stakeholders, regulators, managers, scientists, and engineers is a critical step in the success of an effective vadose zone project. Environmental cleanup and management projects routinely, if not easily, work across scientific, engineering, and management boundaries. But America is a democratic society; citizen groups

"The environment is necessarily shared by us all; but as every littered park reminds us, what is shared by many is typically abused. Reconciling control with the diversity and freedom essential in a democratic society is the task of bounded conflict." K.N. Lee, *Compass and Gyroscope: Integrating Science and Politics for the Environment* (Lee 1993)

are playing an increasingly large role in decisions. Because vadose zone contamination can impact the groundwater resources on which so many depend for drinking water and irrigation sources, stakeholders and regulators often expect their involvement to equal that of the scientists, engineers, and managers. As long as vadose zone managers are in the business of addressing, augmenting, evolving, or creating site-specific cleanup standards, sustained and focused engagement of senior regulatory executives, their technical staff, and their key constituencies is mandatory. While this level of engagement may be necessary, it is never easy.

WHY IS THIS HARD?

To truly engage scientists, engineers, managers, stakeholders, and regulators, in a disciplined manner, in the development and conduct of any technical program can be a challenge. At the most basic level, each of these groups uses a different language and concepts. Language is often a symbol of differences in philosophy, culture, and even values. Among the various stakeholders and participants are undoubtedly those with divergent responsibilities, perspectives, priorities, and learning and communication styles. These differences can make both engagement and communication difficult.

The term "stakeholder" has come to mean the general public. However, a true stakeholder is anyone who perceives themselves as affected by (or having a stake in) a particular issue. Using this broader definition, stakeholders may include regulators, managers, scientists, engineers, environmental groups, Native American tribes, economic interests, and a broad host of others for any specific activity. In this chapter, we use the term stakeholder to mean anyone not directly affiliated with those responsible for managing and regulating the vadose zone.

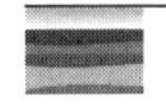

In this day of increasing involvement across multiple groups and diverse viewpoints and backgrounds, roles quickly become unclear and those charged with moving the project forward can find it difficult to determine whose "agenda" has priority. In addition, some managers, scientists, and engineers, the group responsible for the technical rigor of the program, may question the ability and right of stakeholders to make decisions in areas that are not traditionally open to lay participation and interpretation, such as conceptual model development and project management. Without a clear and disciplined process, stakeholder and participant engagement can become a significant (even attractive) diversion. Worse, engagement can fail to yield improvements in the project, such as making the decision process explicit (or transparent), the approach more comprehensive, or the solution more acceptable. Appropriate engagement is, therefore, as necessary as appropriate management and science.

"There is also a need for a formal decision-making framework for future land-use and cleanup standards that will provide an opportunity for consensus-based selection of appropriate data, analysis, and criteria for decision-making. The framework must include an opportunity for stakeholder input at all stages and lead to enforceable agreements that can be modified as further knowledge is gained." National Research Council, *Improving the Environment: An Evaluation of DOE's Environmental Management Program* (NRC 1995b)

While the need for disciplined engagement is rapidly becoming a given in many programs, this critical project component is seldom given the attention paid to other critical project aspects. Planning is often absent, and in its absence, inappropriate methods and processes are used that exacerbate rather than resolve disagreements, thus wasting information and effort. Even when plans are carefully made, failure to proactively manage the process can still derail the effort, just as in other dimensions of program implementation. To fully engage scientists, engineers, regulators, managers, and stakeholders, developing a transparent, disciplined framework is as important as endpoint definition and roadmapping.

BUILDING BLOCKS

Choose a Philosophy

In the last few years, two schools of thought have emerged regarding appropriate ways to engage stakeholders, regulators, and decision makers in scientific and engineering pursuits. One approach contends that each group associated with a decision has a distinct role and these roles should not overlap. Traditionally, those supporting this approach would have scientists plan and conduct science, engineers take the science and apply it to a real-world problem, regulators and facility managers negotiate decisions, and stakeholders review and comment on the decisions. In the best use of this approach, regulators and stakeholders provide input at an early stage to help guide program development and "ratify" proposed decisions before they are finalized. For example, stakeholders and regulators supply suggestions at scoping meetings for an environmental impact statement; scientists and engineers take those comments and build the outline of a document based on their experiences; managers review and approve the outline; and scientists and engineers conduct the impact analysis and present the results to managers, who pass it along to the stakeholders and regulators for review and comment. A common way to look at this approach is the separation of what is traditionally seen as the realm of science and analysis from what is seen as the realm of management and policy making.

The other school of thought contends that stakeholders, regulators, and decision makers can and should have an impact equal to that of scientists and engineers on the development, implementation, and evaluation of programs. Those ascribing to this philosophy would have stakeholders propose alternative models, gather or interpret data, and evaluate results based on their experiences; regulators evaluate alternative remediation technologies and agree to modify endpoints based on their knowledge and experience; and scientists and engineers conduct their work with a clear understanding of 1) the decisions being made with the information they are developing and 2) what stakeholders expect from the analysis. In this approach, science and management functions are more closely fused, and tasks shared, with each group responsible for ensuring that their particular knowledge and perspective is incorporated throughout the process.

Various organizations, agencies, and universities have taken positions advocating one or the other of these approaches. For example, the Engineers' Public Policy Council of the American Association of Engineering Societies (1996) has come out strongly in favor of separating the activities of science and management. On the other hand, a blue-ribbon panel of the National Research Council advocated incorporating stakeholders and managers throughout the scientific process (NRC 1996b). In reality, an effective engagement of the various parties can lie at a variety of locations along the continuum created by these approaches. Those managing the vadose zone need to be aware of these different

Two Sides of the Coin

In 1996, two scientific and engineering organizations came out with differing opinions on the issue of combining science and management. The Engineers' Public Policy Council (EPPC) of the American Association of Engineering Societies (1996), in its statement "Risk Analysis: The Process and Its Application," clearly separated risk assessment (science), risk management, and risk communication (stakeholder engagement). While it recommended "that the regulated community, the scientific and engineering professions, other interested organizations, and the general public be afforded the opportunity to participate in the risk analysis process" it also recommended the divorce of societal value judgments from the scientific process. Relying on National Research Council findings from the 1980s, the statement reinforced the distinction of risk assessment as "the domain of scientific and engineering communities" and risk management as "the responsibility of political decision-makers." Risk communication was simply the method for explaining the process.

On the other side of the coin, the National Research Council (NRC) has been considering how to improve decisions about risks to public health, safety, and the environment for many years. A series of reports, including the one cited by the EPPC statement, focused on ways to improve risk assessment and risk communication. In 1996, the NRC produced a book (NRC 1996b) focused on improving the risk management process. In what was a new approach to many at the time, they advocated combining analysis and deliberation, offering seven principles to achieve this end:

1) The scientific activity should be a decision-driven activity.
2) Those involved in the scientific activity should understand that "coping with a risk situation requires a broad understanding of the relevant losses, harms, or consequences to the interested and affected parties."
3) The success of the scientific activity "depends critically on systematic analysis that is appropriate to the problem, responds to the needs of the interested and affected parties, and treats uncertainties of importance to the decision problem in a comprehensive way."
4) The scientific activity must include early and explicit attention to problem formulation by all parties.
5) The scientific activity must follow a process that is shared, through an iterative process, by all parties.
6) The scientific activity should begin with an analysis of the decision situation that matches the decision-making process to the decision being made, particularly in terms of level and intensity of effort and representation of parties.
7) The organization in charge of the scientific activity should "pay attention to organizational changes and staff training effort that might be required, to ways of improving practice by learning from experience, and to both costs and benefits in terms of the organization's mission and budget."

approaches and make deliberate choices as to which aspects will be most effective given their circumstances.

Determine a Level of Engagement

The appropriate level of engagement will depend, to a large degree, on the level of uncertainty and the stakes of the program. Figure 2-5, which is adapted from Ravetz (1987), lays out the interplay between uncertainty and decision stakes. In general, when stakeholders, regulators, and managers feel the stakes concerning a decision are high (for example, when the safety of the drinking water source is perceived to be at risk, human or environmental health is perceived to be impacted, and cultural factors are perceived to be affected), more involvement will be needed across a broader spectrum of groups. In addition, as uncertainties increase and expand to include gaps in our understanding (for example, the uncertainty as to what governs contaminant movement through the vadose zone), so too will the perceived need for greater engagement in

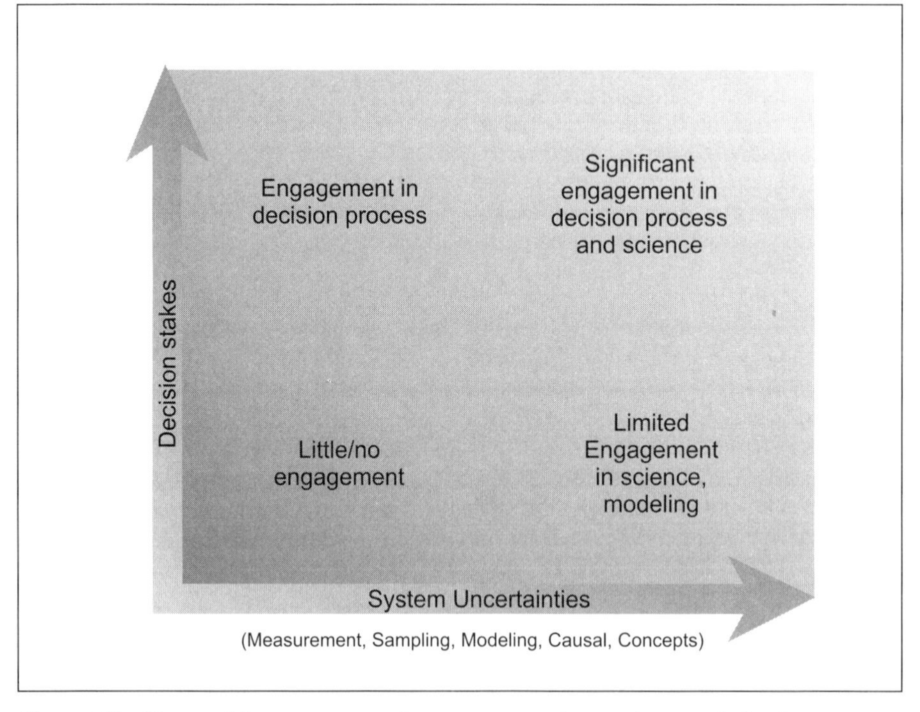

Figure 2-5. The need for engagement increases as stakes and uncertainties increase.

the assessment, and in the underlying science on which decisions will be based.

Additional criteria a project manager might apply to determine the appropriate level of engagement include the following:

- Time-frame of involvement—if a decision is needed within a 2-month window, developing a large stakeholder involvement effort, such as a citizen advisory board, may not be feasible, just as conducting a detailed technical analysis may not be feasible. Activities need to be structured to accommodate both program schedules and public needs. Schedules that are fundamentally inconsistent with the needs for analysis and engagement should be critically evaluated, for they run the risk of failure, as well as wasting time and resources.
- Amount of previous involvement—if stakeholders have recently participated in similar types of projects, and if there is a base of organizational structures and stakeholder understandings that can be built upon, then more in-depth involvement is possible earlier in the project. However, if stakeholder involvement systems are undeveloped or nonexistent, and if stakeholders must reach a level of understanding before they can participate, then considerable time and effort must be devoted to developing common goals and understanding before desired outcomes can be achieved.
- Level of public trust—in circumstances where secrecy or breeches of faith permeate interactions, engagement with stakeholders will be qualitatively different than where there is a history of positive community and regulatory relations.
- Constraints on stakeholder involvement—budgets, schedule, staff availability, facility availability, and stakeholder interest and availability can influence how many stakeholders can be involved, and at what levels, over the life of the project. This constraint affects other portions of the program, as well.

Develop and Manage a Framework

Regardless of which school of thought is embraced or the type of engagement chosen, setting and managing an explicit framework for engagement is critical. Among the many variables in such a framework that must be tailored to a given situation are the definitions of process, problem, and outcomes.

Processes vary widely from strictly parliamentarian rules of order to true collaboration and power sharing. In any case, rules must be established at the beginning. Rules lay out roles and responsibilities of various groups. In addition, rules address questions like how stakeholders, regulators, scientists, and engineers will contribute to the decision-making process; at what points in the process each group will be involved and how; and who is responsible for ensuring that each group is heard, input is used, and that information on its use is communicated back to its provider. Rules also address how the program moves from discussion to decision, what constitutes a resolution of an issue, and what would prompt reopening of an issue. Rules should be firm enough to ensure a transparent process, but flexible enough to account for group learning.

Another aspect to consider in process definition is communication. Is there a shared context and language among groups? Do these need to be built? Is any one group uncomfortable with a given communication style, or are there other ways in which ideas can be shared that span the differences in language and style? Making communication issues concrete is the first step in overcoming them.

While roles and responsibilities, rules, and communication styles may vary across processes, one parameter that does not vary is the perception of equity. For a process to succeed (where success is defined as contributing to problem resolution), all groups must feel engaged. In other words, they must feel that 1) they were heard, 2) their comments and ideas were given weight in the decision-making process (i.e., they made a difference), and 3) they had recourse if items one and two were not satisfied.

Equally important to process definition is problem definition. What question is the vadose zone program addressing? What portion of that problem will the engagement help address, if not all? What is the underlying problem—vadose zone contamination, groundwater protection, receptor protection? Are we deciding how to approach characterization, how much to remediate, how to choose a technology, or all of the above? It would appear obvious that all groups must be working to solve the same problem, but a surprising number of engagements fail because of problem definition discrepancies.

The third key area to address when developing a framework is the definition of outcomes. What can groups legitimately expect when the process ends? Will there be a final report detailing the process, inputs received, and how those inputs were used? (An example might be the comment resolution documents produced after a scoping process in support of an environmental impact statement.) Will all groups sit in on a collaborative meeting to address divergent views? Will groups interact directly to present their issues to management? Will there be mid-points, before the end, when products will also be produced and feedback given? Setting such expectations up front contributes to the transparency and ultimate success of the engagement.

Once a framework has been developed, working with all groups to ensure acceptability, decision makers and managers must take one more step to ensure effectiveness. They must set aside appropriate resources (time, funding, support staff). Depending on the agreed-upon framework, this may include time to compensate heavily involved stakeholder groups for time away from work or for travel expenses (Covello *et al.* 1988; Pacific Northwest Laboratory and Creighton and Creighton 1993; Chrislip and Larson 1994; Bradbury and Branch 1995; Engineers' Public Policy Council 1996; Walker and Daniels 1997).

TOOLBOX

A number of tools and principles have been developed to help scientists, engineers, managers, regulators, and stakeholders work effectively together to reach agreement on technical and risk issues, and to help managers plan and implement effective engagement processes.

Facilitation and Conflict Resolution

Two of the fastest growing areas of relationship management practice are those of facilitation and conflict resolution, although they have yet to be used on a wide scale at federal facilities. In these processes, a neutral party encourages mutual learning and decision making. Depending on the type of process used, the facilitator can focus energies on a particular decision or bring issues into the open for discussion. To allow effective mutual learning and decision making, people must build relationships, trust, and understanding together. An effective dialogue, not simply information transmission, is needed. Relevant examples are hard to find for vadose zone work but an emerging scientific program for the DOE has been studying the use of facilitation to help scientists and the public communicate about vadose zone remediation technologies with some success (Bilyard *et al.* 1998).

The case study "The Columbia River Comprehensive Impact Assessment," by Thomas W. Woods, provides a good example of the importance and benefits of effective dialogue. *See page 127.*

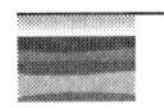

A Balanced Scorecard for Engagement

One tool that may be useful for the managers of vadose zone programs is the concept of acceptability, applied in a manner similar to the balanced scorecard for program management, as described earlier. Acceptability is the level of comfort participants and stakeholders have with a given decision or situation. Research has shown that achieving acceptability requires a multidimensional approach that extends well beyond a specific decision-making process. Planning and managing stakeholder and regulatory engagement is often limited to the decision-making and permitting phases of the project. However, experience has clearly shown that stakeholders and regulators (and program participants) base their judgments on a broader set of considerations, taking into account not only the technical aspects of the program, but the decision making process, mechanisms for providing accountability, and the quality of the relationships among the managers, scientists, and engineers, and between the stakeholders and regulators. Stakeholders need to feel they have an appropriate relationship with the program implementers, that they have standing with program participants, and that their interests have been, are, and will be understood and taken into account by those making decisions. The balanced scorecard for engagement directs management attention to the importance of attending to these four aspects of their engagement among scientists, engineers, stakeholders, and regulators, and helps them recognize that the program's previous record, current behavior, and expected future actions all contribute to stakeholder and regulator judgments of acceptability.

Risk Communication

Most risk communication techniques were developed to help managers and technical staff communicate complex scientific information to members of the public in situations of low trust and high concern, such as often are associated with environmental remediation activities at government and industrial sites. A number of handbooks and guidelines have been developed to help managers, scientists, and engineers understand and apply risk communication principles (for example, Covello and Allen 1988; Hance *et al.* 1988; Chess *et al.* 1989; Lundgren and McMakin 1998).

Some 40 or more factors that can influence how stakeholders perceive scientific and technical work have been identified in the risk communication literature. Those that are most relevant to vadose zone programs include the following:

- Level of understanding—acceptability decreases as the level of scientific understanding decreases.
- Level of personal control—acceptability decreases the more control is vested in someone the stakeholders distrust.
- Timing of effects—acceptability decreases the more immediate the effect.
- Level of trust in institutions—acceptability decreases as trust decreases.
- Amount of media attention—acceptability decreases as the issue is more popularized, particularly if it is popularized in a negative or sensational manner.
- Origin—acceptability is lower for those activities that are human-induced.

When many of these factors combine, the situation is likely to be emotionally charged. Traditional approaches to communication and involvement may not be sufficient to defuse the situation, and more intensive engagement and support, such as facilitation and conflict resolution, will likely be necessary.

Stakeholder Manager

One way to manage appropriate and explicit stakeholder interfaces has been to include as an integral part of the vadose zone program team the person charged with managing stakeholder engagement. This person is focused on identifying and conducting appropriate engagement activities, just as a project integrator is focused on coordinating critical project interfaces, or a manager and scientist partnership is focused on identifying and conducting appropriate science. Ideally, the stakeholder manager has sufficient technical training to understand the program being developed and implemented, exceptional personal communication skills, a trust relationship (either in the beginning or the capability to build one) with stakeholders as well as the other components of the team, and the ability to make decisions regarding levels and kinds of stakeholder engagement. However, all team members must be accessible and willing to answer questions from stakeholders and otherwise participate in the engagement process, as needed. The purpose of an interface person is not to build a fence around the technical team but to allow sufficient focus on the stakeholder effort. It is often true, even in this day of instantaneous e-mail transmissions and Internet opportunities, that person-to-person contact, at the site level, is the most effective form of engagement.

Internet

As an information-sharing tool, the Internet and other computer-based tools are growing in use and popularity. Data that was impossible to share or manipulate easily in print can be placed on the Internet for easy access. Web forums allow online discussions among far-flung stakeholders and the management team (Helie 1999). Electronic mail and Internet feedback forms can encourage stakeholders to provide rapid review and comment (Lundgren and McMakin 1998). Electronic databases make tracking and aggregating voluminous comments easier. Satellite systems can be used to reach remote groups who could not otherwise participate in learning or sharing activities. Group decision software allows all opinions to be heard in a nonthreatening environment (McMakin *et al.* 1995).

One pitfall of Internet use, however, is that simply placing a 1,000-page report on the Web does not guarantee that the information will be read or understood. The old adage "garbage in, garbage out" was never more true than when applied to electronic communication. According to an article in *Consensus*, the quarterly newsletter on conflict resolution and facilitation from the Massachusetts Institute of Technology-Harvard Public Disputes Program, "dispute resolution practitioners do not always think strategically about who is 'talking' to whom electronically or how these communications can affect a consensus building effort." (Helie 1999). Additional pitfalls include the inability to gauge emotion in the written word of an e-mail message and the continued inaccessibility to Internet services for a wide spectrum of the population.

Relevant examples of effective Internet use are still relatively rare in the context of vadose zone management. While the DOE has invested in a number of sites to post documentation (for example, see www.bhi-erc.com/vadose/vadose.htm), few feature opportunities for interaction.

Traps and Gaps

A number of credible methods and tools exist for effectively engaging stakeholders in a scientific and management process. Unfortunately, these methods and tools are seldom applied consistently and with sufficient rigor to achieve the desired results. When funding is required for other project activities, engagement funding is often among the first to be eliminated or reduced. In addition, some managers and scientists express the opinion that "it ain't rocket science" and devote their attentions to what they perceive to be more scientific pursuits, leaving the process to evolve with little management attention. Another difficulty is that credible neutral parties who could lead the engagement process are hard to find.

Management Traps

There are seven significant traps that can prevent effective engagement of stakeholders. Many center around the idea that there is a *finite endpoint to the engagement* that stops short of encompassing the entire project lifecycle. Some managers, for example, focus on the initial phases of work and use stakeholder involvement as an incentive to achieve a budgetary target. In this view, once the stakeholders have effectively lobbied Congress or the agencies to implement or increase that budget, their role in the project has ended. This short-sighted view fails to take advantage of the knowledge, skills, and interests that stakeholders can bring to the work. It also seldom provides a satisfactory role for stakeholders. Prematurely limiting the role of stakeholders in the process often results in stakeholder hostility that operates to the detriment of the project.

"Illusory engagement"—another trap—occurs when projects put the welcome mat out, but forget the invitations and the party. A passive approach to stakeholder engagement that places responsibility for creating and structuring opportunities for engagement on the stakeholders is unlikely to be effective. The process of learning about and joining engagement opportunities sets up too great a barrier for most interested parties to overcome, and few will actually get involved. On the other hand, such passivity may encourage stakeholders to organize to oppose rather than collaborate with the program.

Thinking that if every stakeholder has been contacted and invited to participate, engagement will flow smoothly, or *the "open process" view*, often fails to take into account the needs and desires of stakeholders. It also leaves engagement up to the winds of chance instead of consciously establishing a process and effectively managing that process so that both stakeholders and the program team reach the desired outcomes. This situation can actually engender significant levels of confusion, false expectations, and mistrust. When management does not clearly delineate the "rules of engagement," they do neither themselves nor potential participants (who are often participating at their own time and expense) a favor.

Another trap equates *public availability of project material with engagement.* If all project communications are available on the Internet, in public reading rooms, and mailed directly to stakeholders, that constitutes effective engagement. In some limited cases, in which stakes and uncertainty are low, trust in the agency managing the program is high, and other acceptability issues have been satisfied, an emphasis on disclosure and communication may suffice. These situations are rare! In general, disclosure and effective communication are only part of an effective engagement process. Simply reading materials and providing comments does not satisfy most regulators and other stakeholders.

Another trap is the *reliance on a communication style that is one-way.* Whether written, spoken, or provided interactively on the Internet,

The "Stakeholder List"

Too often scientists, engineers, and managers look to a predefined set of people when determining whom to engage on a particular issue. Many government and industrial cleanup sites have a citizen advisory board that deals with environmental restoration issues. Other regions have vocal activist groups. It is relatively easy to get the ear of such highly visible stakeholder groups. It is much more difficult to ensure that everyone who has a stake in a particular issue has been heard.

Identifying all those who should be involved in a scientific and management process can seem like trying to count the grains of sand on a beach. Any number of groups and individuals may be interested in vadose zone programs. One way to identify potential stakeholders is to look at the various ways in which the program may impact various groups:

- **Organizational or legal responsibility (real or perceived) for activities associated with, influenced by, or otherwise affected by the program.** For example, if other remediation programs have the potential to impact the number and movement of contaminants in the vadose zone, perhaps by flushing water more quickly through the subsurface, representatives of those programs may want to be kept informed or involved with the vadose zone program planning and implementation. Other examples of people with these kinds of interests include the agency charged with making a decision to remediate the vadose zone, regulatory agencies, contractors to these agencies charged with implementing the characterization or remediation, and local government, health, emergency response, and land-use planning organizations.
- **Physical proximity and hence potential exposure to the contamination.** For example, if the nearest city to the contaminant plume uses wells for any part of its municipal water supply, the local government agency will most likely want to be involved in the program, regardless of whether or not the plume has migrated or is likely to migrate in the direction of the city wells. Other examples of people with these kinds of interests include workers who will be near the area or working on the program, anyone who might participate in recreational or other activities in the area, and wildlife management agencies with jurisdiction over the plants and animals near the area.
- **Economic interests that might be affected either positively or negatively by the program.** For example, nearby farmers may want to be kept informed if they perceive the plumes as affecting their land or if the media perceives local crops to be tainted. Other examples are businesses that are impacted by public perception of

continued

particular locations (recreational industry, fishing and hunting industry, tourism industry), businesses bidding on the cleanup effort, and businesses that are using or will be using the land in question.

- **Involvement in an occupation that might be affected by the program.** For example, if the contamination is moving rapidly through the vadose zone, or otherwise has the potential to quickly reach receptor points, emergency response personnel might need to be kept abreast of program activities. Other examples include local workers' unions; professional societies of engineers, scientists, and other technical professions; and health care practitioners.
- **Philosophical or value-related interests.** For example, those with a strong environmental ethic might want to be involved in order to mitigate or remove the contamination before greater damage is done. This type of interest encompasses concerns over quality of life, as well. Examples of people with these kinds of interests include Native American Tribes (who may also share a legal or organizational interest, as well as have certain rights by treaty), environmental activist groups, parent groups, and civic organizations.

This may seem like an exhaustive and unwieldy list; however, while all these groups need to be informed about the planning and activities of the program, not all will want the same level of engagement.

One of the early stages of project development should be the creation of a comprehensive list of stakeholders and a matrix of potential engagement opportunities, ideally prioritized with input from the stakeholders. One such technique is the "onion" diagram (shown in Figure 2-6) developed by Creighton and Associates (various publications over the last ten years; for example, see Pacific Northwest National Laboratory and Creighton and Creighton 1993). This diagram attempts to classify various stakeholders by their level of interest. Each level of interest has an associated strategy for communication or involvement. The key is identifying which portions of the vast number of stakeholders place themselves in each category, and meeting the needs of stakeholders in each category as determined by the stakeholders themselves.

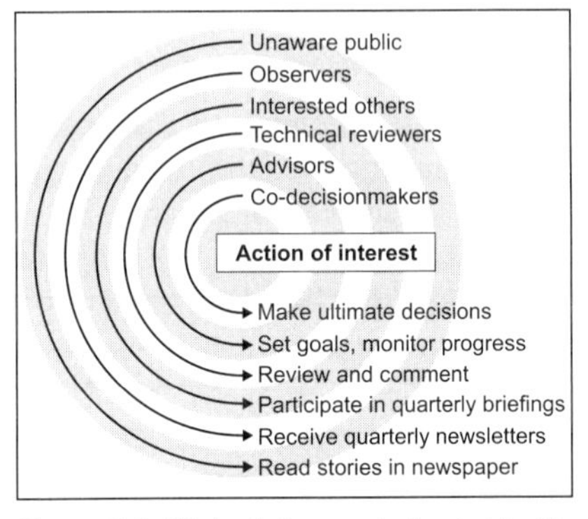

Figure 2-6. "Onion" diagram helps to identify stakeholders interested in a project and methods of engagement and communication.

these information presentations follow a transmission paradigm, otherwise known as one-way communication. Even a well-conducted briefing with a question-and-answer period following does not substitute for a true discussion among stakeholders, regulators, engineers, scientists, managers, and decision makers. This multiparty dialogue builds relationships, trust, and understanding, while a transmission paradigm often, albeit inadvertently, polarizes the situation. Success is rare in such instances.

Assuming that level of engagement is solely a function of schedule and budget is another trap. This mind set assumes that engagement is a luxury that certain projects can do without. For vadose zone programs, this is seldom the case. As noted previously in this chapter, effective engagement is necessary to achieve project endpoints, sometimes even to gain project funding in the first place. It may be better to modify a schedule or budget, however difficult, than to forego effective engagement and run the risk of failure.

A final trap is the mindset that *engagement is effective if no one is complaining*. Silence from a stakeholder group, particularly one that has been previously vocal, can mean that engagement has failed, the doors are closed, and no one is listening. The next step is sometimes a lawsuit or other dramatic statement that is designed to force the agency to pay the attention to the stakeholder process they should have paid in the beginning.

Technical Gaps

One of the major gaps in our ability to effectively engage stakeholders is the lack of organizational and regulatory infrastructure to support innovative methods such as conflict resolution and facilitation. Some of the latest techniques to help groups reach effective, impactful consensus on issues require voluntary sharing of power (Chrislip and Larson 1994). With an increased emphasis on the accountability of government agencies to Congress, many managers feel they alone have the responsibility, and hence the power, to make decisions. In industry, accountability to the Chief Executive Officer and stockholders often forces a similar situation. What vadose zone manager, faced with an inherently uncertain situation, wants to have to point a finger at a group of stakeholders and say "they made me do it"? There is little precedent for delegating responsibility to "outsiders" (Walker and Daniels 1997).

In addition, many managers and decision makers assume that the laws which require stakeholder engagement (for example, the National Environmental Policy Act and its implementing regulations) prescribe the specific type of stakeholder engagement that must be implemented. Unfortunately, this erroneous belief has led agencies, and some stakeholder groups, to opt for hearing-style public meetings that end up as confrontations instead of evaluating options for smaller-scale, more interactive engagements that might encourage listening and group learning. Current laws allow broad discretion in the choice and scope of engagement strategies.

Another glaring gap is the lack of guidance and mechanisms for incorporating stakeholder input into decisions and planning activities, which are typically reserved for management. Natural resource managers, risk assessors, and engineers are learning techniques to identify and incorporate stakeholder values into their activities. Unfortunately, program management practices often fail to adopt or adapt to such techniques.

Another gap is a method for focusing engagement on the topic at hand. This problem arises from the fact that, to be effective (and somewhat efficient), multiple voices, with conflicting opinions, often converge on a single aspect. Complicating this fact is the difficulty of working with stakeholder groups who have developed a conflict of interest with the topic of hand. Such groups may have developed into "cottage industries" that require continuation or expansion of the problem to keep donations flowing, or have developed a long-term commercial interest in the outcome. With the already difficult ebb and flow of interest and opinions among a diverse set of stakeholders, it is hard to manage a disciplined engagement process using traditional stakeholder involvement methods. Conflict resolution and facilitation techniques like those described in Bilyard *et al.* (1998) hold promise.

Some guidance, or a process, for ensuring a balanced representation of perspectives is also needed, particularly when stakeholders must pay their own way. Over the lifetime of a multiyear project, how can all interested groups find a place at the table on an ongoing basis? Sometimes, interested stakeholders must find time outside their day-to-day jobs and family responsibilities to interact with scientists, engineers, and managers whose day-to-day job is the project. Interested scientists and engineers who are not part of the project must also volunteer their time to participate. Yet often if these stakeholders, scientists, and engineers are paid for their involvement (either by the government or other organizations), they are subject to accusations of bias and lack of objectivity.

Taken together, these gaps form a daunting hole over which vadose zone managers hang on a tenuous tightrope of prudent planning and disciplined engagement. Fortunately, because vadose zone managers are in a large company of managers and regulators facing the same or similar challenges, there is an incentive for the market to provide better tools and techniques to address these challenges.

CONCLUSIONS

This chapter has focused on the difficulties, philosophies, tools, traps, and gaps facing those charged with effectively managing a vadose zone project. The basic principles—establish endpoints, develop a roadmap, and practice deliberate and disciplined engagement—detail potential actions a manager can take to improve the chances that vadose zone management will support its ultimate goal, which is protecting our groundwater. Unfortunately, even following these principles will not ensure success. Too many factors outside the manager's control currently play a large role in the management of vadose zone projects.

In summary, we offer three "meta-principles" for integrating the vadose zone into environmental management actions and decisions. These principles are aimed at policy makers rather than project managers—those in a position to affect the drivers and requirements governing vadose zone management.

1. **Establish a consistent, predictable framework to anticipate and guide remediation and site closure decisions.** This principle addresses the problem of case-by-case regulation in the vadose zone, which has led to a dearth of information about the nature and long-term risk of contaminants in the subsurface (not groundwater). Federal cleanup programs operate in two modes: prevention and reaction. The vadose zone offers a key to responsible management and resolution of unanticipated groundwater contamination and of unresolved site closures that dilute national resources directed at the stewardship of one of our most vital national

Management Systems—Wrapping It All Up

We have organized this discussion using a "divide and conquer" approach—break up the problem into manageable parts and then address those parts one at a time. In reality, a significant strain on politically and technically complex programs is the fact that the fundamental issues discussed here are not independent: endpoints are difficult to pin down because managers haven't got the information and don't know what they need; participants often do not agree on an acceptable engagement process; knowledge gaps are difficult to fill until objectives are pinned down, which can't happen until stakeholders and others agree on the important gaps; and so on. When the issues are tangled together so tightly, it is hard to know what string to pull first. While providing a perspective on the pieces may be helpful, we recognize that these parts need to fit within a complete management system and philosophy.

Both the environmental management system and the knowledge management framework offer a coherent, holistic approach that is compatible with the principles presented in this chapter. An *environmental management system* is a disciplined approach to ensuring that environmental activities are well-managed within any project or organization. It does so by focusing on essential management practices and evaluating at each point how operations and supporting procedures impact the environment. The most familiar form of an environmental management system is the ISO 14001 standard, recently established by the International Organization for Standardization (DOE and EPA 1998).

Knowledge management is the continuous process of capturing, creating, packaging, disseminating, and using information. The *knowledge management framework* recognizes more clearly than traditional management systems that a growing number of government programs and businesses are open systems, where the distinction between the program and the external programmatic environment is fairly obscure, control is diffused, and uncertainty and complexity abound. Under these conditions, efforts to apply traditional management systems will frustrate participants because traditional approaches are not aligned with or responsive to the complex and uncertain political and technical environment in which they must operate. For these programs, project management is often synonymous with *knowledge management*. The knowledge management framework offers a philosophy and approach wherein the challenges of setting endpoints, collecting information to refine those endpoints, and engaging stakeholders are all understood in terms of the capture and development of knowledge, and on the progressive integral utilization of that knowledge.

assets—clean water. Some prescription may be necessary to cause us to use that key.

2. **Encourage the disciplined infusion of the best scientific understanding.** Those involved in setting requirements need good data; good data, in many cases, will require exceptional science, and exceptional science requires sustained political and organizational commitment to ensure both its development and use. The time and resources needed to underwrite exceptional science are routinely underestimated. When accurately considered, these resources will often fall outside the bounds of the existing regulations or regulatory guidance.

3. **Recognize and support a variety of roles for stakeholders, Tribal Governments, and regulators in the decision process and in supporting program development**. There is a surprising shortage of theory-based guidance on the topic of how to implement effective engagement. While government and industry have made notable strides in recent years toward a more open process, there appears to be room for major improvements that would benefit most parties and result in more efficient progress. Policies and guidance for deliberate, disciplined engagement should provide a basis for involving stakeholders that minimizes selective influence, but recognizes legitimate special concerns and areas of expertise among stakeholders.

REFERENCES

Bentkover, J.D., V.T. Covello, and J. Mumpower. *Benefits Assessment: The State of the Art*, D. Reidel Publishing Company, Boston, MA (1986).

Bilyard, G.R., J.P. Amaya, S.W. Gajewski, A. Harding, G. Hund, F.B. Metting, T.M. Peterson, S. Underriner, J.R. Weber, and C. Word. *Guidelines - A Primer for Communicating Effectively with NABIR Stakeholders*, PNNL-12041 Rev. 1, Pacific Northwest National Laboratory, Richland, WA (1998).

Bjornstad, D.J., D.W. Jones, M. Russell, K.S. Redus, and C.L. Dummer. *Outcome-Oriented Risk Planning for DOE's Cleanup*, JIEE-98-01, Joint Institute for Energy and Environment, Knoxville, TN (1998).

Bradbury, J.A., and K.M. Branch. "Public Involvement in Chemical Demilitarization," paper presented at the 20th Annual Conference and Exposition of the National Association of Environmental Professionals, Washington, DC (1995).

Chess, C., B.J. Hance, , and P.M. Sandman. *Planning Dialogue with Communities: A Risk Communication Workbook*, Environmental Communication Research Program, New Brunswick, NJ (1989).

Chrislip, D.D., and C.E. Larson. *Collaborative Leadership: How Citizens and Civic Leaders Can Make A Difference*, Jossey-Bass Publishers, San Francisco, CA (1994).

Clemen, R.T. *Making Hard Decisions: An Introduction to Decision Analysis*, Duxbury Press, Wadsworth Publishing Company, Belmont, CA (1996).

Colson, S.D., R.E. Gephart, V.L. Hunter, J. Janata, and L.G. Morgan. *A Risk-Based Focused Decision-Management Approach for Justifying Characterization of Hanford Tank Waste*, PNNL-11231, Pacific Northwest National Laboratory, Richland, WA (1997).

Corbeil, R.C. *Action-Oriented Evaluation in Organizations: Canadian Practices*, Wall & Emerson, Inc., Toronto, Canada (1992).

Covello, V.T., and F.W. Allen. *Seven Cardinal Rules of Risk Communication*, EPA 230/09-88-001, U.S. Environmental Protection Agency, Washington, DC (1988).

Covello, V.T., P.M. Sandman, and P. Slovic. *Risk Communication, Risk Statistics, and Risk Comparisons: A Manual for Plant Managers*, Chemical Manufacturers Association, Washington, DC (1988).

Cummings, R.D., D.S. Brookshire, and W.D. Schulze, (Eds.). *Valuing Environmental Goods: An Assessment of the Contingent Valuation Method*, Rowman & Allanheld, Totowa, NJ (1986).

Dettmer, H.W. *Goldratt's Theory of Constraints: A Systems Approach to Continuous Improvement*, ASQC Quality Press, Milwaukee, WI (1997).

Engineers' Public Policy Council of the American Association of Engineering Societies. *Risk Analysis: The Process and its Application*, American Association of Engineering Societies, Washington, DC (1996).

Fassbender, L.L., M.E. Brewster, A.J. Brothers, S.W. Jajewski, B.L. Harper, T. Eppel, R. John, J.G. Hill, V.L. Hunter, D.A. Seaver, T.W. Wood, D. von Winterfeldt, and J.W. Ulvila. *Application of Value of Information to Tank Waste Characterization: A New Paradigm for Defining Tank Waste Characterization*

Requirements, PNNL-11395, Pacific Northwest National Laboratory, Richland, WA (1996).

Galvin, R. "Science Roadmaps," *Science 280* 803. (1998).

General Accounting Office. *Department of Energy: Management Changes Needed to Expand Use of Innovative Cleanup Technologies*, GAO/RCED-94-205, General Accounting Office, Washington, DC (1994).

General Accounting Office. *Nuclear Waste: Understanding of Waste Migration at Hanford is Inadequate for Key Decisions*, GAO/RCED-98-80, General Accounting Office, Washington, DC (1998).

General Accounting Office. *Performance and Accountability Series, Major Management Challenges and Program Risks—Department of Energy*, GAO/OCG-99-6, General Accounting Office, Washington, DC (1999).

Gilbert, R.B. Think Before Testing, *Practice Periodical of Hazardous, Toxic and Radioactive Waste Management 1(3)* (1997): 90-91.

Hance, B.J., C. Chess, and P.M. Sandman. *Improving Dialogue with Communities: A Risk Communication Manual for Government*, New Jersey Department of Environmental Protection, Division of Science and Research, Trenton, NJ (1988).

Harris, S.G., and B.L. Harper. "How incorporating tribal information will enhance waste management decisions," Waste Management '97 Proceedings (1997).

Harris, S.G., and B.L. Harper. "Using Eco-Cultural Risk in Risk-Based Decision Making," paper and presentation at the American Nuclear Society Environmental Sciences Topical meeting, Richland WA (Proceedings in press) (1998).

Harris, S.G., and B.L. Harper. "Using Traditional Environmental Knowledge As A Risk Characterization Framework," 1999 Society for Risk Analysis annual meeting. (1999).

Helie, J. "Technology Creates Opportunities—and Risks," *Consensus*, January (1999): 5-9.

Holling, C.S. *Adaptive Environmental Assessment and Management*, John Wiley, New York, NY (1978).

Holling, C.S. "Cross-Scale Morphology, Geometry, and Dynamics of Ecosystems," *Ecological Monographs* 62(447) (1992): 5-2.

Kaplan, R.S. and D.P. Norton. "Putting the Balanced Scorecard to Work," *Harvard Business Review*, September-October (1993): 2-16.

Keeney, R.L. *Value-Focused Thinking: A Path to Creative Decisionmaking*, Harvard University Press, Cambridge, MA (1992).

Keeney, R.L., and D. von Winterfeldt. "On the Uses of Expert Judgment on Complex Technical Problems," *IEEE Transactions on Engineering Management 36*, (1989): 83-86.

Keeney, R.L., and D. von Winterfeldt. "Eliciting Probabilities from Experts in Complex Technical Problems," *IEEE Transactions on Engineering Management 38*, (1991): 191-201.

Konikow, L.F., and R.C. Ewing. "Is a Probabilistic Performance Assessment Enough?" *Ground Water* 37(4) (1999): 481-650.

Leary, K.D. *The Use of DQO's in Designing Vadose Zone Monitoring Systems*, U.S. Department of Energy, Nevada Operations Office, Las Vegas, NV (xxxx).

Lee, K.N. *Compass and Gyroscope: Integrating Science and Politics for the Environment*, Island Press, Washington, DC (1993).

Lundgren, R.E., and A.H. McMakin. *Risk Communication: A Handbook for Communicating Environmental, Safety, and Health Risks,* Second Edition, Battelle Press, Columbus, OH (1998).

Matheson, D., and J. Matheson. *The Smart Organization: Creating Value through Strategic R&D*, Harvard Business School Press, Boston, MA (1998).

McMakin, A.H., D.L. Henrich, C.A. Kuhlman, and G.W. White. *Innovative Techniques and Tools for Public Participation in U.S. Department of Energy Programs*, PNL-10664, USDOE Assistant Secretary for Environment, Safety and Health, Washington, DC (1995).

Merkhoffer, M.W. *Decision Science and Social Risk Management*, D. Reidel Publishing Company, Boston, MA (1987).

National Research Council (NRC). *Building Consensus Through Risk Assessment and Management of the Department of Energy's Environmental Management Program*, National Academy Press, Washington, DC (1994).

National Research Council (NRC). *Allocating Federal Funds for Science and Technology*, National Academy Press, Washington DC (1995a).

National Research Council (NRC). *Improving the Environment: An Evaluation of DOE's Environmental Management Program*, National Academy Press, Washington, DC (1995b).

National Research Council (NRC). *Barriers to Science: Technical Management of the Department of Energy Environmental Remediation Program*, National Academy Press, Washington, DC (1996a).

National Research Council (NRC). *Understanding Risk: Informing Decisions in a Democratic Society*, National Academy Press, Washington, DC (1996b).

National Research Council (NRC). *Building an Effective Environmental Management Science Program: Final Assessment*, National Academy Press, Washington, DC (1997).

National Research Council (NRC). *Scientific Opportunity and Public Input: Priority Setting for NIH*, National Academy Press, Washington, DC (1998).

National Research Council (NRC). *Groundwater and Soil Cleanup: Improving Management of Persistent Contaminants*, National Academy Press, Washington, DC (1999a).

National Research Council (NRC). *Improving Project Management in the Department of Energy*, National Academy Press, Washington, DC (1999b).

Olson, R. *The Art of Creative Thinking*, Harper & Row, New York, NY (1986).

Pacific Northwest Laboratory and Creighton and Creighton. *Public Participation for Managers*, U.S. Department of Energy, Washington, DC (1993).

Peterson, G.L., B.L. Driver, and R.S. Gregory. *Amenity Resource Valuation: Integrating Economics with Other Disciplines*, Venture Publishing, State College, PA (1988).

Ravetz, J.R. "Usable Knowledge, Usable Ignorance" *Knowledge Creation, Diffusion, Utilization* 9(1) (1987):87-116.

Rogers, K. "Managing science/management partnerships: A challenge of adaptive management" Conservation Ecology, Vol. 2, No. 2. http://www.consecol.org/vol2/iss2/resp1 (1998).

Ulvila, J., and J. Chinnis. "Decision Analysis for R&D Resource Management," *Management of R&D Engineering*, D.F. Kocaoglu (ed). North-Holland, The Netherlands (publication pending).

U.S. Department of Energy. *Guidelines for Performance Measurement*, DOE G 120.1-5, U.S. Department of Energy, Washington, DC (1996).

U.S. Department of Energy, Richland Operations Office (DOE-RL). *Screening Assessment and Requirements for a Comprehensive Assessment: Columbia River Comprehensive Impact Assessment*, DOE/RL-96-16 Rev. 1, U.S. Department of Energy-Richland Operations Office, Richland, WA (1998).

U.S. Department of Energy, Richland Operations Office (DOE-RL). *Ground Water/Vadose Zone Integration Project Science and Technology Summary Description*, DOE/RL-98-48, Rev. 0, U.S. Department of Energy-Richland Operations Office, Richland, WA (1999a).

U.S. Department of Energy, Richland Operations Office (DOE-RL). *Revised Draft Hanford Remedial Action Environmental Impact Statement and Comprehensive Land-Use Plan.* DOE/EIS-022D, U.S. Department of Energy-Richland Operations Office, Richland, WA (1999b).

U.S. Department of Energy (DOE) and U.S. Environmental Protection Agency (EPA). *Environmental Management Systems Primer for Federal Facilities*, DOE/EH-0573, U.S. Department of Energy, Washington, DC (1998).

U.S. Environmental Protection Agency (EPA). *Data Quality Objectives process for Superfund: Interim Final Guidance*, EPA/540/G-93/071, U.S. Environmental Protection Agency, Washington, DC (1993).

Walker, G.B, and S.E. Daniels. *Collaborative Public Participation in Environmental Conflict Management: An Introduction to Five Approaches*, Proceedings of the Fourth Biennial Conference on Communication and the Environment, State University of New York College of Environmental Science and Forestry's Environmental Studies Department, Syracuse, NY (1997).

Walters, C. "Challenges in Adaptive Management of Riparian and Coastal Ecosystems," *Conservation Ecology* 1(2) (1997): 1.

Wood, T.W., V.L. Hunter, and J.W. Ulvila. *A Value of Information Approach to Data Quality Objectives for the Hanford High-Level Waste Tanks,* PNNL-SA-25649, Pacific Northwest National Laboratory, Richland, WA (1995).

CASE STUDIES

HANFORD GROUNDWATER/VADOSE ZONE INTEGRATION PROJECT "DEPENDENCY WEBS"

Barbara Harper and **Stuart Harris**

Hanford's Groundwater/Vadose Zone Integration Project, together with key stakeholders, has developed a construct called "dependency webs." Dependency webs show which potential impacts from subsurface contamination are important to the key stakeholders, why they are important, and how they are related. Along with a detailed mission statement and a conceptual model, they form an important part of the system description.

As contamination moves through different areas, different resources are affected and different environmental media are impacted. Dependency webs help tell the whole story about what will happen if different locations are contaminated, provide a way to organize metrics, and help manage the risk assessment process. The webs accomplish these three tasks because they:

- Identify resources at risk
- Identify connections among resources
- Provide a conceptual model for an assessment
- Track the relevant impacts to ensure that they are addressed
- Organize the results for risk characterization
- Provide a way to communicate with and receive feedback from stakeholders
- Provide a framework for explaining impacts and risks in a holistic way
- Focus attention on dominant issues by supporting a sensitivity analysis
- Show the full consequences of each decision and how the consequences would change if different decisions were made or if different environmental conditions occurred.

Dependency webs are intended to bound the analytic problem by describing what is "at risk" along the contaminant migration path and what is at stake if critical locations become contaminated. The resources and/or services at a particular location are the attributes that make it important. Among the multiple reasons that a location or resource may be important are aesthetic value, historic significance, the presence of critical habitat, commercial value, the presence of sacred sites or cultural resources, and recreational value. To identify what makes a place or resource important, subject matter experts must be consulted, including cultural experts such as tribal elders, modeling experts, ecologists, civic groups, and advocacy groups (for example, environmental groups or groups representing the "silent voices" of future generations). For instance, the Hanford Reach of the Columbia River provides salmon spawning habitat, tribal subsistence use, migratory bird habitat, native foods and medicines, aesthetics, ceremonial and spiritual resources and areas, recreation, ecotourism income, and public water and agricultural intakes. These elements can be organized into a web showing the linkages between uses and resources and showing which uses depend on or are influenced by which resources.

The process for developing location-specific dependency webs has three basic steps:

1) Identify locations according to contaminant transport results and knowledge of environmental characteristics.
2) Describe the existing habitat quality absent the new contamination or new stressors. Existing habitat quality includes metrics for pre-existing contaminant burdens, existing stressors, and critical ecological characteristics.
3) Identify the critical parameters that need to be evaluated according to the elements that are most important for that location and more likely to be affected by new or additional contamination (that is, draw the web for the location).

The identification of specific web elements is aided by asking the following questions:

- What makes the place important (to anyone)?
- Who lives there? What biota exists there? (What is the existing environmental quality or usability? What environmental quality, functions, or species have already been lost there? What would be expected but isn't there? What trends in environmental quality have been observed there?)
- Who and what uses the location? What happens there (ecological migratory stop, human recreation, and so on)?
- What environmental goods, functions, and services are provided by the location and its natural, cultural, economic, and human resources?
- What is "at stake" there if contamination arrives?
- Who and what is already "at risk" there (biota, cultural activity, economic activity)?
- How are the above factors combined into each locations' descriptive dependency web (influence diagram)?

For more information on dependency webs, see Groundwater/Vadose Zone Integration Project Science and Technology Summary Description (DOE-RL 1999a). For further elaboration, see "Using Eco-Cultural Risk in Risk-Based Decision Making" (Harris and Harper 1998) and "Using Traditional Environmental Knowledge as a Risk Characterization Framework" (Harris and Harper 1999).

THE COLUMBIA RIVER COMPREHENSIVE IMPACT ASSESSMENT

Thomas W. Woods

The Columbia River Comprehensive Impact Assessment (CRCIA) is one of few examples of a truly multiparty, stakeholder-driven assessment of the impacts of subsurface contamination, what such an assessment means, what it requires, and why it is important. The following description of that activity is provided by one of the leading participants in that assessment. It offers an important and insightful perspective on this landmark activity.

The purpose of the CRCIA was to assess the regional effects from the planned final state of hazardous and radioactive wastes at the Hanford Site, a former plutonium production site that is bounded, in part, by the Columbia River. The broad categories of potential impact assessed were human, cultural, economic, and environmental. The geographic area potentially affected, as well as the number of future generations to be assessed, was specified as unlimited. The reason for this specification was because so little was known about the rate of waste movement in the vadose zone and the groundwater, the extent (and significance) of contamination from other sources, the sorbtion and the dilution of wastes, the reconcentration mechanisms, the health risk, and many other variables that determine the level, timing, location, and results of peak concentrations. Populations affected by the Columbia River received the most attention in the development of the CRCIA specification as the river seemed to be the most plausible means of transporting significant quantities of waste to the largest number of receptors.

This regional effects assessment was seen as a critical need by virtually all public groups and a large segment of the scientific community because no other composite impact analysis of Hanford's multiple and extensive nuclear and chemical waste sites had been performed, nor was any planned. As a result, there was no supportable basis for identifying which waste sites to clean up first, the level of clean-up activities needed, and, therefore, how much congressional funding should be provided. In short, there was no credible decision process or technical planning approach for Hanford as a whole.

Congress in particular was unconvinced of the need for billions of dollars based solely on the need to comply with environmental regulations—regulations seriously questioned as to their adequacy as they were not designed for an environmental threat of the magnitude or complexity of the Hanford Site. Ongoing appropriations made by Congress to clean up Hanford were quietly recognized as a tribute to regional leaders and politicians for their porkbarreling skills, skills that were superficially legitimized by a legally binding document endorsing the strategy of "reasonable compliance" with the questionable regulations. That document is known as the Hanford Federal Facility Agreement and Consent Order; it is also called the Tri-Party Agreement for the three parties involved (U.S. Department of Energy [DOE], U.S. Environmental Protection Agency [EPA], and Washington State Department of Ecology [Ecology]). To protect the flow of congressional funding, media spin and hype aimed at salving public and congressional doubts became an intrinsic and vital facet of the Hanford clean-up project. The developers of the CRCIA fervently sought to replace this fragile thread—on which hung continuation of clean-up work (funding) —with an honest, defensible, and robust basis for identifying what work was needed and how well it had to be done.

The most serious limitation that fell upon the CRCIA effort came as no surprise, given the assessment's potential for spearing sacred cows. Indeed, the foundational philosophies of several government environmental agencies might have been brought under public scrutiny if the CRCIA assessment results were to have verified suspected regulatory inadequacies. Consequently, funding was perfunctory causing gross understaffing. Development was done over roughly a three-year period entirely by two or three staff members of public organizations who canvassed most of the potentially effected groups for concerns and assessment requirements. The federal government provided only meeting rooms and publication services. While repeated requests were made for expert assistance from the relevant technical fields, no scientists were made available. As a result, the CRCIA document was conceptually sound but lacked scientific detail on which implementa-tion and conduct of the analysis could proceed. This was a serious blow as it left an open door for unending government delays justified by seemingly sensible questions about the technical feasibility of the specified assessment. While several technical limitations were recognized as needing basic development, most were a simple matter of making changes in local analysis practices.

One of the technical limitations was in learning how to reduce the scope of the impact assessment to fit whatever the funding constraints might be while ensuring that 1) the factors included were always more significant than those excluded, and 2) result uncertainty was always quantified. This unconventional approach required that assessment funding be allocated across all areas of the assessment on the basis of the extent of uncertainty and the relative significance to results. Vadose zone characterization was one of these areas, waste disposal containment performance was another, and groundwater and river water behavior were among others.

The state of understanding of each area varied but each contributed issues that, without basic research and development effort, would limit the assessment.

Among the most daunting limitations was estimating the resultant health risk once peak toxicity levels, locations, and timing were determined. Most existing work in this field focused on estimating instances of cancer in the present generation. The CRCIA, on the other hand, aspired to understand a broader spectrum of health impacts, both to present and to future generations and from direct exposure as well as from inherited genetic damage. Similar multigenerational understanding was sought for plant and animal species.

The process used in developing the CRCIA pivoted on the notion that *requirements* for the assessment were more fundamental than attempting to design the analysis directly. A *requirement* in this context was meant as a minimal constraint on the choices to be made in defining, planning, and conduct-ing the regional effects assessment. A team was formed and first met in August 1995 for the purpose of creating—with DOE—a consensus steering force to define the requirements for the CRCIA. Those represented by the CRCIA team were the Confederated Tribes of the Umatilla Reservation, the Hanford Advisory Board, the Nez Perce Tribe, the Oregon State Department of Energy, the DOE and their Hanford contractors, the EPA, Ecology, and the Yakama Indian Nation. Working to define a common ground on which the three parties (DOE, EPA, and Ecology) and all partici-pants could comfortably stand, the CRCIA team developed and documented assessment requirements through weekly facilitated workshops. Most participants had suggestions, criticisms, issues, and concerns about previous, similar analyses. These were elicited from the participants in a systematic structure that, with some reorganization, became the framework for the CRCIA requirements document. The participants' issues and concerns were translated into the requirements to be met in designing and performing the analysis. In this process, DOE opted only for the role of a host in these workshops and did not develop the document directly or do so through one of their contractors. The team provided its own workshop facilitator and clerical support from among the members of its public organizations. Subsequently, authors of the document also were drawn from this group. Therefore, the CRCIA requirements constituted a definition of acceptability from a cross-section of those people most likely to be adversely affected by Hanford wastes. In April 1998, the 100-plus page CRCIA require-ments document was published along with the results of a screening assessment of the current state of the Columbia River between the Hanford Site and McNary Dam, which is downstream from the Site.

The CRCIA raised many issues and resolved many issues. It was acclaimed by all as the most effective forum to date for meaningful interaction with the potentially affected segments of the public. Inevitably, *control* became the center of the highly contentious atmosphere that eventually engulfed the effort. The CRCIA public organizations wanted the regional effects assessment managed by an organization that was totally independent of DOE. Of course, DOE wanted to retain directive authority. Ironically, the CRCIA team was founded to quell an uprising of resentment in both the technical community and public circles over DOE's handling of analyses of the Columbia River. The open interaction within the team soon quieted that earlier tempest only to raise even more cogent—and heated—questions.

Through the CRCIA, most interested parties were convinced that multiple waste complexes could not be acceptably remediated on the piecemeal basis allowed by the environmental regulations. Interactions of one waste site on another were much too worrisome to ignore. Hypothetically, each one of many waste sites in close proximity could be brought into regulatory compliance while their combined residual wastes could leave a serious environmental problem. Therefore, remediation design had to consider the regional effects of all waste sites in the area, including preexistent contaminant loads in the environment from other, possibly distant, sources. Effects assessment results could then be sensibly compared with limits allowed by the regulations. A closely related issue was that of perpetual waste dilution downgradient of the waste site. Regulatory practice assumed such was the case. As just observed, other waste sites in the area might prove otherwise. Also, reconcentration mechanisms may plausibly reverse this assumed dilution trend. Such has been documented in shellfish in the Columbia River estuary. River sediments have also been shown to accumulate long-lived radionuclides. The CRCIA sought to understand such phenomena.

Use of regional effects assessments like CRCIA as a clean-up project decision tool is an appealing concept. Without question, planning an adequate clean-up effort simply is not credible without gaining insights into the long-term effectiveness of the proposed work. However, insulating the analysts from the fervor and political influence of project planning and budgeting may be essential to achieving the assessment's credible objectivity.

CHAPTER 3 CONTENTS

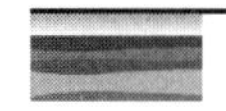

3

Vadose Zone Characterization and Monitoring

Current Technologies, Applications, and Future Developments

Boris Faybishenko

Contributors: M. Bandurraga, M. Conrad, P. Cook, C. Eddy-Dilek, L. Everett, FRx Inc. of Cincinnati, T. Hazen, S. Hubbard, A.R. Hutter, P. Jordan, C. Keller, F.J. Leij, N. Loaiciga, E.L. Majer, L. Murdoch, S. Renehan, B. Riha, J. Rossabi, Y. Rubin, A. Simmons, S. Weeks, C.V. Williams

INTRODUCTION

NEEDS FOR VADOSE ZONE CHARACTERIZATION AND MONITORING

Vadose zone characterization and monitoring are essential for:

- Development of a complete and accurate assessment of the inventory, distribution, and movement of contaminants in unsaturated-saturated soils and rocks.
- Development of improved predictive methods for liquid flow and contaminant transport.
- Design of remediation systems (barrier systems, stabilization of buried wastes *in situ*, cover systems for waste isolation, *in situ* treatment barriers of dispersed contaminant plumes, bioreactive treatment methods of organic solvents in sediments and groundwater).
- Design of chemical treatment technologies to destroy or immobilize highly concentrated contaminant sources (metals, radionuclides, explosive residues, and solvents) accumulated in the subsurface.

Development of appropriate conceptual models of water flow and chemical transport in the vadose zone soil-rock formation is critical for developing adequate predictive modeling methods and designing cost-effective remediation techniques. These conceptual models of unsaturated heterogeneous soils must take into account the processes of preferential and fast water seepage and contaminant transport toward the underlying aquifer. Such processes are enhanced under episodic natural precipitation, snowmelt, and extreme chemistry of waste leaks from tanks, cribs, and other surface sources. However, until recently, the effects of episodic infiltration and preferential flow on a field scale have not been taken into account when predicting flow and transport and developing remediation procedures. The pronounced temporal and spatial structure of water seepage and contaminant transport, which is difficult to detect, poses unique and difficult problems for characterization, monitoring, modeling, engineering of containment, and remediation of contaminants. Lack of understanding in this area has led to severely erroneous predictions of contaminant transport and incorrect remediation actions.

For many years, it was assumed that wastes released or stored in the vadose zone would move slowly, if at all, through the vadose zone. Because of the emerging evidence of waste migration from leaking tanks through the vadose zone to the groundwater, scientists and engineers have begun to develop a strategy to investigate the vadose zone, including a comprehensive plan to assess vadose zone conditions.

Objectives

The overall objective of this chapter is to describe the current status, applications and future developments of vadose zone characterization and monitoring technologies using case-study data from practicing scientists and engineers. Using these data, we will recommend a series of site-characterization and monitoring methods, the development of both expedited and long-term vadose zone characterization and monitoring methods, and future developments for the design, performance, and post-closure of contaminated sites.

Because our understanding of a site is derived from field observations, this chapter describes the basic principles, advantages, and limitations of existing vadose zone characterization and monitoring methods, using case studies from field experiences as well as the American Society for Testing and Materials (ASTM) and Environmental Protection Agency (EPA) standards related to vadose zone studies.

We will present evidence that the central problem of the vadose zone investigation is the preferential fast-flow phenomena and accelerated deep-contaminant transport toward the groundwater that has been observed at several sites. The methods discussed in this chapter can be used for the following purposes:

- Characterization of natural variations of flow and transport processes in vadose zone systems
- Characterization of anthropogenic stresses on vadose zone systems (such as induced point and non-point infiltration, and well injection)
- Design and selection of experimental methods for field and laboratory experiments
- Design of vadose zone remediation systems
- Project planning and data collection.

The methods and efforts required for conceptualization, characterization, and quantification of vadose zone systems for each application will vary with site conditions, objectives of the investigation, and investigator experience. We would like to note that this chapter is not intended to substitute for the thousands of excellent papers and a number of books on vadose zone problems (for example, Everett *et al.* 1984; Jury *et al.* 1991; Kutilek and Nielsen 1994; Wilson *et al.* 1995; Stephens 1996; Selker *et al.* 1999), but rather to present the main directions, advantages, and limitations of vadose zone characterization and monitoring methods.

Conceptualization of Vadose Zone Systems

Conceptualization of vadose zone systems is needed for the integrated qualitative and quantitative characterization of unsaturated flow and transport processes affected by natural behavior and man-induced changes. Conceptualization is provided for any scale of investigation, including site-specific, subregional, and regional applications. Conceptualization involves a step-wise, iterative process of developing multiple working hypotheses for flow and transport process characterization. These hypotheses are then used in selecting a proper combination of monitoring methods, interpretation, and analysis for refinement of flow and transport conceptual models.

A conceptual model is an interpretation or description of the physical system's characteristics and dynamics. The development of a conceptual model is an important step for site characterization because an incorrect model can lead to significant errors in the development of mathematical and numerical models, thus adversely affecting predictions and planning of remediation efforts. The development of a conceptual model is based on the analysis and simplification of data collected during field-monitoring and laboratory experiments, simplification of hydrologic systems, and the representation of hydrogeologic parameters in models (Boulding 1995). In general, conceptual models that describe water flow include a description of the hydrologic components of the system and how mass is transferred between these components.

Without a conceptual model, we do not know what tests to conduct, what parameters to measure, where to place probes, or what probes to use. Conversely, without such data, we cannot develop a conceptual model. This situation requires an iterative approach, in which we conduct a series of observations and tests and, concurrently, develop a conceptual model of water flow to refine our tests. The development of a conceptual model for water flow in the heterogeneous soil and fractured rock of the vadose zone is particularly difficult for three main reasons:

(1) The contrasts in permeability of soils and rocks at different parts of the system may be extreme and localized.

(2) The geometry of water flow depends strongly on the interconnection or connectivity of a preferential-flow-zone network. In a given vadose zone system, many probes may be located within nonconductive zones, which have no significant role in flow. In fractured rocks, fractures may be nonconductive because apertures are closed under the ambient stress state or by mineral precipitation. Additionally, soil and rock hydraulic conductivity may decrease during an infiltration event because of clogging, sealing, or air entrapment.

(3) The design of borehole tests and the interpretation of data in heterogeneous soils and fractured rocks are complicated because the

response in a monitoring well may only be from a single zone of preferential flow or a fracture (Long 1996). Therefore, "point" measurements in heterogeneous soils and fractured rocks cannot reveal complex processes that result from the interaction of features at many different scales.

Conceptualization begins with a theoretical understanding of the entire groundwater-vadose zone-atmosphere system, followed by data collection and the refinement of that understanding. Additional data collection and analysis, as well as the refinement of the groundwater system conceptual model, occur during the entire process of conceptualization and characterization, and during groundwater model development and use (Figure 3-1).

Numerical modeling (forward and inverse) as a means of developing a flow-and-transport-process conceptual model allows us to obtain a better understanding of the level of detail and features needed to improve site-characterization design and monitoring methods.

WATER FLOW AND CHEMICAL TRANSPORT PROCESSES IN DEEP AND SHALLOW VADOSE ZONES

SPATIAL AND TEMPORAL SCALES OF VADOSE ZONE INVESTIGATIONS AND SCALING

Spatial Scales

Heterogeneity of hydraulic processes in soils and sediments occurs in a hierarchy of spatial and temporal scales (Cushman 1986; Faybishenko 1986; Wagenet *et al.* 1994; Wheatcraft and Cushman 1991). Heterogeneity of soils and sediments on different scales and nonuniform areal precipitation and run-off are the main causes of the multiscale flow phenomenon in the vadose zone. The conventional soil-science approach considers flow processes to occur on several scales, shown schematically in Figure 3-2.

This figure illustrates pore (microscopic), Darcian (mesoscopic), and catchment (megascopic) scales. The basic element of soil used for field studies is called a pedon. The pedon is a three-dimensional body having a land surface area of 1 to 10 m^2. The "pedon-scale" investigations are then used to extend the results to a large field scale. Kutilek and Nielsen (1994) proposed the inclusion of two categories within the catchment

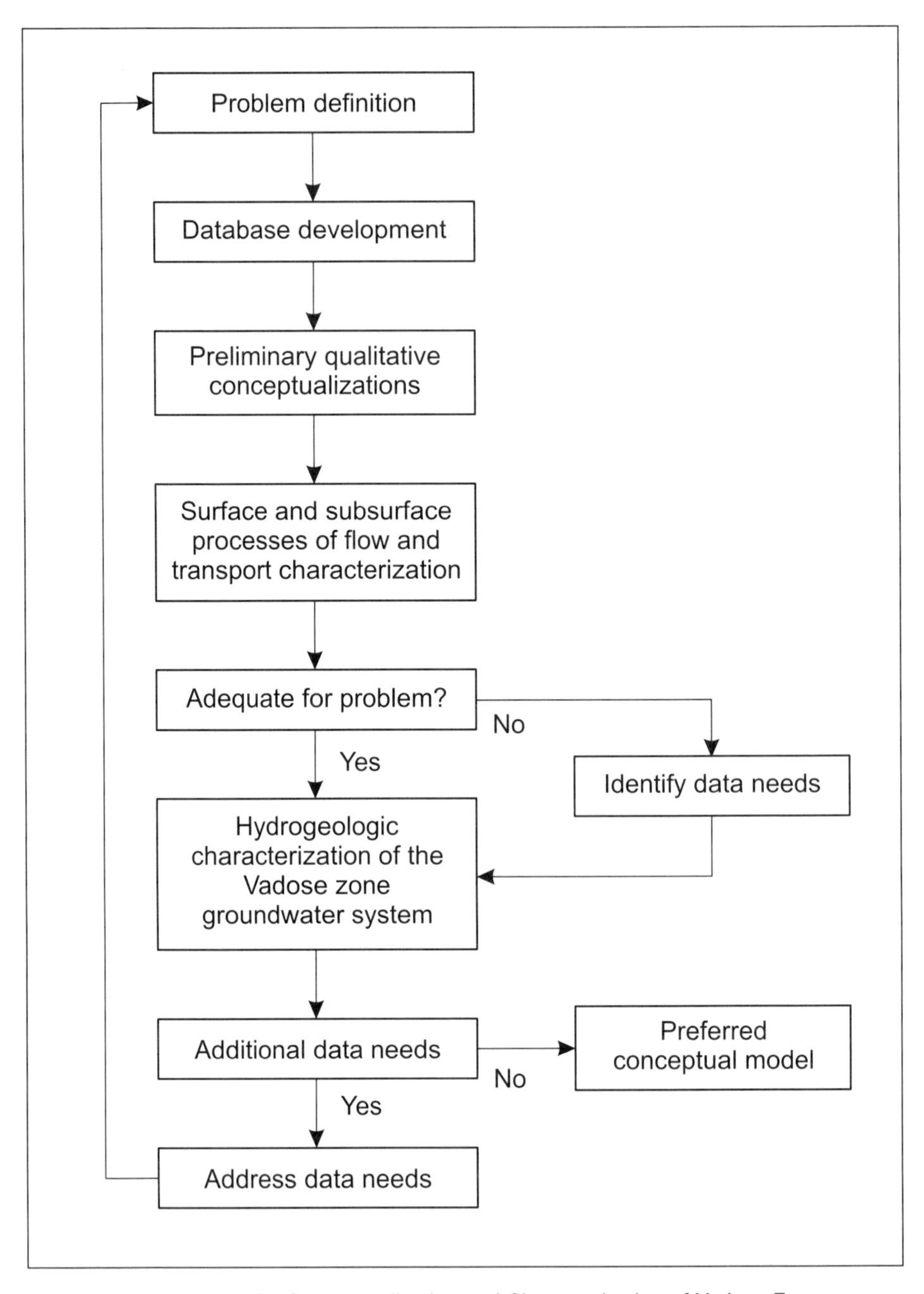

Figure 3-1. Procedure for Conceptualization and Characterization of Vadose Zone - Groundwater Flow Systems (modified from Kolm *et al.*, 1996)

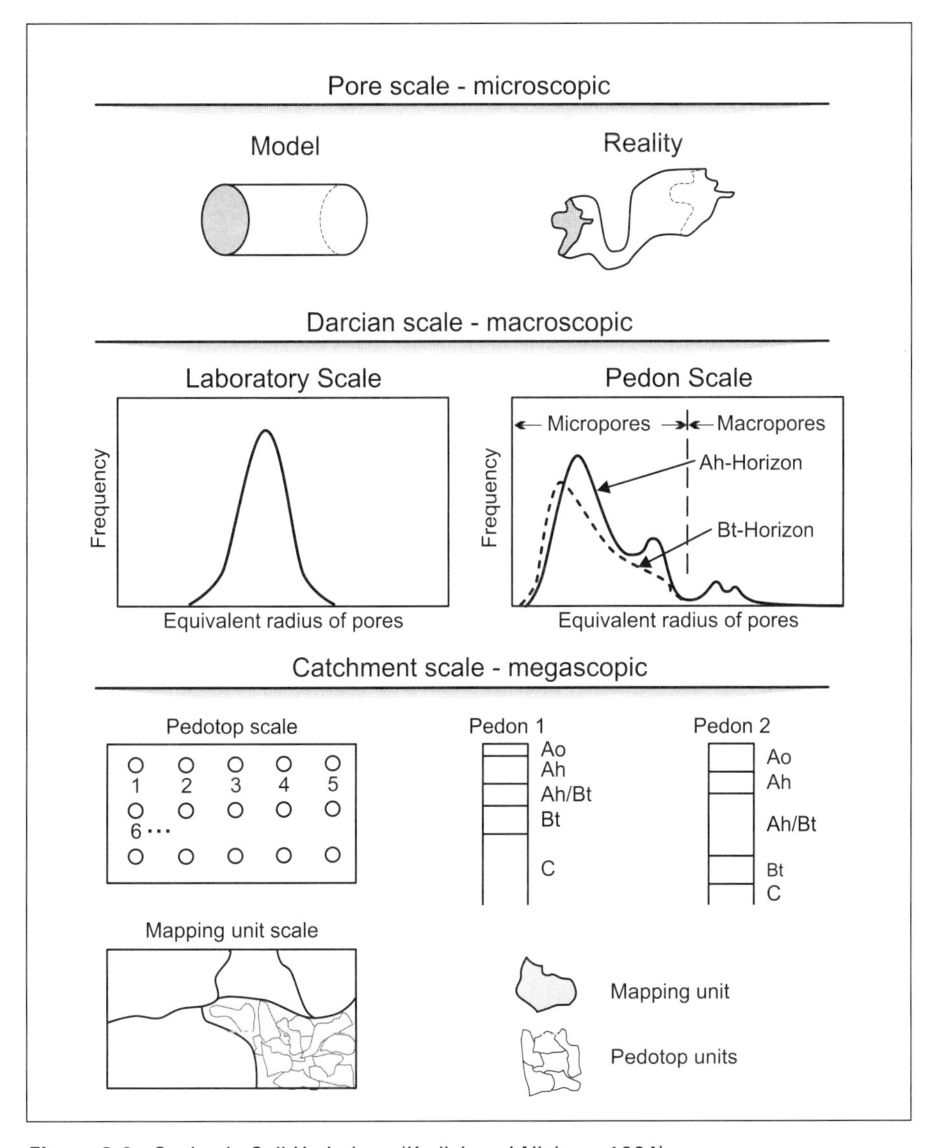

Figure 3-2. Scales in Soil Hydrology (Kutilek and Nielsen, 1994)

scale shown in Figure 3-3: (1) the pedotrop scale, exhibiting a stochastic variability of the infiltration rate (Figure 3-3b), and (2) the mapping unit-scale, exhibiting mostly a deterministic variability (Figure 3-3c) of averaging infiltration rates within each mapping unit.

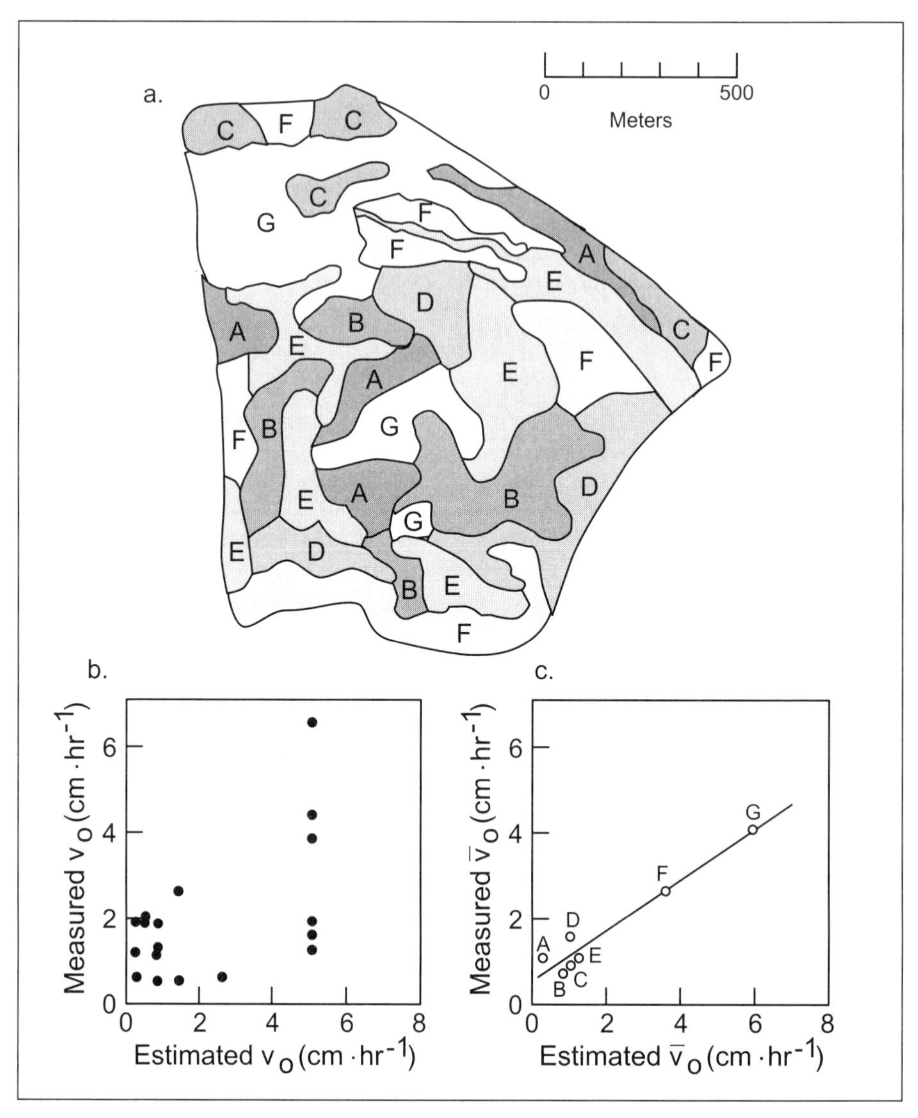

Figure 3-3. **(a)** Pedologic map delineating seven pedotops (designated A through G) within a mapping unit associated with a 100-ha farm. **(b)** Measured quasi-steady state infiltration rates for the seven pedotops illustrated in Figure 3-3a versus those estimated from soil texture. **(c)** Measured mean and estimated mean values $\bar{v}_0$ within each pedotop (Kutilek and Nielson, 1994).

For flow in fractured rocks, Faybishenko *et al.* (1999b) proposed the following hierarchy (Figure 3-4) including elemental-scale, small-scale, intermediate-scale, and large-scale components.

Elemental components of the flow system include a single fracture or a block of porous medium (matrix). Elemental components range in size from a few centimeters to 10-20 cm. Elemental components can be studied in the laboratory using small core samples or larger fracture replicas or in the field using point-size probes. Results of experiments on this scale can be used to describe the details of specific flow and transport processes in fractures, in the matrix, or in fracture-matrix interactions. Some examples of these flow processes are: (1) water dripping from a

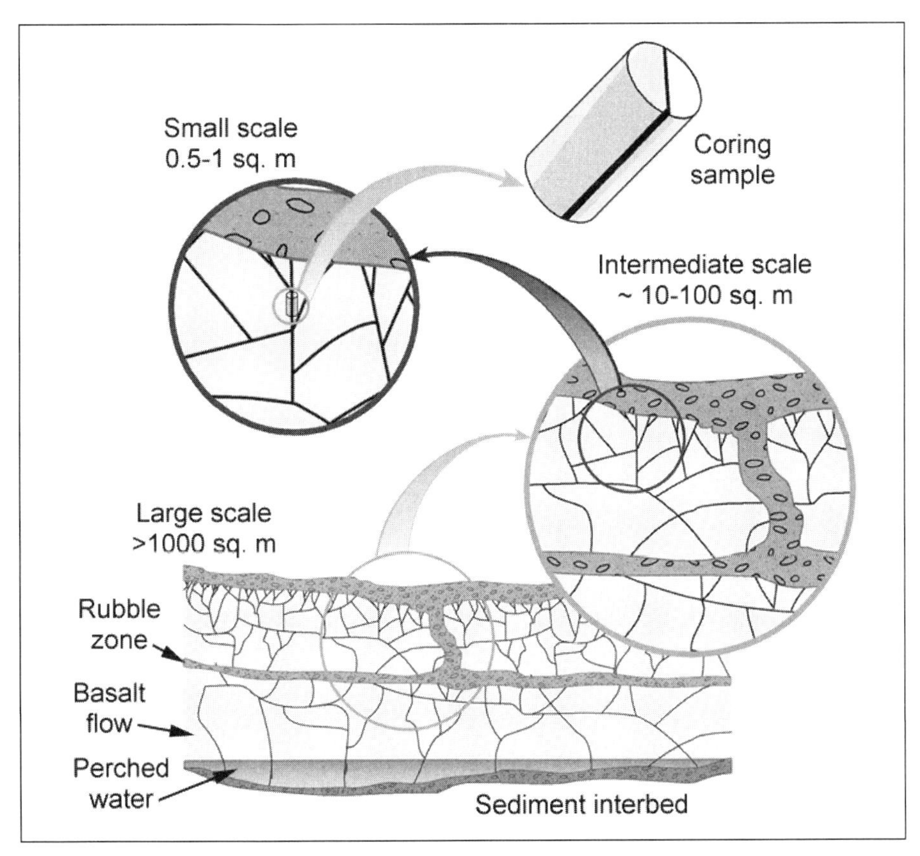

Figure 3-4. A four-level hierarchy of scales of hydrogeological components in fractured basalt (Faybishenko *et al.*, 1999)

fracture under field conditions in boreholes, tunnels, caves, and other underground openings; (2) film flow, or water meandering along a fracture surface; (3) water dripping within flow channels; and (4) intermittent flow along a fracture surface.

Small-scale components include a volume of rock within a single basalt flow with one or a few fractures. The areal extent of small-scale components is approximately 0.5-1 m^2. Results of field experiments on this scale can be used to describe in detail some of the flow and transport processes in a single fracture or a few intersecting fractures. Small-scale infiltration experiments are conducted to take into account fracture-matrix interaction, water-dripping phenomena, and small-scale averaging of flow rates and water pressures measured in fractures and matrix.

Intermediate-scale components include the volume of rock within a basalt flow involving all types of fractures, including the fractured flow top, dense flow interior, the less-fractured flow bottom, and fractures intersecting the basalt flow and rubble zone. The areal extent of intermediate-scale components is approximately 10–100 m^2. The results of field experiments at this scale can be used to describe all flow and transport processes within a single basalt flow. While our prime focus is the study of flow in the fracture network within a single basalt flow, we also study other basalt-flow features such as vesicular or massive basalt, fracture zones in the upper and lower fractured colonnade, and the central fractured zone or entablature.

Large-scale components involve the volume of rock containing several basalt flows and the rubble zones between them. The areal extent of a large-scale component usually exceeds 1,000 m^2. At this scale, we can study flow in the fracture networks and regional hydrogeological processes, which are affected by the network of vertical and horizontal rubble zones, as well as sedimentary interbeds.

Temporal Scales

Water fluxes and moisture profiles in the vadose zone depend on (a) short-term or event-based, (b) seasonal, and (c) long-term or climatic processes (Eagleson 1978; Milly and Egleson 1987). Assuming that the maximum rate of infiltration and exfiltration (evaporation) depends on the initial moisture content of soils, Reeves and Miller (1975), Milly (1986), and Salvucci and Entekhabi (1994) developed the Time

Compression Analysis (TCA) method to estimate the temporal mean moisture profiles of soils. However, Doodge and Wang (1993) demonstrated analytically that the TCA method is not exact.

It is important to determine temporal scales of vadose zone processes because we need to provide long-term predictions (up to hundreds and thousands of years) based on relatively short-term observations of several years. Because of the diurnal, seasonal, and annual variations of moisture content and water pressure at different depths, the vadose zone can be subdivided into several intervals (Figure 3-5). Field observations showed that the depth of these intervals remained approximately the same with time when the cycling of boundary (atmospheric) conditions was the same. Results identical to those of Kutilek and Nielsen (1994), as shown in Figure 3-5, were obtained by Faybishenko (1986), who studied a 44-meter-deep vadose zone in macroporous loam soils. Faybishenko determined that under ambient conditions (natural precipitation and snowmelt) in a semi-arid climate, four distinct zones appear along the vadose zone vertical profile: (1) a near-surface zone affected by episodic rain events, which extends from the surface to depths of 0.5 to 1 m; (2) a zone affected by seasonal infiltration, to depths of 6 to 8 m; (3) a zone of virtually constant moisture content, in which the hydraulic gradient is unity and there is annual downward water flow toward the aquifer; and (4) a capillary fringe zone above the water table. In these conditions, periodic downward and upward flow (as a result of evaporation) were observed in the upper two zones, and annual downward flow occurred below the depths of 6 to 8 m. However, changes in boundary (atmospheric) conditions, may affect the depths of these intervals. Field-observation results, such as these cannot be simulated using a one-dimensional flow model (Salvucci and Entekhabi 1994) because a one-dimensional model cannot describe the three-dimensional flow patterns that result from the deep penetration of water through preferential-flow zones immediately after precipitation, beyond the depth of a zone affected by evaporation. Thus, although the potential evapotranspiration exceeds the precipitation, downward water flow appears in arid and semi-arid climatic conditions (Faybishenko 1986; see also the case study "Near Surface Infiltration Monitoring Using Neutron Moisture Probes, Yucca Mountain, Nevada," by Alan L. and Lorraine E. Flint). The observations summarized above also indicate that water is able to migrate

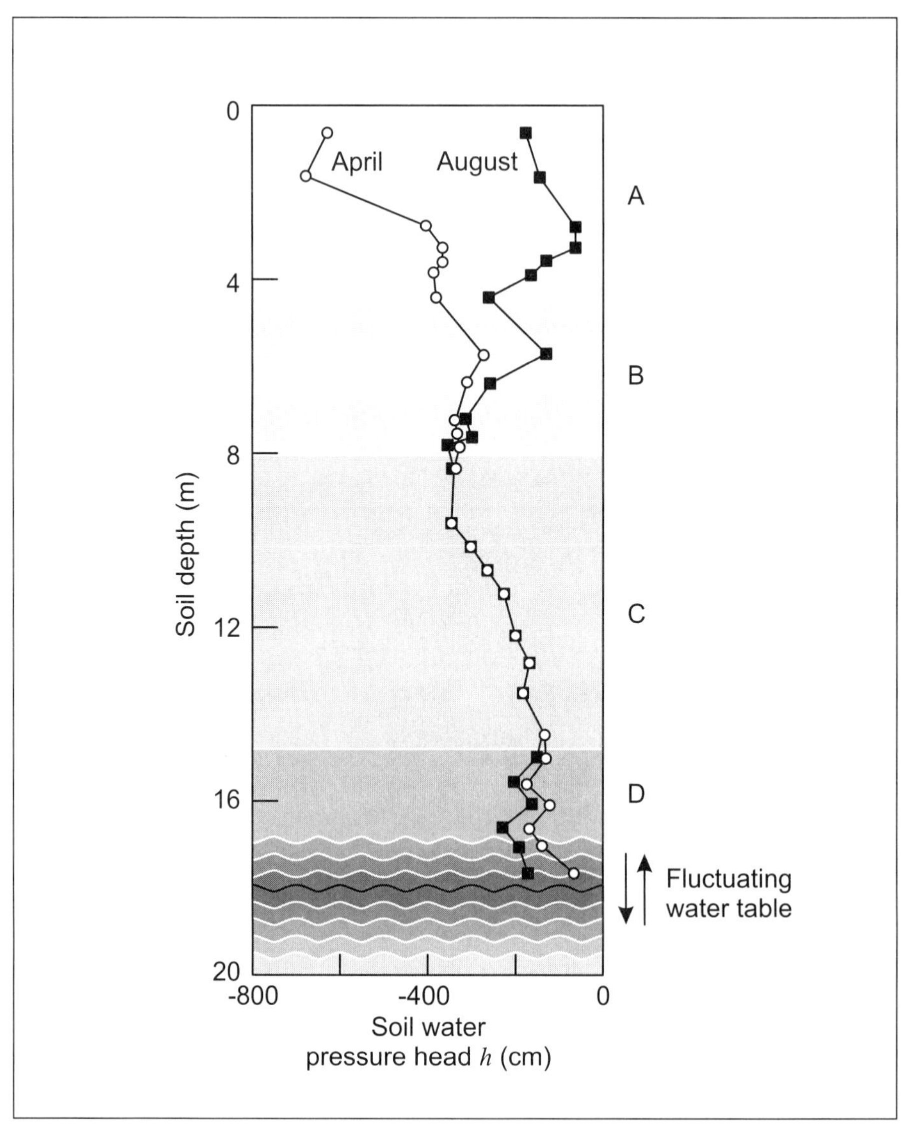

Figure 3-5. Monthly mean values of soil water pressure head h for two months of 1985 measured within the vadose zone of an irrigated soil in a semi-arid region of North China (Kutilek and Nielsen, 1994)

downward rather rapidly, along localized preferential flow paths in partially saturated soils and rocks, without being imbibed into the soil matrix.

Scaling of Hydraulic Parameters

Scaling theories assume that a continuously heterogeneous field is an ensemble of homogeneous domains with geometrical and soil hydraulic-function similarities. Scaling of hydraulic properties of heterogeneous soils was used by several authors (Warrick and Amoozegar-Fard 1979; Milly and Eagleson 1987; Bresler and Dagan 1979, 1983; Kabala and Milly 1991) to calculate water flow. Sposito and Jury (1990) showed that Richards' equation is invariant under the scaling transformation only if $K(\theta)$ is a power or exponential function. To demonstrate the usefulness of a scaling procedure, Figure 3-6 shows the results of water-content field measurements with time at four field plots at eight soil depths. (In total, 608 measurements, 19 times from 32 locations, were taken.)

However, obtaining volume-averaged fluxes by scaling the results of one-dimensional analysis to hydrologically connected and spatially heterogeneous fields is not a trivial task (Salvucci and Entekhabi 1994). Jury and Roth (1990) and Hewett and Behrens (1993) showed that the Miller-similitude assumptions and other scaling techniques may fail to represent heterogeneous soils. In studying fractured basalt, Faybishenko *et al.* (1999b) have found that, at each scale of investigation, different methods and models for flow phenomena must be used to explain observed behavior when no apparent scaling principles are evident.

Two case studies illustrating "Scaling of Soil Hydraulic Properties" by B.P. Mohauty and P.J. Shouse are on the accompanying CD.

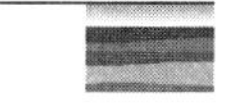

Evidence of Preferential Flow in Heterogeneous Soils and Fractured Rocks

Definition and Main Mechanisms of Preferential Flow

The term "preferential flow" is used to describe the flow that occurs in a non-volume-averaged fashion along localized, preferential pathways, by-passing a fraction of the porous space. Preferential flow in heterogeneous soils may occur along root channels, earthworm burrows,

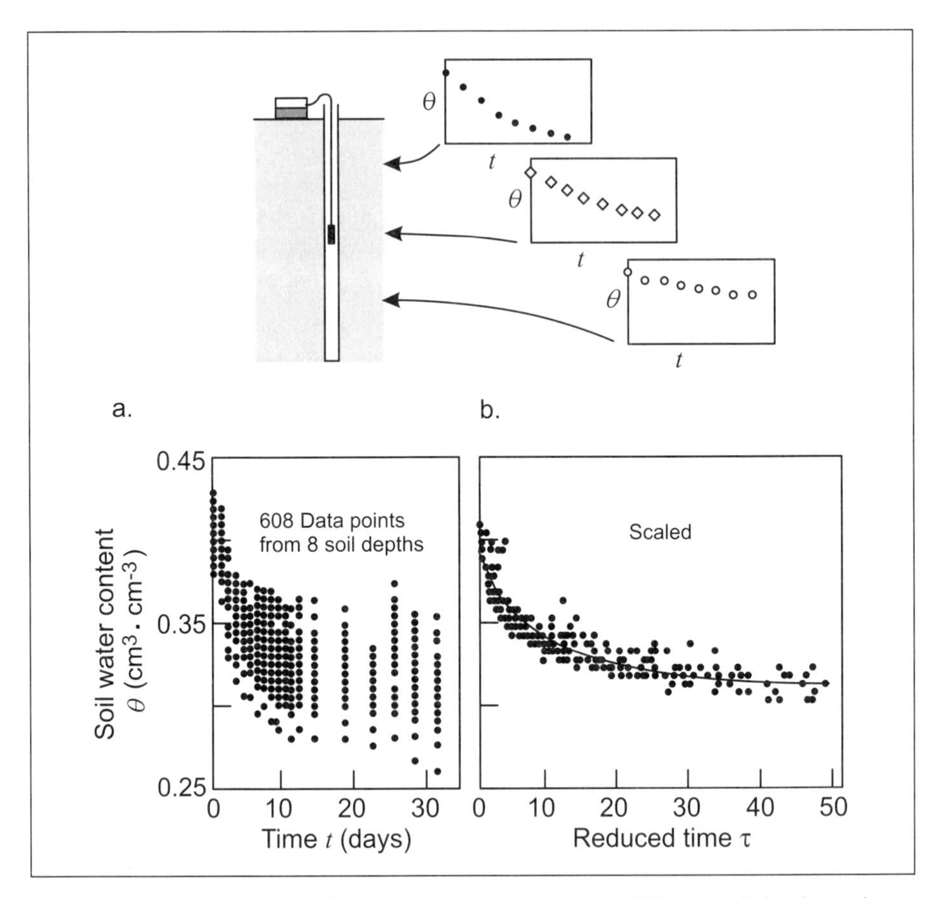

Figure 3-6. Neutron probe soil water contents measured at different soil depths and spatial locations during water redistribution. The solid line in figure c represents the scaled data given in Kutilek and Nielsen (1994, page 265)

and soil fissures or cracks in both fine-textured and coarse-textured soils, as well as at geological heterogeneities such as fractures, clastic dikes, and breaks in caliche layers. Such heterogeneities are created as a result of depositional conditions, diagenesis of sediments, faulting, fracturing, and differential weathering processes. Because flow velocity is higher along the zones of preferential flow than in other parts of the media, preferential flow is also called "fast flow".

Water seepage in the subsurface depends on the state of the land surface, the heterogeneity of the soil profile, and characteristics of the atmospheric and artificial forcing events, which occur on different time scales. Figure 3-7a schematically illustrates several types of fluxes at the

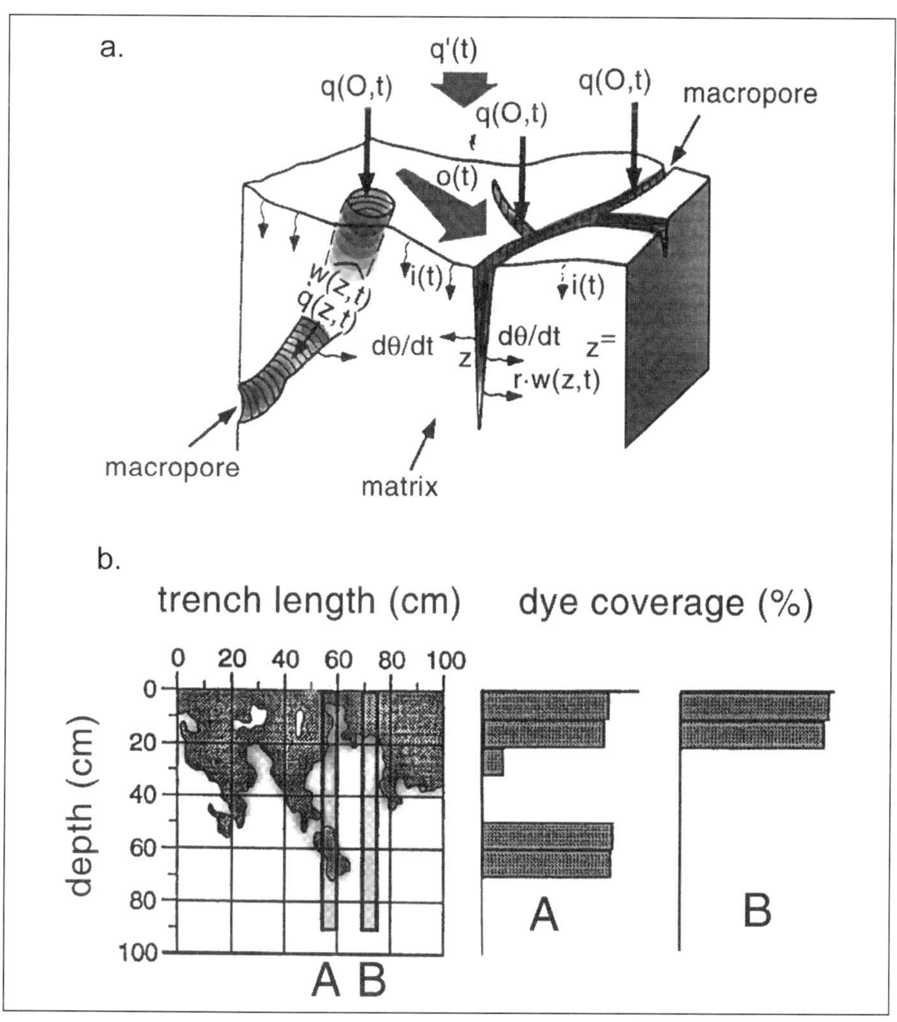

Figure 3-7. Illustration of a concept of preferential flow in heterogeneous soils: (**a**) Schematic of fluxes occurring under infiltration: $q^*(t)$ is overall water input - precipitation and irrigation, i(t) is infiltration into the top soil matrix, $o(t)$ is overland flow (runoff) when $q^*(t)>i(t)$, $q(0,t)$ is volume flux density into the soil mocropores, $q(z,t)$ is volume flux density, $w(z,t)$ is the volumetric soil moisture content, $r.w(z,t)$ is the water sorbance from macropores into the soil matrix (Germann and Beven, 1990). **(b)** Map of the two-dimensional distribution of a chemical at the excavated trench exposure after sprinkling 40-mm dye solution onto the soil surface, and percent of a chemical at different depths determined using cores in vertical wells A and B (Flury, 1996).

land surface causing the phenomenon of preferential flow in macroporous soils and the water redistribution between the macropores and the matrix, and Figure 3-7b depicts a distribution of the dye tracer along the vertical cross-section of the structured, fine-textured soil. Small changes in the water-flow regime of a heterogeneous, deep vadose zone may not be inferred from conventional moisture-content measurements (for example, using neutron logging), because the accuracy of measurements is within range of the moisture-content fluctuations. We can overcome this problem with long-term field measurements of the water pressure using deep tensiometers.

Several mechanisms are assumed to cause preferential flow, including water repellency, cracks, biological effects (such as earthworm or root channels and macropores), air entrapment, small-scale variations in soil hydraulic properties, discrete obstacles, entrapped air behind the wetting front, and confined air ahead of the wetting front. Field infiltration tests with tracers showed that the volume of wetted soils within fingers could occupy from as little as 2% (Kung 1990a,b) to as much as 70% of the total volume of soil (Jury *et al.* 1986; Ghodrati and Jury 1990). This knowledge is important in predicting the fate of contaminants. Small differences in surface topography may significantly affect spatial variations of infiltration, which occur as a result of non-uniform snowmelt on the land surface and changes in soil hydraulic properties in the near-surface zone (for example, resulting from temperature variations). Preferential flow in the vadose zone creates localized groundwater recharge, which may vary with time as a result of changes in the chemical composition of moving and indigenous solutes. For example, sodium concentration, redox conditions, biological transformation of organic materials, and high temperature may significantly affect hysteretic properties of water retention and unsaturated hydraulic conductivity of unsaturated-saturated soils, as well as the processes of water, chemicals, and bio-transformation between the zones of fast flow paths and slow volume-averaged flow.

The capillary characteristics of the heterogeneous sands play a critical role in the displacement of water by a dense, immiscible phase, chlorinated solvent such as a dense nonaqueous phase liquid, or DNAPL (Chen *et al.* 1995; Ewing and Berkowitz 1998). The invading DNAPL may flow laterally and cascade off fine-sand lenses (Kueper *et al.* 1989). Chilakapati *et al.* (1998) showed that the prediction of geochemical transport in heterogeneous systems using a volume-averaged model significantly underestimated oxidation the reaction kinetics, and retardation.

Examples of Preferential Flow in Soils and Fractured Rocks

Heterogeneous Soils and Sediments at the Hanford Site

The sediments beneath waste sites at Hanford are highly heterogeneous (for example, sediments include interbedded sand, silts, gravels, and boulders). Temporal and spatial variations in net water infiltration through current and past liquid discharges, water line leaks, and variable chemical interactions complicate description and understanding of contaminant transport, and lead to uncertainty in the evaluation of transport at contaminated sites. A number of knowledge gaps—including an insufficient understanding of source terms, geological and hydrologic properties, preferential flow, and chemical interactions–make current modeling of contaminant transport in the Hanford vadose zone unreliable.

Figure 3-8 presents an example of three potential types of preferential flow in the vadose zone identified at the DOE Hanford site: (1) fingering, (2) funnel flow, and (3) flow associated with clastic dikes or poorly sealed borehole annular space. According to Ward *et al.* (1997), funnel flow can enhance lateral migration, and horizontal layering will tend to stabilize fingered flow, whereas cross-bedding concentrates and coalesces fingers (Glass *et al.* 1991; Glass and Nicholl 1996; Nicholl *et al.* 1993). Flow through clastic dikes and poorly sealed well-annular spaces could exhibit a hysteretic effect: it may appear during infiltration events, and there may be flow impediments during drying.

A more detailed description of transport beneath Hanford waste tanks is given in "Gamma Borehole Logging for Vadose Zone Characterization Around the Hanford High-Level Waste Tanks " by D.S. Shafer, J.F. Bertsch, C.J. Koizumi, and E.D. Fredenburg. *See page 445.*

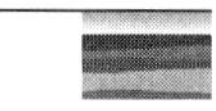

Fractured Basalt at the Idaho National Engineering and Environmental Laboratory (INEEL) Site

Percolation ponds, injection wells, and buried waste sites are primary sources of fractured-basalt vadose zone contamination at INEEL. The

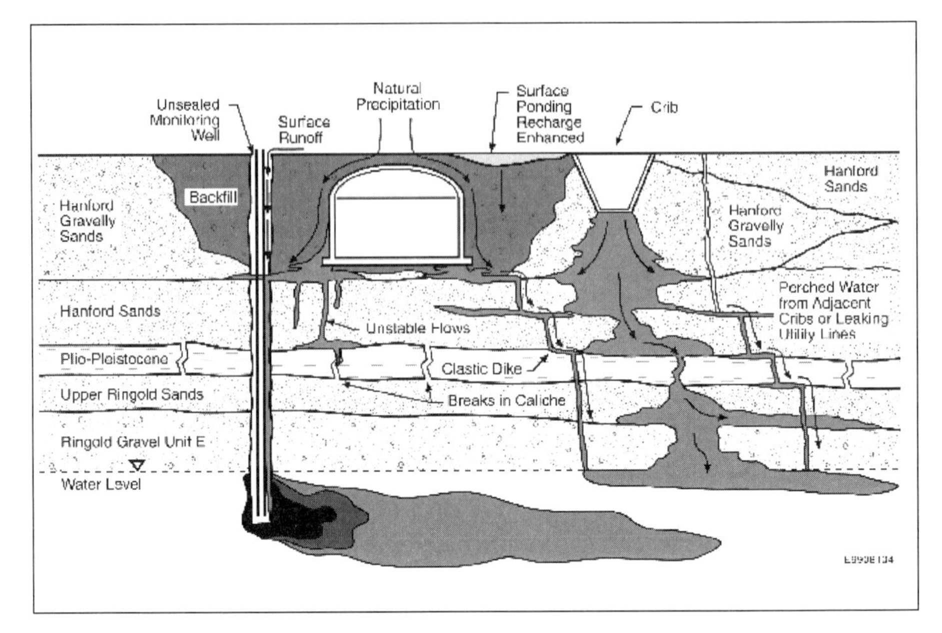

Figure 3-8. Conceptual model of fluid flow beneath single shell tanks at Hanford showing fingering, funnel flow, and flow associated with clastic dikes or poorly sealed borehole annular space (DOE 1999).

Radioactive Waste Management Complex (RWMC) includes one of the largest subsurface waste disposal facilities in the Department of Energy (DOE) complex. In the past, disposal of low-level, mixed, and transuranic radioactive wastes was achieved by direct discharge or burial in shallow, unlined pits and trenches within the surficial sediments. Vadose zone fractures provide a mechanism for deep transport of oxidizing fluid and gas, leading to the release of toxic species, such as heavy metals, that may otherwise be stable in a reducing environment. Flooding of the RWMC has occurred three times in the past, potentially increasing the downward mobility of the subsurface contaminants. The determination of the time required for contaminants to reach the aquifer is important to decision-making regarding remediation options. In 1994, INEEL conducted a Large Scale Infiltration Test (LSIT) 1.4 km south of

RWMC to investigate hydrologic properties of the vadose zone basalts. The vadose zone thickness near the RWMC is about 190 m. The LSIT site consisted of a bermed basin, 183 m in diameter, which contained 32 million L of water. Beneath the basin is a thick sequence of stacked basaltic lava flows, with the first major interbed at a depth of 55 m. Seventy wells were drilled for the test, primarily along four axes extending radially from the basin, with most terminating in the sedimentary interbed. During the LSIT, tracers did not always follow the same pathways as the initial water movement. The tracer distribution could not be modeled as a one-dimensional steady-state flow, but rather as a three-dimensional network of flow paths, which may vary in time (Wood and Norrell 1996; Faybishenko *et al.* 1999a,b).

The case study "Large-Scale Field Investigations in Fractured Basalt in Idaho: Lessons Learned," by Boris Faybishenko and Thomas Wood, discusses the results and lessons learned from Large Scale Infiltration Tests at the National Engineering and Environmental Laboratory. *See page 396.*

Other Sites

The observations at Hanford and INEEL are in agreement with a growing body of field evidence from various sites in semi-arid regions in the U.S.A. and throughout the world (for example, Yucca Mountain [Flint and Flint 1995], Arizona [Wierenga *et al.* 1998], New Mexico [Wierenga *et al.* 1991; Hills *et al.* 1991], Australia [Allison 1988], and Israel [Nativ *et al.* 1995]. These investigations show that water seepage occurs in an episodic manner along localized preferential pathways at depths of several hundred meters beneath the land surface. It is important to note that only portions of fractures carried water, and the chemical composition of water obtained from fractures was substantially different from that of water samples extracted from the nearby rock matrix (Eaton *et al.* 1996). At a field site in the Negev Desert, Israel, man-made tracers were observed to migrate with velocities of several meters per year across an unsaturated zone of fractured chalk 20 to 60 m in thickness (Nativ *et al.* 1995).

CONTAMINANT TRANSPORT

Point and Linear Source Contaminants

Point-source pollutants are associated with leakage from surface and underground tanks, injection of nuclear and organic wastes in boreholes screened at different depths in the vadose zone, surface spills, etc. Pollution is often highly toxic near the point source, but the exact location of the source is often not easily detectable. Linear-source contaminants enter the vadose zone from sewers, trenches, cribs, creeks, and rivers. Contaminant movement from the source often follows zones of preferential flow, which are difficult, if not impossible, to identify in heterogeneous soils and fractured rocks.

Non-Point Source Contaminants

Non-point-source (NPS) pollutants include contaminants in soils, sediments, and surface waters (such as organic and radioactive materials, fertilizers, pesticides, salts, and metals), that are wide-spread over large areas. Contamination of soil and water resources by NPS pollutants is a major global environmental issue, because the pollutant distribution is not limited by geological and physical boundaries such as lakes, rivers, and mountains. Therefore, the extent of NPS contamination and associated chronic health effects are major environmental threats (Corwin 1996).

NPS pollution of surface waters is caused by surface runoff and erosion. The increase in NPS pollutants is usually a result of human activities including agriculture, urban runoff, feedlots, atmospheric pollution, water-resource extraction, and waste storage. The areal extent of NPS contamination in heterogeneous soils and sediments increases the complexity and volume of data required for assessment far beyond that of typical point-source pollutants. Because of the uncertainty associated with the regional-scale assessment of NPS pollutants, the design of site-characterization and monitoring methods poses complex technical problems (Loague and Corwin 1996).

Contaminant Transport Processes

Contaminants can be present in soils in all three phases—liquid, solid, and gaseous. The main transport processes of contaminants in a liquid phase are advection, dispersion, sorption-desorption, ion exchange, and decay reactions. Hydrodynamic dispersion is the process affected by molecular diffusion caused by a concentration gradient, together with dispersion caused by mechanical mixing and fluid advection. The simplest approach to the contaminant-transport investigation is to consider miscible migration of nonvolatile reactive compounds in the liquid phase, which can be sorbed by the solid phase (Figure 3-9).

The liquid flowing through the soil is not pure water but, even with a single chemical dissolved, a complex fluid (Sposito 1981). Sposito (1981) and Nkedi-Kizza *et al.* (1985) demonstrated how to evaluate the enhanced solubility of chemicals affected by the solvent chemical properties. Figure 3-10 shows the impact of the mixed solvents on the retardation factor. It is important to note that the co-solvent may decrease the retardation factor by several orders of magnitude and, therefore, may enhance the migration of toxic chemicals in soils and groundwater. The impact of the co-solvent (for example, methanol or acetone) is greater

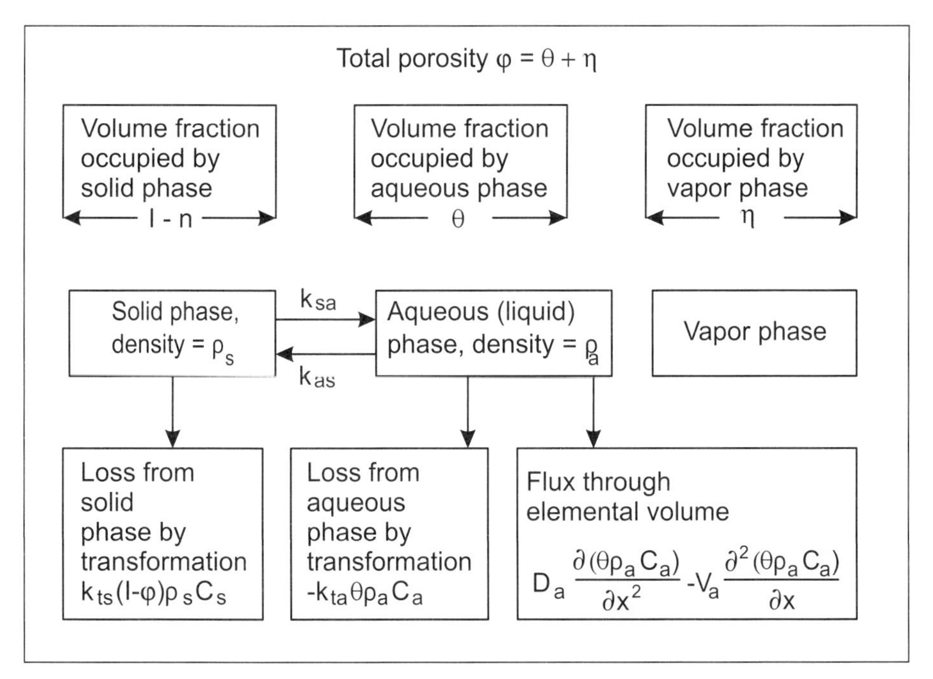

Figure 3-9. Conceptual model showing the transport processes of miscible nonvolatile reactive compounds in soils (Enfield and Yates, 1990)

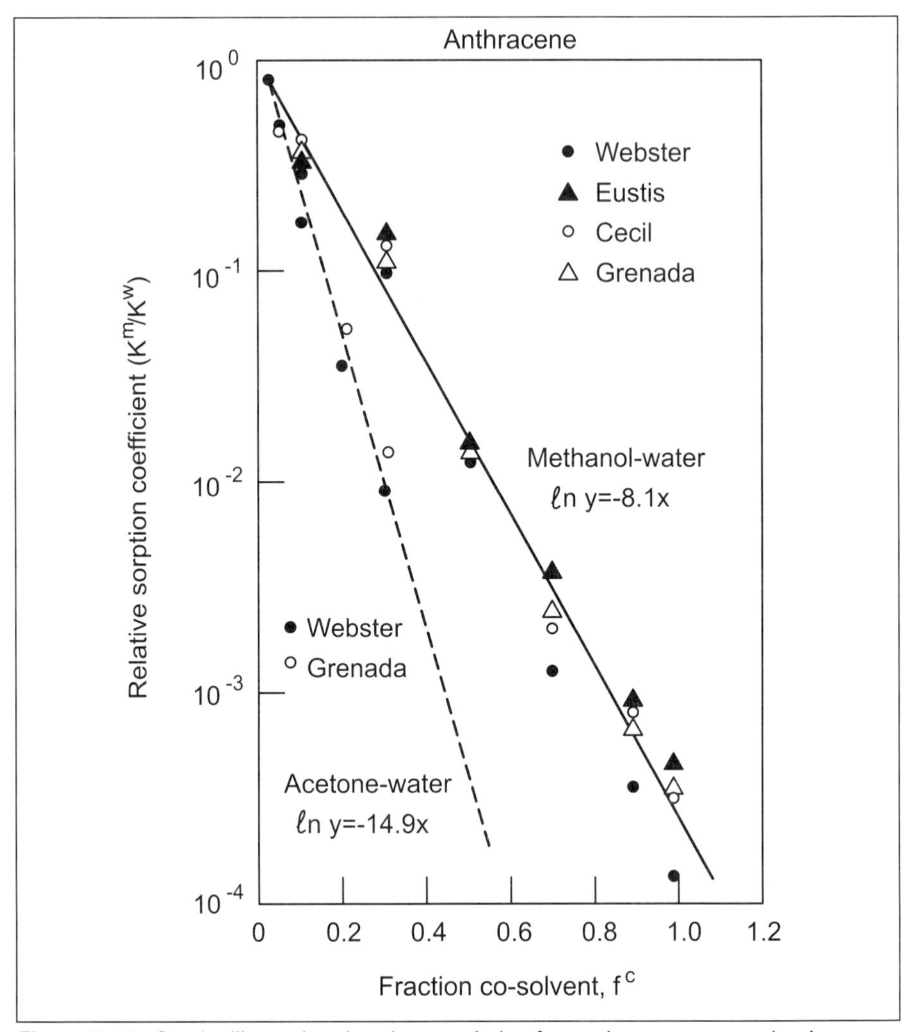

Figure 3-10. Graphs illustrating that the retardation factor decreases several orders of magnitude as the fraction of mixed solvents increases (Nkedi-Kizza *et al.*, 1985)

for more hydrophobic chemicals (Nkedi-Kizza *et al.* 1985). Chemical transport in soils and groundwater can be enhanced by immiscible fluids, or surfactants. Figure 3-11 conceptualizes a number of multiphase flow and transport processes in soils, which one should take into account in designing characterization and monitoring methods for the vadose zone.

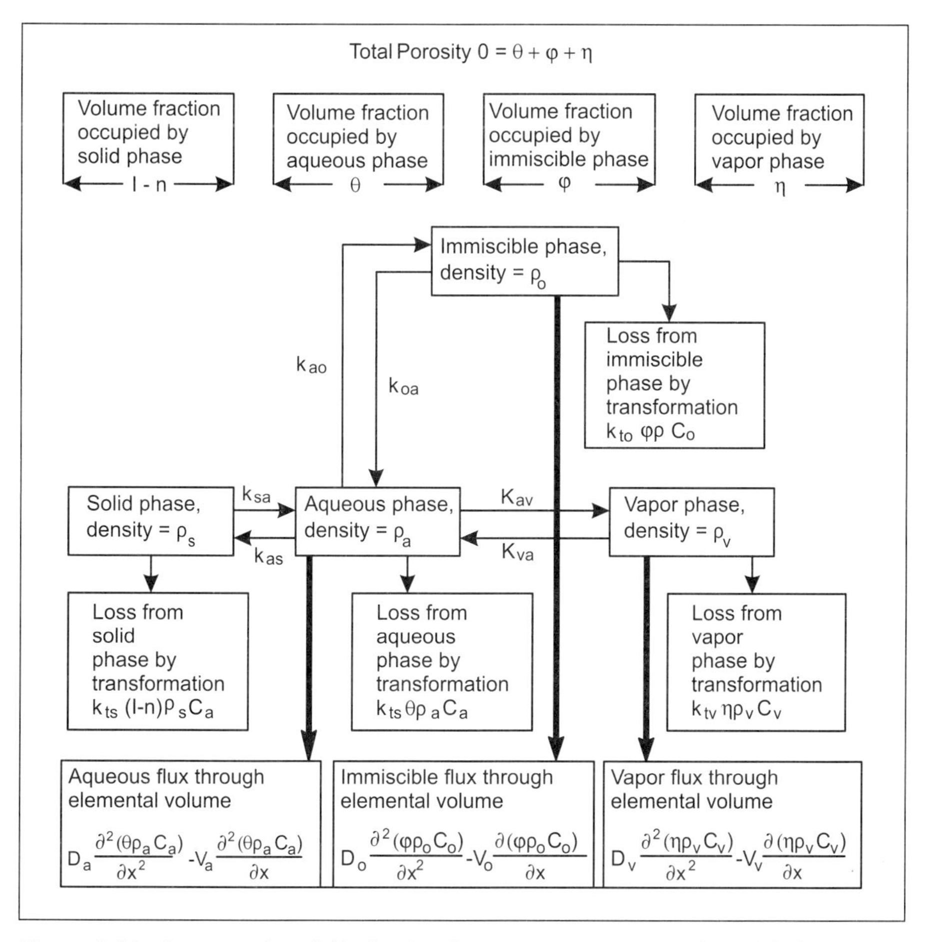

Figure 3-11. Conceptual model indicating the transport processes of multiple fluid phases in soils (Enfield and Yates, 1990)

In structured soils, the macropores can significantly influence movement of volatile compounds because open macropores constitute pathways for vapor-phase movement both downward to the water table and upward from the water table to the atmosphere. Under arid conditions, when the liquid flow in dry soils is insignificant, diffusion and advective transport in the vapor phase may be many times greater than diffusion in the liquid phase.

Figure 3-12 shows a general structure of the system of first-order decay reactions for three solutes (A, B, and C) adopted from Simunek

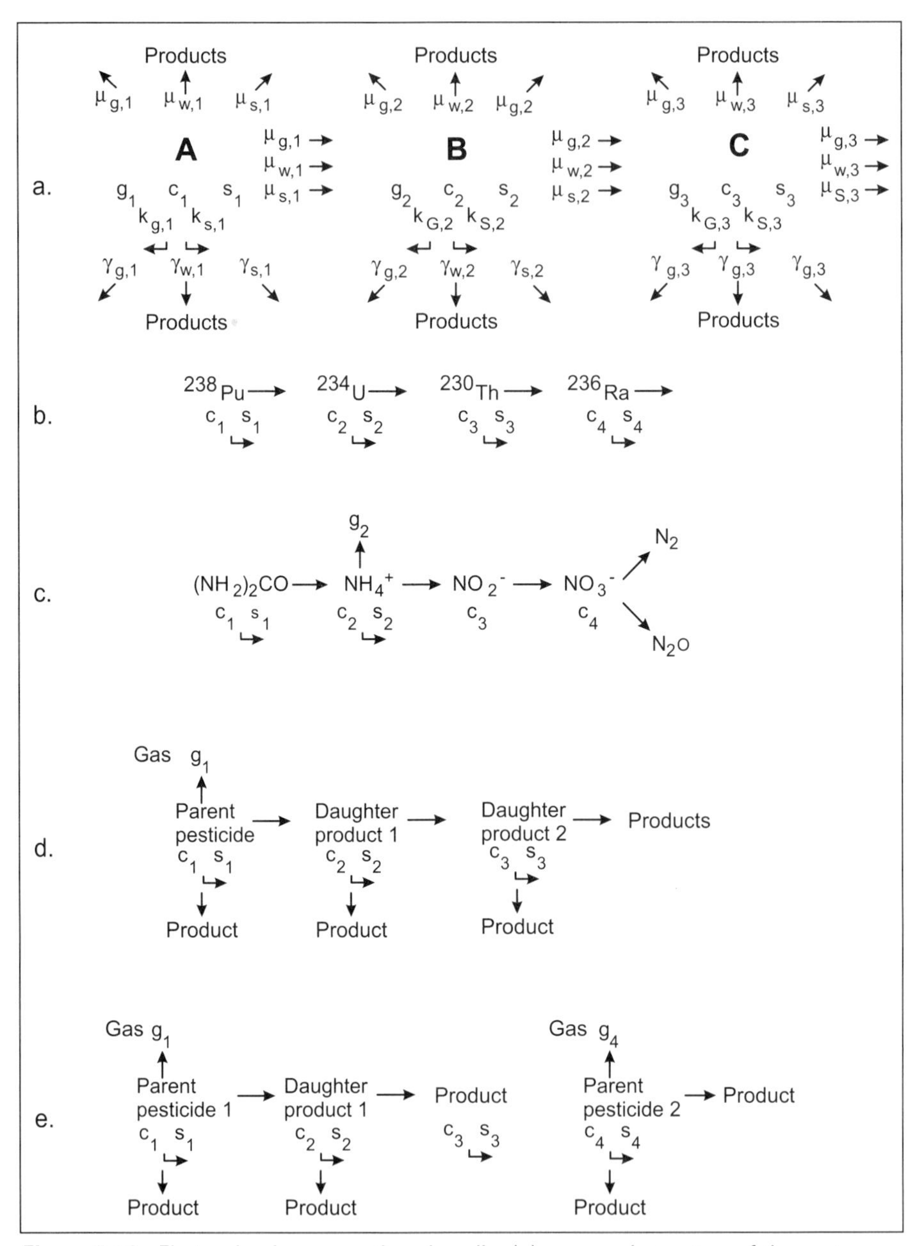

Figure 3-12. First-order decay reactions in soils: (a) a general structure of the system for three solutes (A, B and C), and typical examples of first-order decay reactions for (b) radionuclides, (c) nitrogen, (d) pesticides with interrupted chain (one reaction path), and (e) interrupted chain (two independent reaction paths) (Simunek and van Genuchten, 1995)

and van Genuchten (1995). This figure also shows typical examples of first-order decay reactions for radionuclides, nitrogen, and pesticides. However, the oxidation processes in groundwater may be second-order, depending on the concentration of oxygen (Borden and Bedient 1986). Additional investigations are needed to better understand the second-order reaction processes and determine how chemical reaction coefficients depend on environmental variables such as O_2, pH, temperature, and nutrients affecting the biochemistry.

The assumption of decoupling water flow from chemical transport is not always adequate, especially in geothermal fields where temperature gradients can create density- driven mass transport and affect the unsaturated hydraulic properties of soils. Because of heat generation and significant temperature gradients, the effect of temperature on liquid, vapor, and chemical transport should be taken into account in predicting contaminant transport (for example below Hanford tanks).

Another mechanism affecting contaminant behavior in the vadose zone is colloidal migration of contaminants. Colloids are usually present in soils as suspended substances that facilitate transport of both organic and radioactive contaminants. Mobility of colloids depends on liquid flow velocity and on chemical interactions between colloids and matrix surfaces. Although colloidal transport plays an important role in contaminant migration in partially saturated soils, until now only a few studies addressed this problem on the field scale.

BIOLOGICAL PROCESSES*

Terminology

Bioremediation is the use of biological processes to make the environment less toxic. More specifically, bioremediation, uses enzymes, growth stimulants, bacteria, fungi, or plants to degrade, transform, sequester, mobilize, or contain contaminant organics, inorganics, or metals in soil, water, or air. Thus, characterization and monitoring of bioremediation can be performed by measuring the number and type of microorganism, enzymes, or other biochemical markers, breakdown

*This section was contributed by T.C. Hazen.

products, or metabolic indicators (for example, CO_2). To assist in defining characterization and monitoring opportunities, we summarize bioremedation terminology in Table 3-1.

Engineered Bioremediation Concepts and Objectives

Bioremediation systems can be roughly divided into natural attenuation (or intrinsic remediation) and engineered systems. Monitoring needs are directly related to the approach and objective selected for a site. All engineered bioremediation can be characterized as either biostimulation (that is, the addition of nutrients), bioaugmentation, (that is, the addition of organisms), or as a process that uses both. The problems related to adding chemical nutrients to sediment and groundwater are fundamentally different from those related to adding organisms. Simple infiltration in soils and, subsequently, groundwater is physically quite different in the two processes (Alfoldi 1988). Even the smallest bacteria have different transport properties than chemicals. For example, clayey soils have very low permeability and physically may not allow bacteria to penetrate. These clays also may bind the microbes that are added (for example, as cationic bridges, involving divalent metals and the net negative charge on the surface of the bacteria and the surface of the clay, are formed). In some soils, inorganic chemicals that are injected may precipitate metals, swell clays, and change redox potentials and conductivity, thus having a profound effect on groundwater flow and the biogeochemistry of the environment.

Natural attenuation and biostimulation depend on the indigenous organisms. Thus, these methods require that the correct organisms are present in sufficient numbers. In addition, for biostimulation, we must alter the environment in a way that will have the desired bioremediation effect. In most terrestrial subsurface environments, the indigenous organisms have been exposed to the contaminant for extended periods of time and have adapted (for example, through natural selection). Many contaminants, especially organic compounds, are naturally occurring or have natural analogs in the environment. Rarely can a terrestrial subsurface environment be found that does not already have a number of organisms that can degrade or transform any contaminant present. Indeed, even in pristine environments, bacteria have an increasing number of plasmids (small extrachromosomal bits of DNA that code for

TABLE 3-1 Terminology Related to Bioremediation

Bioaugmentation—The addition of organisms to effect remediation of the environment (for example, the injection of contaminant-degrading bacteria into an aquifer)

Bioavailability—The ability of a compound or element to be used by a living organism. Some compounds lack bioavailability because they are insoluble, strongly sorbed to solids, or, for some other reason, they cannot be utilized as a source of nutrients or energy.

Biocurtain—The process of creating a subsurface area of high biological activity to contain or remediate contaminates

Biodegradation—A biological process of reducing a compound to simpler compounds, which may be either complete (for example, reducing organic compounds to inorganic compounds) or incomplete (for example, removing a single atom from a compound)

Biofilters—Normally used to refer to treatment of gases by passing them through a support material containing organisms, such as soil, compost, or trickle filter; sometimes also refers to treatment of groundwater via passage through a biologically active area in the subsurface

Bioimmobilization—A biological treatment process that involves sequestering the contaminant in the environment, but involves no biodegradation of the contaminant

Biological Treatment—Any treatment process that involves organisms or their products; for example, enzymes

Biomobilization—A biological treatment process that makes the contaminant more mobile in the environment, but involves no biodegradation of the contaminant.

Biopiles—Above-ground mounds of excavated soils that are biologically treated by addition of moisture, nutrients, air, or organisms

Bioreactor—A contained vessel (such as a fermentor) in which biological treatment takes place

Bioremoval—A biological treatment involving uptake of the contaminant from the environment by an organism or its agent

Bioslurping—Soil vapor extraction combined with removal of light nonaqueous phase liquid contaminants from the surface of the groundwater table, thereby enhancing biological treatment of the unsaturated zone and the groundwater, especially the capillary fringe zone

Bioslurry Reactor—Biological treatment of soil-bound and water-soluble contamination by making a thin mixture with water (a "water slurry") and treating the mixture in a contained vessel

Biosparging—Injection of air or specific gases below ground (usually into saturated sediments (aquifer material) to increase biological rates of remediation

Biostimulation—The addition of organic or inorganic compounds (for example, fertilizer) to cause indigenous organisms to effect remediation of the environment

Biotransformation—A biological treatment process that involves changing aspects of contaminants, such as the valence states of metals, the chemical structure, and so on

Bioventing—Originally defined as slow vapor extraction of contaminants from unsaturated soils to increase flow of air into the subsurface via vents or directly from the surface, thus increasing aerobic biodegradation rates; now defined more broadly to include the slow injection of air into unsaturated soils

Composting—Treatment of waste material or contaminated soil by aerobic biodegradation of contaminants in an above-ground, contained, or uncontained environment

Engineered Bioremediation—Any type of manipulated, stimulated, or enhanced biological remediation of an environment

Intrinsic Bioremediation—Unmanipulated, unstimulated, unenhanced biological remediation of an environment (that is, natural biological attenuation of contaminants in an environment)

Land Farming—A process of biologically treating uncontained surface soil, usually by aeration of the soil (tilling) and addition of fertilizer or organisms

Prepared Beds—A contained (lined) area above ground where soil can be tilled or variously manipulated to increase biological remediation; that is, contained land farming

enzymes that can degrade complex compounds like antibiotics) with sediment depth, in response to the increasing recalcitrance of the organics present (Fredrickson *et al.* 1988).

Our ability to enhance bioremediation of any environment is directly proportional to our knowledge of the biogeochemistry of the site. Finding the limiting conditions for the indigenous organisms to carry out the desired remediation is the most critical step. As with surface environments, the parameters that usually limit organisms are required nutrients, inorganic and organic. The most common nutrients are water, oxygen, nitrogen, and phosphorus. In the terrestrial subsurface, water can be limiting, but usually is not. Oxygen is quite often limiting since contaminants can be used as carbon and energy sources by organisms and the contaminant concentration greatly exceeds the oxygen input needed by the organisms. Introduction of air, oxygen, or hydrogen peroxide via infiltration galleries, tilling, sparging, or venting have proven to be extremely effective in bioremediating petroleum contaminants and a variety of other organic compounds that are not particularly recalcitrant (Thomas and Ward 1992). However, if the environment has been anaerobic for extended periods of time and the contaminant has a high carbon content, it is likely that denitrification has reduced the overall nitrogen content of the environment, making this nutrient limiting. Nitrogen has been successfully introduced into the terrestrial subsurface for biostimulation using ammonia, nitrate, urea, and nitrous oxide (EPA 1989). Phosphorus is naturally quite low in most environments and, in terrestrial subsurface environments, even if concentrations are high, the phosphorous may be in a mineral form that is biologically unavailable, such as apatite. Several inorganic and organic forms of phosphate (for example, triethyl phosphate, phosphoric acid, sodium phosphate), have been successfully used to biostimulate contaminated environments (EPA 1989, Hazen 1997). In environments where the contaminant is neither a good carbon nor energy source and other sources of carbon or energy are absent or unavailable, it will be necessary to add an additional source of carbon (Horvath 1972). An additional source of organic carbon will also be required if the total organic carbon concentration in the environment falls below 1 ppm and the contaminant cleanup levels have still not been met. Methane, methanol, acetate, molasses, sugars, agricultural compost, phenol, and toluene have all been added as secondary carbon supplements to the terrestrial subsurface to stimulate bioremediation (National Research Council 1993). Even plants such as poplar trees have been used to biostimulate remediation of subsurface environments (Schnoor *et al.* 1995).

The plants act as solar-powered nutrient pumps that stimulate rhizosphere microbes to degrade contaminants (Anderson *et al.* 1993).

Biostimulation strategies are limited most by our level of ability to deliver the required stimulus to the environment. The permeability of the formation must be sufficient to allow perfusion of the nutrients and oxygen through the formation. The minimum average hydraulic conductivity for a formation is generally considered to be 10^{-4} cm/sec (Thomas and Ward 1989). Additionally, the stimulants required must be compatible with the environment. For example, hydrogen peroxide is an excellent source of oxygen, but it can cause precipitation of metals in soils and lead to such dense microbial growth around the injection site that all soil pores are plugged. It is also toxic to bacteria at high concentrations, for example, above 100 ppm (Thomas and Ward 1989). Ammonia also can be problematic, because it adsorbs rapidly to clays, causes pH changes in poorly buffered environments, and can cause clays to swell, decreasing permeability around the injection point. At some sites, many of these problems can be handled by excavating the soil or pumping the groundwater to the surface and treating it in a bioreactor, prepared bed, land farm, bioslurry reactor, biopile, or compost. In these cases, the permeability can be controlled or manipulated to allow better stimulation of the biotreatment process. It is generally accepted that soil bacteria need a C:N:P ratio of 30:5:1 for unrestricted growth (Paul and Clark 1989). Stimulation of soil bacteria can generally be achieved when this nutrient ratio is achieved following amendment addition. The actual injection ratio used is usually slightly higher (a ratio of 50:5:1), since these nutrients must be bioavailable, a condition that is much more difficult to measure and control in the terrestrial subsurface (Litchfield 1993). It may also be necessary to remove light nonaqueous phase liquid (LNAPL) contaminants that are floating on the water table or smearing the capillary fringe zone, hence bioslurping (Keet 1995). This strategy greatly increases the biostimulation response time by lowering the highest concentration of contaminant the organisms are forced to transform.

Bioaugmentation may provide significant advantages over biostimulation for: (1) environments where the indigenous bacteria have not had time to adapt to the contaminant; (2) particularly recalcitrant contaminants that only a very limited number of organisms are capable of transforming or degrading; (3) environments that don't allow a critical biomass to establish and maintain itself; (4) applications where the desired goal is to plug the formation for contaminant containment, such as a biocurtain; and (5) controlled environments where specific inocula of high-rate degraders will greatly enhance the process (for example, bioreactors, prepared beds, composting, bioslurry reactors, and land farming). Like biostimulation, a major factor affecting the use of bioaugmentation in the terrestrial subsurface is hydraulic conductivity. The 10^{-4} cm/sec permeability limit for biostimulation will need to be an order of magnitude higher for bioaugmentation and may need to be higher yet, depending on the size and adherence properties of the organism being applied (Baker and Herson 1990). Recent studies have shown that the less adherent strains of some contaminant-degraders can be produced, allowing better formation penetration (DeFlaun *et al.* 1994). However, the ability to rapidly clog a formation is a significant advantage of bioaugmentation in applications where containment is a primary goal. The oil industry has been using this strategy for a number of years to plug fluid loss zones and enhance oil recovery (Cusack *et al.* 1992).

Bioaugmentation is indistinguishable from biostimulation in many environments, since nutrients are often injected with the organisms and since dead organisms are an excellent source of nutrients for most indigenous organisms. For many applications it is difficult, if not impossible, to determine if the added organisms *provide* a significant advantage over nutrient stimulation alone. Even some of the best controlled bioaugmentation field studies, such as the caisson studies of polychlorinated biphenyl (PCB) biodegradation in Hudson River sediment, could not show a significant advantage for bioaugmentation over biostimulation alone (Harkness *et al.* 1993). Given the problems and high cost of producing and delivering the organisms, bioaugmentation applications will probably remain limited. However, bioaugmentation may have a very significant advantage when genetically engineered microorganisms (GEMs) are used. It is possible that a GEM could be constructed with unique combinations of enzymes to facilitate sequential biotransformation or biodegradation of a contaminant. Such a microorganism would be particularly helpful for contaminants that are extremely recalcitrant (such as PCBs), or are degraded only under limited conditions (for example, tetrachloroethylene and carbon tetrachloride can only be biodegraded anaerobically). In addition, a GEM could be modified with unique survival or adherence properties that would make it better suited to the environment where it was to be applied.

FIELD VADOSE ZONE CHARACTERIZATION AND MONITORING

Types of Data for Site Characterization and Monitoring Methods

Site Selection

An important step in developing a program for site characterization and monitoring is the selection of the appropriate field site location. The selection of the field site should focus primarily on the detection and characterization of contaminant source areas and existing and potential pathways for contaminant transport in the subsurface. Under the Resource Conservation and Recovery Act (RCRA) of 1976, waste management facilities are required to obtain permits to begin field characterization. The selection of the field site usually starts with a preliminary review of existing information on the facilities in order to identify and characterize existing and potential releases. The next step is a visual inspection of the entire facility for evidence of releases and identification of additional areas of concern. Following the inspection, a plan is developed for the sampling visit, and additional information needed to identify the areas affected by contaminant releases is collected to fill data gaps needed to identify the areas affected by contaminant releases. This information, in combination with basic site characterization data (that is, topography, soils, geology, hydrology, and biota and current and past land use and ownership), is used to determine site boundaries and the monitoring system layout.

While site boundaries may be initially defined by ownership, a broader scale should be evaluated to determine the need for the offsite investigation. For example, investigations of vadose zone and groundwater contamination should include areas of potential release sources located upgradient as well as potential migration paths located downgradient from a site. The boundaries of the area investigated can be changed or extended with time when new information on the extent of contamination becomes available. The case study "Groundwater Contamination in the Perched Aquifer at the DOE Pantex Plant: Successful Characterization Using the ESC Approach," by Caroline Purdy and

Jacqueline C. Burton, gives an example of using new information to manage decision-making about changing the areal extent of contamination at a field site.

Characterization of Natural Conditions

Characterization of natural (ambient background or baseline) conditions involves the following tasks:

- Locating, collecting, and organizing basic types of data from available published and unpublished sources
- Conducting specifically designed field, laboratory, and modeling studies for the sites selected
- Using the information from natural analogue sites.

The basic types of data for the site characterization are summarized in Table 3-2.

When data to characterize the site are limited or unavailable, it is advisable to use information from analog sites as discussed below.

Analog Sites for Vadose Zone Characterization*

Natural analogs refer to natural or anthropogenic (human-produced) systems in which processes similar to those expected to occur at a contaminant site have occurred over long time periods and large spatial scales. Analogs provide an important dimension to the understanding of flow and transport processes. For some systems, natural analogs are the only means of providing the necessary understanding of long-term (thousands of years) and large-scale (kilometers) behavior, which is required to provide the scientific confidence in models used for site characterization and performance assessment. Because analog sites demonstrate the development of natural processes over long time periods, they provide data that cannot be obtained otherwise, or data that are collected more easily and cost-effectively than by means of direct site

*This section was contributed by A. Simmons and M. Bandurraga, Lawrence Berkeley National Laboratory.

TABLE 3-2 **Types of Investigations and Basic Types of Data for Site Characterization** (adapted from ASMT D 5979 - 96).

Types of Investigations	Types of Data
Topography and Remote Sensing	(a) Topography (b) Aerial photography (c) Satellite imagery (d) Multispectral data (e) Thermal imagery (f) Radar, side-looking airborne radar, microwave imagery
Geomorphology	(a) Surficial Surface geology or geomorphology maps (b) Engineering geology maps (c) Surface-water inventory maps (d) Hydrography digital line graphs
Geology	(a) Geologic maps and cross sections (b) Lithologic and drillers logs
Geophysics	(a) Gravity, electromagnetic forces, resistivity, and seismic survey data and/or interpretations (b) Natural seismic activity data (c) Borehole geophysical data
Climate	(a) Precipitation data (b) Temperature, humidity, and wind data (c) Evaporation data (d) Effects of climate change on hydrologic system information
Vegetation	(a) Communities and/or species maps (b) Density map (c) Agricultural species, crop calendars, consumptive use data (d) Land use/land cover maps
Soils	(a) Soil surveys (b) Soil properties determined from field and laboratory analysis
Hydrogeology	(a) Potentiometric head data (b) Subsurface test information (c) Subsurface properties determined from laboratory analyses (d) Previous work regarding modeling studies, hydrogeologic and groundwater system maps

continued

TABLE 3-2 **Types of Investigations and Basic Types of Data for Site Characterization** (adapted from ASMT D 5979 - 96) *(continued)*.

Types of Investigations	Types of Data
Hydrogeology *(cont.)*	(e) Spring and seep data (f) Surface water data (g) Well design, construction, and development information
Hydrochemistry/Geochemistry	(a) Subsurface chemistry derived from well samples (b) Surface water chemistry (c) Rock and soil chemistry (d) Water quality surveys
Anthropogenic Aspects	(a) Planimetric maps (b) Land use/land cover maps (c) Road, transportation, political boundary information (d) Land ownership maps, including historical information, if available (e) Resource management maps

characterization and monitoring. Because the analogs are observable, they also provide an important illustrative function of bolstering public confidence in at least some aspects of environmental restoration and waste disposal (Miller *et al.* 1994).

Analog sites must be carefully selected to exclude those for which initial conditions are poorly known and where important data, such as the source term, are poorly constrained. Chapter 13 of the Yucca Mountain Site Description Report (Rev.1, in press) describes the use of natural analog for applications in the Yucca Mountain Project, concerning the design to building confidence in understanding and modeling natural and engineered barrier system processes associated with the underground high-level nuclear waste disposal facility.

Using Data Quality Objectives in Designing Vadose Zone Monitoring Systems

Successful vadose zone characterization and monitoring programs must efficiently select and combine appropriate technologies to meet the program goals. In Chapter 2, these steps were defined in terms of managers and program directors "setting endpoints" and "developing a roadmap." Over the years, scientists and engineers have used various terms for these activities when applied to characterization and monitoring. Notably, recent literature describes these steps as the Data Quality Objective (DQO) Process. When performed correctly, the DQO process is a valuable tool in designing a program that provides maximum information for a reasonable cost. The DQO process is a mechanism for selecting and specifying consensus standards (for example, ASTM) and screening methods, where appropriate, and incorporating innovative technologies, where appropriate. The DQO process assists in defensibly selecting characterization and monitoring instrumentation, access needs, sample type(s), sampling frequency, and program duration. The DQO process considers technical needs and factors such as uncertainty and ultimate use of the information (for example, identifying critical exposure pathways, potential risks, and so forth).

It is critical that the DQO process is performed jointly by scientists, regulators, people with process and historical knowledge, and other potential technical contributors. The DQO process should be focused on defining a program built around a conceptual model and the stepwise refining of the conceptual model. Importantly, while the DQO process provides discipline and documentation, it should be implemented flexibly and creatively. As discussed below, to be successful, the process must maintain a focus on meeting the technical objectives and not be reduced to a process of defining success or quality by selecting only laboratory analyses from a list of regulatory numbered methods.

A detailed description of the DQO process is given in the case study on the accompanying CD, "The Use of Data Quality Objectives in Designing Vadose Zone Monitoring Systems," by Kevin Leary.

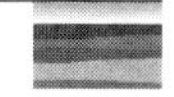

General Approach to Selection of Site Characterization and Monitoring Methods

American Society for Testing and Materials (ASTM) Standards for Site Characterization and Monitoring

The subsurface conditions at a particular site are usually affected by a combination of a variety of natural factors and processes (such as geologic, tectonic, lithologic, topographic, geomorphologic, hydrologic, water quality, hydrogeologic, geochemical, climatic, or microbiologic) and manmade factors and processes (such as waste disposal, contamination, or remediation). A rational site characterization program requires the application of a combination of complementary field, laboratory, and modeling methods to assess the impact of these influences. Consensus standards, such as those produced by ASTM, address both the general and specific procedures, as well as the types of data necessary to conduct qualitative conceptualization and quantitative site characterization and monitoring at any scale, including site-specific, subregional, and regional investigations.

When available, consensus standards are important and useful tools in developing a characterization and monitoring program. Thus, in the following discussion, we will identify ASTM standards related to different aspects of vadose zone characterization and monitoring.

ASTM standard D 5730-98 covers a general approach to improving the consistency of practice and planning a site characterization program for understanding environmental processes in soil and rock systems. This standard lists more than 400 field and laboratory test methods, practices, and guides, as well as a collection of non-ASTM references that describe field methods for site characterization and monitoring. Using this standard, it is possible to obtain an organized collection of information or a series of options for site characterization for environmental purposes. However, the standard does not recommend a specific course of action for each particular site, because ASTM standards cannot replace education and experience. The standard should instead be used in conjunction with professional judgment. Note that the word "standard" used in ASTM documents means that the document has been approved through the ASTM consensus process.

Table 3-3 provides a summary of field and laboratory methods for testing and analyzing soil, water, and waste samples collected during an environmental site investigation. This table also includes methods that are used routinely in field investigations to measure chemical parameters, as well as laboratory methods that are needed to obtain information relevant to the fate and transport of chemical and contaminant constituents in the subsurface.

TABLE 3-3 **Summary of the Field and Laboratory Methods Pertinent to Environmental Site Characterization with References to the ASTM Standards**
(adapted from ASTM 5370-98 with modifications)

	GENERAL
Reports	Indexing papers and reports (D 3584), use of modernized metric system (E 380)
Terminology	Soil, rock, and contained fluids (D 653); atmospheric sampling (D 1356); basic statistics (D 4743); waste and waste management (D 5681, D 5688 sampling and monitoring, D 5689 characterization); water (D 1189)
Objective-Oriented Guides	Acquisition of aerial photography and imagery for establishing historic site use and surficial surface conditions (D 5518)
Contaminated Sites:	Expedited site characterization (PS 85); developing conceptual site models for contaminated sites (E 1689); accelerated site characterization for petroleum releases (PS 3); risk-based corrective action at petroleum release sites (E 1739); short-term measures or early actions for site remediation (D 5745); environmental condition of property area types (D 5746); environmental baseline surveys (D 6008); real estate property transactions (E 1527, E 1528)
Site-Characterization	Environmental (D 5730, D 5995 cold regions; D 6067-ECPT); engineering and construction purposes (D 420)
Septic System Characterization	(D 5879); Subsurface (D 5921); sizing (D 5925)
Environmental Management	Development and implementation of a pollution prevention program (E 1609); lifecycle costing for pollution prevention (PS 14); assessment of buried steel tanks (ES 40); environmental regulatory compliance audits (PS 11); evaluation of an organization's environmental management system (PS 12); development and implementation of a source reduction program (PS 26)
	SAMPLING
General	Collection and preservation of information and physical items by a technical investigator (E 1188); probability sampling of materials (E 105)
Air	Choosing locations and sampling methods for atmospheric deposition at nonurban locations (D 5111, D 5012); guide for laboratories (D 3614); flow rate calibration of personal sampling pumps (D 5337); planning ambient air sampling (D 1357); ambient air analyzer procedures (D 3249); sampling stationary source emissions (D 5835)

continued

TABLE 3-3 **Summary of the Field and Laboratory Methods Pertinent to Environmental Site Characterization with References to the ASTM Standards** *(continued)* (adapted from ASTM 5370-98 with modifications)

Airborne Microorganisms	Sampling at municipal solid waste facilities (E 884)
Sampling Organic Vapors/Toxic Vapors	Charcoal tube absorption (D 3686); canister (D 5466); detector tubes (D 4490); length-of-stain dosimeter (D 4599)
Particulate Matter Determination	Filter absorbance method (D 1704, D 1704M); high-volume sampler (D 4096, D 4536); dustfall (D 1739-settleable particulates)
Worker Protection	Air monitoring at waste management facilities for worker protection (D 4844); air sampling strategies for worker and workplace protection (E 1370); collection of airborne particulate lead during abatement and construction activities (E 1553); activated charcoal samplers (D 4597), liquid sorbent difussional samplers (D 4598); pesticides and PCBs (D 4861); Sampling indoor air quality of building (D 5791)
Biological Materials	Aseptic sampling (E 1287); see also Table A2 in ASTM D 5730-98
Soil/Rock/Sediments	Minimum set of data elements for soil sampling (D 5911)
Drilling Methods	Cable tool (D 5875); casting advancement (D 5872); diamond core drilling (D 2113); direct air rotary (D 5782); direct fluid rotary (D 5783); direct rotary wireline (D 5876); dual-wall reverse circulation (D 5781); hollow-stem auger (D 5784)
Field Sampling and Handling Methods	Auger sampling (D 1452); radionuclides (C 998); ring-lined barrel (D 3550); split barrel (D 1586); thin-wall tube (D 1587); volatile organics (D 4547)
Sediments	Sediments (D 4411-fluvial sediment in motion, D 4823-submerged, D 3213-handling, storing, and preparing soft undisturbed marine soil; E 1391-collection for toxicological testing)
Vadose Zone Water	
Field Methods	Pore liquids (D 4696); soil (D 4700); soil gas (D 5314); purgeable headspace sampling (D 3871); waterborne oils (D 4489); continual online monitoring (D 3864); filterable and nonfilterable matter (D 5907); online sampling/analysis (D 5540-flow and temperature control), water-formed deposits (D 887)
Planning	Water-quality measurement program (D 5612); water-monitoring programs (D 5851)
Groundwater	Sampling methods (D 4448); direct push sampling (D 6001); planning a ground-water sampling event (D 5903)

continued

TABLE 3-3 **Summary of the Field and Laboratory Methods Pertinent to Environmental Site Characterization with References to the ASTM Standards** *(continued)* (adapted from ASTM 5370-98 with modifications)

Surface Water	Dipper or pond sampler (D 5358); closed conduits: equipment (D 1192); sampling (D 3370); laboratory practices: D 3856
Waste/Contaminants	
General Guidance	General planning (D 4687); representative sampling (D 6044); composite sampling and field subsampling (D 6051); heterogeneous wastes (D 5956)
Specific Sampling Procedures	Bituminous materials (D 140); Coliwasa (D 5495); drums general (D 6063 consolidated solids-D 5679, unconsolidated solids-D 5680); pipes and other point discharges (D 5013); scoop (D 5633); unconsolidated waste from truck (D 5658); UST release detection devices (E 1430, E1526); volatile organics (D 4547); waterborne oils (D 4489); oil/water mixtures for oil spill recovery equipment (F 1084); waste piles (D 6009)
Preservation/ Transport	Sample chain of custody (D 4840); estimation of holding time for water samples (D 4515, D4841)
Field Methods	Rock core samples (D 5079); sample containers for organic constituents (D 3694); soil samples (D 4220); sediments for toxicological testing (E 1391); preservation/preparation of waterborne oil samples (D 3325, D 3326); handling, storing, and preparing soft, undisturbed marine soil (D 3213)
	Decontamination of field equipment, nonradioactive waste sites (D 5088); low-level radioactive waste sites (D 5608)
Data Management/ Analysis	
QA/QC	Waste-management environmental data (D 5283); waste-management DQOs (D 5792); precision and bias (E 177); QC specification for organic constituents (D 5789)
Data Analysis	Evaluation of technical data (E 678); outlying observations (E 178); reporting results of examination and analysis of water-formed deposits (D 933)
Geostatistics	Reporting geostatistical site investigations (D 5549); analysis of spatial variation (D 5922); selection of kriging methods (D 5923); selection of simulation approaches (D 5924)
Spatial Data	Digital geospatial metadata (D 5714); see also Groundwater (Data Analysis)

continued

TABLE 3-3 **Summary of the Field and Laboratory Methods Pertinent to Environmental Site Characterization with References to the ASTM Standards** *(continued)* (adapted from ASTM 5370-98 with modifications)

SOIL/ROCK HYDROLOGIC PROPERTIES

Infiltration Rate

Field Methods	Double-ring infiltrometer (D 3385); sealed double-ring infiltrometer (D 5093)

Matric Potential

Field Methods	Tensiometers (D 3404)
Laboratory Methods	Filter paper method (D 5298)

Water Content

Field Methods	Calcium carbide method (D 4944); neutron probe (D 3017-shallow depth, D 5220-depth probe, D 6031-horizontal, slanted, and vertical access tubes)
Laboratory Methods	Direct heating method (D 4959); microwave oven method (D 4643); standard oven-drying method (D 2216); centrifuge moisture equivalent (D 425)

Hydraulic Conductivity

Field Methods	Vadose zone (D 5126)
Laboratory Methods	Granular soils (D 2434 - >1 3 10-3 cm/sec); low permeability soils (D 5084 - <1 3 10-3 cm/sec); rigid-wall compaction-mold permeameter (D 5856); effect of freeze/thaw (D 6035); peat (D 4511)

Other Hydrologic Properties

Laboratory Methods	Air permeability (D 4525); Soil water retention (D 2325-medium/coarse-textured, D 3152-fine-textured)

SOIL/ROCK PHYSICAL PROPERTIES

Particle Size

Soil Laboratory Methods	Analysis (D 422); dry preparation (D 421); <200 sieve (D 1140); wet preparation (D 2217)
Sediment	Selection of methods for fluvial sediment (D 4822)

continued

TABLE 3-3 **Summary of the Field and Laboratory Methods Pertinent to Environmental Site Characterization with References to the ASTM Standards** *(continued)* (adapted from ASTM 5370-98 with modifications)

Soil Density	
Field Methods	Drive cylinder (D 2937); gamma-gamma (D 2922 - <129, D5195 - >129); (D 4531); penetration (D 1586); rubber-balloon method (D 2167), sand-cone method (D 1556); sand replacement method (D 4914); sellve method (D 4564); water-replacement method (D 5030); nuclear method (D 6031)
Pore Volume/Specific Density	
Laboratory Methods	Pore volume (D 4404); specific gravity (D 854, D 5550 - gas pycnometer)
Cone Penetration	
Field Methods:	In situ cone-penetration testing (D 3441, D 5778); CPT stress-wave energy measurements (D 4633); liquification potential evaluation (D 6066); ECPT for environmental site characterization (D 6067)
Classification	
Field Methods:	Field logging (D 5434); noncohesive sediments (D 5387); peat (D 4544-deposit thickness, degree of humification-D 5715); sediments (D 4410); visual-manual procedure (D 2488-unified, D 4083-frozen soils); rock mass classification (D 5878); rock quality designation (D 6032);
Laboratory Methods	Dimension stone (C 199); frozen soils (D 4083); natural mineral aggregates (C 294); peat (D 2607); unified soil classification (D 2487)
Geophysical Properties	
Field Methods	Crosshole seismic testing (D 4428/D 4428M); seismic refraction (D 5777); soil resistivity (G 57 - Wenner 4-electrode method); planning and conducting borehole geophysical logging (D 5753)
Engineering Properties	
In Situ Field Methods	Bearing capacity/ratio (D 1194, D 4429); deformability and strength of weak rock (D 4555); direct shear strength (D 4554, D 5607); erodibility (D 5852), frost heave/thaw, susceptibility (D 5918); extensometers (D 4403); in situ creep (D 4553); in situ modulus of deformation (D 4394-rigid plate, D 4395-flexible plate, D 4506-radial jacking test, D 4729-flatjack method, D 4791-borehole jack); in situ stress (D 4623-borehole deformation gage, D 4645-hydraulic fracturing, D 4729-flatjack method); pressure measurement (D 4719-pressuremeter, D 5720-transducer calibration); vane shear test (D 2573)

continued

TABLE 3-3 **Summary of the Field and Laboratory Methods Pertinent to Environmental Site Characterization with References to the ASTM Standards** *(continued)* (adapted from ASTM 5370-98 with modifications)

Laboratory Methods	California bearing ratio (D 1883); classification (D 2487); compaction (D 698, D 1557, D 5080); compressive strength (D 2166, D 2938); consolidation (D 2435); core dimensional and shape tolerances (D 4543); dispersive characteristics (D 4221-double hydrometer; D 4647-pinhole test); elastic properties (D 2845, D 3148); impact valve (D 5874); linear displacement (D 6027-calibrating transducers); liquid limit (D 4318); moisture content-penetration resistance (D 1558); one-dimensional swell (D 4546); plastic limit/plasticity index (D 4318); point load strength (D 5731); rock hardness (D 5873); shrinkage factors (D 427; D 4943); tensile strength (D 2936; D 3967); thermal properties (D 5334, D 5335); triaxial compression (D 2850, D 2664, D 4406, D 4767, D 5311, D 5407); uniaxial compression (D 4341, D 4405); use of significant digits (D 6026); vane shear test (D 4648);
Miscellaneous	
Field Methods	Geotechnical mapping of large underground openings in rock (D 4543)
Laboratory Methods	X-ray radiography (D 4452)
Peat/Organic Soils	
Laboratory Methods	Bulk density (D 4531); classification (D 2607); hydraulic conductivity (D 4511); pH (D 2976); moisture/ash/organic matter (D 2974)
Frozen Soils	
Field Methods	Description (D 4083)
Laboratory Methods	Creep properties by uniaxial compression (D 5520)
	SOIL/ROCK CHEMISTRY
Basic Chemistry	
Field Methods	Soil pH for corrosion testing (G 51)
Laboratory Methods	Calcium carbonate (D 4373); pH (D 4972); soluble salt content (D 4542); diagnostic soil test for plant growth and food chain protection (D 5435); minimum requirements for laboratories engaged in chemical analysis (D 5522)
Soil Contaminants	Nitroaromatic and nitramine explosives (D 5143); screening fuels (D 5831); PCBs using room temperature phosphorescence (PS 47)
Sediments	Preparation for chemical analysis (D 3975, D 3976)

continued

TABLE 3-3 **Summary of the Field and Laboratory Methods Pertinent to Environmental Site Characterization with References to the ASTM Standards** *(continued)*
(adapted from ASTM 5370-98 with modifications)

Sorption/ Leachability	See fate-related procedures in Table A.1 of ASTM 5730-98
	GROUNDWATER
Characterization/ Monitoring	Assessing aquifer sensitivity and vulnerability (D 6030); conceptualization and characterization (D 5979); existing wells (D 5980); monitoring karst and fractured rock aquifers (D 5717); statistical approaches for groundwater detection monitoring programs (PS 64)
Data Elements	
Field Methods	Minimum set (D 5254); additional identification descriptors (D 5408); additional physical descriptors (D 5409); additional usage description (D 5410); selection of data elements (D 5474)
Data Analysis/ Presentation	Presentation of water level information (D 6000); chemical analysis: diagrams for single analyses (D 5738); trilinear diagrams (D 5754); diagrams based on data analytical calculations (D 5877); use of maps (D 6036)
Monitoring Wells	
Field Methods	Design/installation (D 5092); protection (D 5787); decommissioning (D 5299); casing (D 1785, F 480); grout (C 150-portland cement); water level measurement (D 4750); well development in granular aquifers (D 5521); well discharge (D 5716-circular orifice weir, D 5737-guide to methods); maintenance and rehabilitation (D 5978)
Aquifer Hydraulic Properties	
Field Methods	Packer tests (D 4630, D 4631); aquifer tests with control wells (D 4105, D 4106, D 5269, D 5270, D 5472, D 5473); D 5920 - anistropic unconfined; D 6028; (leaky confining beds); slug tests (D 4044, D 4050, D 4104, D 5785, D 5881, D 5912); constant draw-down for flowing wells (D 5787, D 5855); constant rate pumping (D 6034); partially penetrating wells (D 5850); test selection (D 4043)
Modeling	Site-specific application (D 5447); comparing simulation to site-specific information (D 5490); documenting model application (D 5718); defining boundary conditions (D 5609); defining initial conditions (D 5610); conducting sensitivity analysis (D 5611); simulation of subsurface air flow (D 5719); subsurface flow and transport modeling (D 5880) model calibration (D 5981); developing and evaluating codes (D 6025); describing functionality (D 6033)

continued

TABLE 3-3 **Summary of the Field and Laboratory Methods Pertinent to Environmental Site Characterization with References to the ASTM Standards** *(continued)* (adapted from ASTM 5370-98 with modifications)

Chemistry	
Field Methods	Acidity/alkalinity (D 1067); electrical conductivity/resistivity (D 1125); ion-selective electrodes (D 4127); low-level dissolved oxygen (D 5462); odor (D 1292); pH (D 1293, D 5464); redox potential (D 1498); test kits for inorganic constituents (D 5463); turbidity (D 1889); Extraction Methods: purgeable organics using headspace sampling (D 3871); micro-extraction for volatiles and semivolatiles (D 5241)
Laboratory Methods	Organic carbon (D 2579; D 4129; D 5173; D 6317); minimum requirements for laboratories engaged in chemical analysis (D 5522); see ASTM Volumes 11.01 and 11.02 generally
Microbiology	ATP content (D 4012); iron bacteria (D 932); sulfate-reducing bacteria (D 4412); microbial respiration (D 4478); microscopy (D 4454-total respiring bacteria, D 4455-epifluorescence); plating methods (D 5465); onsite screening heterotrophic bacteria (F 488)
	SURFACE WATER
Geometry/ Flow Measurement	Depth measurement (D 5073, D 5909-horizontal positioning); measurement of morphologic characteristics of surface water bodies (D 4581); operating a gaging station (D 5674)
Discharge	Step backwater method (D 5388)
Open Channel Flow	Selection of weirs and flumes (D 5640); acoustic methods (D 4408); acoustic velocity method (D 5389); broad-crested weirs (D 5614); culverts (D 5243); developing a stage-discharge relation (D 5541); dye tracers (D 5613); electromagnetic current meters (D 5089); Palmer-Bowles flume (D 5390); Parshall flume (D 1941); rotating element current meters (D 4409); slope-area method (D 5130); thin-plate weirs (D 5242); velocity-area method (D 3858); width contractions (D 5129)
Open Water Bodies	Water-level measurement (D 5413)
Other Characteristics	Suspended sediment concentration (D 3977); environmental conditions relevant to spill control systems (F 625)
Chemistry	See Groundwater above

continued

TABLE 3-3 **Summary of the Field and Laboratory Methods Pertinent to Environmental Site Characterization with References to the ASTM Standards** *(continued)*
(adapted from ASTM 5370-98 with modifications)

	WASTE / CONTAMINANTS
Waste Properties	
Field/Screening Methods	Compatibility (D 5059); cyanides (D 5049); flammability potential (D 4982); oxidizers (D 4981); pH (D 4980); physical-description screening analysis (D 4979); radioactivity (D 5928); sulfides (D 4978); waste specific gravity/bulk density (D 5057)
Laboratory Methods	Waste bulk density (E 1109); biological clogging of geotextiles (D 1987); coal fly ash (D 5759); solid waste freeze-thaw resistance (D 4842); stability and miscibility (D 5232); wetting and drying (D 4843)
Extraction Methods	Single batch extraction methods (D 5233); sequential batch extraction with water (D 4793- water, D 5284-acidic extraction fluid); soxhlet extraction (D 5369); total solvent extractable content (D 5368); solvent extraction of total petroleum hydrocarbons (D 5765); shake extraction of solid waste and water (D 3987)
Contaminant Fate	See fate-related procedures in Table A.2 of ASTM 5730-98
Radioactive Materials	
Monitoring	Detector calibration (E 181); radiation measurement/dosimetry (E 170); radiation protection programs for decommissioning operations (E 1167)
Sampling/ Preparation	Sampling surface soil for radionuclides (C 998); soil sample preparation for determination of radionuclides (C 999)
Asbestos	Screen analysis (D 2947)
	OTHER SITE CONDITIONS
Field Atmospheric Conditions	Atmospheric pressure (D 3631); conversion unit and factors (D 1914); determining comparability of meteorological measurements (D 4430); humidity: dew-point hygrometer (D 4030); psychrometer (E 337); terminology (D 4023)
Wind	Anemometers (D 4480, D 5096, D 5741); surface wind by acoustic means (D 5527); wind vane (D 5741, performance -D 5366); see Volume 11.03 generally
Solar Insolation	Pyranometers (E 824, E 913, E 941); pyrheliometers (E 816)

Characterization and Monitoring Technologies at DOE Sites*

It is instructive to summarize characterization and monitoring technologies that have been used at Department of Energy (DOE) sites. Table 3-4 presents a summary of current technologies that have been used at Idaho National Engineering and Environmental Laboratory (INEEL), Oak Ridge National Laboratory (ORNL), and Savannah River Site (SRS) for site characterization and monitoring. This table is based on the eight survey responses that were obtained by Loaiciga, *et al.* (1997). These site-characterization technologies are mostly conventional rather than innovative. Innovative technologies for soil and groundwater sampling and probing, such as the Geoprobe push-down sampler and the Site Characterization and Analysis Penetrometer System (SCAPS), were not reported in use at INEEL, while both are used at SRS. Geoprobe is only used at ORNL. INEEL personnel explained during the survey that site characterization is currently performed in a definitive-level mode rather than in a screening-level mode. This poses restrictions on the type of technologies that can be deployed *in situ*, according to regulator-approved standard operating procedures, to collect and analyze data used in risk analysis decisions.

According to Table 3-4, only a few vadose zone water and gas monitoring technologies were used. Only conventional suction lysimeters were in use at all three sites. Evidently, vadose zone sampling for air and water relies heavily on traditional soil coring devices, such as the split-spoon sampler to retrieve soil samples, which are then shipped to a laboratory for full analytical characterization.

Remote sensing is applied at all three sites primarily for topographic mapping and surface environmental reconnaissance of temperature, vegetative status, and gamma activity, as seen in Table 3-4. Remote sensing mapping allows fast, regional-scale assessment of surface properties, typically conducted on a preliminary screening level.

It is also evident from Table 3-4 that numerical models showed strong acceptance and a wide range of applications at all three sites. The main reason is that environmental restoration has progressed from screening- and definitive-level characterization to risk analysis, cleanup, contain-

*This section was contributed by H. Loaiciga, S. Renehan, and S. Weeks.

ment, and remediation. As a result, models have become useful tools for creating and analyzing a variety of scenarios in a very cost-effective manner. For example, a mass transport numerical model can simulate the fate and transport of a contaminant (such as benzene) in groundwater that is being pumped, treated, and recharged according to a specific pump-and-treat scheme. A vadose zone model such as SESOIL (Seasonal Soil Compartment Model, Bonazountas and Wagner 1984) may be implemented to assess the effect of soil capping on long-term metal vertical migration in the vadose zone. Modelers typically are part of the risk-analysis group at the surveyed sites.

One aspect of site characterization that was overlooked initially by the survey relates to ecological monitoring. ORNL and SRS maintain active monitoring of vegetation, fish, mammals, and other species, as well as of surface-water bodies. Living organisms are tested primarily for radionuclides and metals (such as cesium and strontium isotopes, and mercury) that accumulate in their tissues. Ecological monitoring is done by capturing and/or sampling of specimens and testing parts or tissue in the laboratory according to standard protocols.

TABLE 3-4 **Summary of Current Site Characterization and Monitoring Technologies Used at INEEL, ORNL, and SRS.**

Technology	INEEL	ORNL	SRS
Remote Sensing			
Remote sensing/aerial photography	+	+	+
Surface Geophysics			
Electrical resistivity	+	+	+
Electromagnetic conductivity	+	+	+
Seismic methods	Past use	+	+
Ground-penetrating radar	+	+	+
Magnetometer surveys		+	+
Borehole Geophysics			
Resistivity surveys	+	+	+
Cross-borehole tomography	+		+

continued

TABLE 3-4 Summary of Current Site Characterization and Monitoring Technologies Used at INEEL, ORNL, and SRS. (*continued*)

Technology	INEEL	ORNL	SRS
Nuclear Logging			
Density logging	+	+	+
Nuclear logging (natural gamma, neutron logging, gamma-gamma radiation)	+	+	+
Drilling			
Geoprobe®-type penetrometer		+	+
Large site characterization and analysis penetrometer system (SCAPS) platform			+
Standard methods (hollow-stem auger, rotary, and so on)	+	+	+
Direct sonic drilling	Past use		+
Rotosonic drilling	Past use	+	+
Horizontal drilling	Past use	+	+
Groundwater Sampling			
Sampling (bladder, dedicated pumps, and so on)	+	+	+
Sampling bailers (such as thief sampler)	+	+	+
Soils Characterization			
Sampling technologies (discrete, continuous, and so on)	+	+	+
Vadose Zone Water and Gas Monitoring			
Lysimeter (suction, pressure/vacuum, and so on)	+	+	+
Electrical resistivity blocks	Past use		
Soil-gas monitoring (probes, chambers, and so on)	+		+

TABLE 3-4 Summary of Current Site Characterization and Monitoring Technologies Used at INEEL, ORNL, and SRS. *(continued)*

Technology	INEEL	ORNL	SRS
Time domain reflectometry	+		
Electronic leak detection system			
Thermocouple psychrometers			
Tensiometers	Past use		+
Frequency domain capacity probes			
Automatic VOC collection/gas chromatography			
Analytical Technologies			
Gas chromatography	+	+	+
High-performance liquid chromatography			+
Thin-layer chromatography			
Super-critical fluid chromatography			
Gas chromatography/mass spectrometry	+	+	+
Mass spectrometry	Past use	+	+
Ion-mobility spectrometry			
Atomic-absorption spectrometry	Past use	+	+
Atomic-emission spectrometry		+	+
Laser-induced breakdown spectrometry			
Infrared spectrometry (Fourier transform, and so on)	Past use	+	+
Near IR reflectance/transmission spectrometry			
Raman spectroscopy			
UV-visible spectrometry (fluorescence, synchronous luminescence, etc.)		+	+
Fluorescence spectrometry	+		+
X-ray fluorescence	Past use		+
Gamma spectrometry	+		+
Radiation detectors (Geiger counter, solid/liquid scintillator, semi-conductor detector, and so on)	+	+	+

continued

TABLE 3-4 Summary of Current Site Characterization and Monitoring Technologies Used at INEEL, ORNL, and SRS. (*continued*)

Technology	INEEL	ORNL	SRS
Nuclear magnetic resonance			+
Photo-ionization detector	+	+	+
Electrical conductivity sensor		+	+
Electrochemical techniques			+
Explosive sensor			+
Free product sensor			+
Fiber optics sensor (solid, porous, etc.)			+
Piezoelectric sensors			+
In situ chemical probes (chlorine, pH/ORP, TDS, DO, and so on)	+	+	+
Membrane-based testing devices (RDX, TNT, PCBs, and so on)		+	+
Environmental test kits (color testing, titrimetric testing, immunassays)	+	+	+
Detector tubes	+		+
Numerical/Spatial/Statistical Models			
Geostatistical/statistical	+	+	+
Flow and transport models	+	+	+
Geographic/expert/decision support systems	+	+	+

Legend:
A plus sign in Table 3-4 means that the technology is currently used.
A blank space in the table indicates neither current nor past use of a specific technology.

Selection of Drilling and Soil Sampling Methods*

Drilling

Selection of an appropriate drilling method is the primary step in site characterization and monitoring (Driscoll, 1986; Nielsen and Schalla, 1991). The decision should be based on the specific characteristics of each site, including, but not limited to, the geologic, hydrogeologic, topographic, climatic, and anthropogenic conditions of the site. Selection of a drilling method must also consider the goal of the drilling (for instance, soil and fluid sampling requirements, and/or monitoring equipment installation). Further, horizontal drilling and boring machines are emerging as useful adjuncts to vertical drilling for environmental work, especially methods minimizing the use of drilling fluids.

The ASTM D 6286-98 Standard Guide for Selection of Drilling Methods for Environmental Site Characterization summarizes most drilling methods available (Table 3-5), and provides the advantages and disadvantages of each method. The two main classes of drilling methods are auger drilling and fluid-rotary drilling. Other, less common methods are vibratory drilling, cable-tool drilling, and jet drilling. Casing-drive systems, which are typically combined with fluid-rotary, cable-tool, or jet drilling techniques, provide another drilling method. Vibratory drilling is best understood as a type of highly efficient casing-drive system. Casing-drive systems are less commonly used alone for borehole drilling. Direct-push methods, such as the cone penetrometer (CPT), are discussed in the section "Cone Penetrometer and Direct Push Tools for Vadose Zone Characterization," below.

Of the drilling methods listed in Table 3-5, water-based fluid-rotary methods and jet-drilling methods can introduce large quantities of water to the subsurface. This result may be unacceptable, because the water will disturb all of the parameters of interest, including saturation, permeability, microbial community structure, and the concentrations and distributions of the chemical constituents. Therefore, water-based fluid-rotary methods and jet-drilling methods generally should not be used for vadose zone investigation.

*This section was contributed by B. Faybishenko and P. Jordan.

TABLE 3-5 Summary of well Well-Drilling Methods (ASTM 6286-98)

Drilling Method	Drilling Fluid	Casing Advance	Type of Material Drilled	Typical Drilling Depth, in ft[A].	Typical Range of Borehole Sizes, in In.	Samples Obtainable[B]	Coring Possible
Power auger (Hollow-stem)	none, water, mud	yes	soil, weathered rock	<150	5-22	S, F	yes
Power auger (Solid-stem)	water, mud	no	soil, weathered rock	<150	2-10	S	yes
Power bucket auger	none, water (below water table)	no	soil, weathered rock	<150	18-48	S	yes
Hand auger	none	no	soil	<70 (above water table only)	2-6	S	yes
Direct fluid rotary	water, mud	yes	Soil, rock	>1000	2-36	S, R	yes
Direct air rotary	air, water, foam	yes	soil, rock	>1500	2-36	S, R, F	yes
DTH hammer	air, water, foam	yes	rock, boulders	<2000	4-16	R	yes
Wireline	air, water, foam	yes	soil, rock	>1000	3-6	S, R, F	yes
Reverse fluid rotary	water, mud	yes	soil, rock	<2000	12-36	S, R, F	yes
Reverse air rotary	air, water, foam	yes	soil, rock	>1000	12-36	S, R, F	yes
Cable tool	water	yes	soil, rock	<5000	4-24	S, R, F (F-below water table)	yes
Casing-advancer	air, water, mud	yes	soil, rock, boulders	<2000	2-16	S, R, F	yes
Direct-push technology	none	yes	soil	<100	1.5-3	S, F	yes
Sonic (vibratory)	none, water, mud, air	yes	soil, rock, boulders	<500	4-12	S, R, F	yes
Jet percussion	water	no	soil	<50	2-4	S	no
Jetting	water	yes	soil	<50	4	S	no

[A]Actual achievable drilled depths will vary depending on the ambient geohydrologic conditions existing at the site and size of drilling equipment used. For example, large, high-torque rigs can drill to greater depths than their smaller counterparts under favorable site conditions. Boreholes drilled using air/air foam can reach greater depths more efficiently using two-stage positive-displacement compressors having the capability of developing working pressures of 250 to 350 psi and 500 to 750 cfm, particularly when submergence requires higher pressures. The smaller rotary-type compressors only are capable of producing a maximum working pressure of 125 psi and produce 500 to 1200 cfm. Likewise, the rig mast must be constructed to safely carry the anticipated working loads expected. To allow for contingencies, it is recommended that the rated capacity of the mast be at least twice the anticipated weight load or normal pulling load.
[B]Soil = S (Cuttings), Rock = R (Cuttings), Fluid = F (some samples might require accessory sampling devices to obtain).

Soil Sampling

Soil samples from the vadose zone are collected during drilling and are to be used for various aspects of site characterization, including stratigraphic description, measurements of moisture content and matric potential, testing of hydraulic conductivity and water retention, geotechnical testing, soil-gas analyses, microbiological investigations, and chemical analyses of pore liquid and soils (ASTM 4700-91). Soil samples are also used for chemical analyses of liquids, solids, and gases to determine the presence, possible source, migration route, and physical-chemical behavior of contaminants in the vadose zone.

Two types of vadose zone sampling devices have been designed: (1) samplers used in conjunction with hand-operated devices, and (2) samplers used in conjunction with multipurpose or auger drill rigs. These devices are included in the ASTM Guide 4700-91. During drilling, encased and uncased soil samples can be taken from specific depths according to requirements of the analyses.

A major disadvantage of hand-operating samplers (such as screw-type augers, barrel augers, tube-type samplers, and hand-held power augers) is the limited depth of sampling. Sampling devices used in conjunction with hollow-stem augers and in holes advanced by solid-stem augers include thin-walled tube samplers (also called Shelby tubes), split-barrel drive samplers (also called split spoons), ring-lined barrel samplers, continuous-sample tube systems, and piston samplers. These samplers are either pushed (or driven) in sequence with an increment of drilling, or are advanced simultaneously with the progression of a hollow-stem auger column. It is necessary to take into account that, in general, the soil-sampling methods are destructive, and multiple sampling at the same location is not possible. During drilling, sampling, and sample preparation for the analysis, some portions of soil gas and liquid are usually lost, which may lead to erroneous results, as shown in the case study "Comparison of Vadose Zone Soil and Water Analytical Data for Characterization of Explosives Contamination."

The case study, "Comparison of Vadose Zone Soil and Water Analytical Data for Characterization of Explosives Contamination," by Wilson S. Clayton, Ph.D., P.G., and Peter Wirth, P.E., compares the analytical results obtained from soil samples with those acquired from permanently installed suction lysimeters. *See page 423* .

Disturbed soil samples obtained by backhoe, bucket auger, or other destructive techniques may lose volatile components during soil sampling. Special procedures to restrain biodegradation of VOCs using methanol, and solution of cupric sulfate, are described in the ASTM Standard D 4547-98, "Sampling Waste and Soils for Volatile Organic Compounds."

The reliability of the sampling procedure for site characterization can be significantly increased using an innovative adaptive-sampling approach, which is described in the case study on the accompanying CD, "Adaptive Sampling Approach to Environmental Site Characterization," by Grace Bujewski, Sandia National Laboratories. This approach was used for soil sampling to detect radiation, organic compounds, and metals, and showed a significant cost saving compared to the conventional RCRA site characterization.

Cone Penetrometer and Direct Push Tools for Vadose Zone Characterization*

Over the past twenty years, one of the most important technological developments for vadose zone characterization of unconsolidated sediments is the direct-push method for accessing and probing the subsurface. The cone penetrometer and related direct-push technologies, such as the Geoprobe®, have been increasingly used for geologic and chemical characterization at sites throughout the United States and elsewhere. In addition to the standard suite of sensors (that is, tip pressure, sleeve friction, and capillary pressure), the cone penetrometer has been used with innovative sensors and samplers to perform contaminated-site assessments, and has also been used to install wells. By integrating geologic information from the standard cone-penetrometer sensor with the depth-discrete chemical and physical information obtained from cone-penetrometer-based samplers and sensors, we can perform an accurate, rapid, and cost-effective characterization. Using the added capability of cone-penetrometer-installed wells, the placement of

*This section was contributed by J. Rossabi.

targeted remediation systems can be initiated during the characterization. Cone-penetrometer tests (CPT) provide high-resolution, high-quality data, are minimally invasive, and produce a minimum of investigation-derived waste. These attributes are critical to investigative and cleanup operations at large hazardous waste sites with heterogeneous sediments.

Background

Most environmental professionals would prefer to use noninvasive techniques for characterizing the subsurface. However, although promising surface geophysical methods are being developed, subsurface characterization at the resolution needed for most environmental site assessment currently requires accessing the subsurface. One of the least invasive ways of achieving subsurface access is to use the small-diameter (less than 5 cm) cone penetrometer. A cone-penetrometer test is performed by pushing an instrumented steel rod into the ground to determine the properties of the penetrated subsurface materials. The standard array of instruments on a cone penetrometer includes tip pressure, sleeve friction, and pore pressure sensors. This ensemble is commonly called a piezo-cone configuration. Geotechnical properties, stratigraphy, and soil type of the subsurface materials can be estimated using the data from these sensors (Lunne *et al.* 1997).

The cone penetrometer was first developed in the Netherlands in 1932 as a manually deployed instrument to determine hard soil zones and for measuring pile-bearing capacity, and has since developed into an automated system used throughout the world (Barentsen 1936). Currently the CPT is performed using a hydraulic pressure system to deploy the rods and a heavyweight truck to supply the inertial mass. Sensor data are collected and processed electronically with a typical resolution of approximately 2 cm. Depths of penetration vary, depending on the subsurface materials. For example, at the Savannah River Site (where there are coastal plain sediments), depths greater than 35 m are routine, with occasionally pushes beyond 85 m; however, refusal of penetration has also occurred at depths of less than 10 m (Rossabi *et al.* 1998). Conditions that cause refusal include the presence of gravel, cobbles, rock or highly consolidated-cemented strata. Sites with these attributes are less suited to CPT methods.

Predominately used for geotechnical applications such as bearing capacity and liquefaction, the CPT only recently has been used for environmental characterization. The fast and inexpensive access to the subsurface provided by the CPT makes it an ideal tool for contaminated-site investigations. In addition, the development of chemical and other sensors, combined with the geologic sensing of the CPT, has hastened the development of new protocols that substantially reduce characterization times and increase characterization accuracy.

The Department of Defense and Department of Energy have led the way in sponsoring the development, deployment, and testing of new sensors for the cone penetrometer. Significant developments and innovations in CPT technology have come from both government and private industry.

Direct Push Tools

As mentioned previously, the standard suite of devices on the cone penetrometer includes tip resistance, sleeve friction, and pore pressure sensors. This ensemble is commonly called a piezo-cone configuration. Electrical resistance measurement capability has been recently included as a standard tool in several cone penetrometers. (This measurement has a long history as a standard borehole-logging tool but only lately has been commonly implemented with the cone penetrometer.) In Figure 3-13, plots from the three standard cone penetrometer sensors (tip pressure, sleeve friction, and pore pressure) are displayed. Tip pressure or cone resistance is a measure of the normal force felt by the cone in a direction opposite to that of the push, approximately perpendicular to the ground surface. Sleeve friction is a measure of the frictional force on the outer cylindrical surface of the penetrometer rod related to the "stickiness" of the formation material. Pore pressure is a measure of the instantaneous pressure on the cone in a direction perpendicular to that of the push. In addition to these plots, the ratio of sleeve friction to tip pressure and electrical conductivity are plotted. The sleeve friction to tip pressure ratio is a useful construct for evaluating the behavior of the sediments. The electrical conductivity is generally measured directly through electrodes located on the outside of the cone and insulated from one another. Each of these measurements offers corroborating or complementary information about the penetrated sediments.

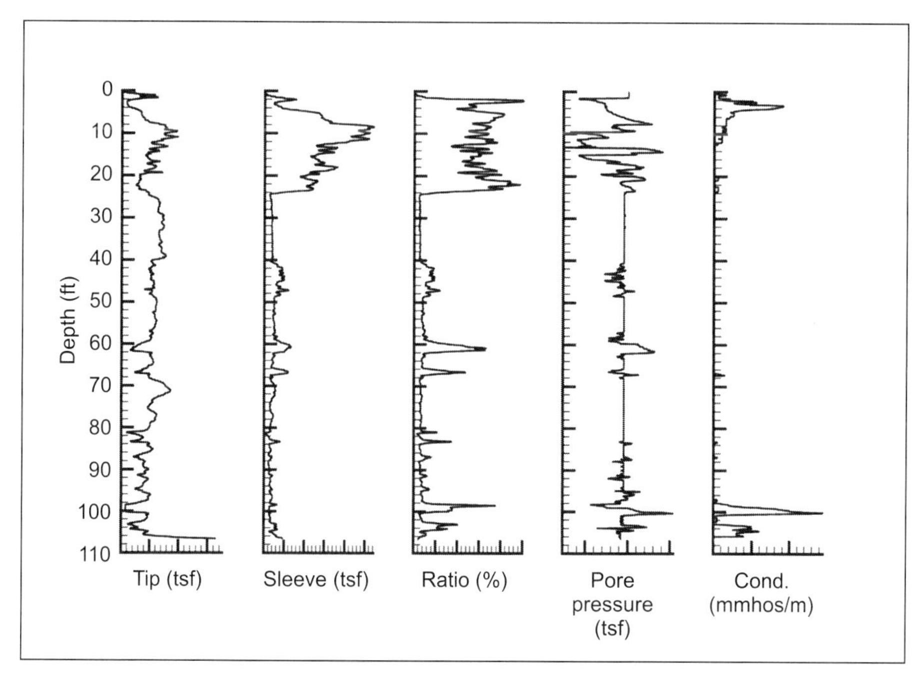

Figure 3-13. Cone penetrometer data from the vadose zone at the Savannah River Site

The sleeve-friction-to-tip-pressure ratio is often used as the most basic soil-behavior-type indicator (Lunne *et al.* 1997). From these data, the soil type can often be inferred (but not absolutely identified) according to grain-size distribution. In general, the lower the ratio, the higher the sand content of the soils, with a nominal value of 6 percent chosen as the demarcation line for clay. Unless a site-specific correlation is calculated, however, it is prudent to use the log in relative terms with a higher ratio corresponding to more clayey soils, and a lower ratio to sandier soils. Soil-behavior type is often defined using a more formal basis for soil-type classification, an empirically developed correlation chart based on tip pressure, friction ratio, and data from several sites (Douglas and Olsen 1981). Other soil classification methods have been developed from inclusion of the pore-pressure data with tip and sleeve information. Three examples of these charts are shown in Figure 3-14. The charts may have to be adjusted for more precise soil-behavior-type determination at specific sites, but they generally provide an accurate description of the subsurface materials.

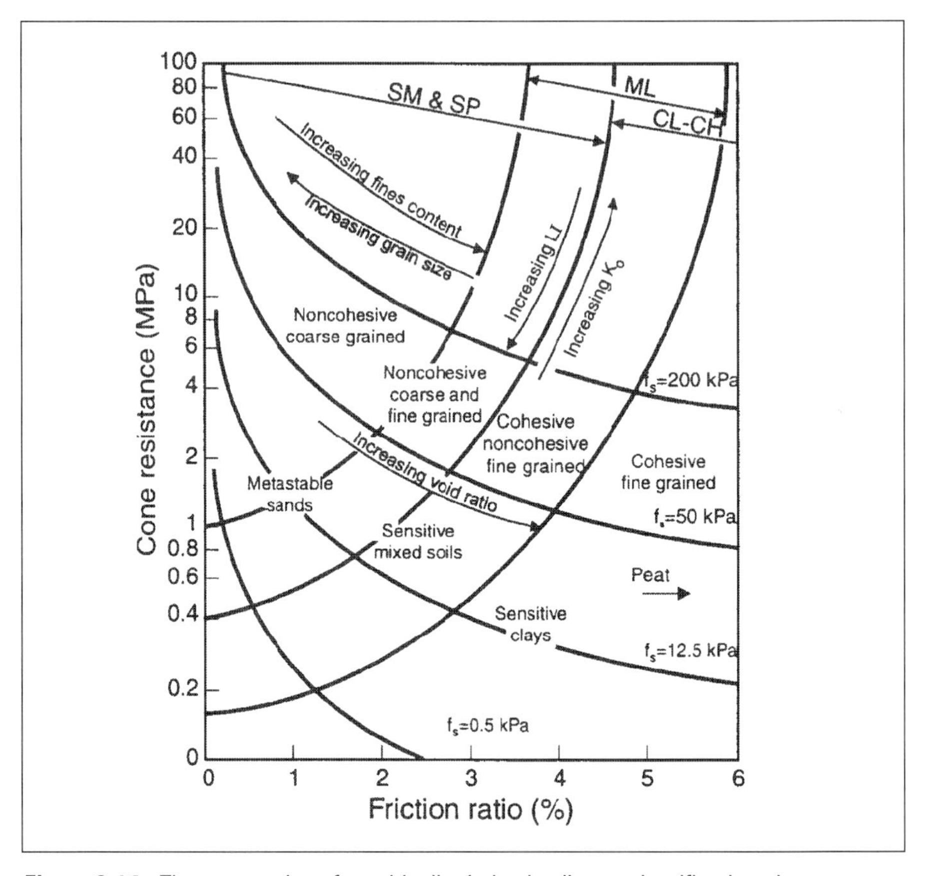

Figure 3-14. Three examples of empirically derived soil type classification charts *(continued)*

The most compelling aspect of cone penetrometer data is its vertical resolution. The penetration rate for cone penetrometers is 2 cm/sec, and data acquisition systems accompanying these tools acquire data at a rate of approximately 1 Hz. The 2 cm resolution contrasts with conventional borehole logging tools with typical resolution of approximately 1 m. Even recovered core is rarely described at a resolution of less than 30 cm. The value of high-resolution data has been felt most strongly in the characterization of contaminated sites. Although subsurface data on the order of centimeters is rarely necessary for site structural assessments or production-well characterization, the flow and transport of subsurface contaminants is often strongly influenced by thin, discontin-

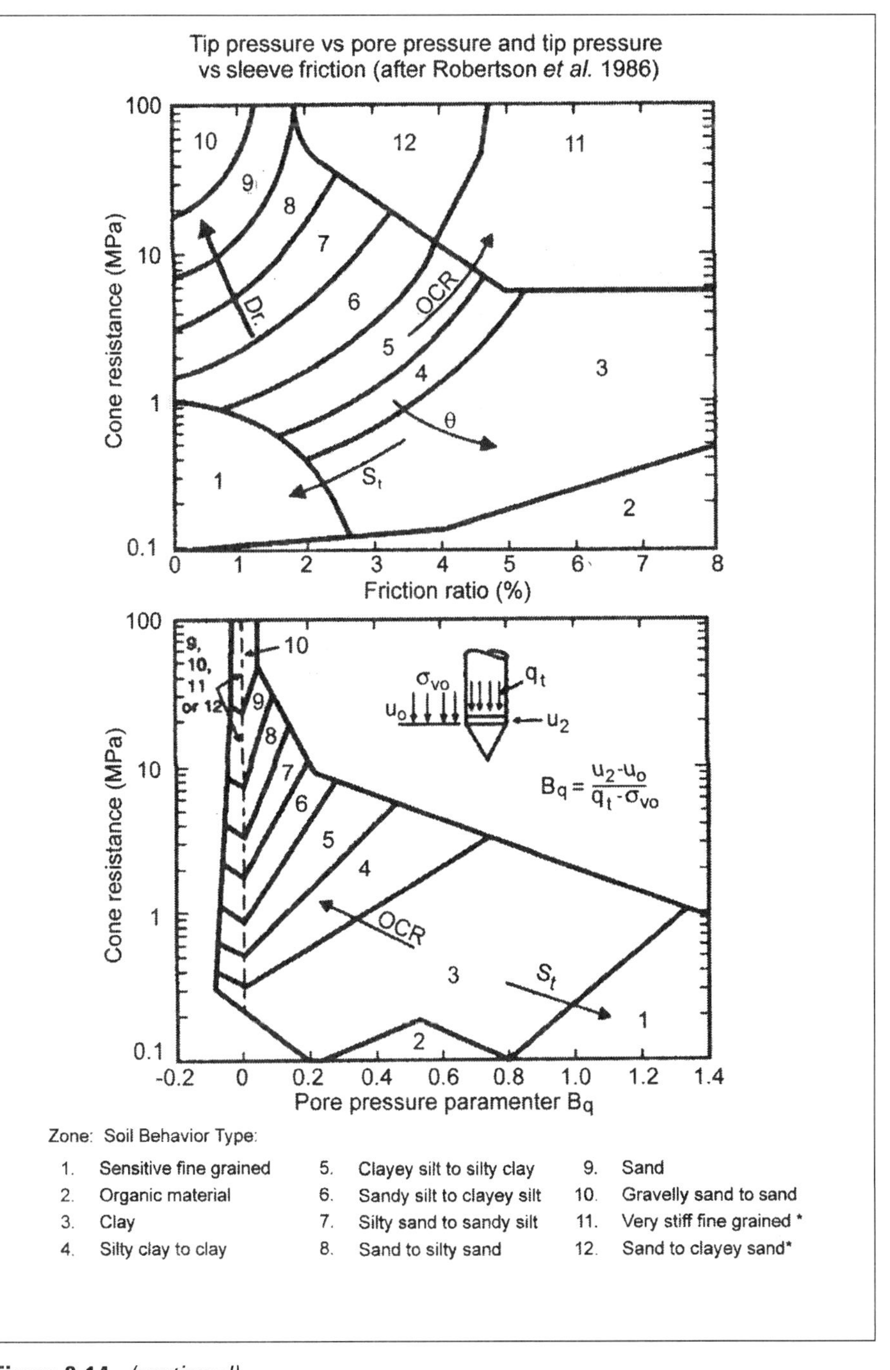

Figure 3-14. *(continued)*

uous layers. For example, a 2-cm layer of nearly saturated clay may prevent the downward migration of a nonaqueous phase fluid (NAPL). In addition, partially saturated materials with high capillary suction forces will absorb and retain nonaqueous phase fluids. Their slow dissolution into surrounding pore and infiltrating water permits the contaminants to act as a long-term source for years.

In Figure 3-13, the friction ratio indicates a clayey material from the ground surface to a depth of 24 ft below ground surface (bgs). The pore-pressure response corroborates these data. The electrical conductivity plot shows that the upper 12 ft of the clayey material is wet, with higher saturation values near the ground surface, as from a recent infiltration event. The plots show dry sand from 24 ft bgs to 60 ft bgs with a brief interval of fine sand or silt 41 to 49 ft bgs. Clayey layers are evident at 60, 68, and 83 ft bgs with very thin clayey laminates appearing between layers. The pore pressure sensor detects these laminates more clearly than the ratio log because the sleeve friction sensor (10 to 20 cm in length) tends to smear the ratio data resolution. The clay at 99 ft marks the beginning of an interbedded sand and wet clayey zone that continues until the saturated zone at a depth of 130 ft bgs.

These data collected at a site contaminated by volatile organic compound (VOC) releases near the surface have been important for targeting contaminant investigations. Figure 3-15 shows the friction ratio plot of Figure 3-13, as well as the plot of results of tetrachloroethylene (PCE) analyses performed on soil samples taken from two cores located near the cone penetrometer push. In this case, soil sampling was performed at the deeper, clayey zones, rather than at prescribed intervals, in order to minimize the chances of missing contamination and to reduce the expense of collecting and analyzing non-detects. Soil-gas sampling performed in this area identified high gas concentrations just above the 99-ft clay. It is clear from this plot that high residual concentrations of PCE are associated with the clayey zone. Further characterization activities near the release area would logically focus on the clay. For example, the cone penetrometer might be used to determine the topographic gradient of the 99-ft clay and track the migration of a dense nonaqueous phase liquid (DNAPL) source down the gradient surface.

One of the most important characteristics of the vadose zone is the amount of water in the predominately two-fluid phase system. The pore water is crucial to many subsurface processes (such as chemical and

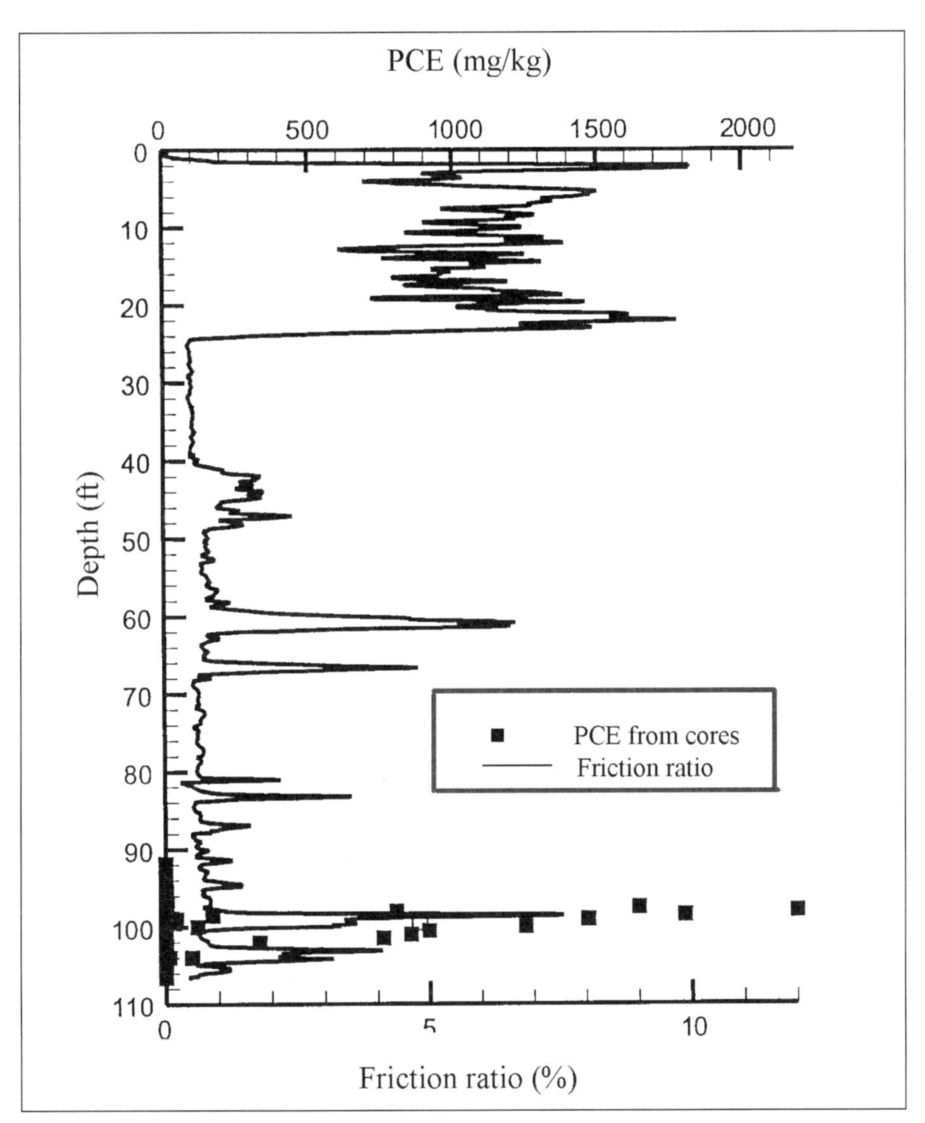

Figure 3-15. Comparison of friction ratio data with PCE concentration from soil cores

biological events), but it is especially important to the advective and diffusive transport of contaminants. The degree of saturation of subsurface sediments directly affects the relative permeability of potentially mobile fluids throughout the vadose zone. Accurate models of the subsurface for contaminant fate and transport rely on the intrinsic permeability values of individual strata within the model domain. These relatively time-insensitive values provide the basis for determining the contaminant conductivity. The actual conductivity, however, is determined by the relative permeability of the contaminant fluid. This value depends on the presence and amount of other fluids in the system. For example, the transport of gases and nonaqueous phase liquids (NAPL) will be significantly reduced when they encounter a nearly water-saturated stratum of fine-grained sediments. A less water-saturated system will allow free transport of the gas, and, often, increased receptivity of the NAPL by imbibition through capillary suction. An important complication of the soil-moisture parameter is that it changes over time depending on the infiltration events to which the system is exposed. Therefore, knowledge of the soil moisture through time by multiple measurements or constant logging is desirable for the most accurate subsurface representation.

Several *in situ* methods of determining the soil-moisture characteristics of the unsaturated subsurface have been devised. The most commonly used methods are the *in situ* point techniques (such as the tensiometer and time domain reflectometer) and the borehole logging methods (neutron probes, for example). The point methods are reliable and accurate but can only provide information about the material immediately adjacent to the location of the fixed probe. The neutron log is valuable because it provides depth-discrete information over the full length of the borehole. The method has its drawbacks, however, in that it requires a radioactive neutron source, special well design and installation, and has a vertical resolution of approximately 30 cm or more. With lithology changes and associated water retention capacities that often occur at smaller scales than one foot, information impacting contaminant transport may be lost.

Recently, innovative tools for evaluating soil moisture have been developed for deployment with the cone penetrometer. These tools have many of the advantages of the *in situ* point measurement methods, but they are advanced through the depth of the vadose zone. They can therefore provide contiguous high-resolution measurements of the soil moisture properties of the vadose zone. Three commercially available probes for the CPT have been tested at the Savannah River Site. Two of the probes use a measurement of the dielectric properties of the subsurface in the frequency domain at a given point to determine the moisture content. The other, developed by Sandia National Laboratory personnel, uses a time domain reflectometer in a cone section. All three methods provided very accurate results when compared to the baseline results obtained from Shelby tube samples and laboratory analysis.

"Case Study of Cone Penetrometer (CPT)-Based Soil Moisture Probes," by Joe Rossabi, describes the application of a cone penetrometer to the problem of measurements of moisture content in soils. *See page 428.*

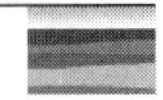

Cone penetrometer data might be used for specification of boundary conditions and heterogeneity in models, and for targeting zones for installation of tensiometers, lysimeters, or other equipment for which operation depends on soil type. Essentially, the standard cone penetrometer provides data that can be used in the same way as core description data, that is, to identify stratigraphic intervals and geologic features that are contextually related to any specific subsurface investigation.

An application of cone penetrometers for DNAPL detection is examined in the case study "Cone Penetrometer-Based Raman Spectroscopy for DNAPL Characterization in the Vadose Zone," by J. Rossabi, B.D. Riha, C.A. Eddy-Dilek, A. Lustig, M. Carrabba, K. Hyde, and J. Belo. *See page 431.*

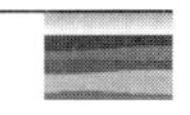

In many cases, boreholes can be grout-sealed upon removal of the rods, further reducing the chances of cross-contamination and exposure. Wells (currently up to 5 cm in diameter) can also be installed directly into the formation using direct-push methods. Another use of direct-push technologies is the injection of fluids for either reaction with contaminants or creation of fractures in low-permeability soils. Tools that have been implemented with the cone penetrometer at the Savannah River Site or other sites are listed in Table 3-6, along with their use, the type of data they produce, their features, and their commercial status. Although this list is not exhaustive (new sensors and techniques are constantly being developed), it does show the amount and variety of tools that are compatible with direct-push methods.

TABLE 3-6 Sampling and Deployment Capabilities.

Tool	Use	Data/Output	Features	Status
		SOLID, LIQUID AND GAS SAMPLING		
FLUTe Membrane Installation	Sampling, chemical (NAPL), installation	Recovered sample, visual indicator, ports pressure, and so on	Cheap, fast, simple	CA-P
MOSTAP Sampler	Sediment samples	Tube or split spoon samples	Targeted depth	CA-P
Cone Sipper	Multi-depth liquid and gas samples	Liquid or gas samples	Multiple samples on single push	CA-P
BAT	Liquid samples	Septum-sealed sample	Targeted depth	CA-P
Bailing in Rods	Liquid samples	Liquid samples	Targeted depth	CA-S
Multilevel Gas	Gas samples	Gas samples	Targeted depths	CA-P
Wireline Sampler	Multi-depth sediment samples/cutting tool	Soil samples	Multiple samples on single push	UD
Well Point Samplers–Frits	Shallow groundwater, gas	Liquid or gas samples	Targeted depths	CA-P
Well–Inside Rod	Small diameter wells	Liquid or gas samples	Direct contact with the formation	CA-S
Well–Outside Rod	2-in. diameter wells	Liquid or gas samples	Direct contact with the formation	CA-S
Lysimeter	Vadose zone soil pore water	Liquid samples	Direct contact with the formation	FT
ERT (Electrical Resistivity) Probes	3D resistivity field	Relative soil moisture in plane or volume	Plane or volumetric differencing information	CA-S
Soil Moisture Probes	Volumetric soil moisture—single depth	Quantitative dielectric-based soil moisture values	Installed at a single depth	FT
		PHYSICAL SENSING CAPABILITIES		
Standard	Lithology, soil properties	Soil behavior type classification	High resolution	CA-S,P
Temperature Sensor	Temperature	Temperature	Identify anomalous temperature perturbations	CA-P

continued

TABLE 3-6 Sampling and Deployment Capabilities. *(continued)*

Tool	Use	Data/Output	Features	Status
Science and Engineering Associates (SEA) Cone Permeameter	Pneumatic relative permeability, hydraulic conductivity	Depth-discrete	Targeted depths permeability	CA-S
Soil Moisture TDR	Volumetric soil moisture log	Quantitative time domain reflectometry-based soil moisture values	High vertical resolution	CA-S
Fiber Optic Probe	Capillary pore pressure	Relative humidity	*in situ* measurement of soil properties	FT
Soil Moisture Dielectric	Volumetric soil moisture log	Quantitative dielectric-based soil moisture values	High vertical resolution	CA-S
POLO (Subsurface Position Locating System)	Subsurface position relative to entry point at surface	Incrementing three-dimensional data on distance from datum	Physical position	FT
GeoVis Video Microscope	Grain size distribution, soil type, fluid behavior, contaminant identification	High -resolution color video	Very high resolution	CA-S,P
Inclinometers	Boring deviation	Deflection from normal	Infers physical position	CA-S
Vibratory Cone and Resonant Sonic Cone	Liquefaction of soils	Liquefaction parameters	Structural parameters and increased depth capability	CA-S
Seismic Cone	Density change, soil type, liquefaction	Geophone data	Structural parameters, identifies units	CA-S
Index of Refraction	Fluid phase change	Change of fluid	Identifies sharp phase changes	CA-P
Hyperlog	Soil color, contaminants	Munsell color charts and three-wavelength fluorescence	Augments soil type classification, identify fluorescence contaminants	CA-S

continued

TABLE 3-6 Sampling and Deployment Capabilities *(continued)*

Tool	Use	Data/Output	Features	Status
		CHEMICAL SENSING CAPABILITIES		
Raman Spectroscopy	Nonaqueous phase compound identification	Inelastic scattering spectrum of compounds	Uniquely identifies nonaqueous phase contaminants *in situ*	CA-S,P
LIF (Laser-Induced Fluorescence)	Locate fluorophores	Fluorescence spectrum of compounds	Identifies nonaqueous phase compounds *in situ*	CA-S
ROST™ (Rapid Optical Screening Tool)	Locate fluorophores	Time domain fluorescence spectrum of compounds	Identifies nonaqueous phase compounds *in situ*	CA-S,P
FFD	Locate fluorophores	Fluoresence intensity of compounds	Detects nonaqueous phase compounds *in situ*	CA-S,P
Laser-Induced Breakdown	Identify inorganic compounds	Emission spectroscopy of compounds	Detects some nonaqueous phase inorganic compounds	FT
Hydrosparge	*In situ* measurement of aqueous phase compounds	Headspace measurement of volatile and semivolatile compounds	*In situ* aqueous measurement	CA-S
***In Situ* Soil Sensor**	*In situ* thermal desorption and analysis of VOCs/ SVOCs	Heated headspace from *in situ* soil sample	*In situ* soil sample measurement	CA-S
Gas Analyzers at the Surface (Includes Chromatographic, Solid State and Other Detection Methods)	Analyze gas sample stream during penetration	Gas phase analysis of volatile compounds	Measures at the surface in real time	CA-S,P
Other In-Cone Gas Analyzers (such as RCI 5000, PAWS)	VOC measurement in the gas phase	Gas phase analysis of volatile compounds	Measures at the surface or in cone	FT
X-ray Fluorescence	*In situ* detection of metals/rads	X-ray fluorescence spectrum	*In situ* detection of metals/rads also clay components (such as titanium)	FT
Gamma Spectroscopy	*In situ* detection of Cs-137 and other gamma-emitters	Gamma spectrum	*In situ* detection of gamma emitters	CA-S

continued

TABLE 3-6 Sampling and Deployment Capabilities *(continued)*

Tool	Use	Data/Output	Features	Status
MIP (Member Interface Probe)	Detection of aqueous and nonaqueous phase volatiles	Membrane mediated organics in gas phase	Multiple depths, source identifier	CA-S,P
PiX (Precision Injection/ Extraction of Alcohols)	Identify DNAPL	Increase concentration in injected and recovered solvent	NAPL identifier, probes volume around cone	FT
Fiber Optic TCE	Gas phase TCE *in situ*	Reagent-based gas and aqueous sensor	Sensitive at maximum containment levels (MCLs) for drinking water	FT

Legend:

CA-PS—Commercially Available for Purchase or Service
FT—Field Tested
UD—Under Development

Advantages and Limitations of Direct Push Methods

The cone penetrometer has many advantages over conventional methods of accessing the subsurface. Some of these are listed below.

- Rapid penetration (the normal push rate is 2 cm/sec)
- Minimally invasive (the diameter of the cone penetrometer rod is normally 4.445 or 3.175 cm)
- Minimal investigation-derived waste (no drill cuttings or other potentially hazardous waste)
- Can acquire multiple data sets simultaneously (three to five different kinds of measurements, as well as sampling, can be performed on a single push)
- Produces data in real time
- High vertical resolution (data are taken at least every 2 cm, and sensors are designed to exploit this capability)

- Direct formation contact (particularly useful for electrical resistivity and spectroscopy)
- Cost-effective (multiple logs and formation samples can be obtained in a single push, and wells can be installed in a second push at less than one-third the cost of conventional drilling and logging)

It is important to keep in mind the following limitations of the cone penetrometer:

- Used only in unconsolidated sediments (limits the number of sites that are accessible)
- Depth limitations (limited push capacity and formation resistivity limit the total depth achievable by the cone penetrometer to between depths of 3 and 100 m [10 and 300 ft])
- Small diameter (limits the types of tools that may be deployed in the cone penetrometer and also the size of wells that can be installed)
- Skin effects (created by the displacement and compaction of soils during the cone-penetrometer push—generally not significant)
- Inferred measurements (not practical to use for obtaining continuous core, so most cone-penetrometer logs, like traditional borehole logs, are inferred measurements from properties of the formation)
- May not be acceptable for long-term monitoring.

The use of direct-push technologies for vadose zone site characterization can greatly enhance the quality of the investigation. Daily operating costs of the cone penetrometer are comparable to those of conventional drilling, but the speed of penetration, low associated waste, minimal invasiveness, and the ability to deploy *in situ* and downhole sensors make the direct-push methods more cost-effective in many cases. The best method of applying direct-push methods and tools depends on the needs and resources of the specific site, and the application of these procedures and instruments will often serve to complement conventional drilling and other site-assessment methods. As with other characterization techniques, the direct-push method should be incorporated in the environmental professional's toolbox and used when appropriate.

Sources on Direct Push Technology

In addition to the references cited, private companies offering cone-penetrometer or direct-push services, direct-push tool and instrument developers and manufacturers, and government sources (Department of Defense, Department of Energy, Environmental Protection Agency, and Department of Transportation) can supply information on direct-push technology. There is also information available on the following Internet sites:

- Geoprobe Systems Inc. http://geoprobesystems.com/
- Applied Research Associates, Inc. http://www.ara.com/
- Gregg In Situ Inc. http://www.greggdrilling.com/INSitu.html
- Fugro Geosciences http://www.geo.fugro.com/
- Department of Energy CMST program http://www.cmst.org/
- Liquifaction, the Web Site for the CPT Industry http://www.liquefaction.com/index.html.

ENVIRONMENTAL MEASUREMENT-WHILE-DRILLING SYSTEM FOR REAL-TIME SCREENING OF CONTAMINANTS*

One of the disadvantages of current characterization technologies is the lack of reliable observations and measurements conducted during borehole drilling. This drawback is especially significant for contaminated sites, because the soil and rock conditions can be altered by drilling tools and atmospheric air and water. Information on environmental conditions and drill-bit location and temperature during drilling is required in many environmental restoration operations. An inexpensive data collection system for identifying and tracking contaminant

*This section is based on the case study "Environmental Measurement-While-Drilling (EMWD) System for Real-Time Screening of Contaminants," by Cecelia Williams, on the accompanying CD.

concentrations and monitoring drill-bit conditions is needed for many waste-site procedures.

The Environmental Measurement-While-Drilling (EMWD) system (Figure 3-16) represents an innovative blending of new and existing technology in order to obtain real-time data during drilling. The objective of this method is to distinguish contaminated from non-contaminated areas in real time while drilling at hazardous waste sites. In EMWD, downhole sensors are located behind the drill bit and are linked by a high-speed data transmission system to a computer at the surface. As drilling is conducted, data is collected on the nature and extent of contamination, enabling on-the-spot decisions regarding drilling and sampling strategies. The EMWD system has been adapted by the integration of a gamma ray spectrometer (GRS) in place of the original simple gamma radiation detector. The GRS consists of a sodium iodide-thallium activated crystal coupled to a photomultiplier tube (PMT). The GRS output feeds to a multichannel analyzer (MCA). The MCA data, as a 256-channel gamma spectrum (100 KeV-1.6 MeV), is transmitted to the surface via a signal-conditioning and transmitter board. The system

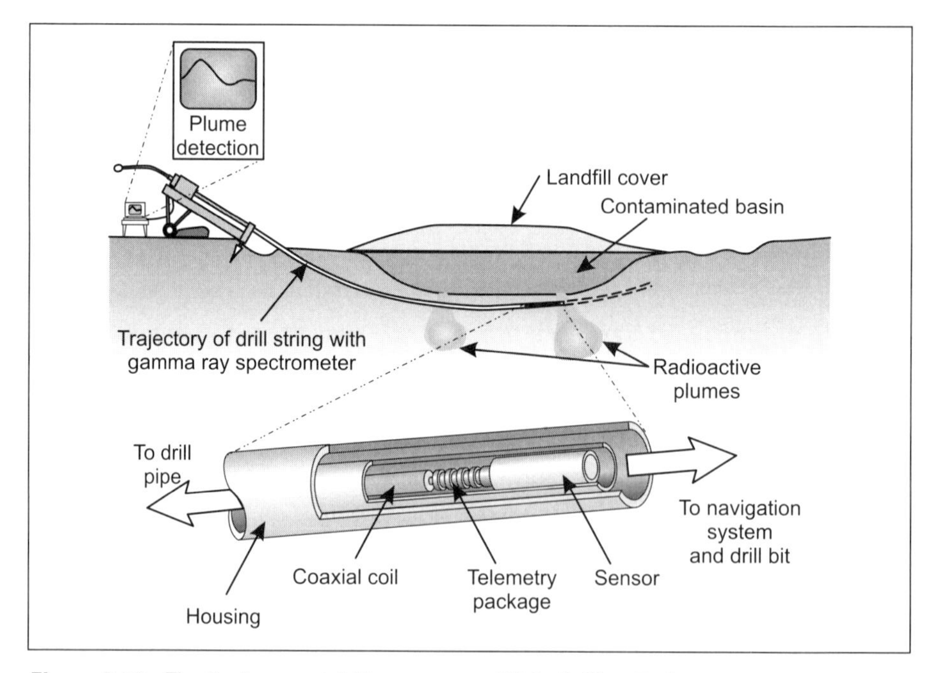

Figure 3-16. The Environmental Measurement-While-Drilling Tool.

also monitors the uphole battery voltage as measured downhole and the temperatures associated with the detector and instrumentation. The design includes data-assurance techniques to increase safety by reducing the probability of giving a "safe" indication when an unsafe condition exists.

The system provides real-time data on an eight differential/single analog multiplexer and any number of digital channels. Sampling speed from the analog channels can reach 100 kHz. The telemetry system is firmware-programmable to easily support many different data formats and additional data channels. The data transmission format (Digital FM Bi-phase, 4800 baud) provides excellent noise rejection for jumping the wireless connection between the rotating drill pipe and the stationary receiver. A Sandia-designed receiver removes the FM carrier, generates the data clock, and buffers data to be used by an IBM or IBM-compatible personal computer. A 28V rechargeable battery pack can supply downhole instrumentation power for more than 18 hours of drilling. The battery pack remains topside for easy maintenance and/or recharging.

The system is compatible with directional drilling techniques that use minimal drilling fluids and generate little-to-no secondary waste. The current system includes a continuous read-out-non-walk guidance and location system for use with the EMWD system. The orientation sensor package was integrated with the EMWD-GRS system without significant modification. In addition to the existing techniques, sensors are needed for the detection of heavy metals, volatile organic compounds, and natural gas. Technology developers are currently working with the EPA to obtain certification.

The EMWD system has many practical applications, including site characterization for contaminant detection and delineation. This system will guide sampling activities and borehole emplacement options. For example, a drill operator can back out of contaminated soils and redirect

The results of field testing of the EMWD are illustrated in the case study on the accompanying CD, "Environmental Measurement-While-Drilling (EMWD) System for Real-Time Screening of Contaminants," by Cecelia Williams.

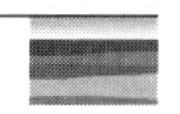

a drilling operation around the contamination. Other potential users of EMWD include utility emplacement and petroleum industries.

WELL COMPLETION AND INSTRUMENTATION

Permanent Well Completion

Wells drilled in hostile environments (such as weak, unconsolidated soils and sediments, fractured rocks, and contaminated sites) should be cased, and the annulus between the casing and the formation should be backfilled or grouted. Such procedures are intended to seal off unstable zones to prevent the collapse of the well, allow the installation of monitoring probes and periodic insertion of logging tools, and prevent cross-contamination of different layers along the borehole length.

Permanent well completion can be performed using borehole casings and backfilling the annulus. Steel or plastic casings, which usually span the entire length of the well, are often used to provide borehole wall stability (Rahman and Chilingarian 1995). Materials commonly used to seal the borehole annulus are bentonite, cement, or a mixture of both.

It is common practice to install an array of single monitoring probes in boreholes at different depths within fine-sand layers, separated by impermeable bentonite or cement. However, these materials may settle unevenly in the borehole, creating air pockets (Everett *et al.* 1984a,b). Water used to harden cement and swell bentonite pellets can be absorbed by the surrounding formation and thereby change the formation moisture content. If a water-conducting pathway intersects the sand layer in the borehole, water may accumulate in the sand layer around the monitoring probe or may perch on top of a bentonite seal (Figure 3-17). The presence of water may lead to erroneous measurements in boreholes. Moreover, field observations showed that the well-casing composition and bentonite may affect measurements conducted with logging tools and may contribute solutes to the groundwater (Houghton *et al.* 1984; Remenda and van der Kamp 1997).

During drilling at Hanford, it was assumed that the borehole annulus (the space between the soils and the casing) in unstable soils could remain ungrouted immediately after drilling, because the soils would ultimately collapse and seal the borehole annulus. However, observations at Hanford showed that the unstable soils did not seal off the

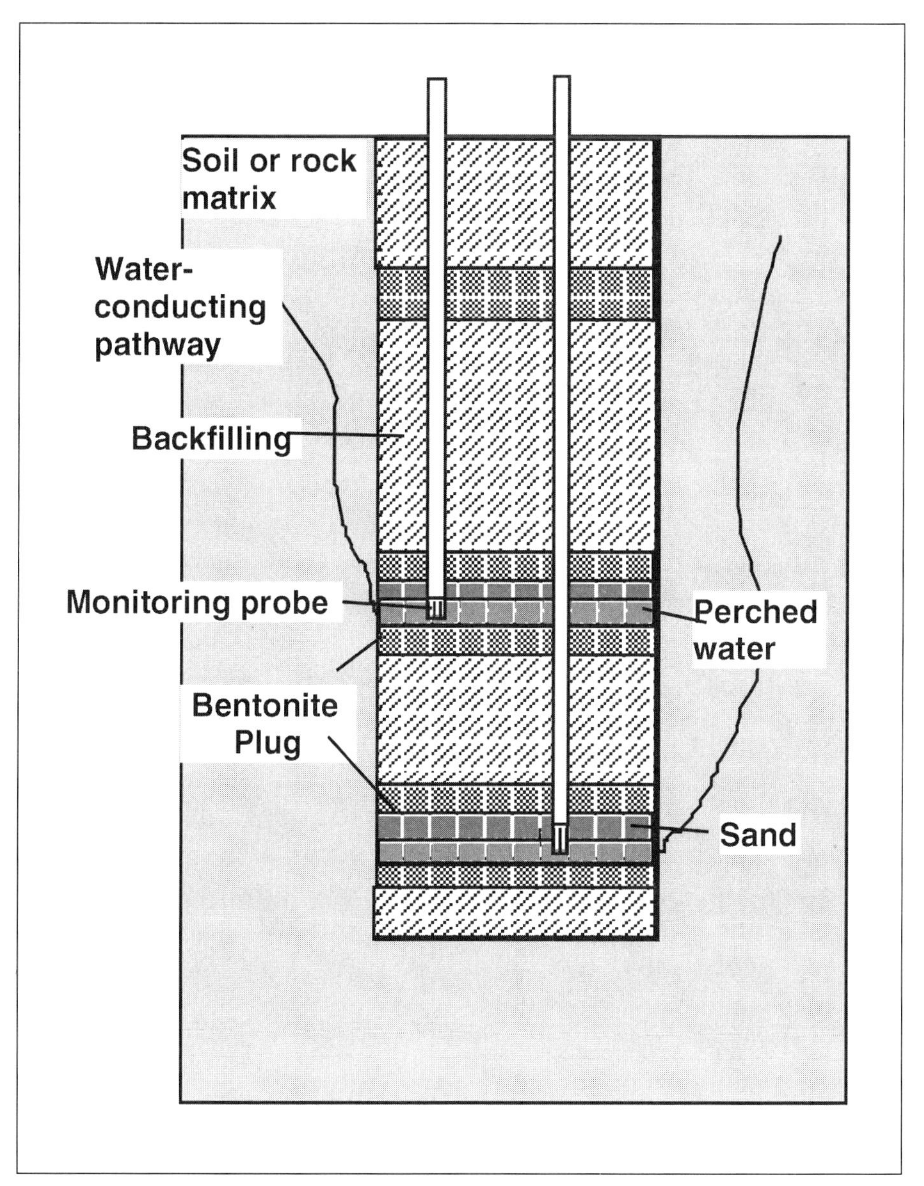

Figure 3-17. Conventional method of installing instrumentation in boreholes (Faybishenko *et al.*, 1998)

The case study ""Gamma Borehole Logging for Vadose Zone Characterization Around the Hanford High-Level Waste Tanks," by David S. Shafer, Desert Research Institute; James F. Bertsch and Carl J. Koizumi, MACTEC-ERS; and Edward A. Fredenburg, Lockheed Martin Hanford Company, describes the application of the Spectral Gamma Logging System(SGLS) at Hanford. *See page 445.*

borehole annulus, which led to the accelerated migration of radioactive and organic contaminants toward the groundwater.

An innovative technology of borehole sealing and instrument installation involves the use of polyurethane grout, which is schematically shown in Figure 3-18. For this method, monitoring probes are installed in contact with natural soils or rocks exposed on the borehole walls using permanently inflatable packers. The procedure is as follows:. Plastic packers are attached to a special 2-in. PVC pipe, and monitoring probes (including tensiometers, suction lysimeters, thermistors, electrical-resistivity sensors, and TDR probes) are taped to the outer surface of the plastic packers. The pipe, assembled with packers and instrumentation, is lowered into a borehole at a field site. Pipes 10 to 20 ft long can be lowered manually, and pipes 40 to 60 ft long are lowered using a boom-truck. After the pipe and string of monitoring probes are placed in a borehole, a water-activated polyurethane resin is injected in the plastic packers. When polyurethane is foamed inside the packer, the packer forces the probes toward the borehole wall. Then, polyurethane is injected in the space between the packers to fully grout the well and to prevent water from entering it.

This technology was tested in fractured basalt with positive results (Faybishenko *et al.* 1999a,b). Based on the testing of the polyurethane resin on Idaho soils and basalts, polyurethane has the following attributes favorable for permanent well completion:

- It adheres well to dry and wetted materials including soils, sediments, and rocks, and will not affect contaminant transport.
- It comes in liquid form, and therefore can easily be injected through a small-diameter tube to infill an annular space of boreholes. The polyurethane set time can be regulated.

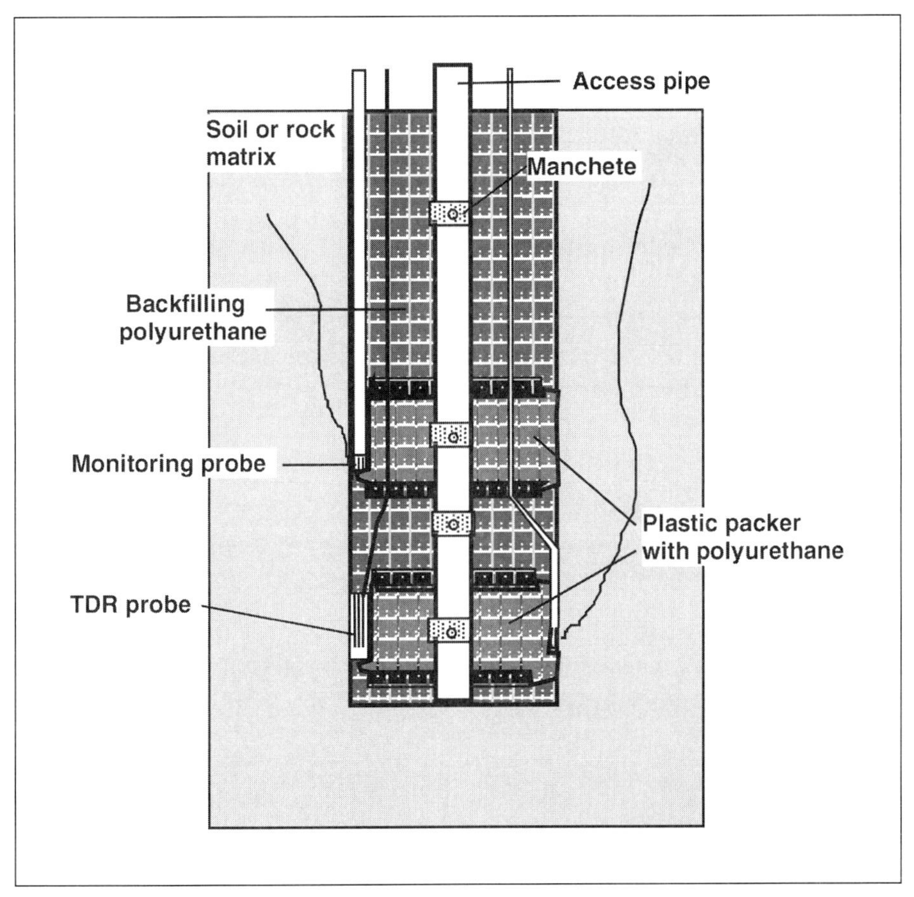

Figure 3-18. Schematic of installing instrumentation and grouting a borehole using polyurethane (Faybishenko *et al.*, 1998)

- It has very low bulk density and electrical properties, which are favorable for geophysical logging.
- Unlike bentonite and cement, it will not change the moisture content of surrounding rocks.
- The use of acrylic casings grouted with polyurethane in boreholes may increase the accuracy of neutron logging (Zawislanski and Faybishenko 1999).

Removable SEAMIST (a. k. a. FLUTe™ Systems) Liners*

SEAMIST is an innovative technology that can function both as a borehole casing and as a support platform for sampling devices and instrumentation. SEAMIST consists of the borehole liner of a tubular, impermeable membrane, and a tether that gathers the bottom of the membrane, extends up the center of the hole, and travels to a reel in a canister at the surface (Keller 1991). SEAMIST employs an everting/inverting, flexible liner to seal and support an open borehole while carrying instruments into place and isolating them one from another (Figure 3-19). The everting liner is driven into the hole by air

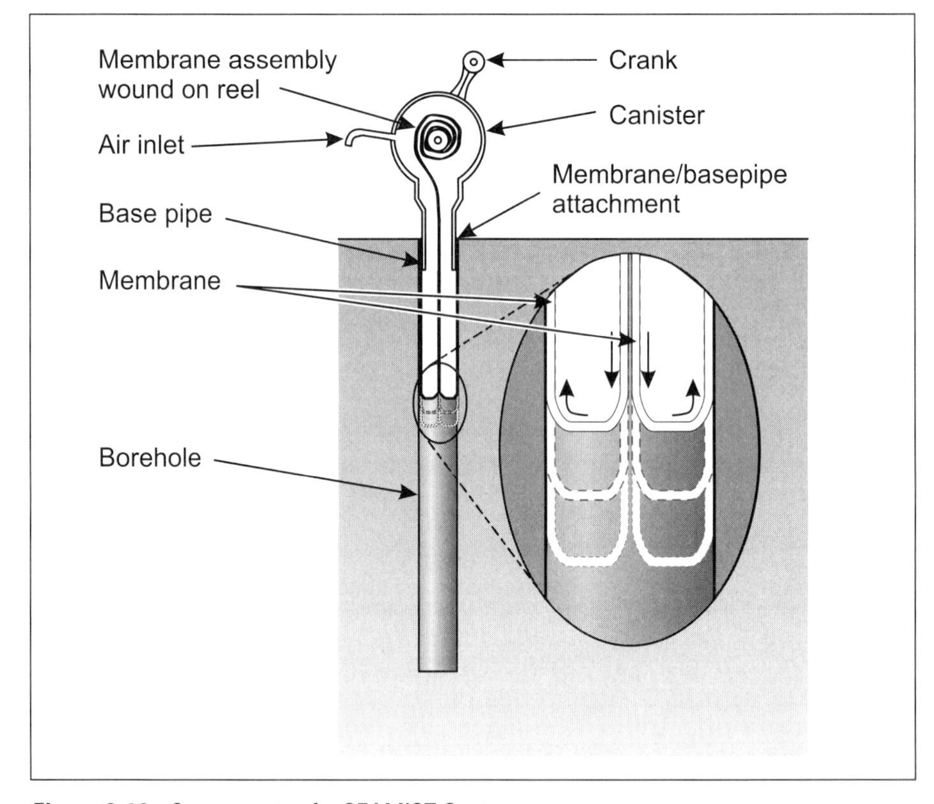

Figure 3-19. Components of a SEAMIST System

*This section was contributed by Carl Keller and Cecelia V. Williams.

pressure. As the liner is propelled into the hole, it carries a variety of "instruments" into place, and seals the holes between the instruments.

Using the apparatus shown in Figure 3-19, the impermeable membrane is emplaced downhole at a speed between approximately 20 to 50 ft/min. Positive pressure is supplied to maintain the integrity of the borehole once the liner is emplaced. Grout or a filler material such as sand may be used instead of air for long-term installations. Because the filler material is located inside the liner, contamination of the geologic medium by the filler is prevented. If the system is later removed from the subsurface, the membrane is wound onto the reel and into the canister by inversion. Therefore, no cross-contamination of absorbent pads or sensors can occur as the membrane is removed. Liners can be made of a wide variety of impermeable materials, including plastic tubular films and laminates. Liner materials are selected on the basis of cost, durability, fabrication ease, impermeability, and chemical compatibility.

SEAMIST liners have been used in horizontal, vertical, enlarged, constricted, and curved holes, and in open and cased boreholes. They can line the borehole temporarily or permanently, preventing the collapse of the borehole, limiting movement of air into the subsurface, and blocking fluid flow into and within the borehole. Sampling ports with attached tubes extend to the surface, while larger instruments (such as gamma logs, neutron logs, resistance logs, and television cameras) can be carried into the hole on the tether. The sample collection instruments that have been emplaced downhole include:

- Absorbers pressed against the hole wall for wicking of liquid samples
- Absorbcnt matcrial that completely covers the SEAMIST liner from top to bottom of the borehole, allowing a continuous map of the subsurface to be obtained at that location
- Gas-sampling ports and tubing for vapor collection in discrete intervals of the borehole
- Electrical contacts for resistance measurements.

Many of the problems with conventional vadose zone monitoring techniques are eliminated or minimized by the SEAMIST design. These include: borehole instability, single-point sampling with screened wells, the inability to retrieve and repair instrumentation buried in backfill, and cross-contamination of samples as a result of inadequate backfill seals. These advantages are achieved with a portable, lightweight, and robust emplacement system that is fast and relatively inexpensive. The primary limitation of the technology is that the borehole must remain open long enough after drilling to allow deployment of the membrane (usually less than 30 minutes). If regions of swelling clays are encountered in the lithology, the air-pressurized membrane may not prevent closure of the borehole, unless the liner is sand-filled. The seal of the interface between the membrane and the borehole wall may not be as absolute as in a grouted hole, but appears to be adequate for most applications.

The basic everting liner method has evolved into a large family of techniques for vadose zone measurements. The Duet™ uses two liners: the first liner supports and seals the hole, and the second, instrumented, liner, is deployed in parallel in the same hole, and is used to provide measurements at any place or time without violating the seal or the borehole support. The LAHD™ (Liner Augmentation of Horizontal Drilling) application uses a propagating liner to support a horizontal hole while the hole is being reamed by a horizontal drill rig. The result is an exceptionally clean, sealed hole with little mud invasion, no mud cake, and complete support against hole collapse.

The liner is also used to case and/or instrument the horizontal hole while fully supported. The everting liner can be deployed in cone penetrometer holes and through cone penetrometer rods to emplace the reactive ribbon. The multilevel sampling in the vadose zone has been extended to ground water sampling with a downhole pump for each port. Table 3-7 summarizes SEAMIST applications at various field sites.

Removable Automatic Pneumatic Packers*

Automatic pneumatic injection packers can be used to provide controlled, site-to-site, and borehole-to-borehole testing of stable soils and rocks using the same packer design for injection, pumping, and observations. This approach is amenable to the automation and remote control

*This section was contributed by P. Cook.

TABLE 3-7 **Summary of SEAMIST Field Applications**

APPLICATION	DESCRIPTION	SITE	PERIOD OF USE
Tritium Plume Monitoring	Two systems are used for tracking the movement and concentrations of a t ritiated water plume (both vapor and liquid water sampling to 40 ft).	Lawrence Livermore National Laboratory(LLNL), CA	May 1991 to present
Carbon Tetrachloride Monitoring	Two systems with disposable membrane liners are in use in a carbon tetrachloride plume. The membranes pack off the cased borehole while sample tubing to the bottom draws the vapor sample.	Hanford, WA	1992 to present
Fracture Flow Mapping and Rate Measurement	Membranes coated with liquid-indicating and wicking layers mapped and measured brine flows (grams per day) underground at Waste Isolation Pilot Project, NM.	Waste Isolation Pilot Project, NM	January 1992
Tritium and VOC Sampling	Vapor sampling tubes and absorbent collectors 230 ft were installed horizontally underneath an old radwaste landfill.	Los Alamos National Laboratory (LANL)	April 1992
Sandia National Laboratory - Integrated Demonstrations	• Gas sampling and permeability measurements in two boreholes (11.5 in. in diameter and 110 ft deep) • Three 110 ft borehole liners to support holes during logging.	The Chemical Waste Landfill Sandia National Laboratory (SNL)	Spring through Fall 1992
Vapor Sampling/ Permeability Measurements	Three membranes were instrumented and installed for sampling soil vapor, vapor pressure and permeability measurements. Maximum depth is 130 ft with ten sampling elevations per membrane.	Savannah River Site (SRS)	July 1992 (Continuing for up to 2 years)
Neutron Logging Tool Transport	Neutron moisture logging in horizontal boreholes underneath waste landfill in four boreholes (200 to 250 ft long)	LANL	August 1992
Vapor Sampling	Vapor-sampling with nine sampling points installed to 90 ft deep.	Tucson	September 1992
High Pressure Borehole Liners	Two Kevlar-reinforced membranes were installed to a depth of 155 ft, then filled with water inside cased wells to prevent collapse of the PVC casino during steam injection remediation experiments.	LLNL	October 1992
Borehole Liner	A SEAMIST liner was installed to support/ seal hole while long-term monitoring system is designed.	Utica, NB	October 1992

continued

TABLE 3-7 **Summary of SEAMIST Field Applications** *(continued)*

APPLICATION	DESCRIPTION	SITE	PERIOD OF USE
Sandia National Laboratory–Integrated Demonstration	Installed gas-sampling tubing to 393 ft in a horizontal borehole of 4 in. in diameter.	SNL	Fall 1993 to Fall 1994
Monitor Fuel Oil Plume Position	Monitor the position of the plume via closelyspaced soil vapor sampling ports in a vertical hole.	Swedish Hospital	May 1993 to present
Stabilization of Contamination	SEAMIST is being used to apply strip coat to the interior of ducting to immobilize hazardous dust (such as U and Pu)	Old plutonium facility	Spring 1994
Monitoring of Soil Vapor Extraction	Nine vertical SEAMIST systems with 10 vapor sampling ports to 80 ft used for pressure, permeability, and concentration monitoring.	Sacramento Army Depot, CA	March 1994 to present
Mapping of Contamination at Radioactive Waste Site	Installation, in a mole hole 4.5 in. in diameter.,of an absorbent covering on SEAMIST	Oak Ridge WAG6	August 1994 to present
Monitoring of a Radioactive Waste Landfill	Use of SEAMIST in tunnels built in trenches below low-level radioactive waste landfill to monitor for leachate migration.	TA-54, Pit 39	September 1994 to present
Sampling at Discrete Levels below the Water Table	Installation of SEAMIST with water and vapor sampling ports. The water table is at 40 ft below the surface.	March AFB, Riverside, CA	Being installed
Permeability Measurements	Use of SEAMIST in 60 ft deep hole for measurement of permeability and changes during thermal enhanced vapor extraction.	TA-3, Chemical Waste Landfill	July 1994 to present
Tritium Plume Measurement	Installation of absorbers on SEAMIST membrane to 300 ft for wicking of water in vadose zone.	TA-33	August 1994 to present
Vapor Sampling	Multilevel water sampling to 80 ft below SLW.	Saegertow, PA	September 1997
Vapor Pressure	Pore pressure monitoring to 800 ft.	Nevada Test Site, NV	1995 to present
Monitoring	Landfill monitoring in horizontal holes.	Los Alamos, NM	April 1998 to present

continued

TABLE 3-7 **Summary of SEAMIST Field Applications** *(continued)*

APPLICATION	DESCRIPTION	SITE	PERIOD OF USE
Vapor Sampling	Installation in Sonic Casing to 300 ft.	Asuza, CA	October 1997 to present
Liquid Sampling	Groundwater Sampling (6 ports in 3 in. hole)	Milford, NH	November 1998 to present
Vapor Sampling and Absorber Installations	Duet™ double liners	Yucca Mt., NV	September 1997 to present
Vapor Sampling	Gas sampling in CPT holes	Savannah River Site, SC	September 1997
DNAPL Location	Installation of color-reactive ribbon	Savannah River Site, SC	August 1998
Liquid Monitoring	Absorbers for fracture flow	Yucca Mt., NV	1998

Trademarks: SEAMIST, Robert Alpert Companies
FLUTe, Duet, LAHD, Flexible Liner Underground Tech.

necessary to establish consistent testing regimens and to accommodate the large number of tests. The packers developed at LBNL for air injection tests (Cook 1999) include inflatable rubber sealing bladders on a packer string, which can be manipulated independently and can divide a borehole into a number of separate zones over the length of the string (Figure 3-20). Each zone is serviced by one 3.2-mm-diameter port for pressure measurement and one 6.4-mm-diameter port for air injection. Several boreholes may be instrumented at one time. A modular design allows partial dismantling of the packer strings in the field for repair or work in tight quarters.

If all the bladders were to be inflated at once, the packer string would seal the entire borehole section that it occupied. However, by inflating every other bladder and allowing the remainder to remain deflated, an alternating sequence of open and closed (sealed) intervals is produced. Depending on the injection control valves, an open interval becomes a pressure monitoring zone, or the injection zone where air is injected during a test. Once tests have been performed with these open zones, the

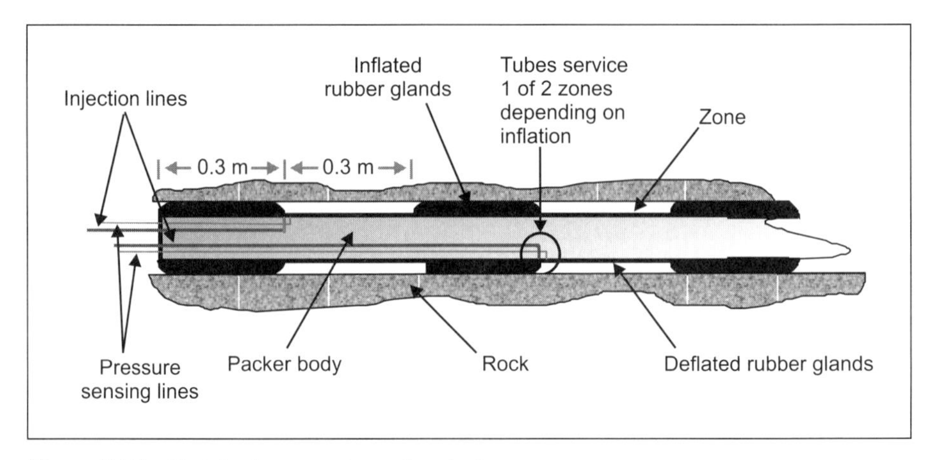

Figure 3-20. Sketch of automatic packer design

inflated bladders are deflated and deflated bladders are inflated, opening those zones that were once sealed, and sealing those that were originally open. In this manner, nearly the entire length of the packer string is usable for testing every 0.3 m without having to move the string. If the zones on the injection packer are changed independently from those on the observation packers, there will be four possible zone configurations available during a given packer installation. Permutations of these injection and observation positions are used to ensure that all positions within each observation borehole are allowed a chance to respond to a given injection zone. The observation packer zones are usually changed in unison, because the locations of the observation zones are thought not to perturb the flow field significantly, and permutations between them would cause only second-order effects in the response system.

Installation of Embedded Sidewall Sensors in Boreholes*

Sensors that are installed in vertical and slanted boreholes at different depths are used to determine matric potential and moisture content, as well as to collect water samples for chemical composition. Sensors placed in boreholes need to be in contact with the formation, as the

*This section was contributed by L. Murdoch and FRx Inc. of Cincinnati.

results of measurements depend on the contact area of the sensors. To address such limitations, Murdoch *et al.* (1999) developed a method for accessing the sidewall of a borehole. The method uses a device that embeds sensors or sediment samplers into the sidewall to distances of about 15 cm (Figure 3-21). The device can also be used to obtain a core sample 15 cm long and 4 cm in diameter, and then to insert a permeable sleeve for extracting samples of water or gas. This device can be used to install several types of electrode sensors, including devices for measuring water content (TDR waveguides), redox potential (platinum-tipped electrodes), or electrical resistivity (a four-conductor, Wenner-type electrode). Horizontally oriented TDR waveguides can be inserted at virtually any depth, thereby extending the TDR technique to the study of deep vadose zones. The borehole is completely sealed to minimize the possibility of cross-contamination after the sensors have been emplaced. The sidewall technique currently can place as many as 60 sensors in a single boring, and offers the potential for markedly increasing the spatial resolution with which processes in the vadose zone can be monitored. The method can be used for embedding sensors in the sidewalls of horizontal or directional boreholes. This application could provide a platform for placing sensors beneath sensitive structures, such as the tanks containing high-level waste at Hanford.

The case study, on the accompanying CD, "Embedded Sidewall Sensors," L. Murdoch, Clemson University, provides the results of field validation of moisture-content measurements using TDR probes installed at different depths in a vertical borehole sidewall with those from conventional soil sampling. The case study also provides the results of measurements of the redox potential using Eh electrodes installed in the borehole, which would be impossible to do otherwise in the field.

HYDROGEOLOGICAL CHARACTERIZATION USING GEOPHYSICAL METHODS*

Goals of Geophysical Methods and Acquisition Geometry

Geophysical investigations complement hydrogeological methods of site characterization by providing a denser spatial resolution for sub-

*This section was contributed by S. Hubbard, Y. Rubin, and E.L. Majer (LBNL), based on Chapter 10 of *The Handbook of Groundwater Engineering*, by Y. Rubin, S. Hubbard, A. Wilson, and M. Cushey.

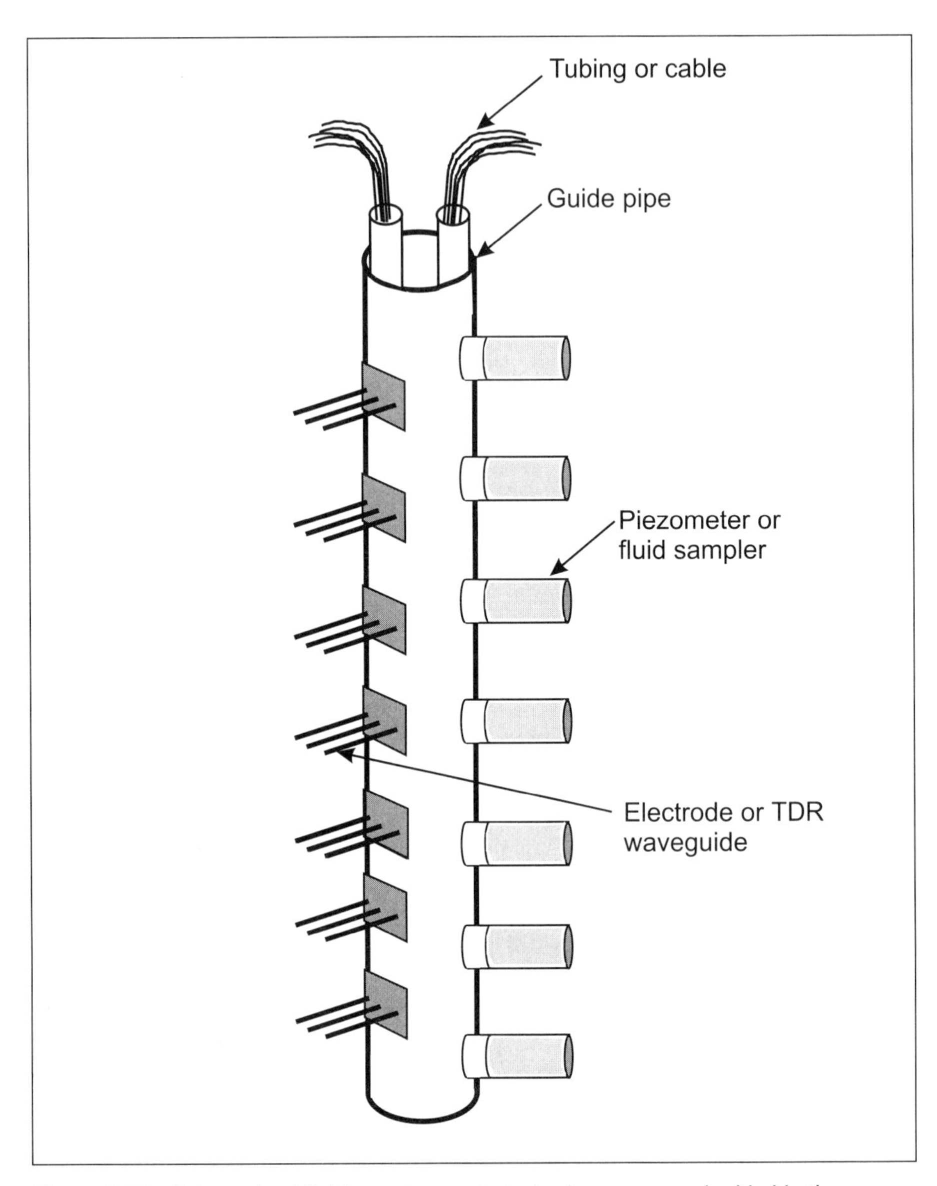

Figure 3-21. Schematic of fluid samplers and electrode sensors embedded in the sidewall of a borehole (after Murdoch *et al.*, 1999)

surface measurements. The main advantage of geophysical investigations is collection of data in a non-invasive manner, which reduces the number of direct measurements needed to fully characterize a site. Figure 3-22 illustrates the resolution and support scale of different types

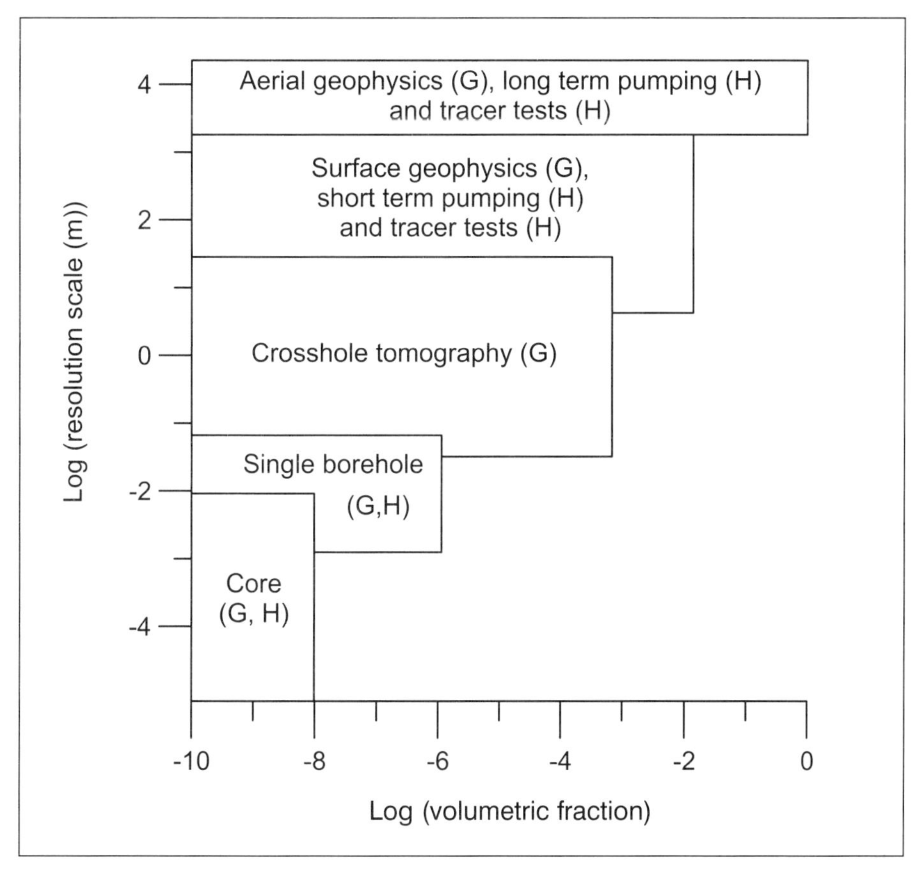

Figure 3-22. Resolution and volumetric fraction of subsurface sampled using hydrological (H) and geophysical (G) characterization methods

of geophysical and hydrogeological measurement techniques. This figure shows that some geophysical techniques have, on average, higher resolutions than others, and that for each method there is a range of possible resolutions. Figure 3-22 also shows that, in terms of both resolution and volume of soil sampled, geophysical data bridge the information gap between the more traditional site characterization techniques of core analysis and borehole testing.

The purpose of this section is to present the main features of geophysical methods that are currently used for near-surface site characterization. The dimensionality and volume of sampled subsurface soil are in part governed by the acquisition geometry of the geophysical methods. Common acquisition geometries include surface, cross-hole tomographic, and single borehole modes.

Surface geophysical methods are used to detect variations in subsurface properties in one, two, or three spatial dimensions. In electrical-method terminology, the term 'profiling' refers to measuring lateral changes in electrical properties over a constant subsurface depth, and the term 'sounding' refers to a collection of measurements that are associated with a single surface location and are made as a function of depth. Surface-seismic and ground-penetrating-radar data are displayed as 'wiggle-trace' profiles with distance on the horizontal axis and arrival time (which can be converted to depth) on the vertical axis. The vertical and lateral variations in arrival time, amplitude, and phase of the wiggles that comprise the vertical cross-sections are indicative of subsurface physical property changes. Figure 3-23a illustrates how surface radar data are collected by moving the transmitting and receiving antenna across the ground surface and recording the reflected arrivals. Seismic and ground-penetrating-radar profiles yield two-dimensional information about physical property changes.

Cross-borehole tomographic acquisition geometries are used with electrical, seismic, and radar geophysical methods for detailed site investigation. Tomographic data can produce high-resolution images. A typical tomographic geometry consists of two vertical boreholes separated by an interwell region of interest. With this acquisition geometry, sources and detectors are located in separate boreholes. This geometry is illustrated for seismic methods in Figure 3-23b, where direct energy from a source or transmitter in one borehole travels to and is recorded by geophones that are connected by a cable and located in the other borehole. The source position is changed and the recording repeated until the source has occupied all positions in the source borehole. Travel times and amplitudes of the recorded wave are picked and inversion algorithms are used with these data to estimate velocity and attenuation at each cell in the interwell area (Peterson *et al.* 1985). With electrical methods, electrodes occupy the wellbores; with electromagnetic (EM) data, borehole induction coils occupy the wellbores; and with radar methods, antennas occupy the boreholes when collecting tomographic data.

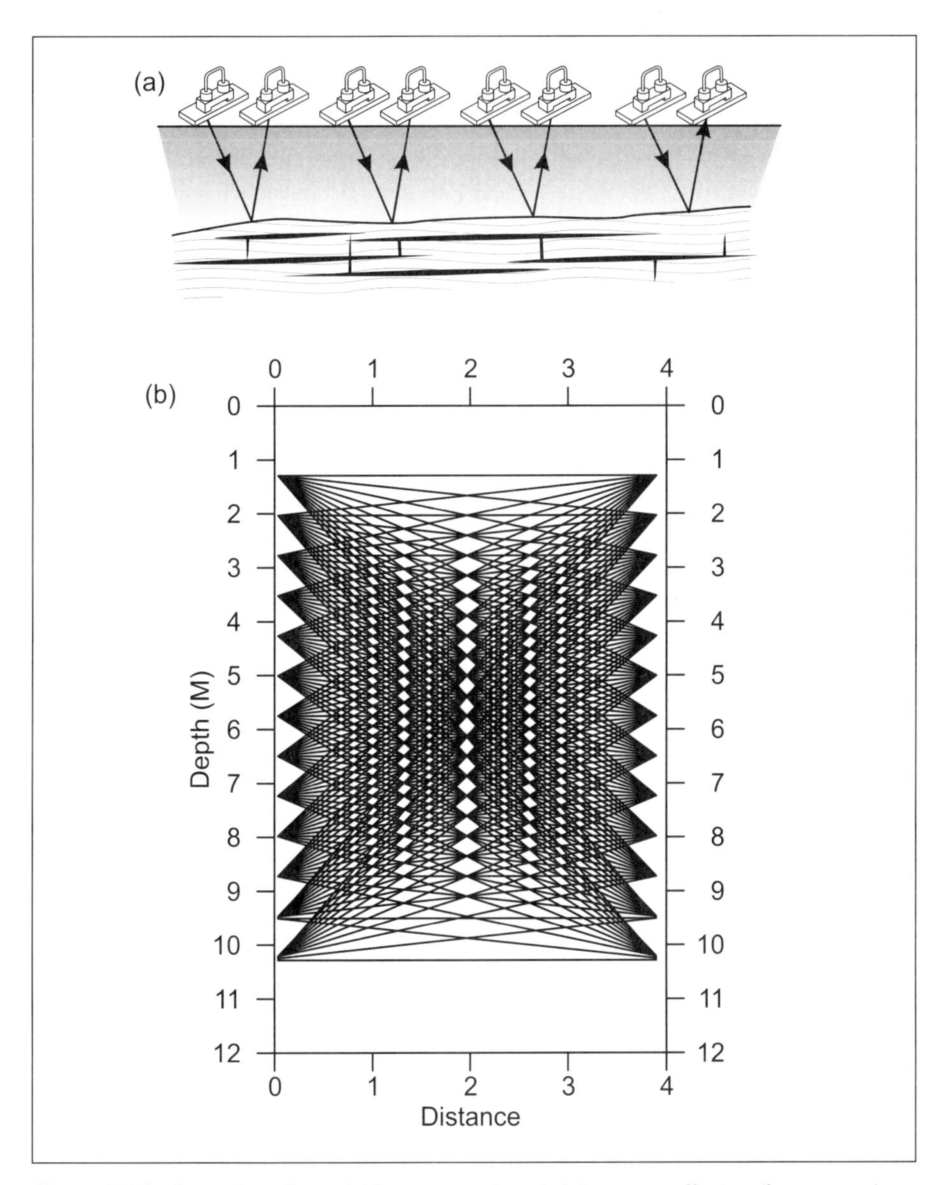

Figure 3-23. Examples of acquisition geometries: (a) Common-offset surface ground penetrating radar acquisition geometry to collect one trace per surface location from a transmitter-receiver pair, and (b) Cross hole tomographic acquisition geometry used for seismic and radar methods. Sources or transmitters and receivers are located in separate boreholes, and energy from each source is recorded at all receivers

Lastly, single downhole acquisition geometries are used with logging tools to sample the physical properties in the vicinity of the wellbore.

Geophysical methods can be used with all of these different acquisition geometries, including surface, tomographic, and single borehole, to collect data at a single point in time as well as over a period of time. Data using "time difference" displays (a data set collected at an earlier time subtracted from a data set collected at a later time) enhances the image of subtle geophysical attribute changes, for example, caused by steam flooding, hydraulic fracturing, or the spread of contaminant plumes.

The particular geophysical method and acquisition geometry used for a given investigation is chosen by considering the investigation target, the necessary level of resolution, conditions of the site, funds available for the investigation, and the availability of other data. In the following sections, we present the concepts of several geophysical methods, including electrical, seismic, gravity, and magnetic techniques. All of these methods can be employed using surface-acquisition geometries, and it is also becoming more commonplace to collect electrical and seismic data using tomographic geometries. Tables 3-8 and 3-9 summarize potential applications and limitations of the different surface or tomographic techniques, as well as the frequency of use, relative cost, and relative resolution of each method. Table 3-8 also lists the geophysical attribute that is commonly measured using each technique. For hydrogeological characterization, it is common to transfer these geophysical attributes into hydrogeological parameter estimates using petrophysical relationships (Rubin *et al.* 1998). For more detailed information about these individual geophysical methods, refer to Telford *et al.* (1990) and Robinson and Coruh (1988). Single wellbore logging methods are discussed separately below, and Tables 3-10 and 3-11 provide a summary of potential applications and limitations of these borehole techniques.

Electrical Methods

Electrical methods involve the detection of the effects of electric current flow in the soils. The two properties of primary interest are (1) the ability of a material to conduct electrical current, or the electrical conductivity, and (2) the displacement current, which occurs when an electrical current is passed through the material (that is, the dielectric

TABLE 3-8 Surface and Crosshole Geophysical Methods, Parameters, Potential Applications, and Limitations

Method	Parameters Commonly Inferred from Measurements	Potential Applications	Limitations	Relative Cost*1	Relative Resolution*2	Frequency of Use for Site Characterization
ELECTRICAL						
DC Resistivity	Electrical Resistivity	• Mapping of: gross stratigraphy, faults, depth to bedrock fresh/salt water interfaces, landfills, and some contaminant plumes	• Can not be used in paved areas • Wires can be cumbersome, and arrays can be long for deep investigations	Moderate-high	Moderate	Very common
EM	Electrical Conductivity	• Mapping of: gross stratigraphy, salt/fresh water interfaces, depth to bedrock, faults, and some contaminant plumes • Detection of buried tanks and pipes	• Difficult to resolve resistive targets, noise from power lines, and fences and pipes	Low-moderate	Moderate	Very common
GPR	Dielectric Constant	• Mapping of: detailed stratigraphy, some contaminant plumes, cavities, depth to bedrock, and water table • Detection of buried tanks and pipes • Estimation of hydrogeologic properties using petrophysical relationships and tomographic or CMP acquisition geometries	• Ineffective in highly electrically conductive environments, due to interference from electrical noise such as power lines and fences • Hydrogeologic property estimation entails more sophisticated data acquisition and processing procedures	Moderate	High	Very common (surface GPR) Uncommon (tomographic or CMP GPR)
SEISMIC						
Reflection	P-wave reflectivity and velocity	• Mapping of: gross and detailed stratigraphy, faults, and water table	• Acquisition often difficult in unconsolidated environments • Sophisticated acquisition and processing system required • Sensitive to cultural noise	Moderate-high	High	Common

continued

TABLE 3-8 Surface and Crosshole Geophysical Methods, Parameters, Potential Applications, and Limitations *(continued)*

Method	Parameters Commonly Inferred from Measurements	Potential Applications	Limitations	Relative Cost*1	Relative Resolution*2	Frequency of Use for Site Characterization
Crosshole	P-wave velocity and attenuation	• Mapping of: detailed stratigraphy, faults, cavities, and some contaminant plumes • Estimation of porosity and permeability using petrophysical relations	• Sophisticated acquisition and processing necessary • Sensitive to cultural noise • Best in saturated sections	High	High	Uncommon
Refraction	P-wave velocity	• Mapping of: gross stratigraphy and velocity structure, depth to bedrock, water table, and significant faults	• Gross feature identification only • Can not resolve low layers that have lower velocities than overlying layers • Sensitive to cultural noises	Low	Low	Common
GRAVITY AND MAGNETICS						
Gravity	Density	• Mapping of: depth to bedrock, faults, landfills, and cavities	• Gross feature identification only • Requires extensive data reduction and accurate elevation information	Low-moderate	Low	Uncommon
Magnetics	Presence of magnetic materials	• Mapping of: depth to magnetic basement, locating buried drums and pipes, and landfill delineation	• Interference from industrial and near surface magnetic features	Low	Low-high depending on application	Very Common

*1 Using acquisition parameters that sample similar subsurface volumes.

*2 Resolution is a function of geophysical method, acquisition parameters, and site conditions. This column presents relative resolutions for acquisition in environments favorable for each method using parameters optimal for a near-surface investigation for that method.

TABLE 3-9 Summary of Possible Applications of Surface and Cross-Borehole Geophysical Methods for Site Characterization

Surface/Crosshole Method: Application	Resistivity	EM Induction	GPR	Seismic Reflection	Crosshole Seismic	Seismic Refraction	Gravity	Magnetics
Depth to Water Table	2	2	4	2	0	4	1	0
Fresh/Salt Water Interface	4	4	2	1	0	2	0	0
Depth to Bedrock	4	4	4	4	0	4	4	2
Gross Hydro-stratigraphy	4	4	4	4	1	4	1	1
Detailed Hydro-stratigraphy	2	3	4	4	4	1	1	0
Significant Fault Detection	4	4	4	4	4	4	4	4
Cavity Detection	2	1	3	2	3	0	2	1
Porosity, Permeability Estimation	3	3	3	3	3	1	1	0
Water Content Estimation	3	3	3	1	0	1	1	0
Contaminant Detection	3	3	3	1	3	0	0	0
Detection of Buried Metallic Objects	2	4	4	2	1	0	1	4
Landfill Delineation	4	4	4	2	2	2	2	4

Key:
0 = not considered applicable
1 = limited use
2 = used, or could be used, but not the best approach or has limitations
3 = excellent potential but not fully developed
4 = generally considered an excellent approach, techniques are well developed

TABLE 3-10 Summary of Physical Property Measured, Applications and Limitations of Borehole Geophysical Methods

Borehole Tool	Measurement	Applications	Limitations and Sources of Error	Frequency of Use for Groundwater Characteristic	Borehole Conditions	Radius of Investigation
Caliper	Borehole diameter	• Borehole diameter measurement, fracture detection, and lithology inference		Common	• Cased or uncased • Saturated or unsaturated	Borehole wall only
Electrical						
Induction	Electrical Conductivity	• Lithological identification, clay content, and lithology inference	• Works best for delineating electrically conductive targets, affected by resistivity of borehole fluid	Uncommon	• Uncased or PBC cased • Saturated or unsaturated	¼ m to several m from borehole wall
Nuclear						
Gamma	Natural gamma	• Lithological identification, clay content, and permeability estimation	• Affected by casing, cement, mud, borehole diameter, and position of probe in well	Common	• Cased (PVC or steel) or uncased • Saturated or unsaturated	15-30 cm from borehole wall
Gamma Spectrometry	Natural gamma radiation	• Lithological identification • Identification of radioisotopes in groundwater	• Affected by borehole diameter, fluid, casing, annular material, and instrument drift • Requires computer analysis	Uncommon	• Cased (PVC or steel) or uncased • Saturated or unsaturated	15-30 cm from borehole wall

continued

TABLE 3-10 Summary of Physical Property Measured, Applications and Limitations of Borehole Geophysical Methods

Borehole Tool	Measurement	Applications	Limitations and Sources of Error	Frequency of Use for Groundwater Characteristic	Borehole Conditions	Radius of Investigation
Gamma-Gamma or Density	Electron density	• Lithologic and depth to bedrock identification, density, porosity, and moisture content estimation	• Affected by borehole diameter, gravel pack, casing, mud cake, cement, and position of probe in hole	Common	• Saturated or unsaturated • Optimal in uncased	15 cm from borehole wall
Neutron	Hydrogen content	• Moisture content, saturated porosity, lithology, and depth to water table	• Affected by borehole diameter, thickness of mud cake,casing, cement, mud weight, temperature, and pressure • Source requires license	Common	• Saturated or unsaturated • Optimal in uncased	15-25 cm from borehole wall
Acoustic						
Acoustic, Sonic or Velocity	P-wave transit time	• Lithology and fracture identification • Depth to bedrock, estimation of grout integrity and porosity	• Spikes occur when amplitude of P-wave is less than detection level of tool • Dry hole sondes can be used in vadose zone	Common	• Conventional–Uncased and • Dry hole sondes–Uncased and unsaturated	25 cm for sediments, 120 cm for rocks

TABLE 3-11 **Summary of Possible Applications of the Most Common Borehole Geophysical Methods for Site Characterization**

Borehole Method: Application:	Caliper	Induction	Gamma	Gamma-Gamma (Density)	Neutron	Borehole Imager
Lithology	4	4	4	4	4	4
Fracture Detection	4	2	0	0	0	4
Porosity Estimation	0	3	1	4	4	3
Moisture Content	0	2	0	4	4	3
Permeability Estimation	0	3	3	2	2	2
Water Quality	0	3	2	0	2	2

Key:
0=not considered applicable
1 = limited use
2 = used, or could be used but not the best approach or has limitations
3 = excellent potential, but not fully developed
4 = generally considered an excellent approach, techniques are well developed

properties). In the following paragraphs, we review the most commonly used electrical methods for near-surface investigations of electrical resistivity, electromagnetic techniques, and ground-penetrating radar.

Resistivity

Resistivity is a measure of the ability of current to flow through a given material, and is an intrinsic property of the material. Resistivity methods involve the introduction of a direct current (DC) or very low frequency (less than 1 Hz) current into the ground between two current electrodes. In a two-electrode system, current flows from the positive current electrode to the negative current electrode. These currents establish equipotential surfaces, and current flow lines are perpendicular to these surfaces. To deduce the subsurface resistivity, we place two potential electrodes between the current electrodes to measure the difference in potential or voltage. These measurements, together with the known current and a geometric factor (which depends on the particular electrode configuration), can be used to calculate resistivity following Ohm's law.

No other physical property of naturally occurring material displays such a large range of values as electrical resistivity, which commonly varies over 12 orders and has a maximum range of 24 orders of magnitude (Zohdy *et al.* 1974; Telford 1990). This wide range has rendered electrical resistivity a useful tool for mapping subsurface structure and stratigraphy, and for estimating hydrogeological parameters. In general, the electrical resistivity is reduced with an increase in water content and salinity, an increase in clay content, and a decrease in grain size (Burger 1992). Because of the myriad of factors affecting electrical resistivity measurements, it is often difficult to directly correlate resistivity with lithology without other constraining information. However, general statements can be made regarding electrical resistivity values, such as: (1) resistivity is sensitive to moisture content, and thus, unsaturated sediments usually have higher resistivity values than saturated sediments, (2) sandy materials generally have higher resistivity values than clayey materials, and (3) granitic bedrock generally has higher resistivity values than saturated sediments and often offers a large apparent resistivity contrast when overlaid by these sediments. Reviews of the resistivity method are given by Ward (1990), Van Nostrand and Cook (1966) and Zohdy *et al.* (1974).

Electromagnetic Induction

The use of electromagnetic techniques for environmental site assessments has increased dramatically in recent years. Controlled-source inductive EM methods use a transmitter to pass a time-varying current through a coil or dipole on the earth's surface. This alternating current produces a time-varying magnetic field, which interacts with the conductive subsurface to induce time-varying eddy currents. These eddy currents give rise to a secondary EM field. Attributes of this secondary magnetic field, such as amplitude, orientation, and phase shift, can be measured by the receiver coil. By comparing these attributes with that of the primary field, information about the presence of subsurface conductors, or the subsurface electrical conductivity distribution, can be inferred. Electrical conductivity is the inverse of electrical resistivity, which is measured using resistivity techniques. As such, electrical conductivity measurements are also affected by material texture, porosity, presence of clay minerals, moisture content, and the electrical resistivity of the pore fluid. EM methods tend to require less time, and achieve greater penetration depth with shorter arrays than DC resistivity methods. However, the EM equipment can be more expensive, and the interpretational methods necessary to extract qualitative information can be more complicated than those used with resistivity methods. In general, EM methods are best suited for use when attempting to detect the presence of high-conductivity subsurface targets such as salt-water saturated sediments. Reviews of the instrumentation available for EM induction systems, their applicability for environmental site characterization, and EM interpretational methods are given by Hoekstra and Blohm (1990), Goldstein (1994) and McNeill (1990).

Ground-Penetrating Radar

Ground-penetrating radar (GPR) is a relatively new geophysical tool that has become increasingly popular with the growing need to better understand hydrogeological conditions. GPR methods use electromagnetic energy at frequencies of 10 to 1000 MHz to probe the subsurface. At these frequencies, dielectric properties (that is, the separation, or polarization, of opposite electric charges within a material that has been subjected to an external electric field) dominate the electrical response. GPR systems include an impulse generator, which repeatedly sends a particular voltage and frequency source to a transmitting antenna. A signal propagates from the transmitting antenna through the earth and is reflected, scattered, and attenuated by variation in subsurface dielectric contrasts. Subsequently, the modified signal is recorded by the receiving antenna. In general, GPR performs better in unsaturated coarse or moderately coarse textured soils. GPR performance is often poor in electrically conductive environments such as saturated, clay-rich, or saline soils.

Estimation of the dielectric constant is necessary to infer quantitative hydrogeological information from GPR data. For the high frequency range of interest for GPR methods, the propagation phase velocity (V) in a material with low electrical conductivity can be related to the dielectric constant (κ) as:

$$\kappa \approx \left(\frac{c}{V}\right)^2 \tag{3.1}$$

where c is the propagation velocity of electromagnetic waves in free space, or the speed of light, of 3×10^8 m/sec (Davis and Annan 1989). Equation 3.1 enables the estimation of dielectric information from radar signal propagation velocity estimates, which can be extracted from radar data collected with common midpoint (CMP) or tomographic acquisition geometries. Dielectric constants are affected by material saturation, porosity, material constituency, temperature, and pore fluid composition. The dielectric constant of air and water are 1 and 80, respectively; most dry materials have dielectric constants of 3 to 7; most wet materials have dielectric constants of 15 to 30. Surface radar profiles are commonly used for mapping stratigraphy and structure. Dielectric constant values obtained from radar data in surface CMP or tomographic modes can be used to estimate characteristics such as soil saturation.

Seismic Methods

Seismic reflection, cross-hole transmission, and refraction methods use artificially generated, high-frequency (100 to 5000 Hz) pulses of acoustic energy to probe the subsurface. These disturbances are produced at a point and propagate outward as a series of wavefronts. The passage of the wavefront creates a motion that can be detected by a sensitive geophone and recorded on the surface. According to the theory of elasticity upon which seismic wave propagation is based, compressional, shear, and surface waves are produced by a disturbance, and each wave travels with a different propagation velocity. Because of the relative ease of detecting the compressional (or P-wave) energy, the most common surface seismic acquisition modes of reflection and refraction have been designed to provide information about this wave.

Seismic data are also collected using cross-borehole tomographic acquisition modes. Because of the high resolution offered by seismic tomographic methods, this technique is ideal for detailed stratigraphic and hydraulic characterization of interwell areas.

Reflection

The surface reflection technique records the return of reflected compressional waves from boundaries where acoustic contrasts exist. Seismic reflection data are usually collected as common-shot or common-receiver gathers, which are sorted during processing into common-midpoint gathers. These gathers of traces represent reflections from a subsurface location (the midpoint) that has been sampled by several source-receiver pairs. Due to the lack of well-defined velocity contrasts in unconsolidated and unsaturated materials, seismic reflection data acquisition is often difficult in the vadose zone. Computer-based processing of seismic reflection data generally produces a wiggle-trace profile that resembles a geologic cross section. In addition to obtaining structural and stratigraphic information about the subsurface from the wiggle traces, information about seismic velocity is available through seismic data processing (Yilmaz 1987).

Seismic compressional wave velocity is affected by porosity, permeability, pore fluid type, depth of burial, consolidation, and temperature. However, unique relations between seismic P-wave velocities and lithology generally do not exist. Some generalities can be made regarding the relationship between seismic velocities and lithology (Burger 1992), including:

- Unsaturated sediments have lower velocities than saturated sediments
- Unconsolidated sediments have lower velocities than consolidated materials
- Velocities are very similar in saturated, unconsolidated sediments
- Weathered rocks have lower seismic velocities than unweathered rocks of the same type

Refraction

With refraction methods, the incident ray is refracted along the target boundary before returning to the surface. The refracted energy arrival times are displayed as a function of distance from the source. The arrival times and distances can be used to directly obtain velocity information. Refraction techniques are useful when there are only a few shallow (less than 50 m in depth) targets of interest, or for identifying gross lateral velocity variations or changes in interface dip (Lankston 1990). Seismic refraction methods yield a much lower resolution than seismic reflection and crosshole methods. Because refraction methods are cheap and acquisition is sometimes more successful in unsaturated and unconsolidated environments, these methods are often used to detect the depth to the water table and top of bedrock, the gross velocity structure, and the location of significant faults. A review of the refraction method is given by Lankston (1990).

Gravitational Methods

Measurements of the changes in gravitational acceleration can be used to obtain information about subsurface density variations. As density is a bulk property of rocks and tends to be consistent throughout a geological formation, gravity methods are used to identify gross features based on density variations. Because of the lower resolution afforded by this method, it is not commonly used for detailed site characterization. It does, however, provide a cheap way to detect some targets, such as the interface between sedimentary overburden and bedrock, or the locations of significant faults. Gravity methods have also been used to detect sinkholes and other subsurface voids, and to establish landfill boundaries. The common measuring device is a gravimeter, which is a portable and easy-to-use instrument. A spring balance inside the gravimeter measures differences in the weight of a small internal object from location to location, which are attributed to changes in the acceleration of gravity resulting from lateral variations in subsurface density. Measurements can be collected at a regional or local scale depending on the station spacing. The station spacing is usually less than half of the depth of interest. Reviews of the gravity technique and applications to environmental studies are given by Hinze (1990) and Butler (1991).

Magnetic Methods

Magnetic methods are used to detect the direction, gradient, and intensity of the earth's magnetic field. The intensity of the magnetic field at the earth's surface is a function of the location of the observation point in the primary earth magnetic field, as well as local or regional concentrations of magnetic material. Magnetometers are used to measure the total geomagnetic field intensity or relative values of the vertical field intensity; magnetic gradiometers measure the horizontal and vertical gradient of this magnetic field. After correcting for the effects of the earth's natural magnetic field, magnetic data can be presented as total-intensity, relative-intensity, and vertical-gradient-anomaly profiles or contour maps. Magnetic solutions are non-unique, and interpretation generally involves forward modeling or mapping of the anomalies and correlating the results with additional geologic information. Magnetic methods are generally used to identify gross features at a resolution similar to that of seismic refraction and gravitational methods. In near-surface studies, magnetic data are commonly used to map the depth to the basement, providing provided that the basement rock contains sufficient magnetic minerals. Sedimentary materials, which are most common in aquifers, are essentially non-magnetic. Magnetic methods with much finer station spacing and higher lateral resolution (a few meters) are now among the most commonly used geophysical methods for site investigation, because of their ability to locate shallow metal objects, such as drums and abandoned drill-hole casings. A review of magnetic methods as applied to environmental problems is given by Hinze (1990).

Borehole Geophysical Methods

Borehole geophysics refers to the process of recording and analyzing physical property measurements made in holes or wells. One-dimensional borehole data can be correlated to extrapolate the information between the wellbores, and can also be used to calibrate surface geophysical data. The volume of investigation for downhole logs is related to log type, source-detector spacing, and subsurface material, and thus varies with the well-site conditions and the logging parameters employed. Interpretation of the recorded log data often involves comparing several different logs displayed side-by-side, or by cross-plotting data from one log against data from other logs, core analysis, or tests. For hydrocarbon exploration, the decision to test and complete a well is largely based on geophysical log information, and, as a result, most of the interpretation guidelines for borehole geophysics have been developed for borehole and rock environments encountered in petroleum exploration. An excellent reference for borehole geophysics applied to groundwater investigations is given by Keys (1989), which is the reference for the following logging information unless otherwise cited. Another one-dimensional sampling tool that is useful for vadose zone characterization is the cone penetrometer (CPT). A complete overview of the CPT method and applications is given by Lunne *et al.* (1997). Both borehole logging devices and CPT tools will be discussed, briefly, below.

Borehole Logging Tools

Borehole log measurements are made by lowering a sonde into the borehole on the end of an electric cable. The sonde is a probe, 2.5 to 10 cm in diameter and 0.6 to 9 m in length, which encloses sources, sensors, and the electronics necessary for transmitting and recording signals. Measurements made in the borehole are recorded on the surface in digital form, or in analog form on chart paper. The following discussion focuses on the underlying physical principles of only those logging methods that are either currently used in vadose zone applications or that have potential for aiding these investigations, including caliper, electric, nuclear, acoustic logging and borehole imaging tools. Table 3-10 lists the applications, limitations, and borehole conditions required for each method, as well as the frequency of use of the method for vadose zone applications. Although some borehole techniques have potential for aiding vadose zone studies, they are nevertheless not commonly used, due to the sophistication, and, thus, higher cost of using the method. Table 3-11 summarizes the applications of some of these borehole geophysical methods. The table refers to information available from individual logs only. More information about fault displacement can be obtained by correlating several logs or by integrating the well logs with surface geophysical methods or other data.

Caliper Logging

Caliper logs are mechanical or acoustic tools that measure the diameter of the borehole. The mechanical caliper tool includes between one and six caliper probes connected to a single arm. The probes are pressed against the borehole wall by spring pressure. As the tool is pulled up the borehole, the mechanical caliper probes move in response to changes in borehole diameter, and the acoustic calipers measure the reflection transit time of an acoustic signal from the borehole wall. Changes in the diameter of the borehole affect the response of all geophysical tools, and therefore, a caliper log is generally collected in conjunction with all logging suites to aid in interpretation and correction of the other logs. Changes in wellbore diameter can be related to casing design as well as to fracturing or caving along the borehole wall, which can sometimes be indicative of the lithology.

Electric Logging

Electric logs measure potential differences resulting from the flow of electric current in and adjacent to the well. There are many different types of electric logs including single-point, resistivity, dipmeter, and spontaneous potential. These logs are often used in groundwater applications to investigate subsurface properties such as lithology, water quality, fracture locations, and porosity. However, because these logs must be run in saturated boreholes prior to casing, their application is limited for vadose zone investigations. An electric log that can be run in unsaturated and uncased borehole environments is the induction log. This tool has two coils: one for transmitting and alternating current into the surrounding formation and a second for receiving the returned signal. The transmitted alternating current induces eddy-current flow in surrounding conductive materials. These eddy currents set up secondary magnetic fields that induce a voltage in the receiving coil; the magnitude of the received current is proportional to the electrical conductivity of the surrounding material.

Nuclear Logging

Nuclear logging entails the detection of unstable isotopes near the borehole. The considerable advantage of nuclear logs over electric logs is that they can be run after casing has been installed. As isotopes decay, they emit radiation, usually from the nucleus. Of the radiation emitted, gamma photons and neutrons are often used in borehole applications because of their ability to penetrate dense material such as rock and casing. Borehole geophysical tools that measure radioactivity of nearby formations may be classified as either those that detect natural gamma radiation, those that employ controlled gamma rays to induce radiation, or those that use neutron sources to induce nuclear processes. The radioactivity is measured as electronic pulses, and the quantity and amplitude of the pulses yield information about the surrounding formation. Logging tools that use artificial radioisotopes as sources are regulated by governmental agencies and require a license for use.

The gamma log uses a scintillation detector to measure the amount of naturally occurring gamma radiation of the material penetrated by the borehole within a selected energy range. The three most common naturally occurring radioactive materials that affect the gamma log are potassium-40, uranium-238, and thorium-232. The utility of this log lies in the fact that these isotopes are generally more abundant in shales and clays, that the isotopes are less common in sands and calcareous materials, and that reliable measurements can be made above the water table. The gamma log reveals the characteristically high gamma-log count rate associated with silts and clays. In addition to the count rate that is measured with the gamma log, the gamma-spectrometry method records the amplitude of the pulses over a wide energy range. Analysis of this energy yields more diagnostic information on lithology, and also permits estimation of the type and quantity of radioisotopes that may be contaminating the groundwater. Gamma-gamma or density logs record gamma radiation that originates from an artificial gamma source in the well and that is backscattered by the borehole and surrounding material. The count rate of the backscattered gamma rays can be related to the electron density of the material, which is in turn proportional to the bulk density of the material. If the fluid and grain densities are known, the bulk density measured with the gamma-gamma log can be used to calculate porosity. Because moisture content affects the bulk density of materials, gamma-gamma logs can also be used to record changes in moisture above the water table.

Neutron tools consist of an artificial low-energy wellbore neutron source and one or two neutron detectors. The neutrons emitted by the source lose energy upon collision with other elements in the vicinity of the borehole. Because hydrogen has a mass similar to the neutron, it is the element that is most effective at slowing the neutrons. The quantity of slowed neutrons is thus interpreted to be proportional to the quantity of hydrogen present, which is in turn interpreted to be proportional to the moisture content or saturated porosity. For most materials, resistivity and neutron logs have a similar log character because of the relationship between saturated porosity and pore fluid resistivity.

Acoustic Logging

Acoustic (sonic or velocity) tools transmit an acoustic pulse through the fluid and material near the borehole, from a source to the detector. These tools emit an acoustic source at frequencies of 10 to 35 kHz, which creates compressional or P-waves. As the waves travel, some of the energy is refracted back to two receivers located on the sonde. The difference in travel time between the receivers is used to calculate interval velocity, which is recorded as a function of depth in the wellbore. Acoustic waveform logging entails the recording and interpretation of the entire waveform rather than just the travel time. Because conventional acoustic logs require saturated borehole conditions for signal transmission, acoustic logs are not commonly used in the vadose zone. In some circumstances, use of a water-filled casing with an end-cap may allow sufficient coupling for conventional acoustic logs to be used in the vadose zone. Alternatively, dry-hole acoustic sondes exist that can be used in uncased and unsaturated boreholes. Dry-hole sondes can not be used while moving the tool in the borehole, and because there is no contacting fluid with this tool, the signal-to-noise ratio is typically smaller than that for conventional acoustic tools (Hearst and Nelson, 1985).

Borehole Imagers

Borehole cameras or scanners provide very high-resolution video or digital images of the borehole wall. An unfolded borehole image can be thought of as a cylinder that has been opened along a side and flattened to provide a 360-degree picture of the borehole wall. Planar horizontal features that intersect the well appear horizontally, and planar dipping features that intersect the well appear as sinusoids with the lowest point of the curve in the direction of dip.

FIELD MEASUREMENTS OF WATER POTENTIAL

One of the challenging problems of vadose zone investigations is the determination of water potential in unsaturated-saturated heterogeneous soils and fractured rocks. The water potential is a characteristic of the energy status of water in the subsurface. Because the soil matric potential in unsaturated soils and sediments varies in significant range from 0 to -15 bar, there is no single technique to measure over the entire range of the matric potential. The detailed description of field and laboratory methods of measuring the water potential is given in a monograph by Klute (1986). Gee and Ward (1999) presented a thorough review of innovations in two-phase measurements of soil hydraulic properties. Figure 3-24 shows schematically the water pressure ranges for laboratory and field measurements using tensiometers, heat dissipation probes, and thermocouple psychrometers.

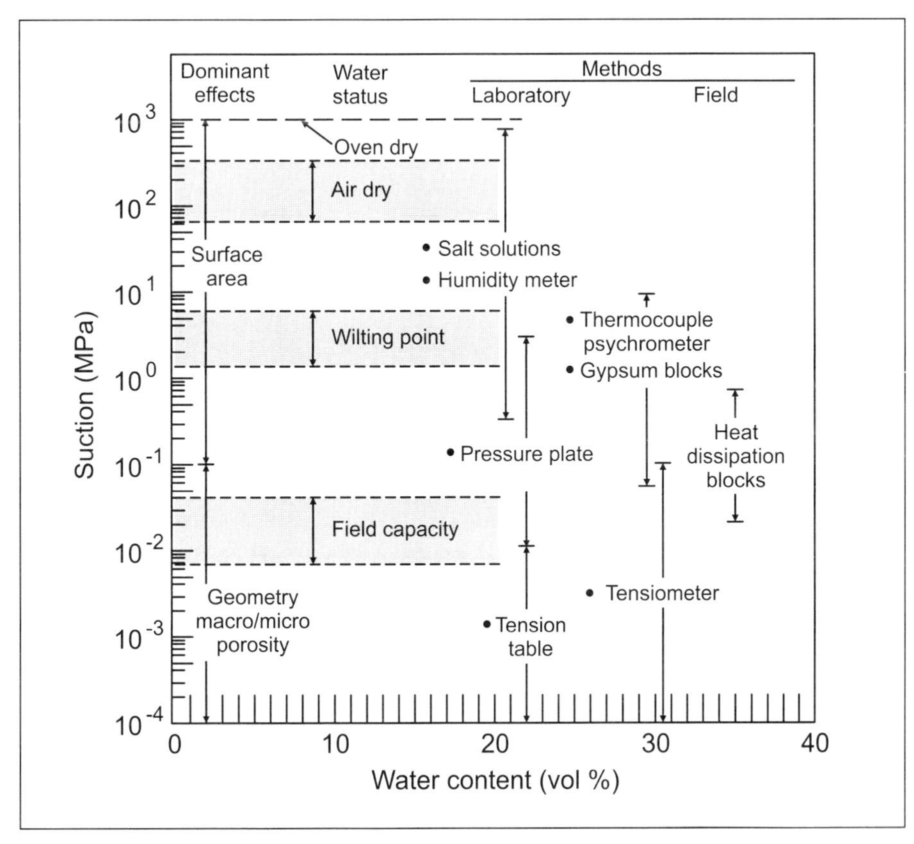

Figure 3-24. Schematic showing the operational ranges of field and laboratory methods used in monitoring the matric suction in the vadose zone for soil water physical processes (Gee and Ward 1999)

General Design and Applications

Tensiometers have been used from as early as the 1920s for a variety of specific applications (Kornev 1924; Richards and Gardner 1936), including water-pressure measurements at hazardous-waste sites (Healy *et al.* 1983; McMahon *et al.* 1985; Ripp and Villaume 1985; Ryan *et al.* 1991); recharge areas (Lichtler *et al.* 1980; Sophocleous and Perry 1985); irrigation land (Richards *et al.* 1973); and for civil-engineering projects (McKim *et al.* 1980; Richards *et al.* 1938). Tensiometers have been used under both laboratory and field conditions in variably saturated porous media to assess the soil-water potential, which is then used to determine the hydraulic gradient, direction of water flow, and the water flux in soils, as well as to indirectly estimate the soil-water content using water-retention curves (Richards *et al.* 1938; McKim *et al.* 1980; Sposito 1981; Stannard 1986; ASTM 1998; Kutilek and Nielsen 1994).

Although tensiometer design has undergone various changes, its basic components have remained unchanged. A tensiometer is comprised of a porous tip (usually a ceramic or metal cup) connected to a water-filled tube and a pressure sensor. It is advisable to use deaerated water to fill in tensiometer tubes. Pressure sensors commonly used in tensiometers include vacuum gauges, mercury manometers, and pressure transducers (Marthaler *et al.* 1983). Pressure transducers can be connected to tensiometers either remotely (Klute and Peters 1962; Healy *et al.* 1986) or directly (Bianchi 1962; Watson 1967; Hubbell and Sisson 1998).

An expanded cross-sectional view of the interface between a tensiometer porous tip and soil is shown in Figure 3-25. The porous cup is buried in the soil and transmits the soil-water pressure to a pressure sensor. During normal operation, the saturated pores of the cup prevent bulk movement of soil gas into the cup. Water held by the soil particles is under tension; absolute pressure of the soil water is less than atmospheric. This pressure is transmitted through the saturated pores of the cup to the water inside the cup. Conventional fluid statics relates the pressure in the cup to the reading obtained at the manometer, vacuum gage, or pressure transducer, which are shown in Figure 3-26.

The tensiometers with water-filled tubes cannot be used at depths greater than approximately 5 to 7 m. This restriction arises because, in addition to the soil suction, a water column creates an extra negative pressure in the tensiometer, leading to the degassing of water and the accumulation of water vapor and air in the connecting tube extending to the surface. When the water pressure drops to -30 to -40 kPa, air nucleates in the tensiometer water and diffuses from the atmosphere and unsaturated soils through the connecting tubes and the ceramic tip, thus creating air bubbles. It is virtually impossible to avoid the formation of air bubbles even if deaerated water is used to fill the tensiometer. If the inner diameter of the tube is less than 4 to 5 mm, air may stick to the tube walls, gradually accumulate, and create air plugs along the water connecting tube. In tubes with larger diameters, air can move up and accumulate at the top of the tube. As the air volume increases, it forces

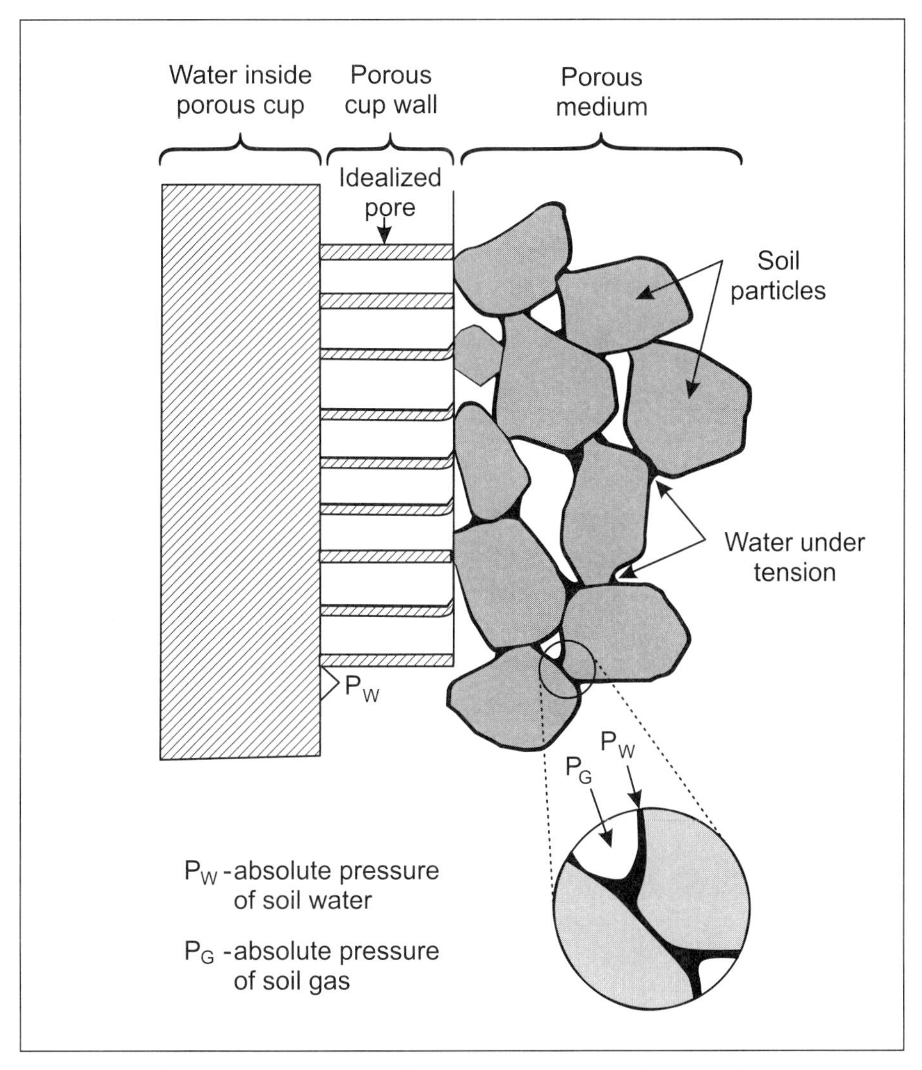

Figure 3-25. Enlarged cross section of porous cup-porous medium interface (ASTM, 1998)

water to discharge from the tensiometer into the surrounding soils, causing the water level in the water-filled tube to drop.

The porous tip of the tensiometer can have a cylindrical shape (Everett *et al.* 1984a), a conical shape (Dzekunov *et al.* 1987), or any other shape (for example, rectangular, ellipsoidal, circular or conical)

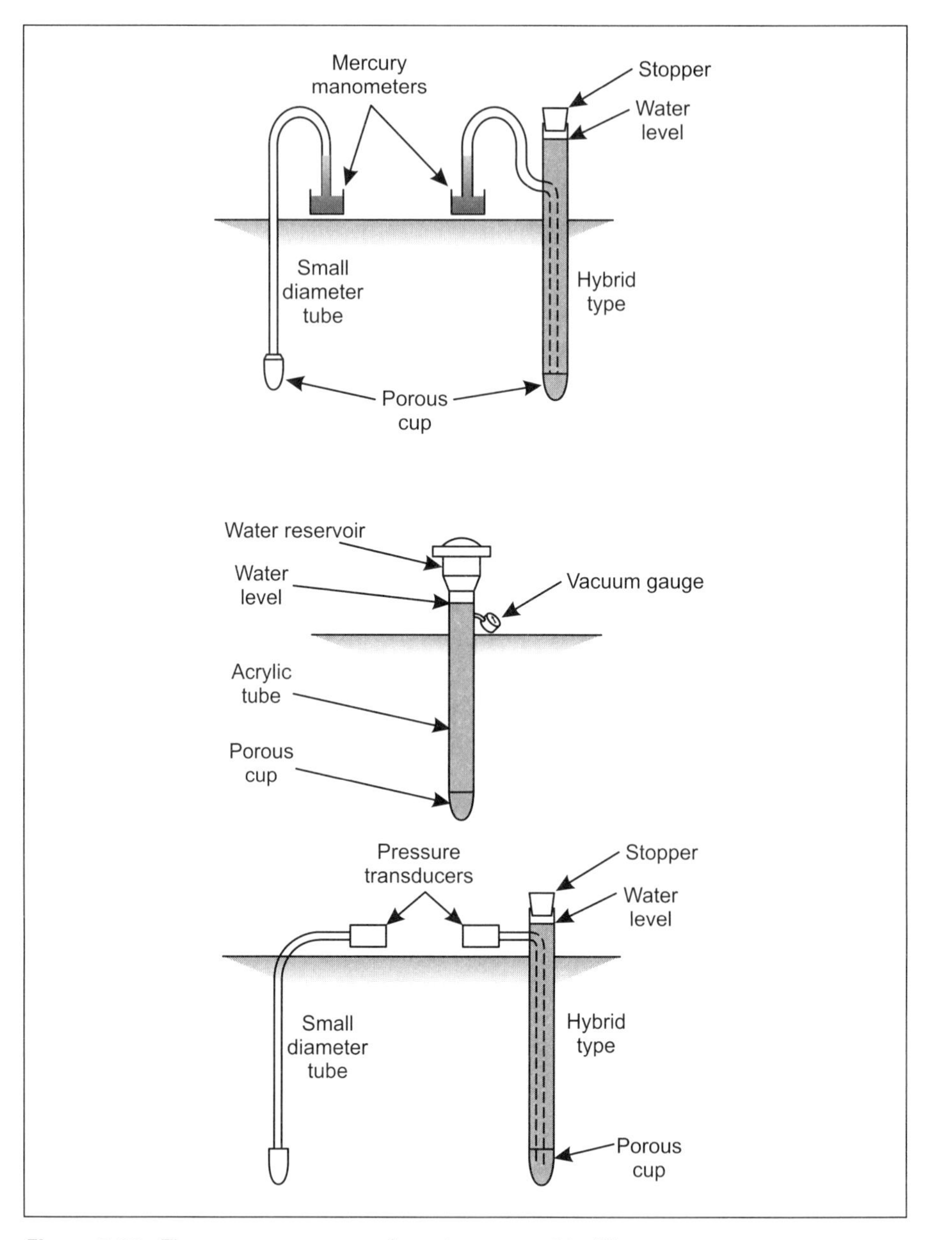

Figure 3-26. Three common types of tensiometers with different pressure sensors: (a) Manometer; (b) Vacuum Gage; and (c) Pressure Transducer (ASTM D 3404-91)

that enhances contact with the soil in a particular geological setting. For example, the upper portion of the tip may be cylindrical and the lower portion conical. The cylindrical portion of the porous tip maximizes surface contact with soil, whereas the conical shape may penetrate some soils more easily than the cylindrical shape, and thus create a better contact with soils. The cylindrical shape has been found advantageous in soft material, such as soil slurry, silica flour, or fine sand. The porous tip may be made from any water-permeable material that does not plug easily. Both ceramic material (for example, as fabricated by SoilMoisture Inc., of Santa Barbara, California) and stainless steel porous tubes (such as those manufactured by Soil Measurement Systems Inc., of Tucson, Arizona) can be used to make the tensiometer. Stainless construction of the porous tip is particularly useful when soils containing volatile organic compounds are to be investigated.

A better contact between the porous tip and surrounding soils can be achieved by inserting the tensiometer tip into a flexible, porous bag filled with soils taken during borehole drilling. Note that when the tensiometers are used to measure water pressure in contaminated soils, both the water and air may be mixed with impurities such as volatile organic compounds (VOCs), which may change the air pressure above the liquid level in the cells of the tensiometer.

Tensiometers can also be used for field measurements of water pressure in unfrozen soils at air temperatures below zero degree centigrade. In this application, the tensiometer cup and connecting tubes are filled in with an antifreeze solution (Gorden and Veneman 1995).

Problems with tensiometers designed for measuring water in the vadose zone include uncontrolled water-level changes when water from the soil enters the tensiometer, or when air enters the tensiometer and water discharges into the soils during soil drying.

Air-Free Tensiometers

In order to remove air accumulated in the tensiometer, Miller and Salehzadeh (1993) developed an air stripper. The stripper consists of an air-permeable tube that is inserted between the tensiometer fluid and a wet vacuum (containing some water droplets). When the vacuum is applied, air molecules diffuse from the tensiometer fluid into the stripper tube and are vented outside the tensiometer. By maintaining the wet vacuum, the diffusion of water droplets from the tensiometer into the air stripper is practically eliminated. Ward *et al.* (1998) successfully coupled this method with time domain reflectometry (TDR) to obtain direct, continuous measurements of the matric potential in undisturbed cores, and then use the simultaneously determined moisture content and matrix potential to determine water retention curves.

Air-Pocket Tensiometer

Villa Nova *et al.* (1989) designed a tensiometer with an air pocket at the top of a water-filled tube connected to a porous tip. This tensiometer included only one tube, and the water level was required to be above the ground surface. Consequently, it was limited to measurement depths of approximately 5 m. If the water level in the tube were below the surface, it would vary in a manner that cannot be observed directly or controlled at the surface (Stephens 1996). Using the calculations of the volume of air above the water level based on Boyle's Law, Tokunaga (1992) determined the water level in a water-air access tube connected to a ceramic cup. In his apparatus, the same tube is used for both water and air feed.

Deep Tensiometers

Hubbell and Sisson (1998) developed an advanced tensiometer in which the porous tip is connected to a 1-in. polyvinyl chloride (PVC) tube through a central hole in the rubber stopper. The porous tip of the tensiometer is filled with water from the surface, and a pressure transducer is then lowered through the PVC access pipe and inserted into the rubber stopper hole. Thus, a water column between the tensiometer and the land surface was eliminated. However, this tensiometer has no control over the presence of water in the tensiometer. In order to either refill the porous cup with water or to check the tensiometer calibration, the pressure transducer (installed at the top of the porous cup) must be pulled to the surface. The total number of tensiometers installed in a single well is limited to the total cross-section of access pipes in this borehole. These tensiometers were tested under field conditions and showed good performance at several sites (Hubbell and Sisson 1998).

Morrison and Szecsody (1987) developed a solenoid transducer tensiometer that is installed at depths greater than 10 m and that automatically recirculates fluid at a given frequency (up to 42 days). However, like Hubbel and Sisson's model, this design has no control of the water level in the tensiometer. Therefore, the frequency of water replenishing is arbitrary.

Faybishenko (1999b) developed a two-cell tensiometer capable of (a) measuring water pressure in soils and rocks in the vadose zone and groundwater at any depth by measurement of a single parameter, namely, the pressure of an isolated volume of air in the tensiometer; and (b) maintaining a constant water level above the porous cup. Figure 3-27 depicts the design of the tensiometer. If the lower-cell water level drops below the bottom of the connector, water from the upper cell flows into the lower cell and reestablishes the water level in the lower cell. Thus, the lower cell maintains an essentially constant water level, above which a small air volume is isolated. Because the air volume in the lower cell is isolated, and because the lower-cell water-level height is kept essentially constant and is in equilibrium with the surrounding liquid, the pressure of the isolated lower air volume is directly proportional to the water pressure of the soil surrounding the porous tip of the tensiometer. Air pressure measured in both cells is stored remotely using a data acquisition system. The difference in pressure in both cells is used to determine the presence of water in the tensiometer. The tensiometer can be used in both drying and wetting conditions in the vadose zone and below the water table. Several tensiometers can be installed in a borehole at different depths to measure water pressure of soils or rocks above and below the groundwater table. The tensiometer can be used in both vertical and slanted boreholes with practically no limitations in depth. The tensiometer can also be designed as a suction lysimeter to collect water samples from surrounding soils. For this purpose, the water access tube may be extended to the bottom of the porous tip. Alternatively, a fourth tube can be used for this function.

Heat-Dissipation Method

The heat-dissipation probe is a small, porous, ceramic block containing a heater and a thermocouple embedded in a ceramic block (Reece 1996). The ceramic block is inserted in soils, and when a pulse of heat is applied to it, the thermal response is determined by measuring the volt-

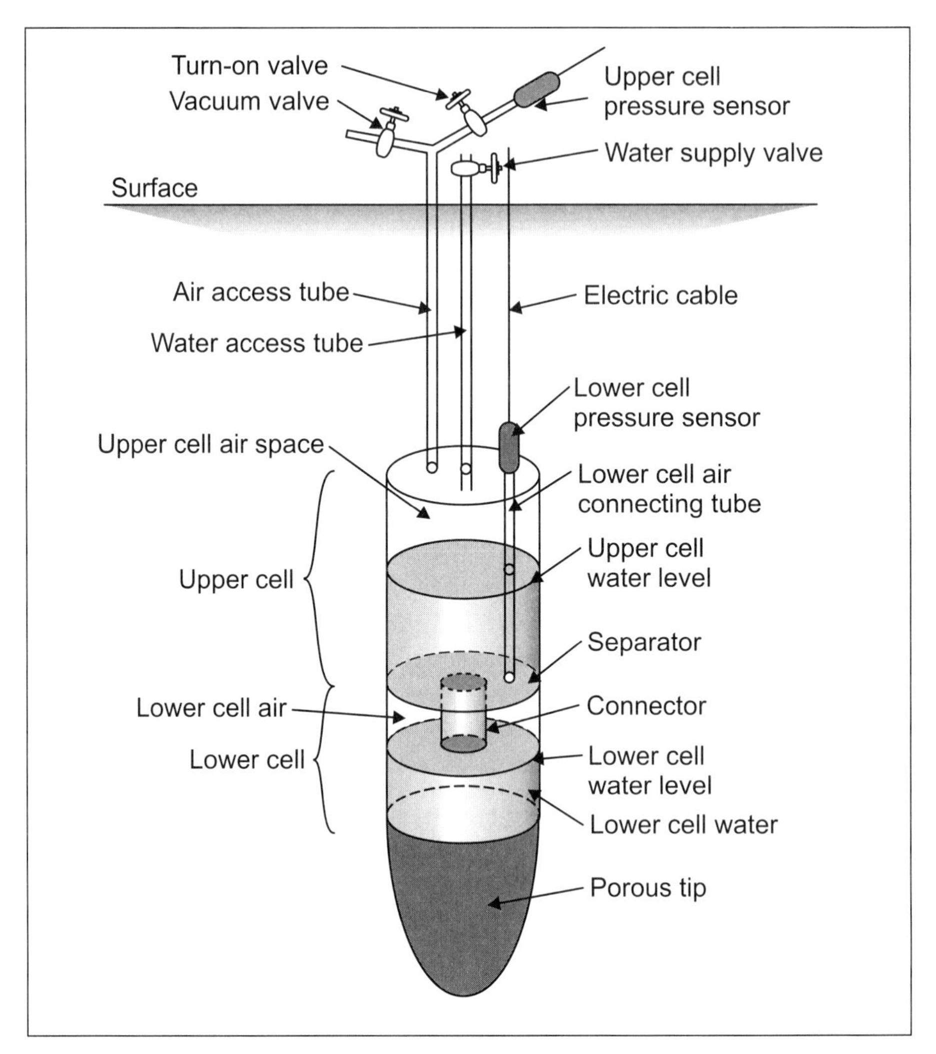

Figure 3-27. Schematic of the two-cell tensiometer (Faybishenko, 1999b).

age. This response is caused by heat dissipation, which is related to the water content of the ceramic block. It is assumed that the matric potentials of the ceramic block and the surrounding soils are in equilibrium. Therefore, if the heat dissipation probe is calibrated in terms of the matric potential, the soil matric potential can be determined. The practical range of the matric potential for the heat dissipation method is from the ceramic air entry value of about -0.1 bar to approximately -15 bar (Phene *et al.*

1992). A nearly linear relationship between the voltage and the matric potential exists in the range of -0.2 to -3.5 bar (Kutilek and Nielsen 1994). An advantage of the heat-dissipation probe over tensiometry is that water filling is not required, and therefore the system can operate in unsaturated soils with minimal maintenance. Heat-dissipation probes can also be connected to a data acquisition system. However, the heat-dissipation method does not work properly near the full water saturation and below the water table (if there is a need to determine the pressure distribution below the water table) because the analysis of the measurements is based on the assumption of a constant power dissipation (for all values of soil moisture content) at the heating element. The variations in the applied power may create temperature fluctuations, causing errors in the estimated matric potential (Bilskie 1999).

Granular Matrix Sensors

Electrical-resistance sensors (such as gypsum, nylon, or fiberglass matrix) for measuring matric suction (Eldredge *et al.* 1993, Shock *et al.* 1998) have not been very effective because of problems caused by salt effects, hysteresis, and degradation of the sensor material, particularly gypsum blocks (Campbell and Gee 1986). A new model of electrical-resistance sensors (Watermark™ Irrometer Co., Riverside, CA) utilizes a gypsum block imbedded in a granular fill material. These new sensors have been used for agricultural applications. They were calibrated against neutron probes and tensiometers (Eldredge *et al.* 1993). The sensors were used in alkaline soils and were practically insensitive to low salt concentrations. However, they are temperature-sensitive, so that temperature corrections are required. The matric potential range of these sensors is from the air-entry value of granular material to about -10 bar or lower.

Filter-Paper Technique

The filter-paper method has been known for over 60 years, but its practical applications are limited. In this method, a small filter-paper disk is placed in contact with soil and is equilibrated for several days. The filter paper is then removed from the soil, and the water content of the filter paper is determined with a conventional gravimetric method. Using the known water-retention curve of the filter paper, the soil matric potential is estimated. Advantages of this method include its simplicity, low cost, and its wide range of measurements (from near saturation to oven dry). The disadvantages are that the paper has to be removed from the soils and cannot be replaced at exactly the same location, and that a long equilibration time (up to a week or more) is needed for drier soils. Additionally, Deka *et al.* (1995), who compared the filter-paper method with tensiometry and psychrometry data, determined that for the dry materials, the filter-paper method overestimates the matric suction, because longer equilibrium times are required for drier samples.

Electro-Optical Methods

An electro-optical method can be used for measuring the water content of a filter paper (or any thin porous material, such as cellulose, nylon, or hydrophilic plastic filter paper) in a continuous fashion. Cary *et al.* (1991) developed a method of using a small light source, such as an infrared emitter, to beam light through the filter. The light is captured by a small photo-detector that measures output voltage as a characteristic of the attenuation of the light beam. The photo-detector voltage is a function of the water content of the filter. It is assumed that a unique, nonhysteresis relationship exists between the water content and matric suction of the filter. For dry soils, the method is insensitive to small changes in water content.

Thermocouple Psychrometry (TCP)

Comprehensive discussions on the use of TCP to measure soil-water potential and to estimate soil matric suction is provided by Rawlins and Campbell (1986) and Kutilek and Nielsen (1994). Thermocouple psychrometers infer the water potential of the soil liquid phase from measurements in the vapor phase. Use of thermocouple psychrometry for matric suction measurements is advantageous primarily in dry soils (-2 bar to -6 bar and lower). With TCPs, no direct hydraulic connection is required and only microscopic quantities of vapor are exchanged between soil and sensor. The main difficulty in using psychrometers arises because the range of the soil gas relative humidity is relatively narrow (from 0.99 to 1) in comparison to the wide range of the moisture content and water pressure (from 0 to –15 bar) for soils, which complicates the calibration procedure of psychrometers. Andraski (1997) showed that although frequent calibrations of probes are required, TCPs could be used for at least five years.

Water Activity Meters

Chilled-mirror psychrometers (commercially available as water activity meters) have been used to measure total and matric soil suctions (Gee *et al.* 1992). Typically, these measurements are made under laboratory conditions on disturbed soil samples collected in the field with minimal water loss. The water activity meter can be used to estimate matric potential in soils with a low–salinity pore solution (less than 1 dS m^{-1}). Measurements in dry soils can be made quickly (in less than 5 minutes). A disadvantage of the water activity meter is that it becomes insensitive to changes in the matric potential (higher than -0.5 bar) in wet soils.

TDR-Tensiometer

An important innovation is the combination of a tensiometer and a TDR probe to obtain simultaneous measurements of the matric potential and moisture content of soils, which can be used to determine the water-retention curve under field conditions. In this method, tensiometer tips constructed from hollow metal tubes serve as the TDR wave guides. The TDR-tensiometer systems recently reported by Baumgartner *et al.* (1994) and Whalley *et al.* (1994) have two major differences. The Baumgartner *et al.* (1994) design shown in Figure 3-28 is based on a two-rod TDR system, which requires a balancing transformer. The tensiometer tubes are made of stainless steel tubes with porous steel cups welded or threaded to the distal end. In contrast, the Whalley *et al.* (1994) design is based on a three-rod TDR system with no balancing transformer. The tubes are made of aluminum, with porous ceramic cups glued to the distal end, and are filled with deaerated water. The tensiometer readings are taken using either a needle-tensiometer or a pressure transducer attached to the proximal end of each tube.

FIELD MEASUREMENTS OF WATER CONTENT IN UNSATURATED SOILS

One of the main goals of site characterization is to conduct rapid, reliable, and cost-effective monitoring of soil water content of soils needed

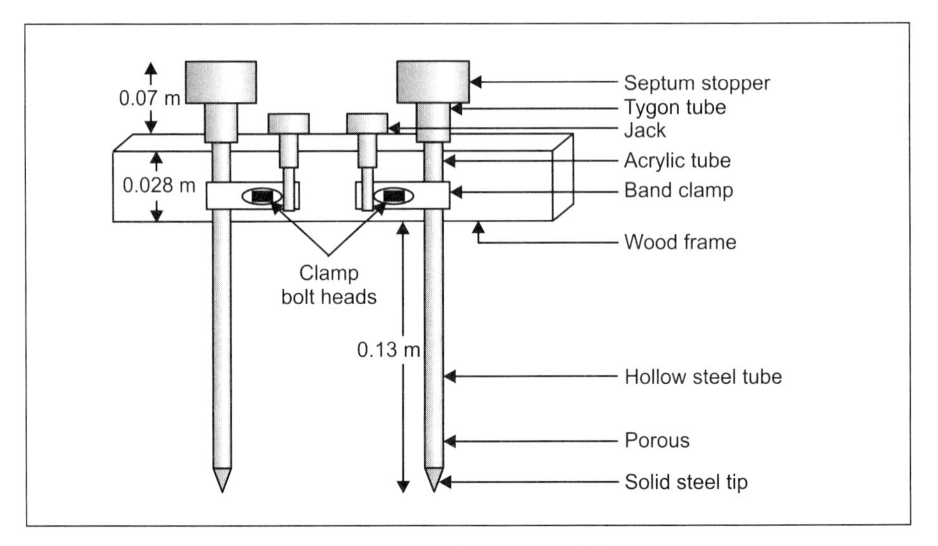

Figure 3-28. Schematic of a combination Tensiometer-TDR system (Baumgartner *et al.* 1994)

in practically every aspect of site characterization. The moisture content, θ, along with the matric potential, ψ, are main parameters characterizing the spatial and temporal variations of water flow in the unsaturated-saturated soils and sediments of the vadose zone. These data are also used to estimate the main hydraulic parameters characterizing water flow, such as: the unsaturated hydraulic conductivity as a function of the water content $K(\theta)$; or the matric head, $K(\psi_m)$; or the water-retention function, $\theta(\psi_m)$, which is the relationship between the water content and matric potential.

Determination of the soil and sediment water content in soils and sediments can be conducted using direct and indirect methods. Direct methods involve removing water from a soil sample by evaporation, leaching, or chemical reactions, and measuring (or inferring) the amount of water removed (Gardner 1958). A key problem in water-content determination using direct methods is the difficulty in defining a dry soil, that is, determining when to stop water removal. This problem does not have a unique answer, because different soils are comprised of various amounts of colloidal and noncolloidal mineral particles, organic matter, volatile compounds, water, and dissolved chemicals. In particular, drying at temperatures in excess of 70°C can cause the decomposition of organic matter. On the other hand, soils with colloidal minerals may require temperatures between 165 and 175°C, while the standard range is 100 to 110°C. Examples of conventional direct methods of determining moisture content are gravimetry with oven drying, gravimetry with microwave oven drying, and alcohol leaching. Direct water-content measurements are destructive. Soil samples are taken at different places during borehole drilling.

Indirect methods of determining soil moisture content are based on measurements of some soil properties affected by water content, and are discussed next.

Neutron Logging

The neutron logging method is used extensively in soil sciences and civil and environmental engineering for determination of water content of soils, sediments, and rocks. The principle of neutron logging is based on the use of a radioactive source, such as americium-beryllium, that emits fast neutrons into the surrounding formation. The radioactive source is inserted in a borehole access tube, and neutron thermalization caused by hydrogen atoms in soil water and hydrocarbons is measured. The neutron logging procedure is approved by ASTM Standard D3017-96, and is described in detail in many texts and papers (such as Klute 1986; Jury *et al.* 1991; Kutilek and Nielsen 1994). A schematic of the neutron-logging principle is presented in Figure 3-29.

Neutron logging is a rapid and nondestructive method of *in situ* measurements in boreholes. It can be used for long-term, repetitive monitoring of the water balance and infiltration in the vadose zone.

The use of a neutron probe in environmental and environmental remediation applications usually requires a frequent recalibration of the neutron probe. This recalibration is necessary because of changes in the hydrogen concentration, especially in the presence of hydrocarbon contaminants. Neutron-probe measurements within the top 15 to 20 cm of the soil surface are not accurate because of neutron loss from the probe and the topsoil layer into the atmosphere.

The neutron logger provides measurements within the distance of 16 cm in wet soils to about 70 cm in dry soils (Kutilek and Nielsen 1994). Because the neutron logger is more sensitive to water near the

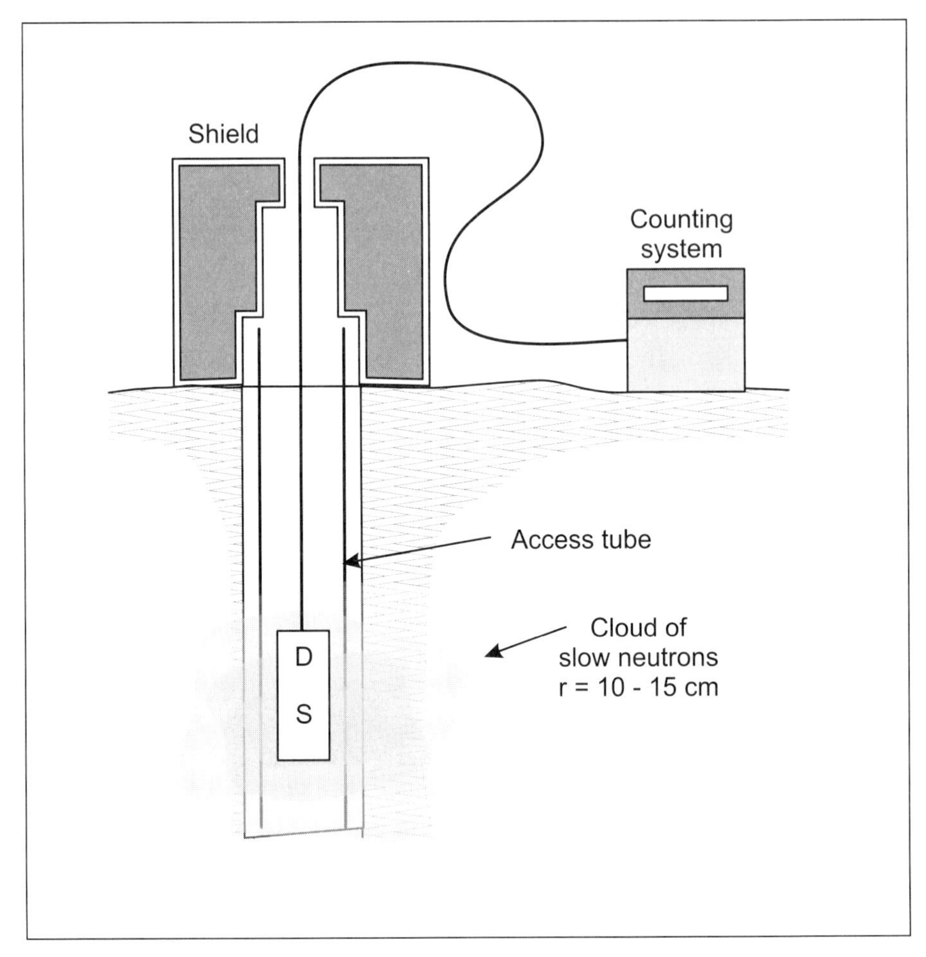

Figure 3-29. Schematic of a neutron probe used for measurement of soil water content in a soil profile (Kutilek and Nielsen 1994)

borehole than to water farther away, and because the soil volume involved in measurements varies as moisture content changes, the water content that is inferred from neutron logging is not a volume-averaged value. The low spatial resolution of neutron logging does not allow precise detection of changes in the moisture content caused by discontinuities in fractured rocks and heterogeneous soils. In addition, chemical composition, density, and heterogeneity of the formation may affect neutron-logging measurements.

The case study "Near Surface Infiltration Monitoring Using Neutron Moisture Logging, Yucca Mountain, Nevada," by Alan L. Flint and Lorraine E. Flint, U.S. Geological Survey, describes a long-term monitoring program at Yucca Mountain. *See page 457.*

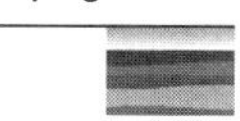

The calibration of neutron logging is needed to determine the absolute moisture content of soils. The calibration under field conditions is usually based on neutron logging with simultaneous sampling of soils to determine the moisture content by gravimetric method. The calibration under laboratory conditions involves measurements in large containers of soils with regulated moisture content. Laboratory calibration curves may be different from those obtained in the field because the soils in laboratory containers might have different textures than those under natural field conditions. The calibration curves may also depend on the neutron probe "aging" as well as the concentration of elements such as iron, boron, molybdenum, and cadmium. The errors of neutron logging can be reduced by determining the changes in the moisture content rather than the absolute values of the moisture content. The restrictions for using neutron probes require that the user have special training and a license for transport, ownership, and use of a radioactive source.

Time Domain Reflectometry Methods

General Considerations

The time domain reflectometry (TDR) technology has been employed successfully in geotechnical and mining industries for monitoring rock mass deformation and subsidence, and for measuring water level, water pressure, and soil moisture content (Huang and Dowding 1994; O'Connor and Wade 1994). The TDR method was first applied to soil-water investigations by Topp *et al.* (1980). This method is based on the determination of the propagation velocity of an electromagnetic wave along a transmission line (that is, waveguides inserted in the soil). Because of the large difference in dielectric constants of the soil components (Table 3-12), the TDR method is practically insensitive to soil solid-phase composition and texture. However, at contaminated sites, the dielectric constant may vary as the contaminant volume changes.

TABLE 3-12 **Dielectric Constants of Fluids and Solid Materials in the Range of Temperature from 20 to 25°C**
(CRC Handbook of Chemistry and Physics 1993; von-Hippel 1954)

Fluids	Dielectric Constant	Solids	Dielectric Constant
Water	78.5 -80.4	Ice	3.7-4.1
Ethanol	24.3	Fused quartz (Si02)	3.78
Acetone	20.7	Sandy soil (dry)	2.55
Ammonia	16.9	Loamy soil (dry)	2.51
Benzene	2.29	PVC	2.89
C02 (liquid)	1.6	Polyethylene	2.2.5
C02 (gas)	1.001	Teflon	2.1
Air	1.0	Wood	1.9–1.95

Two basic approaches have been used to establish the relationships between the soil-bulk dielectric constant (ε_b) and volumetric soil-water content (θ). The first approach is empirical and is based on fitting observed data of ε_b and θ using a polynomial function. Topp *et al.* (1980) described the observed relationship between ε_b and θ using a third-order polynomial given by:

$$\theta = -5.3\text{x}10^{-2} + 2.29\text{x}10^{-2}\,\varepsilon_b - 5.5\text{x}10^{-4}\,\varepsilon_b^{\,2} + 4.3\text{x}10^{-6}\,\varepsilon_b^{\,3} \quad (3.2)$$

This equation describes the relationship $\varepsilon_b(\theta)$ for the water content $\theta < 0.5$, with an error of approximately 0.013. However, this equation is not valid for soils with large concentrations of organic matter.

The second approach to describing the relationship $\varepsilon_b(\theta)$ is based on assessing the contributions of dielectric constants of soil physical components, namely solid particles, gas, and water (Roth *et al.* 1990). According to Roth *et al.* (1990), a model for the bulk dielectric constant of a three-phase system is given by equation 3.3.

$$\varepsilon_\beta = (\theta\varepsilon_\omega^{\,B} + (1-\phi)\,\varepsilon_\sigma^{\,B} + (\phi-\theta)\,\varepsilon_\alpha^{\,B})^{(1/B)} \quad (3.3)$$

In this equation, ϕ is the soil porosity; B is an exponent that depends on the geometry of the medium in relation to the axial direction of the wave guide (B = 1 for an electric field parallel to soil layering, B = -1 for a perpendicular electrical field, and B = 0.5 for an isotropic two-phase mixed medium); $(1 - \phi)$, θ, and $(\phi - \theta)$ are the volume fractions of solid, liquid, and gaseous phases; and ε_σ, ε_w and ε_a are the dielectric constants of the solid, liquid, and gaseous phases, respectively. Roth *et al.* (1990) showed that for $B = 0.5$, equation 3.3 produces a relationship similar to that of equation 3.2.

A variety of TDR probes were developed for different practical applications. Topp and Davis (1985) and Yokuda and Smith (1993) determined vertical moisture-content profiles using rods of different diameters at fixed intervals, which created impedance discontinuities. Ward *et al.* (1994) used horizontally installed probes. Ferre *et al.* (1994) developed a field multilevel probe consisting of two water-filled PVC access tubes (located in pilot holes) and a pair of target rods that are moved manually.

DeRyck *et al.* (1993) and Redman and DeRyck (1994) developed a multilevel TDR device that includes several horizontal rods attached every 2.5 cm to a PVC access tube at one end, and to a PVC support tube at the other. Each pair of rods forms a waveguide. A spring-loaded contact device moves through the access tube to provide measurements of changes in the apparent dielectric constant, which are related to changes in the moisture content. This approach was also used to determine the changes in concentration of kerosene and tetrachloroethylene (TCE). However, this technique was used only in repacked flow cells.

The assumption that the dielectric properties of soil water are similar to those of bulk water does not hold for sediments with high surface charge (Dirksen and Dasberg 1993; Roth 1990; White *et al.* 1992). Significant deviations from a linear calibration curve occur at low water contents because of the effect of bound water. Herkelrath *et al.* (1991) recommended determining the TDR calibration between ε_b and θ for particular types of soils investigated. The main advantages of the TDR method over other direct and indirect methods of soil-water content measurement are listed below:

- Accuracy may be up to 1 or 2 percent of the volumetric water content, if a proper calibration is made.
- Calibration may not be needed for non-saline soils.
- There are no radiation hazards associated with neutron-probe or gamma-attenuation techniques.

- TDR equipment (manufactured by Tektronix, Inc., and Campbell Scientific, Inc.) is available for telemetric and multiplexing enhancement of remote measurements in different geotechnical and environmental engineering applications (Huang and Dowding 1994).

Mallants *et al.* (1996) showed that the success or failure of TDR for monitoring solute transport depends on the quality of the calibration procedure. These authors recommended calibrating TDR probes using undisturbed soils taken from a location adjacent to the measurement point.

Bilskie (1999) showed that the error of the TDR method for measuring water content is caused by reflection distortion, which is caused either by attenuation in the coaxial connecting cables, or by attenuation of the signal at the probe via free ions in the soil solution. These processes are not taken into account in mathematical models used for evaluation of the reflection waveform. The TDR measurements can be improved by taking into account the resistance of the coaxial cable and connectors.

Remote-Shorting TDR Method

Hook *et al.* (1992) developed a remote-shorting method using a probe consisting of a conducting strip line, which is segmented using positive-intrinsic-negative diodes. Figure 3-30 shows a schematic of a seven-segment, eight-diode probe 1.85 m long that is used to monitor water storage in a hazardous-waste landfill cover at Hanford (Gee and Ward 1999). The diodes activated by the external voltage are used to apply a short circuit across the waveguides. A positive voltage shorts the first diode and opens the second diode. At the same time, a negative voltage reverses the process. It is assumed that the transit time of the electromagnetic (EM) pulse traveling along the segment is a linear function of the moisture content in this segment (Hook *et al.* 1992). Because this method is based on using the differencing technique, it can be used to obtain measurements with long cables and in saline soils.

Measurements at Hanford using cables 50 m long showed that the linear relationship between the water content and transit time holds for sands, however, it does not hold for silt loams, for which it becomes nonlinear (Gee and Ward 1999).

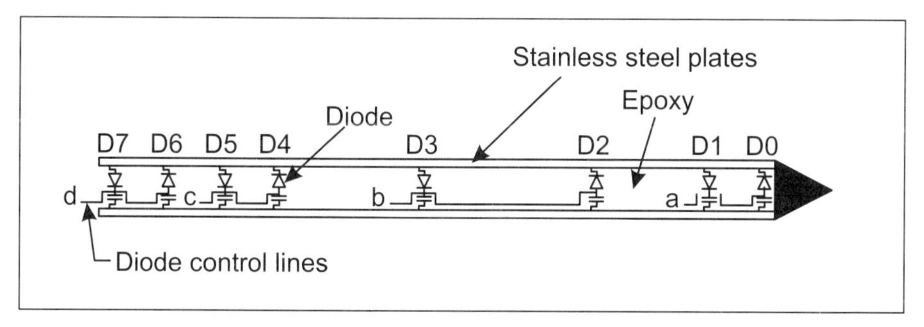

Figure 3-30. Schematic of a 7-segment, 8-diode probe, 1.85 m long with a 19x12.5 mm cross section. The space between the stainless steel plates is filled with electrical casting epoxy. The EM pulse generated by the TDR is introduced at diode D3 and propagates toward both ends. The diodes are switched with ±5V DC using the control lines a, b, c, and d (Gee and Ward 1999)

To obtain small-scale, high-resolution TDR measurements, Nissen *et al.* (1999) proposed using a printed circuit board TDR probe that can increase the travel time of the electromagnetic waves in the waveguides.

TDR Measurements in Saline Soils

Dalton and van Genuchten (1986) determined that in saline soils with high bulk electrical-conductivity values, the attenuation of the TDR waveform leads to an overestimation of moisture content. To overcome this problem, Dalton and van Genuchten (1986) and Malicki and Skierucha (1989) proposed increasing the TDR voltage from the usual 250 mV.

Another approach to overcoming the problem of saline soils is the use of coated probes. These probes are now commercially available from most conventional TDR probe suppliers. Ward *et al.* (1992) tested TDR probes (coated with different nonconductive materials, namely, paint, epoxy resin, PVC heat-shrink tubing, ceramic, and a silicone-based compound) in high-salinity and high-temperature soils. The coated probes were tested in soil columns saturated with KCl solutions (0 to 0.4 M). Tests demonstrated that these probes can be used to determine travel time in soil under temperatures as high as 80°C. However, the disadvantage of coated probes is their variable sensitivity, which depends on the soil-moisture content (Ferre *et al.* 1998).

Ward *et al.* (1996) and Ward (1998) tested a prototype of a probe with waveguides 0.9 m long under laboratory and field conditions using the remote-shorting method and analyzed the results using a differential waveform analysis, which confirmed that this method is able to reliably measure transit time, dielectric constant, and electrical conductivity in saline soils.

Thermo-TDR Probe

The exchange of thermal energy in a soil is influenced by the soil-heat capacity and the thermal conductivity, both of which depend on moisture content. A thermo-TDR probe provides simultaneous measurements of the soil thermal properties and the bulk electrical conductivity. This probe combines a conventional TDR probe with a dual-heat-pulse probe (Noborio *et al.* 1996; Ren *et al.* 1999). The TDR probe comprises three hypodermic needles (20 gauge, 70 to 100 mm long) spaced 10 mm apart. A line heater and a thermocouple are enclosed in the center needle, and an additional thermocouple is located in one of the outer needles (Noborio *et al.* 1996). During the operation, the first step is to conduct a TDR probe measurement, and the second step is to apply heat to the center electrode (the heater) and to monitor the temperature in the outer electrode. Recorded TDR waveforms are used to determine the moisture content, while temperature measurements are used to calculate the thermal properties. Laboratory tests conducted for different soils in a range of the water content from 0.05 to 0.36 m^3 m^{-3} showed that the probe provides quick, continuous, and accurate measurements of the soil thermal properties and water content. Field testing of this probe is needed. Because of the probe's small size, its application may be limited to non-consolidated soils into which it can be carefully installed.

TDR-Cone Penetrometer Method

A TDR-based moisture sensor was combined with the Navy's Site Characterization and Analysis Penetrometer System (SCAPS) to allow rapid, minimally-intrusive measurements of soil moisture content to depths of 45 m (Ward *et al.* 1996). In the first design, the TDR waveguide consisted of a pair of ring electrodes 2.5 cm thick and 40.6 mm in diameter that were electrically isolated from each other and from the steel mandrel using an ultra-high molecular-weight insulator. The probe was tested in the laboratory to obtain the correlation between the moisture content and the dielectric constant. Several field tests were conducted to depths of 10 m at a coastal site in San Diego that was affected by seawater intrusion. To overcome problems caused by ionic solutes, Ward *et al.* (1996) and Ward (1998) used a helical waveguide with remote shorting and differential waveform analysis. The probe consisted of a pair of parallel stainless steel helices separated from each other and from the cone penetrometer mandrel by an insulator.

TDR-Tensiometer

A combination of a tensiometer with TDR probes, to obtain simultaneous measurements of the matric potential and moisture content of soils reported by Baumgartner *et al.* (1994) and Whalley *et al.* (1994), was described above. These probes were used in field-infiltration experiments by Si *et al.* (1999).

There are several vendors who manufacture TDR hardware. They are listed on the following Web site: http://iti.acns.nwu.edu/clear/tdr/.

Capacitance Methods

This borehole-logging method is based on using frequency-domain reflectometry (FDR) to determine the changes in the dielectric constant (capacitance) of a material. This method works well in homogeneous soils. Capacitance sensors have been used to measure soil-moisture content for many years (Thomas 1966; Dean *et al.* 1987). The results of measurements depend on the type of capacitance probe used. Evett and Steiner (1995) determined that a commercial probe (the Sentry 200 CP, developed by Troxler Electronics Laboratories, Raleigh, NC) produced a lower resolution in θ compared to the standard neutron-logging probe. Note that the soil volume for a capacitance probe is limited to a sphere of no more than a few centimeters around the probe access tube, which is considerably smaller than the sphere of influence of a neutron probe.

Paltineanu and Starr (1997) and Starr and Paltineanu (1998) successfully tested multisensor capacitance probes under laboratory and field conditions. The probes operate at a frequency in excess of 100 MHz (and are commercially available from EnviroSCAN, Sentek Pty. Ltd., South Australia).

Because the capacitance method uses a single-frequency approach, simultaneous measurement of soil temperature and the dielectric constant allow one to determine the dielectric loss, which is needed to calculate the moisture content of saline soils (Hilhorst and Dirksen 1994). The response of FDR to changes in the moisture content is non-linear, and therefore, a careful calibration of the probes is necessary. This problem becomes more complex if the soil properties are changed over time, which may be the case for many environmental remediation sites. Under such conditions, the probes must be recalibrated.

Phase Transmission Methods (VIRRIB® method)

VIRRIB® method is based on measurements of the propagation of electromagnetic (EM) waves through the medium. This method was developed and successfully used for irrigation control and measurement of deep percolation in the Czech Republic (Litschmann 1991). Unlike TDR and capacitance methods, the VIRRIB method relies on the determination of a phase shift of a sinusoidal EM pulse, relative to the phase at the origin, after traveling a fixed distance along a conductor. According to Litschmann (1991), who extensively tested this method, the VIRRIB sensor consists of two stainless steel concentric circles (that must be inserted into the soil). These are connected in the body of the sensor, where the electronics are located. The sensor is embedded in a cover, which prevents water penetration into the electronic part. The diameter of the outer ring is 280 mm. The soil volume affected by measurements is approximately 15,000 to 20,000 cm^3. The sensor uses DC current with a voltage of 12 to 20 volts supplied from an external source. The output current is directly proportional to the moisture content of the soils. Horizontally installed probes can measure the moisture content in a soil layer 12 cm thick, and vertically installed probes can measure the moisture content in a soil layer 30 cm thick. The primary advantage of these probes is that they can be connected to cables up to 1 km in length. However, very little data exist that document this promising method. Major disadvantages are the difficulty of installation of the sensor and its sensitivity to ionic solutes.

Electromagnetic Induction (EMI)

The spatial distribution of bulk electrical conductivity can be measured using Electromagnetic Induction (EMI) methods. Because a dry soil layer is a poor conductor, the EM measurements depend on the water content, the solute concentration, and the amount and type of clay in the soil. Innovative applications of EMI for measuring water content were demonstrated for unsaturated soils (Kachanoski *et al.* 1988; Sheets and Hendrickx 1995), as well as radioactive sludges (Crowe and Wittekind 1995). Sheets and Hendrickx (1995) used a Geonics EM-31 ground-conductivity meter to monitor moisture content along a transect 1,950 m in length near Las Cruces, New Mexico. According to Sheets and Hendrickx (1995), changes in moisture content could be more accurately determined by changes in the EM measurements than by neutron logging. A disadvantage of this method is the ground-conductivity meter's sensitivity to metallic objects (such as fences, high-voltage power lines, and buried metallic objects), soil salinity, and temperature.

Thermal Probes

Campbell *et al.* (1991), Bristow *et al.* (1993), Tarara and Ham (1997), Bristow (1998), and Song *et al.* (1998, 1999) imposed changes in the temperature and thermal properties of soils, which are affected by the soil-moisture content. The probe used to affect these changes consists of two small needles separated a small distance, one of which is a heater, the other a thermal sensor. The heater periodically emits a pulse, and the thermal sensor needle monitors thc temperature decay. The temperature decay is used to determine the soil volumetric heat capacity, which is a function of the soil-water content. Thermal probes are commercially available from Thermal Logic, Pullman, WA. The advantage of this method is that the probe is insensitive to the soil salinity. The main disadvantage is that only a small volume of soils is involved in measurements.

Fiber Optic Sensors

Fiber optic cables contain thin strands of glass that carry light. Fiber optic cables are lightweight and do not change the physical, chemical, or biological properties of soils. Moreover, they do not interfere with the electrical fields that may be generated by geophysical techniques. Fiber optic sensors are currently used in many scientific and industrial applications for real-time monitoring of a number of processes (Griffin and

Olsen 1992; Lieberman *et al.* 1990, 1991). The primary function of fiber optic sensors is to determine the changes that physical or chemical processes impose on a constant beam of light (Udd 1991). Fiber optic sensors have also found an application in environmental sciences (Lieberman *et al.* 1990; Rogers and Poziomek 1996).

Alessi and Prunty (1985) determined the moisture content of soils by measuring the attenuation of light transmitted through an optical fiber embedded in soil. The light received from the sensor was converted to a voltage. The resulting measurements showed a linear relationship between the voltage and the moisture content of a silt loam soil. An important advantage of this method is its independence of the salinity of soils.

According to the DOE Characterization, Monitoring, and Sensor Technology-Cross-Cutting (CMST) Program, the design, construction, and evaluation of fiber optic, laser-induced breakdown spectroscopy (LIBS) probes in conjunction with a cone penetrometer are in progress at Science and Engineering Associates, Inc. (Stephen Saggese, Principal Investigator). The purpose of these instruments is to quantify the concentration of heavy metals such as Cr, Pb, and other DOE-specified elements. Sandia National Laboratories is fabricating, testing, and evaluating a new cone penetrometer for characterizing hydrogeologic parameters and subsurface contaminant concentrations with fiber optic probes and a Time Domain Reflectometry (TDR) sensor.

CHEMICAL DISTRIBUTION AND TRANSPORT MONITORING

Monitoring of chemical transport in the vadose zone involves determining the concentration of chemicals in the pore liquid. Various types of pore-liquid samplers were discussed by Morrison (1983), Wilson (1980, 1981, 1982, 1983, 1990), Everett (1981), Everett *et al.* (1982, 1984), Robbins and Gemmel (1985), Merry and Palmer (1986), U.S. EPA (1986), and Ball (1986).

Suction Lysimeters*

Suction lysimeters are devices that are used for collecting pore-liquid samples from unsaturated and saturated soils and fractured rocks. The

*This section was contributed by L. Everett and B. Faybishenko.

description of suction lysimeters is given below based on recommendations summarized in the ASTM D 4696-92 Standard Guide.

The selection of the sampler type for site characterization and monitoring should be based on consideration of several criteria, including sampling depths and required sample volumes for a particular type of analysis (volatile organic compounds [VOCs], major ions, or radioactive elements). Table 3-13 summarizes the features of various types of suction samplers.

Vacuum lysimeters are samplers that can be used at depths up to 6 or 7.5 m. *Pressure-vacuum lysimeter*s are samplers that can be used at depths up to 10 or 15 m. *High-pressure-vacuum lysimeter*s (also known as pressure-vacuum lysimeters with transfer vessels) can be used from the surface to depths at least 91 m. (Installations as deep as 91 m were reported by Bond and Rouse [1985].) *Suction lysimeters with low bubbling pressures* can be used to maximum depths from about 7.5 to 46 m (U.S. EPA 1986; Johnson and Cartwright 1980).

Suction lysimeters consist of a hollow porous tip attached to a vessel or a body tube. Samples are obtained by applying a vacuum to the sampler and collecting pore-liquid in the body tube (Figure 3-31). When suction greater than the soil pore-liquid capillary pressure is applied to the sampler, liquid moves into the sampler. Ceramic porous segments are hydrophilic, and the maximum pore sizes are small enough to allow meniscuses to withstand the entire range of sampling suctions. If the maximum pore sizes are too large, the meniscuses are not able to withstand the applied suction. When the meniscuses break down, hydraulic contact between the water in the porous tip and soils is lost, and soil air enters the sampler, leading to the release of suction.

There are several methods for retrieving liquid collected in the sampler. For depths up to 6 or 7.5 m, liquid samples can be brought up to the surface using suction (Figure 3-31a). For depths greater than 6 or 7.5 m, samples may be retrieved by pressurizing the gas above the liquid in the sampler through a gas-access line, which pushes the liquid sample up to the surface through a second line (Figure 3-31b).

Pressure-vacuum lysimeters (Parizek and Lane 1970) for deep sampling (Figure 3-31b) have two lines that are forced through a two-hole stopper sealed into the upper end of the body tube. The discharge line extends to the base of the sampler, and the pressure-vacuum line terminates just below the top stopper. At the surface, the discharge line

TABLE 3-13 Suction Sampler Summary (ASTM D 4696-92)

Sampler Type	Porous Section Material	Maximum[A] Pore Size (μm)	Air Entry Value (cbar)	Operational Suction Range (cbar)	Maximum Operation Depth (m)
Vacuum lysimeters	Ceramic	1.2 to 3.0 (1)[A]	>100	<60 to 80	<7.5
	PTFE	15 to 30 (2)[A]	10 to 21	<10 to 21	<7.5
	Stainless steel	NA[B]	49 to 5	49 to 5	<7.5
Pressure-vacuum lysimeters	Ceramic PTFE	1.2 to 3.0 (1)[A]	>100	<60 to 80	<15
		15 to 30 (2)[A]	10 to 21	<10 to 21	<15
High pressure-vacuum lysimeters	Ceramic PTFE	1.2 to 3.0 (1)[A]	>100	<60 to 80	<91
		15 to 30 (2)[A]	10 to 21	<10 to 21	<91
Filter tip samplers	Polyethylene	NA[B]	NA[B]	NA[B]	None
	Ceramic	2 to 3 (1)	>100	<60 to 80	<7.5
	Stainless steel	NA[B]	NA[B]	NA[B]	None
Cellulose-acetate hollow-fiber samplers	Cellulose Acetate	<2.8	>100	<60 to 80	<7.5
	Non cellulosic Polymer	<2.8	>100	<60 to 80	<7.5
Membrane filter samplers	Cellulose Acetate	<2.8	>100	<60 to 80	<7.5
	PTFE	2 to 5	NA[B]	NA[B]	<7.5
Vacuum plate	Alundum	NA[B]	NA[B]	NA[B]	<7.5
	Ceramic	1.2 to 3.0	>100	60 to 80	<7.5
	Fritted glass	4 to 5.5	NA[B]	NA[B]	<7.5
	Stainless steel	NA[B]	49 to 5	49 to 5	<7.5

[A]Pore size determined by bubbling pressure (1) or mercury intrusion (2).
[B]NA = Not available

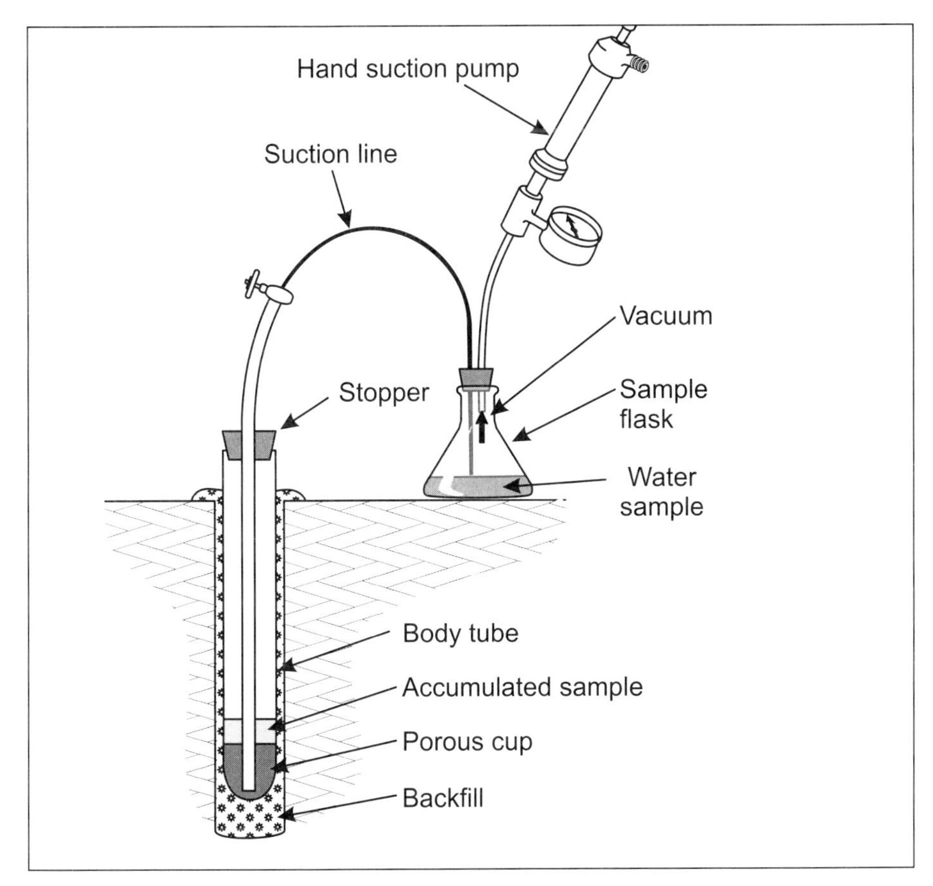

Figure 3-31a. Schematic of vacuum lysimeter (ASTM D 4696-92)

connects to a sample bottle and the pressure-vacuum line to a pressure-vacuum pump. These samplers can retrieve samples from depths greater than 7.5 m because pressure is used to retrieve liquid samples. During pressurization, however, some of the sample is forced back out of the cup. At depths greater than 15 m, the volume of sample lost in this fashion may be significant. In addition, at these depths, pressures required to bring the sample to the surface may be high enough to damage the cup or to reduce its hydraulic contact with the soil (Young 1985).

High-pressure-vacuum lysimeters operate in the same manner as pressure-vacuum lysimeters. These instruments include a built-in check valve between the chambers, which prevents both sample loss through

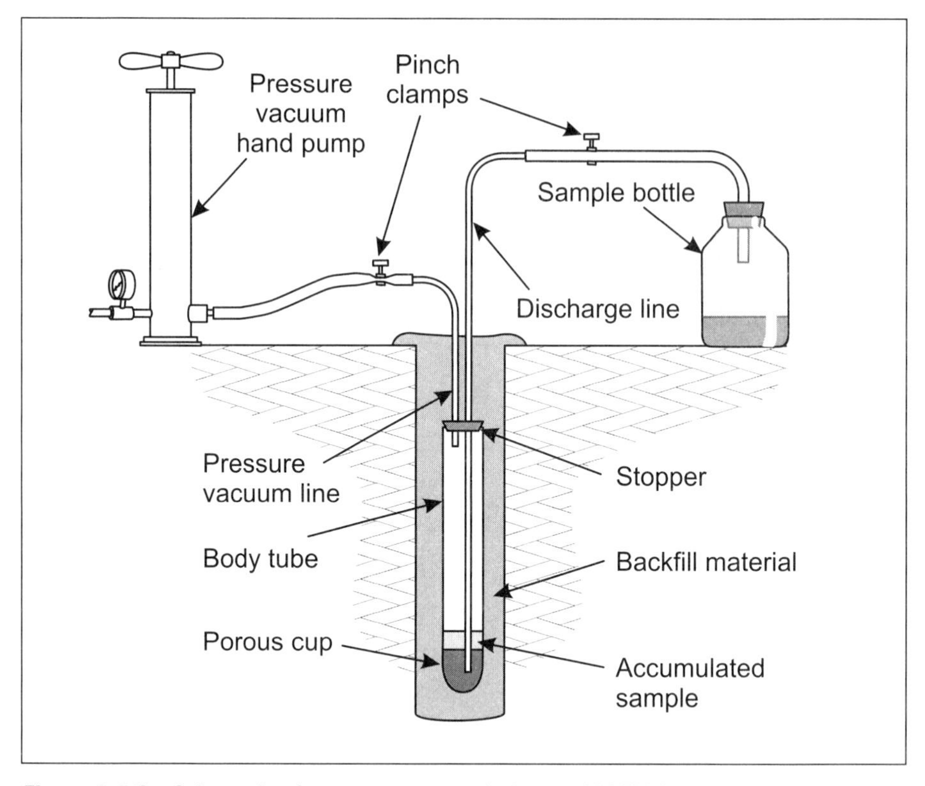

Figure 3-31b. Schematic of pressure vacuum-lysimeter (ASTM D 4696-92)

the porous tip during pressurization, and possible cup damage caused by over-pressurization.

Filter-tip samplers consist of two components: a permanently installed filter tip, and a retrievable glass sample vial. The filter tip includes a pointed end to help with installation, a porous section, a nozzle, and a septum. The tip is threaded onto extension pipes that extend to the surface. The sample vial has a second septum, and when in use, the vial is seated in an adaptor with a disposable hypodermic needle. The needle penetrates both septa, allowing the sample to flow from the porous segment into the vial. The inside diameter of the extension pipes varies from 2.5 to 5.1 cm. Vial volumes range from 35 to 500 mL (Knighton *et al.* 1981). The body of the filter tip is made of thermoplastic, stainless steel, or brass, and the attached porous tip is constructed from either high-density polyethylene, sintered ceramic, or sintered stainless steel. The septum is made of natural rubber, nitrile rubber, or fluororubber (Knighton *et al.* 1981). Filter-tip samplers collect liquid samples by lowering an air-evacuated sample vial through an access tube to a permanently installed porous tip. The vial is connected to the porous tip, liquid flows from the porous tip into the vial, and the vial is then retrieved.

Various materials have been used for the porous tip, including nylon mesh (Quin and Forsythe 1976), fritted glass (Long 1978), sintered glass (Starr 1985), stainless steel (Mott Metallurgical Corp. 1988), and ceramics (SoilMoisture Equipment Corp. 1988). The sampler body tube has been constructed with PVC, acrylic, and stainless steel (Smith *et al.* 1986; Caster and Timmons 1988; SoilMoisture Equipment Corp. 1988).

A cellulose-acetate, hollow-fiber sampler (Jackson 1976; Wilson 1981) consists of a bundle of flexible, hollow fibers (maximum pore size of less than 2.8 μm) pinched shut at one end and attached to a suction line at the other end. The suction line leads to the surface and attaches to a sample bottle and the source of suction in the same manner as a vacuum lysimeter. The fibers, which are analogous to the porous sections of vacuum lysimeters, have outside diameters of up to 250 μm (Levin and Jackson 1997). Levin and Jackson (1977) described similar fibers made from a noncellulosic polymer solution (maximum pore size of less than 2.8 μm). Those fibers have dense inner layers surrounded by open-celled, spongy layers with diameters ranging from 50 to 250 μm.

A membrane-filter sampler (Stevenson 1978; Morrison 1983; Everett *et al.* 1982; U.S. EPA 1986) consists of a membrane filter of polycarbonate, cellulose acetate (maximum pore size of less than 2.8 μm), or cellulose nitrate, that is mounted in a "swinnex" type filter holder (Stevenson 1978; Wagemann *et al.* 1974; Wilson 1983). The filter rests on a glass fiber prefilter, the prefilter on a glass fiber "wick," and the glass fiber "wick" on a glass fiber collector. The collector is in contact with the soil and extends the sampling area of the small-diameter filter. A suction line leads from the filter holder to the surface. At the surface, the suction line is attached to a sample bottle and the suction source in a manner similar to vacuum lysimeters.

To reduce chemical interference from substances on the porous tips, the U.S. EPA (1986) recommended preparation of ceramic units prior to installation, following procedures originally developed by Wolff (1967),

Wood (1973), and Neary and Tomassini (1985). The process involves passing hydrochloric acid (HCl) through a porous tip.

Morrison and Szecsody (1985) found that the radius of sampling influence is maximized if the borehole diameter is only slightly larger than that of the sampler, and if a silica flour pack is used. The U.S. EPA (1986) recommended that the hole have a diameter at least 5 cm larger than the sampler, which will facilitate installation of the silica flour. The silica flour slurry is usually emplaced using the tremie-pipe method. The excess water from the sampler and silica slurry should be removed immediately after installation (U.S. EPA 1986). To allow the samplers to equilibrate with the surrounding soil, Litaor (1988) recommended their installation a year before sampling begins.

Cellulose-acetate hollow-fiber-sampler installation procedures were described by Everett *et al.* (1984); membrane-filter-sampler installation procedures were described by Stevenson (1978), Everett *et al.* (1984), and Morrison (1983); and vacuum-plate-sampler installation procedures were described by Everett *et al.* (1984) and Morrison (1983). Because the fibers are thin and fragile, to be positioned in the soil, they must be placed in a predrilled vertical or horizontal hole or installed in a perforated, protective PVC tubing filled with soil slurry (Silkworth and Grigal 1981).

Membrane-filter samplers are placed in a hole to the top of the selected sampling depth. First, sheets of the glass fiber "collectors" are placed at the bottom of the hole. These collectors develop the necessary hydraulic contact between the sampler and the soil, and because they cover a larger area than the filter holder alone, they also extend the area of sampling. Second, two or three smaller discs are placed on the collectors. Third, the filter holder fitted with a glass fiber prefilter and the membrane filter is placed on top of the wick disks. The suction line leads to the surface. Finally, the hole is backfilled (Morrison 1983; Everett *et al.* 1984).

The major causes of sampler failure are line damage and leaks (caused by freezing, installation, rodents, and so on), connection leaks, and clogging of the porous material. Freeze damage to the lines can be minimized if the lines are emptied of sample prior to applying a vacuum. Care must be taken to prevent the tubing line from freezing.

Biofilm growth and plugging by colloids may clog porous tips (Morrison 1983; Quin *et al.* 1976; Debyle *et al.* 1988). However, such clogging may not affect the composition of liquid sampled, but will may only increase the time of sampling (U.S. EPA 1986; Johnson and Cartwright 1980).

Morrison and Szecsody (1987) described devices that could be used as tensiometers and then converted to pressure-vacuum lysimeters. However, they found that gases entering these devices prevented accurate measurement of pore-liquid tensions. Baier *et al.* (1983) also discussed methods of converting tensiometers to pressure-vacuum lysimeters. Suction lysimeters may be used as tensiometers, but the volume of water drawn from the soil through a lysimeter may significantly affect natural pore-liquid tensions (Taylor and Ashcroft 1972).

Operational lifetimes of suction samplers depend on installation, subsurface conditions, maintenance, and sampling frequency. Some samplers are operational for as long as 25 years (Baier *et al.* 1983).

Because vacuum lysimeters and experimental samplers use suction to retrieve samples, the maximum sampling depth is limited by the maximum suction lift of water, about 7.5 m (U.S. EPA 1986). In practice, these samplers are generally used to approximately 2 m below the surface (U.S. EPA 1986). They are primarily utilized to monitor near-surface movement of pollutants, such as those from land disposal facilities or from irrigation return flow.

Ordinarily, pressure-vacuum lysimeters are not used deeper than 15 m below ground surface. At greater depths, sample loss and overpressurization problems are considered significant enough to warrant the use of high pressure-vacuum lysimeters that do not have these limitations. High-pressure-vacuum lysimeters are not preferred at shallower depths because they are more expensive than pressure-vacuum lysimeters. In addition, high-pressure-vacuum units have more moving parts than pressure-vacuum units, and as a result, the possibility of failure is higher.

Factors affecting the volume of a pore-liquid sample are the suction applied, the schedule of suction application, the spatial distribution of pore-liquid, the soil texture and structure, and the porous tip design. Samples collected with lower suctions (approximately 10 cbar or less) usually come from liquids migrating through soil macropores (Morrison 1983). Samples collected with higher suctions (greater than about 10 cbar) also include fluids held at higher tensions in micropores. The case study illustrates the application of liquid samplers with different suctions.

"Characterization and Monitoring of Unsaturated Flow and Transport Processes in Structured Soils," by Philip M. Jardine, R.J. Luxmoore, J.P. Gwo, and G.V. Wilson, illustrates the application of liquid samplers with different suctions. *See page 473.*

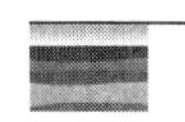

The sampler may disrupt normal flow patterns as a result of the applied suctions. Although the area nearest the sampler undergoes the most disturbance, the effects may extend several meters from the sampler (Bouma *et al.* 1979; Warrick and Amoozegar-Fard 1977; Van der Ploeg and Beese 1977). Because of this disturbance, samples are averages of the affected flow area rather than point samples (Morrison 1983). Morrison and Szecsody (1985) found that (under the conditions of their study) the radii of influence for suction lysimeters ranged from 10 cm in coarse soils up to 92 cm in fine-grained soils.

Sampling with falling suction produces samples with compositions that are "averages" of the liquids held at the range of tensions applied. Because suctions, and, therefore, inflow rates, decrease with time, these "averages" are weighted toward the portions of the samples obtained in early times. Samples collected over prolonged periods are "averages" of the liquids flowing past the sampling region.

During wet periods, samplers affect a small volume of soil and pull liquids from a sequence of pores that may include macropores. During dry periods, samplers affect a larger volume of soil, draw from micropores (because the macropores have been drained), and collect less liquid (Anderson 1986). The net result of these trends is that sampled soil solutions are "averaged" over different volumes and derived from different pores as a function of the soil-moisture content and distribution.

Soil textures and pore-liquid tensions control a sampler's radius of influence, and the amount of liquid that it can remove. The slope of the pore-liquid release curve for a sand is greater than that for a clay at low pore-liquid tensions. This indicates that, for an equal change of pore-liquid tension at these low tensions, a larger quantity of pore-liquid will be released from a sand than from a clay. At higher tensions, the slope of a clay pore-liquid-release curve is greater than that of sand. This indicates that more pore-liquid will be released from a clay than from a sand for an equal change in pore-liquid tension at higher tensions. Therefore, suction samplers may not obtain samples from coarse-grained soils at higher pore-liquid tensions.

Hansen and Harris (1975) demonstrated that intake rates may vary substantially because of variability in the ceramic sections from one manufacturer's batch to another. The intake rate of a sampler is also a function of the degree of clogging. The range of pore-liquid tensions over which a sampler can operate is a direct function of the maximum pore size of the porous section.

Nagpal (1982) recommended that several consecutive extractions of liquids be taken during a sampling event, and that only the sample from the last extraction be used for chemical analyses. The purpose of this is to flush out cross contaminants from previous sampling periods, and to ensure that any porous segment/soil solution interactions have reached equilibrium. Debyle *et al.* (1988) also suggested discarding the first one or two sample volumes when sampling dilute solutions with newly flushed (HCl method) and installed samplers. The purpose of this is to allow the cation exchange between the porous segment and the pore-liquid (caused by the HCl flushing) to equilibrate.

Other Methods

Electrical-Conductivity Imaging

Electromagnetic-conductivity imaging is a relatively rapid and cost-effective field method to investigate the spatial and temporal variability in soil salinity, and it is widely used all over the word. Several field tools were developed (GEONICS EM31, EM34, EM38) that replaced the traditional four-electrode resistivity traversing techniques. The advantage of electromagnetic-induction tools is that ground-contacting electrodes are not required, and consequently, these tools can be used in open boreholes in the vadose zone. Williams and Baker (1982) showed that EM meters can be used for measurements in saline soils. To provide measurements in saline soils, Rhoades and Oster (1986), Cook and Walker (1992), and Acworth (1999) determined the empirical correlation between the soil salinity and the measured electrical conductivity (EC). The EC of a homogeneous porous material depends on many variables, including porosity, saturation, clay content, grain size, and the total dissolved solids in the soil water.

Using inversion, a two-dimensional vertical section of true electrical conductivity can be created from field images of apparent electrical conductivity (Acworth 1999). The distribution of different electrical-conductivity zones can be correlated against borehole and other field data.

Fiber Optics

Ghodrati (1999) developed a method for and demonstrated successful applications of the fiber optic sensors for *in situ* point-type measurements of the breakthrough curves. (The description of the principle of fiber optics is given in the section "Field Measurements of Water Content in Unsaturated Soils," above).

TDR Probes

TDR probes that were initially designed to measure moisture content also can be used to determine breakthrough curves. There are numerous examples showing the performance of TDR probes to monitor the chemical transport. (The description of the performance of TDR probes in saline soils is given in the section "Field Measurements of Water Content in Unsaturated Soils," above).

Pore Water Extraction by Refractometer

Soil pore solution can be extracted using the refractometer method. Pore water is squeezed from fine-grained soil samples (ASTM D 4542-95 Standard Test Method) and this water is used to determine the soluble chemicals in the extracted pore water. This method is applicable for soils with water content equal to or greater than approximately 14 percent, by volume. An extensive summary of methods used to extract pore water from soils has been presented by Kriukov and Manheim (1982). The refractometer method cannot be used to extract liquids from coarse-grained soils, such as sand or gravel.

SEAMIST Absorbent Pads

Soil pore solution of low-moisture-content soils can be extracted using absorbent pads attached to the outside surface of SEAMIST liners, which are pressed against the walls of the borehole. (See the section "Removable SEAMIST™ Liners," above for the details of installation.)

SOIL GAS CHARACTERIZATION AND MONITORING

Goals of Soil-Gas Monitoring

Soil-gas monitoring is a widely used and effective method of subsurface investigations in several industries, including agriculture (Boynton and Reuther 1938), petroleum and mineral exploration (Horvitz 1969; Ullom 1988), underground coal gasification (Jones and Thune 1982), and environmental investigations (Roffman *et al.* 1985; Wittmann *et al.* 1985). Soil-gas monitoring is used to assess the extent of groundwater contamination by volatile organic contaminants (Wittmann *et al.* 1985), and to detect discharges to the vadose zone and groundwater from underground storage tanks (Scheinfeld and Schwendeman 1985; Wittmann *et al.* 1985). The RCRA and the Comprehensive Environmental Response, Compensation, and Liability Act (CERCLA) recommend soil-gas monitoring for soil and groundwater remedial actions (Karably and Babcock 1989).

In addition to measuring the soil-gas concentration along the soil profile, there is a need to measure soil gas flux, especially at the soil surface (Rolston 1986). These measurements are conducted to assess the soil-atmosphere gas exchange and to determine, directly, the gas flux. One of the methods developed for this purpose is the use of closed chambers installed over the soil surface.

It is important to note that soil-gas monitoring cannot be used as a stand-alone method, but only in combination with other methods. The success of soil-gas monitoring is strongly dependent upon the effects of geologic variation and moisture content in the sampling horizon, as well as the physical properties of contaminants. Note also that the soil gas sampled actually represents a mixture of soil gas and atmospheric air entering the soil at the time of sampling. Atmospheric air can enter soils through macroporous or soil cracks, which are located far from the sampling device.

Soil-Gas Characterization and Monitoring for VOCs*

Soil-Gas Processes Affecting VOC Concentrations

Soil-gas processes and monitoring methods are described below using the ASTM Standard 5314-92. The soil-gas composition depends

*This section was contributed by L. Everett and B. Faybishenko.

on a combination of several processes, such as migration, partitioning, and degradation of chemicals. Chemicals can enter the soils either in the liquid or vapor phase from the surface, from groundwater, or from leaking underground tanks (see Chapter 1).

Partitioning is the process that controls contaminant movement between phases. The unsaturated zone has four phases of interest: the water phase, the soil mineral and organic particle phase, the gas phase, and the nonaqueous phase liquid (NAPL) phase. In unsaturated soils, partitioning depends on air-filled porosity, water content, the presence and composition of NAPL, and the presence of clay and discrete inorganic soil particles.

Partitioning includes dissolution, adsorption, volatilization and evaporation. Dissolution is the partitioning of contaminants between the NAPL and water phases. It is impacted by the presence of liquid-phase co-solvents (such as gasoline additives) even at low concentrations in liquid-phase mixtures. The change in dissolution equilibrium can affect certain liquid-phase components in water, often enhancing the solubility of the components beyond what is indicated by partitioning coefficient data generated under laboratory conditions.

The effects of temperature upon dissolution are generally insignificant for aliphatic hydrocarbons between 15 and 50°C (Price 1976), which is typical for most vadose zone soils under natural conditions. However, temperature effects upon dissolution equilibrium can be significant for other contaminants (Owens *et al.* 1986). Dissolution equilibrium is also affected by changes in water salinity. The rate of dissolution depends upon the partitioning coefficient of the particular contaminant of interest and the degree to which the NAPL phase and water have been mixed. For example, frequent water-level fluctuations accelerate partitioning contaminants accumulated within the capillary fringe.

Volatilization and evaporation are the processes of evaporating volatile contaminants moving from either the NAPL or the liquid phase water phase to the surrounding gas phase. However, the soil-gas composition may not be similar to that of the liquid phase, because of the lack of constituents with the lowest vapor pressures. The rate of volatilization is expected to be higher in macroporous soils as affected by soil-gas convection currents. Manos *et al.* (1985) demonstrated that the organic matter and clay content in the soil impact the volatile organic-compound emission from soils. Soil-gas contaminants with high sorption properties cannot be efficiently sampled. Knowledge of the presence of barriers to vertical or horizontal migration (such as foundations, buried pavement, or perched water zones) and the existence of preferential pathways for contaminant migration (such as backfill rubble, utility vaults, storm sewers, or soil cracks) can assist in designing the soil-gas monitoring system.

Degradation is the process of contaminant attenuation by oxidation or reduction, either through biogenic or abiogenic processes. Degradation by oxidation usually occurs in shallow soils. Biodegradation is caused by the presence of microorganisms capable of using the contaminant as a substrate. Populations of various microorganisms that naturally occur in soils can degrade petroleum products (Dragun 1985). Contaminant biodegradation is known to occur in groundwater and in soils prior to contaminant partitioning into a vapor phase (Davis 1969). Contaminant biodegradation rates are highly variable and are controlled by a number of kinetic factors influencing the distribution of microorganisms (White *et al.* 1985; Jensen *et al.* 1985).

Contaminants can degrade to compounds that may or may not be detectable in soil gas. For example, while aerobic degradation can produce carbon dioxide, which is easily detectable and is an indirect indicator of the presence of contaminants (Diem *et al.* 1987), this process can also generate organic acids and phenols (Dragun 1985), which are not routinely detectable using whole-air soil-gas sampling because of their low Henry's constants. Anaerobic degradation can produce compounds such as methane, ethylene, propylene, acetylene, and vinyl chloride, which also can be monitored as indirect indicators of the presence of contaminants.

Biodegradation of contaminants in the vadose zone can also occur naturally. Natural biodegradation can result from indigenous microbial populations adapting to metabolize contaminants as primary substrate, or by introducing foreign populations that have been preconditioned to metabolize contaminants of interest. Certain compounds may not be present in soil gas because of the effect of biodegradation (Kerfoot 1987; Chan and Ford 1986). High clay content, organic matter, water content, and the processes of degradation can reduce the efficiency of soil-gas sampling and therefore, cause contaminant concentrations to drop below detection limits.

The vadose zone is a highly complex soil-air-water-hydrocarbon system in which contaminants can move from one phase to another when affected by chemical, physical, and microbiological processes. In addition, the soil-gas sampling procedure itself causes disruption in the soil-gas equilibrium condition. As a result, subsequent soil-gas sampling very often produces different results and may not be comparable. Therefore, the data obtained can be compared only qualitatively.

Sampling Method

The selection of a soil-gas sampling method involves consideration of the type and the methodology of the sampling-system application, as well as the QA/QC protocol. There are approximately 100 soil-gas sampling systems in existence (Kerfoot and Sanford 1986; Eklund 1985; Mayer 1989; Spittler and Clifford 1985). According to ASTM Standard 5314-92, a soil-gas sampling method should be selected based on the consideration of the site-specific conditions. The six basic soil-gas sampling methods are listed below:

(1) The whole-air-active method, which involves the sampling of a mixture of contaminant and noncontaminant vapors.

(2) The sorbed-contaminants-active method of placing a sampling device in the subsurface and withdrawing soil gas through the device.

(3) The whole-air-passive method of continuously injecting a carrier gas of a known composition and determining the contaminant concentration in a carrier gas-contaminant mixture.

(4) The sorbed-contaminants-passive method of placing a collection device in the soil and allowing the device to equilibrate with the soil atmosphere.

(5) Soil sampling and subsequent sampling of gas from a headspace atmosphere.

(6) Soil-liquid sampling and subsequent sampling of gas from the headspace atmosphere of a liquid sample.

Whole-air-active methods of soil-gas collection involve the forced movement of soil gas from the soil to a collection device through a probe (Devitt *et al.* 1987; Kerfoot 1987). Contained samples of soil gas are then transported to a laboratory for analysis, or the sampling device is directly coupled with an analytical system. Whole-air-active sampling is best suited to soil-gas monitoring where contaminant concentrations are expected to be high and the vadose zone is vapor-permeable. Probes are installed into pre-existing holes or are driven into the vadose zone. The volume of a gas sample can vary from a few milliliters to several liters, depending on the sampling rate through the probe, the vapor storage capacity of the soil, and the soil's ability to deliver vapor to a probe under a vacuum.

The active approach may not be effective in clayey and moist soils. Driven probes tend to compact natural soils around the probe. In very dry soils, driven probes can create cracks that can enhance soil-vapor permeability and create pathways to the atmosphere. Under conditions of low soil permeability and low contaminant concentration, purging of the probe prior to sampling may lower contaminant levels below the limits of analytical detection. Discussions of numerous whole-air-active sampling systems can be found in Devitt *et al.* (1987), Boynton and Reuther (1938), and Nadeau *et al.* (1985).

Sorbed-contaminants-active methods involve forcing bulk soil gas from the soil to an apparatus designed specifically to extract and trap gas contaminants by adsorption. The adsorptive material used includes charcoal or a carbonized molecular sieve adsorbent (US EPA 1988b), porous polymers, silica gel, and activated alumina (Devitt *et al.* 1987). This system is well-suited to sites where the soil may be highly permeable to vapor and where the contaminant concentration may be lower than required for successful whole-air surveys. This approach is especially useful for the detection of nonpolar volatile organic compounds. Sorbent collection devices are commercially available or can be specially prepared with an appropriate sorbent material that concentrates desired compounds for future analysis. Colorimetric detector tubes are available that provide an indication of the presence of target compounds at the time of sampling.

These devices are limited in application by the high concentration requirements for many compounds and by the tube's compound-specific nature. The effectiveness of the sorbed-contaminants-active method is limited in soils with high clay and water contents. Some sorbents may be affected by high humidity in soil gas. For example, for some chemicals, humidity greater than 60% (very common for soil gas) can reduce the adsorptive capacity of activated charcoal to 50%. The presence of condensed water in the sample tube may also indicate the reduced amount of adsorbed chemicals. Organic compounds that are reactive, oxygenated, or gaseous at room temperature are either not adsorbed by, or not efficiently desorbed from, charcoal tubes (US EPA 1988a).

Whole-air-passive methods involve continuously injecting a gas of known composition to create a small vacuum in a collection chamber, which results in the diffusion of soil-gas components from the soil into the chamber. The resultant carrier gas-contaminant mixture is collected for analysis. This method is effective in monitoring contaminant emissions from soil or water, and in assessing the health risk of such emissions to the general public. The air-passive method is limited in application because the injected gas greatly dilutes the contaminants in the sample stream. High water, clay, and organic matter content will restrict the rate of contaminant flux to the chamber.

Sorbed-contaminants-passive methods of soil-gas collection involve the passive movement of contaminants from the soil to a sorbent collection device. The main mechanisms of contaminant migration to a passive-sorbent device are (1) the diffusion of gas molecules from soil regions with a high concentration of gas to a sorbent device with a low concentration of gas, and (2) the advective transport of gas through the vadose zone intersecting the sorbent device. Passive samplers that have been used include occupational-health volatile-organic-compound monitors and a sampler originally developed to detect hydrocarbons in petroleum exploration (Spittler and Clifford 1985; Wesson and Armstrong 1975). These devices use containers that are several inches in diameter and that utilize charcoal as a sorbent. The end of the instrument is left open and the device is placed in a borehole at the desired depth. The borehole is then backfilled (Mayer 1989), and the sorbent device is usually left in the soil for two to ten days, or in some cases, up to 30 days or more. This method can be used even in frozen and high-water-saturation soils (Wesson and Armstrong 1975).

Soil sampling for subsequent headspace atmosphere or extraction sampling methods are used to determine contaminants present in a headspace atmosphere above a contained soil sample. Note that this headspace atmosphere is not gas extracted from the soil, rather it is an artificial atmosphere that volatilized from a potential contaminant source, that is, the soil sample. Contained atmosphere methods do not yield representative samples for *in situ* vadose zone investigations, because the sampled soil gas is not the same gas as the *in situ* vadose zone gas. During sampling, large amounts of the vapor phase can be lost, which reduces the applicability of this method. Uncontrolled volatilization can be reduced using two methods: (1) recovering small soil cores with polypropylene syringes, or (2) adding buffering solutions or sodium sulphate and phosphoric acid to the vial prior to sealing, which will shift the activity coefficients of the subject contaminants to favor the vapor phase. To reduce contaminant degradation (especially biodegradation) in the container, samples should be stored in the dark at approximately 4°C.

Soil pore-liquid headspace gas methods involve the collection of soil gas that has accumulated above the soil pore liquid in suction lysimeters, pan lysimeters, or free-drainage glass block samplers. After a lysimeter has been installed for some period of time, vapor is sampled from a soil pore liquid sampler. In dry soils, when the lysimeter cannot recover a pore liquid sample, the soil gas can freely pass through the porous cup of the lysimeter, and the suction lysimeter serves as a soil gas sampler. Soil-gas samplers can be installed at different depths in vertical or slanted boreholes. The space between the samplers should be properly backfilled in order to avoid cross-contamination.

Soil-gas sampling should involve the collection of field blanks, travel blanks, and sample probe blanks to test for residual contamination in the sampling system, and to examine sample integrity during handling and transport. Field replicates can be demanded by a client or dictated by a particular situation. The number of replicates is usually 10% of the total number of soil-gas samples. In general, the time between sample collection and analysis should be minimized. Investigators should also protect samples against light and heat, and exercise precautions against leaks (see ASTM Practice D 1605).

Problems of sample handling and transport can be minimized by the integration of sampling and analytical systems. For example, a whole-air-active-sampling system can be coupled directly to a portable VOC (volatile organic compound) analyzer. The sample stream is fed directly to the intake port of the analyzer and passed through the detector.

Small volume samples are commonly recovered by syringe for immediate injection into an analyzer or small-volume container. Glass gas-tight chromatography syringes are employed when rigorous QA/QC protocol is required and samples are injected into the analyzer immediately upon recovery. These syringes must be decontaminated prior to the recovery of each sample aliquot. Disposable syringes are employed when samples are to be transferred to a small-volume container for transport.

Hand pumps are also used to transfer samples into tedlar bags or glass bulbs. Hand pumps are preferably installed behind the analyzer or container in the sample train to avoid contribution from or loss of contaminants to the hand pump. Hand pumps commonly contain petroleum-based lubricants, which will contribute to the hydrocarbon content of soil gas. These devices must be either placed at the end of the sample train or abandoned.

For low-level detection, tubing can cause a cross-contamination if it is not replaced in the sampling train prior to sampling at a new location. Vacuums can be employed to transfer soil gas from a sampler to a container. Evacuated glass bulbs, some containing adsorbents or absorbing liquids (see ASTM Practice D 1605), can be affixed to an in-place, purged sampling device and allowed to come to pressure equilibrium. Care must be exercised in recovering the gas sample from a vacuum cylinder to avoid contaminating the sample with atmospheric air.

Sorbent traps are commonly self-contained. A trapping device should be compatible with the properties of the target compounds and the technique of desorption chosen. Good practice for use of these devices, including handling and desorption procedure, is required for successful implementation of sorbent traps when sampling organic compound vapors (see ASTM Practice D 3686). Containers for soil samples should be preserved for a subsequent headspace analysis. The choice of container for soil headspace determination is dependent upon the method of sampling chosen.

Soil-gas samples have limited holding time depending on the degradation of the VOCs in the container, which may be caused by exposure to light or heat, agitation during shipping, as well as the air diffusivity of the container material. Sample biodegradation may occur in containers if the water vapor condensed in a container contains microorganisms capable of metabolizing contaminants.

On-Site Measurements

Flame Ionization Detectors (FIDs) generate electric current when gases containing carbon atoms are oxidized to carbon dioxide in a hydrogen flame and potential is applied across the flame. The magnitude of the electric current generated is termed the detector response. FIDs are responsive to hydrocarbon contaminants in soil gas and are therefore commonly employed as contaminant detectors. These detectors are durable for field application, have a wide linear range, and generally respond uniformly to organic gas species. FIDs are usually unresponsive to inorganic gases and water vapor, which are common constituents in soil gas. FID performance can be evaluated independent of the chromatographic column (see ASTM Practice E 594). Although highly versatile, these detectors are not selective for halogenated compounds. They require supplies of fuel gas, which necessitate careful safety practices in handling and flame ignition.

Photoionization Detectors (PIDs) employ ultraviolet radiation to ionize organic molecules. Positive ions and free electrons are formed, which migrate to the detector electrode(s), resulting in an electric current that is proportional to the contaminant concentration at the detector. PIDs are extremely sensitive to aromatic hydrocarbons. The range of detectable contaminants can be extended by using lamps of different energies, which will cause a change in the response of contaminants with different ionization potentials.

Figure 3-32 summarizes the data illustrating the relationship between PID readings taken in the field and the total VOC concentrations determined using sampling of soil gas into 1L SUMMA canisters and the TO-14 EPA analysis of soil gas. Data from all available soil gas probes within LBNL are included in the figure. A nearly 1 to 1 linear relationship between the PID reading and the total VOC concentration can be used to describe the data at VOC concentrations that exceed 1000 ppbv. However, when the actual VOC concentrations were lower than 1000 ppbv, the PID equipment could not accurately measure the concentration. In this case, the PID either registered a null sample (Group 1) or over-predicted the actual concentration (Group 2).

Electron Capture Detectors (ECDs) are highly sensitive to and selective for compounds with electronegative functional groups such as chloro-fluorocarbons (CFCs). The sensitivity of the detector is propor-

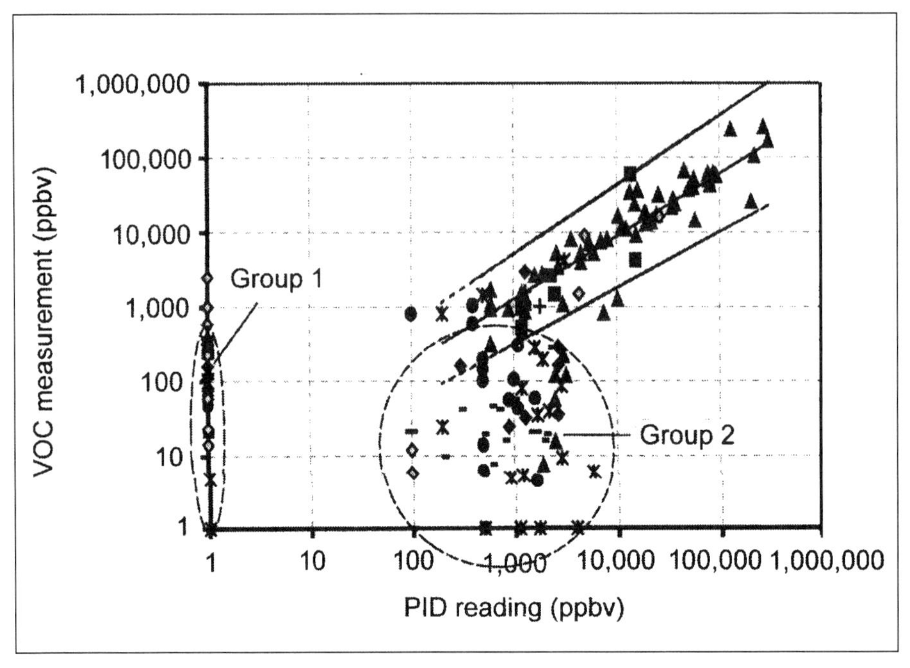

Figure 3-32. Relationship between PID and VOC measurements taken at all available gas probes at LBNL

tional to the number of these groups on a compound, resulting in a detector response that is unique to each compound. The ECD comprises a source of thermal electrons inside a reaction chamber (a radioactive source emits β radiation, which ionizes the carrier gas to produce electrons). The device detects compounds with electronegative functional groups capable of reacting with thermal electrons to form negative ions. Such reactions decrease the concentration of free electrons, and the detector is designed to measure such changes in electron concentration inside the chamber (see ASTM Practice E 697). Calibration of the ECD is therefore linked to each compound and is determined by the detector. ECDs are also sensitive to water, oxygen, and other common components of soil gas, which may cause problems in method performance. Because ECDs emit radiation, the space should be properly vented, and under Federal regulation, licensing is required for ECD operation.

Other detectors that provide soil-gas analysis are the argon ionization detector (a nondestructive device similar in operating design to the ECD), the flame photometric detector (FPD) used to determine organic compounds containing sulfur and phosphorus, and the hot-wire (pyrolyzer) detector used to identify compounds containing nitrogen.

The selection of a particular soil-gas monitoring technique should always be based upon the physical properties of the vadose zone, the chemical and physical properties of the compounds of interest, and the type of the remediation system. A case study on the accompanying CD shows three applications of an innovative *in situ*, real-time automated soil-gas detection and monitoring system, MultiScan™. The uses of the Multi Scan System demonstrated in this case study "MultiScan™ Case Study #1: Los Alamos National Laboratory Vapor Extraction System, and MultiScan™ – Case Study #2: Subsurface Barrier Verification," are listed below:

(1) Monitoring and performance assessment of the Pilot Vapor Extraction Test was conducted at Los Alamos National Laboratory. The soil gas probes were embedded in monitoring boreholes using the SEAMIST™ system. A photoacoustic gas analyzer was used to detect TCA, TCE, Freon, CC_{14}, CO_2, and water vapor. Barometric pressure and temperature were measured.

(2) Subsurface barrier verification was used in combination with the SEAtrace™ to locate leaks in a colloidal silica permeation-grouted barrier at Brookhaven National Laboratory and a thin-wall jet-grouted barrier at the Dover Air Force Base.

(3) Monitoring and performance assessment of a passive vadose zone remediation system (BERT™) was completed at the Radioactive Waste Management Complex of the Idaho National Engineering and Environmental Laboratory.

Soil-Gas Sampling for Radon*

Radon is a radioactive gas that is generated naturally by the radioactive decay of radium, an element that is contained in all rocks and soils.

*This section was contributed by A.R. Hutter and B. Faybishenko.

Recent investigations have found that radon (^{222}Rn) concentrations measured in soil gas and groundwater can be used to detect tectonic structure and the weak zones in the near-surface zones (Choubey and Ramola 1997). Hutter (1996) identified the factors affecting radon migration in soils, which are humidity, porosity, barometric pressure, wind, precipitation, temperature, and soil permeability. Each parameter strongly correlates with the radon concentration. The factors controlling the spatial and seasonal variations in radon concentration are the local soil conditions, moisture content, and temperature (Rose *et al.* 1990; Hutter 1996). Radon migration along deep cracks to the surface can be caused by a combination of several processes such as diffusion, barometric pumping, and thermal convection (Rose *et al.* 1990).

The isotope ^{220}Rn is known as thoron. It has a half-life of approximately 55 seconds. Hutter (1995) showed that the ratio of ^{220}Rn to ^{222}Rn can be used to assess the soil-gas processes. A case study on the accompanying CD, "Investigation of Fast Migration in the Vadose Zone for Assessment of Groundwater Contamination by Chernobyl Radionuclides," by V.M. Shestopalov, V.N. Bubilas, and D.U. Kukharenko, Radioecological Center, NAS of Ukraine (based on the results of investigations in the Chernobyl area vadose zone) demonstrates that the increased ratio of ^{220}Rn to ^{222}Rn indicates a zone of potential preferential flow in the vadose zone. Isotopic studies were confirmed using radar measurements.

The methods used for soil-gas sampling typically involve emplacing tubes into the ground permanently or temporarily, depending upon the requirements of the sampling program. Generally, permanently emplaced tubes help to ensure more accurate analytical results because of possible changes of the soil structure around the tubes during installation. Soil gas is then extracted from these tubes and analyzed for the radon isotopes ^{220}Rn and ^{222}Rn, using scintillation cells and gross alpha counting equipment or solid-state alpha spectroscopy instruments. For instance, a typical soil-gas-sampling set-up is shown in the DOE *Environmental Measurements Laboratory's Procedures Manual* (EML 1997).

Methods of determining ^{220}Rn and ^{222}Rn contamination using with soil-gas samples have been widely published (Hutter 1995; Lahti *et al.* 1998). Using these methods to measure typical soil gas ^{220}Rn and ^{222}Rn concentrations (greater than 5 kBq m^{-3} [130 pCi L^{-1}]), Hutter (1995) showed an uncertainty of 20.7 percent and 10 percent (90 percent confidence levels), respectively, from analyses of duplicate field measurements.

An international comparative study of soil-gas radon measurements was conducted in 1995. The results, based on pooling of the participant data and use of the ratio of the standard deviation (SD) to the arithmetic mean, showed an agreement of approximately 27 percent at depths greater than 0.75 m (Hutter 1998). From this study, it was concluded that sampling errors are two to three times that of analytical errors.

Many factors may introduce significant errors during soil-gas sampling. For instance, if a loosely fitting sample tube is inserted into a drilled hole with only a plug at the surface, the soil gas that is drawn is probably not from the depth of the tube bottom, but rather, from some unknown and varying depth along the length of the tube. Using "packers," or very small diameter probes, helps to ensure proper depth determination (Tanner 1988; Reimer 1990). Inherent in this consideration is the "target" volume of the sampled soil gas. Even though sampling depths may be similar, the measured ^{222}Rn concentration may be influenced by the volume of the sample, as a result of vertical variation in the soil gas ^{222}Rn concentration. Samplers that draw a large volume of soil gas are likely to introduce a greater uncertainty because of vertical variation in soil-gas radon. For example, a 1 L soil-gas sample drawn from the bottom of a tube is likely to come from several centimeters above and below the tube end (assuming packers are not used). On the other hand, small sample volumes of extracted soil gas are more likely to have been derived from a smaller zone. The data from the comparative study described above indicate that, for the test site, a greater error was introduced by poor estimation of the soil-gas extraction depth than by any variations in ^{222}Rn concentrations caused by soil heterogeneities (Hutter 1998).

Pneumatic Pumping and Injection Experiments

Pneumatic pumping and injection experiments using boreholes screened in the vadose zone are important for many environmental and engineering applications including soil vapor extraction (SVE) systems. SVE systems are widely used for remediation of volatile and semi-volatile organic compounds accumulated in the vadose zone from leaking surface and underground tanks, pipelines, cribs, spills, and other sources, at many contaminated sites (Massmann 1989; Pedersen and Curtis 1991). The design of soil vapor extraction systems is based upon the results of pneumatic (gas) pumping and/or injection experiments conducted to determine the soil air permeability and porosity, which are then used in assessing the well spacing, well configuration, and blower or pump specifications (Pedersen and Curtis 1991; Massmann and Madden 1994). Pneumatic injection and pumping experiments can be conducted independently or simultaneously using different wells.

Single-and cross-borehole air-injection interference tests can be used to characterize unsaturated heterogeneous soils and fracture networks. To conduct the field tests in fractured rocks, strings of straddle packers fitted with injection and monitoring ports are employed to seal off discrete depth intervals in different boreholes (Cook 1999). Field experiments in unconsolidated sediments are typically conducted by connecting a blower up to the top of the well casing and injecting or extracting air from the well screened interval. Figure 3-20 shows a schematic of the automatic packer design. Ambient-temperature air or a gas-tracer can be injected at a constant mass rate into a given depth interval in an injection well, and pressure responses are measured at different intervals in monitoring wells. The first type of test represents zero-offset profiles (ZOP), in which the injection packer string and monitoring strings are moved together along the length of the borehole by the same increment. The second type of test represents multiple-offset profiles (MOP), with monitoring locations in other wells, which are held fixed while the injection zone is moved incrementally along the length of the borehole. These tests can be repeated systematically for different combinations of wells, and injection and monitoring intervals. The changes in the injection mass flow and pressure response are used to assess the three-dimensional pneumatic connectedness of the formation (Benito *et al.* 1999; Cook 1999). This type of information can be combined with geophysical, geologic, and other hydrologic data to improve the understanding of flow paths and develop an adequate conceptual model of the site.

Air-injection tests have been successfully used in determining the air permeability and the geometry of flow in relatively dry soils and fractured rocks at several sites (Rasmussen *et al.* 1995; Granovsky and McCoy 1997; Illman *et al.* 1998; Benito *et al.* 1999). For example, air-injection tests have been used successfully at the Yucca Mountain Site in Nevada, which is a potential site for a geologic high-level nuclear waste repository. To date, an estimated 3,500 separate air injections have been undertaken, and nearly a quarter of a million pressure-response curves have been logged in the study (Cook 1999).

Pneumatic pumping experiments are conducted by extracting gas from a borehole at an open (or screened) interval and measuring spatial and temporal distributions of the decrease in the gas pressure using pressure sensors installed in the injection and observation wells. For example, Figure 3-33 shows schematics of the vacuum extraction from a soil profile open to the atmosphere. The flow pattern developed around the

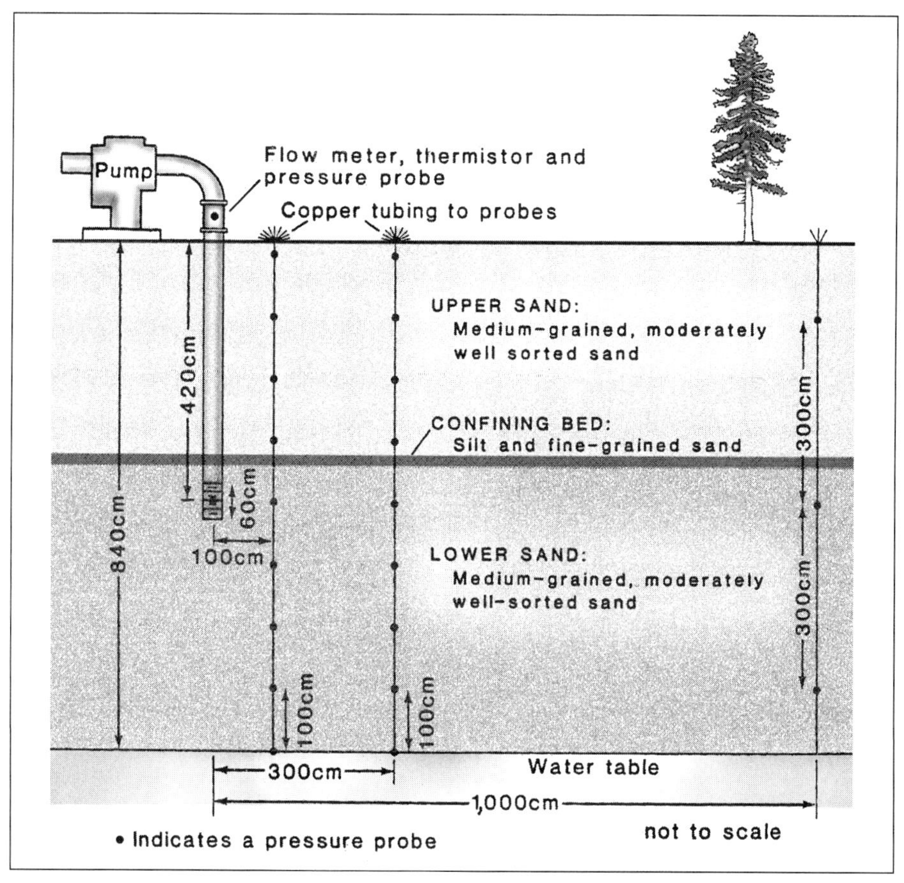

Figure 3-33. Schematic of site instrumentation for conducting pneumatic tests in the vadose zone (Baehr and Hult 1989).

injection well depends on the boundary conditions. The design and performance of the pneumatic pumping tests are described in detail by, Cho and DiGiulio (1992), Edwards and Jones (1994), Massmann (1989), and Massmann and Madden (1994).

Models of Airflow

Contrary to liquid flow in porous media, the gas velocity at the pore wall cannot always be assumed to be zero. This is because of an additional flow component termed the "slip flow" or "drift flow" (Klinkenberg 1941), also known as "Klinkenberg's effect." Klinkenberg's effect leads to the total gas flow exceeding the viscous flow predicted by Darcy's law. The ratio of the slip flow to the viscous flow increases as the average pore radius decreases. For example, the effect of slip flow is important in silt and clay sediments when the pore radius is less than approximately 10^{-3} mm. The effect of slip flow is negligible in sand and gravel materials when the pore radius is from 10^{-2} to 10^{-1} mm and larger (Massmann 1989).

Assuming the validity of Darcy's law for flow of ideal gas of a constant viscosity and composition in isothermal and constant volumetric gas content media, and neglecting slip flow, transient gas flow is described by a nonlinear equation (Bear 1972):

$$nS_g\mu \frac{\partial P}{\partial t} = \nabla(\mathbf{k}_g P \nabla P) \tag{3.4}$$

where n = porosity, S_g = gas saturation, μ = gas viscosity, P = gas pressure, and $\mathbf{k}_g$ = gas-permeability tensor (including gas phase relative permeability effects). This nonlinear equation can also be written in terms of the pressure squared:

$$\frac{nS_g\mu}{P} \frac{\partial P^2}{\partial t} = \nabla \mathbf{k}_g \nabla P^2 \tag{3.5}$$

Table 3-14 summarizes the assumptions and corresponding limitations used in deriving a model of gas flow in porous media described by equations (3.4 and 3.5). For small pressure gradients, a linear approximation of equation (3.4) can be used (Massmann 1989)

TABLE 3-14 Summary of Assumptions and Limitations (After Massmann 1989)

Assumptions	Limitations
The equation of motion for gas transport can be approximated using equations similar to Darcy's law	In fine-grained materials, Darcy's law underestimates discharge by neglecting slip flow; however, this assumption is likely a valid approximation for flow in sand and gravel
Effects of diffusional flow are negligible	Valid assumption for predicting pressure distributions
Vapor behaves as ideal gas	Valid approximation of temperature and pressure conditions typical for vapor-extraction systems
Constant and uniform porosity	Porosity will generally vary with time and with location due to natural variations in geologic materials and temporal and spatial variations in moisture content
Molecular weight is uniform	Molecular weight will vary with gas composition; variations will generally be small for typical applications of methane control and organic vapor recovery
Gravitational effects are negligible	Valid assumption for vapor extraction applications
Compressibility of the porous media is negligible	Valid assumption because compressibility is essentially small as compared to that of vapor
Gas transport can be modeled using the equation for incompressible flow	Valid assumption for pressure variations on order of one-half an atmosphere and less

$$nS_g \mu \frac{\partial P}{\partial t} = \nabla(\mathbf{k}_g P_o \nabla P) \tag{3.6}$$

where P_o = initial or static gas pressure. For larger pressure gradients, it is more accurate to linearize equation (3.5):

$$\frac{nS_g \mu}{P_o} \frac{\partial P^2}{\partial t} = \nabla \mathbf{k}_g \nabla P^2 \tag{3.7}$$

Equations (3.6 and 3.7) show that five parameters are needed to simulate gas flow, including viscosity, average gas pressure, porosity, gas saturation, and the gas phase permeability (the product of the intrinsic permeability and the gas phase relative permeability) of the porous media. Note that equations (3.6 and 3.7) assume a constant volumetric gas content (the product of gas-phase saturation and the porosity) and a constant gas composition. It should be noted here that while analytical gas flow models are usually based on equations (3.6) or (3.7), several numerical simulators are capable of solving the more general problem with mixtures of nonideal gases, slip flow and variable gas saturation (see Chapter 5).

Analytical and Numerical Solutions and their Limitations

Analytical solutions of equation (3.7) for air permeability are usually based on assumptions of either one-dimensional radial (McWhorter 1990) or two-dimensional radial and vertical axisymmetric flow of compressible air toward a partially penetrating well screened in the vadose zone (Baehr and Hult 1989, 1991; Massmann 1989; Shan *et al.* 1992). Falta (1996) developed a program called GASSOLVE for analyzing several types of transient and steady-state soil gas pump tests, which are listed in Table 3-15. In developing his program, Falta (1996) used analytical solutions from Hantush (1964), Baehr and Hult (1989), and Falta (1993).

Illman *et al.* (1998) analyzed the data of a series of multistep, single-hole pneumatic tests conducted by Guzman *et al.* (1994) and Guzman and Neuman (1996) that used transient pressure and pressure-derivative type-curve methods. They determined that airflow around the 1-m long borehole injection intervals appeared mostly to be three-dimensional; air compressibility led to a significant borehole storage effect, rendering the first step of a single-hole test to be unreliable; borehole skin effects were minimal; and air permeabilities determined from steady-state and transient-type-curve methods were practically the same. The accuracy of estimates of air-filled porosity and air permeability increases if multi-step injection experiments are complemented by air-pressure recovery experiments (Vesselinov and Neuman 1999). However, an application of the type-curve approach is limited because the subsurface formation around the injection interval is assumed to be homogeneous, a linear equation of airflow is used, and the borehole storage effect is taken into

TABLE 3-15 Types of Gas Pump Tests Analyzed by GASSOLVE (Falta 1996)

Type of soil gas pump test	Parameters to be determined
Transient, open to the atmosphere	k_r, k_z (θ_g)
Steady-state, open to the atmosphere	k_r, k_z
Transient, fully confined, fully penetrating well	k_r, (θ_g)
Transient, fully confined, partially penetrating well	k_r, k_z, (θ_g)
Transient, leaky confining layer, fully penetrating well	k_r, (k'/h'), (θ_g)
Transient, leaky confining layer, partially penetrating well	k_r, k_z, (k'/h'), (θ_g)
Steady-state, leaky confining layer, fully penetrating well	k_r, (k'/h')
Steady-state, leaky confining layer, partially penetrating well	k_r, k_z, (k'/h'),

Legend: k_r = radial air permeability,
k_z = vertical air permeability,
k' = permeability of the leaky confining layer,
h' = thickness of the leaky confining layer.

account only for a single monitoring interval at a time, neglecting the air storage effect in the rest of the injection borehole.

Numerical inverse modeling has several advantages over analytical methods. For example, Vesselinov and Neuman (1999) considered airflow through a nonuniform, locally isotropic, porous continuum system, including all boreholes with multiple packed-off intervals, and simultaneously measured pressure data at several monitoring intervals, taking into account atmospheric pressure fluctuations. Their method also provided kriged estimates of spatial variations in air permeability and air-filled porosity throughout the tested fractured rock volume.

Analytical and numerical solutions of airflow equations often assume constant parameters; however, parameters may vary with time because of a variety of processes affecting airflow. If one neglects these processes, one may obtain erroneous prediction results and then design ineffective remediation systems. For example, gas flow in the subsurface can induce volatilization and evaporation of liquid water and nonaqueous phase liquids (NAPLs). These changes in the liquid saturation can have a significant effect on the gas phase relative permeability. Vacuum pumping involves both radial air movement from the soils surrounding the well and vertical flow from the atmosphere through an uncovered land surface or cracks in the cover (Edwards and Jones 1994). Neglecting the leakage through a soil cover (for example, through cracks in asphalt) toward the pumping well is a source of errors in the determination of air permeability (Massmann and Madden 1994).

The values of air permeabilities determined from field tests may be several orders of magnitude higher than those determined from laboratory core experiments and the permeability determined from hydraulic tests (Guzman *et al.* 1996). This is mainly because air permeability determined using laboratory cores represent the rock matrix and air permeability determined using field experiments represent a soil/rock system affected by zones of fast, preferential flow. Such zones are usually absent in small laboratory cores. However, the complexity of the flow field in a heterogeneous formation leads to a poor correlation between the values of air permeability and the fracture geometry parameters—such as density, trace length, orientation, aperture, and roughness (Neuman 1987). Because the directions of airflow depend on a combined effect of the geometry of the injection interval, flow paths in the fractured-porous system, and boundary conditions, the flow dimensionality may not be an integer (1, 2, or 3). It can be a noninteger or fractional (Barker 1988).

Massmann and Madden (1994) demonstrated that the air conductivity determined using horizontal wells was about twice that determined using vertical wells. Cho and DiGiulio (1992) demonstrated that horizontal permeabilities determined from air-injection tests are slightly higher than those determined from vacuum extraction tests. The difference can be caused by subtle layering or soil stratification at the sites. Because the range of air permeability determined from small-scale core measurements and single-hole experiments varies randomly in space by orders of magnitude and exceeds the range of air permeability from the variations of applied pressure, one can use geostatistical methods to analyze the spatial distribution of flow parameters.

Air permeability determined from air-extraction tests can be as much as 20% larger than that from injection tests (Massmann and Madden 1994). The reason for this is that the injection of warmer air in the formation (which usually has a lower temperature) results in water condensation in the near-well zone. To avoid this problem, it is recommended to use N_2 gas or run the ambient air through a desiccant before injecting it into the well.

Because air permeability is a function of the soil moisture content, an increase in the moisture content of soils leads to a reduction in air permeability (Stonestrom and Rubin 1989a, b; Guzman *et al.* 1994; Guzman and Neuman 1996). Because the matrix water saturation is usually higher than that of fractures, the matrix air permeability may become insignificant compared to that of unsaturated, air-filled fractures. In this case, the air permeability can be determined using numerical inverse modeling by treating air as a single mobile fluid phase in a continuum medium represented primarily by interconnected air-filled fractures (Vesselinov and Neuman 1999).

Air injection into a formation containing some moisture causes pressure buildup followed by a slight drop, as water is pushed away from the near-borehole zone or it evaporates. The amount of time required for pressure in the injection interval to stabilize typically ranges from 30 to 60 minutes. However, as the flow rate increases, the nonsteady flow regime may last for 24 hours or longer, as a result of water displacement from air-conducting fractures. Consequently, the gas permeability increases. Thus, short-term air-injection tests may lead to an incorrect estimation of air permeability. In large open fractures (or fracture zones), the inertial effects grow as the applied pressure increases. Experimental observations in fractured rocks have shown that the relationship between air-pressure change and the airflow rate, which takes into account both viscous flow and inertia effects, is given by (van Golf-Racht 1982)

$$\Delta(P^2) = AQ + BQ^2 \tag{3.8}$$

where A and B are coefficients representing the effects of viscous (laminar) and inertial (turbulent) flow, respectively. The values of A and B can then be used to calculate the rock permeability (van Golf-Racht 1982). Equation (3.3.11.4-5) can be represented as

$$Q/\Delta(P^2) = 1/(A + BQ) \tag{3.9}$$

where the ratio $Q/\Delta(P^2)$ is called the gas deliverability index (van Golf-Racht 1982), which characterizes the ability of soils/rocks to transmit gas.

An alternative to pneumatic pumping and injection experiments is the determination of the vertical air permeability of the unsaturated soils and rocks using the analysis of the time trend of atmospheric pressure and air pressure at different depths of the vadose-zone profile (Weeks 1978; Shan 1995).

Thus, because a variety of environmental processes and factors affect air pumping/injection and the test design, one needs to study airflow processes and determine air permeability by taking into account site-specific conditions and using the same well design as planned for the SVE system (Cho and DiGiulio 1992). It is important to conduct air-injection experiments using single-hole or cross-hole air-injection tests at several applied flow rates and pressures.

MONITORING FOR NONAQUEOUS PHASE LIQUIDS*

The removal of residual solvents, primarily dense nonaqueous-phase liquids (DNAPL), is currently the most significant challenge for the successful completion of many large groundwater and soil-cleanup efforts. Slowly dissolving DNAPL may provide a major source of vadose zone and groundwater contamination for hundreds of years. The problem is further complicated by the fact that DNAPLs are present as dispersed blobs at many sites and are therefore very difficult to characterize in the subsurface. At waste sites where DNAPL contamination is suspected, robust characterization of the nature and extent of the contamination is an essential component of any comprehensive remediation strategy.

Traditional sampling approaches usually are unsuccessful in locating DNAPL. Many of the current methods used for characterizing DNAPL-contaminated sites are described by Cohen and Mercer (1993). These methods generally consist of inferred measurements of DNAPL (such as soil-gas analysis and geophysical methods), rule-of-thumb empirically developed methods from aqueous well samples, and direct measurements using invasive methods such as drilling and soil sampling. Most

*This section was contributed by Joe Rossabi, Carol Eddy-Dilek, and Brian Riha.

geophysical techniques do not have the resolution needed to detect DNAPL present at scales smaller than one cubic meter. Conventional soil and liquid sampling are too costly to be used for detailed DNAPL characterization. However, precise delineation of DNAPL-contaminated areas will facilitate the design of appropriate remediation strategies and prevent the escalation of cleanup costs.

Because of the complexity of spatial distribution of DNAPL in the subsurface, several characterization methods should be used in an ensemble approach. The techniques described in this section were designed specifically for implementation with the cone penetrometer (CPT). This takes advantage of the high-resolution geologic information obtained with the CPT.

Above the water table, DNAPL resides in intergranular pores held by capillary forces. Below the water table, DNAPL behaves in a complex fashion, moving downward as an immiscible phase and accumulating in highly concentrated discrete and dispersed ganglia. Because of the physical and chemical characteristics of DNAPL, characterization and remediation methods that minimize unnecessary waste generation are prudent.

Many strategies and tools target the refractory case of DNAPL occurring in thin, highly discrete zones, which are typical of most sites. The innovative DNAPL characterization tools that have proved to be most successful in field tests include: (1) hydrophobic sorbent ribbon on FLUTe™ (also known as SEAMIST™) membrane, and (2) Laser-Induced Fluorescence (LIF), Raman, and Optical Cone Penetrometer Test (CPT) probes. The case study "Cone Penetrometer-Based Raman Spectroscopy for DNAPL Characterization in the Vadose Zone," by J. Rossabi, B.D. Riha, J. Haas, C.A. Eddy-Dilek, A. Lustig, M. Carrabba, K. Hyde, and J. Belo, provides a detailed description of the cone penetrometer-based Raman spectroscopy.

These technologies have been successfully demonstrated and will complement tools currently used or proposed by industry, DOE, the U.S. Environmental Protection Agency (EPA), and the U.S. Department of Defense (DoD). The innovative characterization technologies (such as CPT-based Raman and FLUTe™) build on the baseline DNAPL characterization techniques and generally strive for direct detection of DNAPL with minimal invasion and minimal investigation-derived waste (IDW).

In addition, several other promising technologies have been tested, such as alcohol micro-injection/extraction through CPT, differential-partitioning gas tracer tests, and measurement of radon partitioning to DNAPL. However, these technologies require additional development.

At the DOE Savannah River Site, a package of innovative DNAPL characterization tools is being developed and deployed to:

- Unambiguously identify DNAPL in the subsurface
- Minimize secondary waste
- Eliminate undesirable gravitational movement of DNAPL
- Minimize IDW
- Mitigate similar types of collateral environmental damage inherent in addressing this complex environmental need.

Rapid Hydrophobic Sampling

The rapid hydrophobic sampling system is fast and easy to deploy with a cone penetrometer system, and it yields depth-discrete samples from boreholes. For this technique, hydrophobic sorbent ribbons (that is, ribbons that preferentially absorb non-polar liquids) are attached to the liner of small-diameter FLUTe membranes (see the subsection "Removable SEAMIST™ Liners," above). The ribbons are designed to collect DNAPL samples (Figure 3-34), and are impregnated with a DNAPL indicator dye for immediate assessment of the presence of DNAPL's at a specific depth. The ribbon is pressed against the formation on the walls of the borehole, and the hydrophobic material preferentially collects organic liquids. The liner is then retrieved from the borehole and is rapidly scanned both visually and with a volatile organic compound (VOC) analyzer. After screening, the depth-discrete sorbent pads can be analyzed in more detail in the laboratory. The FLUTe can also be deployed using drilling methods.

Laser-Induced Fluorescence (LIF), Optical, and Raman-Cone Penetrometer Methods

The cone penetrometer is particularly suited for characterization of DNAPL-contaminated sites because of its ability to deploy a variety of

Figure 3-34. Picture of FLUTe™ membrane recovered from a CPT borehole. Dark marks indicate DNAPL at that depth.

sensors as well as its capacity to delineate depth-discrete lithology and contaminant distribution with relative ease. Because chlorinated alkanes do not fluoresce at standard excitation wavelengths, LIF sensors cannot measure chlorinated alkanes directly. However, fluorescent intensities are found to increase one to three orders of magnitude over the background in zones known to contain DNAPL. The large increase may be due to the leaching of natural organic matter or the incorporation of other likely fluorophores into the DNAPL. Co-disposed lubricants, hydraulic oils, and cutting oils are also potential candidates for fluorescence probing. Thus, the fluorescence measurements can be used to infer the presence of DNAPL. Used in concert with Raman spectra, the presence of DNAPL in a particular location can be confirmed. For details of this method, see the case study "Cone Penetrometer-Based Raman Spectroscopy for DNAPL Characterization in the Vadose Zone, by J. Rossabi, B.D. Riha, J. Haas, C.A. Eddy-Dilek, A. Lustig, M. Carrabba, K. Hyde, and J. Belo."

Raman spectroscopy is one of the few direct detection-characterization technologies for DNAPL. Each compound has a unique Raman spectrum that can be probed through the optics deployed in a cone penetrometer. Thus, specific DNAPL compounds can be identified. Unfortunately, the Raman technique is inherently weak, and the spectra must be separated from the fluorescence spectrum, which often dominates over it. Other optical techniques such as CPT video microscopy (for example, the Navy GeoVis system) have also helped to identify DNAPL in the subsurface. Specific formations can be visually identified for DNAPL potential for precise targeting by spectroscopy. If co-constituents color the DNAPL, DNAPL may be directly identified.

Small-Scale Alcohol Micro Injection/Extraction Test

The single-well, alcohol injection-extraction test uses a cone penetrometer delivery system through which less than one gallon of a solution of alcohol is injected. The injected fluid permeates into an area of the size of a small cylinder around the CPT, and can solubilize DNAPL without mobilizing it. A small volume of water is injected a small distance into the formation (less than 1 ft) and is then extracted. The extracted water samples are analyzed to determine the concentration of organic contaminants. Then, a small volume of an alcohol and water solution is injected a similar distance into the formation and is subsequently extracted. The extracted solution is sampled and analyzed. DNAPL is significantly more soluble in the alcohol/water solution than in water alone. A large increase in the concentration of DNAPL components is an unequivocal indicator of the presence of residual DNAPL. The test provides clear confirmation of the presence of DNAPL without the drilling of additional holes. The test has been used to target specific strata that were thought to contain DNAPL (that is, above clay in the saturated zone).

Limitations

The use of CPT technologies is limited to unconsolidated sediments and to depth refusal of the cone penetrometer. Comparison of the results of the DNAPL detection using core sampling, Raman spectroscopy, and FLUTe data shows that the data complement each other. However, we cannot obtain a unique result about the DNAPL distribution from single-borehole measurements (Figure 3-35). At the same time, the design of remediation requires knowledge of the total volume of DNAPL in the subsurface. For this purpose, cross-borehole, partitioning interwell tracer tests can be used.

The case study "The First Vadose Zone Partitioning Interwell Tracer Test (PITT) for NAPL and Water Residual," by Paul E. Mariner, Minquan Jin, James E. Studer, and Gary A. Pope, describes a method of estimating the total volue of DNAPL in subsurface. *See page 491.*

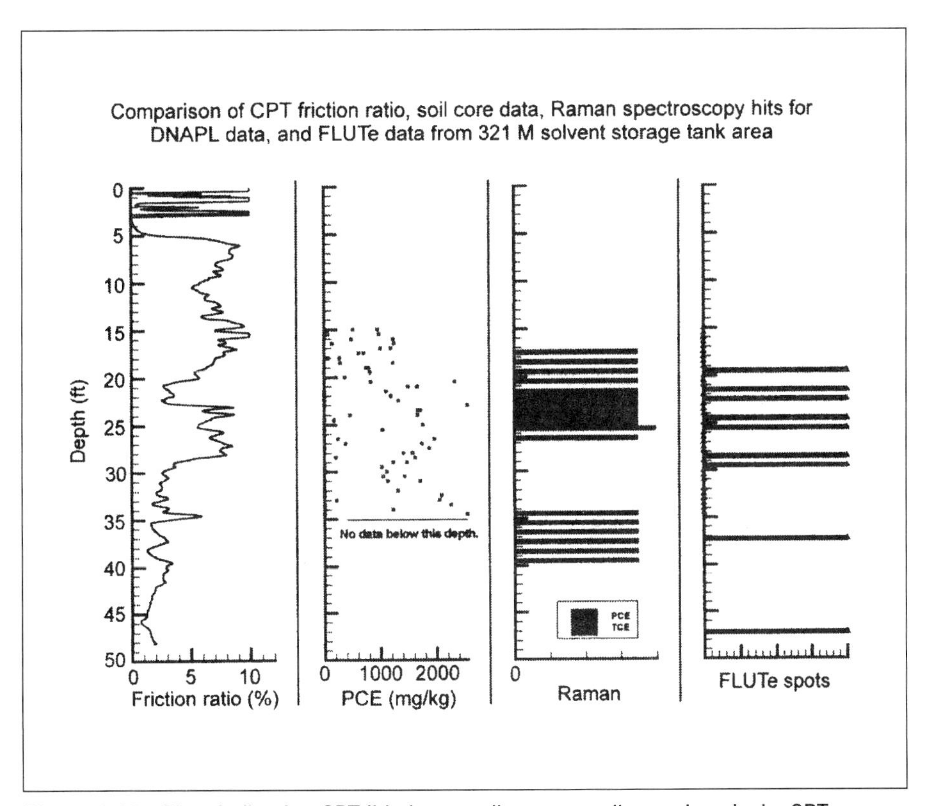

Figure 3-35. Plots indicating CPT lithology, sediment sampling and analysis, CPT Raman spectroscopy identification of PCE and TCE, and FLUTe membrane identification of DNAPL contamination at SRS solvent storage tank area. Note the benefits of complementary data.

ISOTOPIC TRACERS OF FLOW AND TRANSPORT THROUGH THE VADOSE ZONE*

Applications of environmental isotope tracers are complementary to conventional water-flow and solute-transport investigations. This is especially true in arid and semi-arid regions that are commonly characterized by extensive zones of unsaturated rock. Natural variations in the hydrogen and oxygen stable isotope ratios of water were first noted in the early 1950s (Friedman 1953; Epstein and Mayeda 1953). Since that time, a number of different isotopic methods have been developed for studying hydrologic problems, including groundwater flow and mixing, recharge areas of aquifers, subsurface residence times, and multi-phase reactive exchange of chemicals (Phillips 1995). In this section, a brief overview of several of the more commonly used isotopic techniques for tracking the movement of water through the vadose zone will be presented.

Hydrogen and Oxygen Isotope Ratios of Water

Craig (1961) found that the hydrogen (δD) and oxygen ($\delta^{18}O$) isotope ratios of most rainwater lie on or near a line of slope 8 on a plot of δD versus $\delta^{18}O$. This line is commonly referred to as the Global Meteoric Water Line (GMWL) and is shown in Figure 3-36. The relationship between δD and $\delta^{18}O$ for rainwater from a specific area relative to the GMWL is a function of a variety of factors, including the distance from the ocean, altitude, and temperature. In regions where there are large variations in the δD and $\delta^{18}O$ values of rainwater (for example, in the western United States), differences in the isotopic compositions of groundwater have been used to distinguish sources of recharge, flow directions, and groundwater mixing (see for example, Stahl *et al.* 1974; Hearn *et al.* 1989; Criss and Davisson 1996; Davisson *et al.* 1999).

The δD and $\delta^{18}O$ values of rainwater can also exhibit significant seasonal variations. This difference is most striking in arid or semi-arid regions, which are generally characterized by large temperature variations between summer and winter rainy seasons. These natural, seasonal changes in the isotope compositions of rainwater have been used in a number of studies to estimate infiltration rates (see for example, Phillips

*This section was contributed by Mark Conrad and Boris Faybishenko.

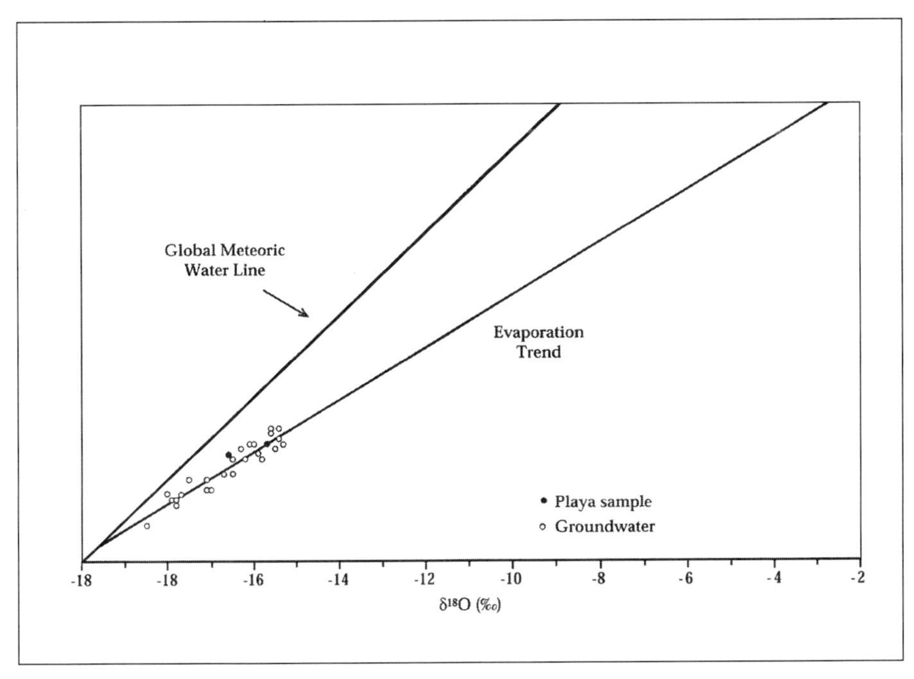

Figure 3-36. $\delta^{18}O$ and δD data for groundwater samples collected from monitoring wells at the TAN site of the Idaho National Engineering and Environmental Laboratory and from playas adjacent to the TAN site. Also shown are the global meteoric water line and a best-fit line through the data, which is equivalent to a typical trend of data observed for evaporated waters (Conrad *et al.*, 1999).

1994; Allison *et al.* 1994; Liu *et al.* 1995; Newman *et al.* 1997). However, water-vapor diffusion and isotopic exchange can attenuate these seasonal signals over relatively short time -scales, limiting the utility of these studies to relatively recent infiltration events.

The hydrogen and oxygen isotope ratios of surface waters can be altered if they undergo significant evaporation before they infiltrate into the subsurface. As water evaporates, lighter isotopes are preferentially separated into the vapor phase, and the residual water becomes enriched in D and ^{18}O. Depending on the humidity and temperature of an area, the slopes of evaporation trends on plots of δD versus $\delta^{18}O$ are generally between 3 and 6. The result is that highly evaporated waters will lie significantly off of the GMWL on δD-$\delta^{18}O$ plots. The distinctive isotopic compositions of evaporated surface waters can be used to track infiltration of those waters through the vadose zone (Gat *et al.* 1969; Gonfiantini *et al.* 1974; Komor and Emerson 1994; Gaye and Edmunds 1996). An example of this is given in Figure 3-36. At this site at the Idaho National Engineering and Environmental Laboratory in eastern Idaho, the isotope compositions of the groundwater have been significantly shifted by infiltration of strongly evaporated water from playa lakes. In addition to the natural evaporative process, waters with isotope compositions shifted by evaporation during industrial processes can also be used to identify and trace input of water from leaking tanks and pipes (Ingraham 1994).

It is also possible to use water artificially enriched in the heavy isotopes (D, ^{18}O) as tracers for studying infiltration through the unsaturated zone (Swenson 1997; Anderson *et al.* 1997). An advantage of using D_2O and $H_2{}^{18}O$ is that they serve as direct tracers of water flow and are not affected by processes that can influence introduced solute tracers. Until recently, tracer studies using D_2O and $H_2{}^{18}O$ have been limited by relatively time-consuming and expensive analytical procedures. However, recent advances in stable-isotope analytical techniques have greatly reduced the time and cost per analysis, and should lead to more widespread use of these isotopic tracers.

Dating Water Using Cosmogenic Radionuclides

Cosmogenic radionuclides are a special group of isotopes that have gained widespread use for dating soil water and estimating net infiltration rates through the vadose zone. These isotopes are naturally produced at low levels by interaction between cosmic rays and the atmosphere. The most widely used are ^{3}H, ^{14}C, and ^{36}Cl. Other cosmogenic radionuclides that deserve mention, but will not be discussed here, include ^{39}Ar (Loosli 1983) and ^{129}I (Fabryka-Martin *et al.* 1985).

In addition to natural production, above-ground testing of nuclear bombs during the 1950s and 1960s also produced ^{3}H, ^{14}C, and ^{36}Cl. For ^{3}H and ^{36}Cl, these increases were very large and dominate the natural signals in recent waters. Since above-ground testing was halted, the atmospheric levels have been dropping back to normal (pre-bomb testing) atmospheric levels. These signals have provided additional opportunities for tracking water movement in the vadose zone, especially on short time-scales.

^3H (tritium) has a relatively short half-life (12.4 years). Concentrations are measured in tritium units (TU) that correspond to 1 ^3H atom in 10^{183} hydrogen atoms (Libby 1971). The normal background concentration is low (between 2 and 10 TU). For these reasons, under normal circumstances, tritium is only useful for identifying an input of very recent water. However, tritium levels from nuclear testing reached >2000 TU in the northern hemisphere and >50 TU in the southern hemisphere, peaking during 1963-64. In the southern hemisphere, ^3H concentrations have already decayed to near pre-bomb atmospheric levels and are of limited use (Cook *et al.* 1994). In the northern hemisphere, the bomb pulse can still be clearly distinguished and has been used in numerous studies to date infiltration rates of recent waters (see for example, Chapman *et al.* 1992; Scanlon 1992). As with stable hydrogen isotope ratios, however, tritium signals can also be attenuated by vapor exchange that can smear out the bomb-pulse signal in the unsaturated zone.

^{14}C has a significantly longer half-life (5730 years) than ^3H and can be used to date much older events. The natural ratio of $^{14}C/C$ is 1.175 ××x 10^{-12}. During nuclear bomb testing, the concentration of ^{14}C only reached approximately twice its natural levels (Lehman *et al.* 1993). This level of ^{14}C does not overwhelm the natural signal, but should be considered when used to interpret transport of recent waters.

Using ^{14}C concentrations to date groundwater assumes that water infiltrating through the root zone will pick up dissolved inorganic carbon (DIC) from sources in equilibrium with the atmosphere (for example, from root respiration or decay of recent organic matter). Then, as the water infiltrates deeper into the soils, the DIC will remain with the water, allowing dating by radioactive decay. The most significant problem with using ^{14}C for studying transport in the vadose zone is its reactive nature. Shallow soils often contain relatively high concentrations of organic matter, some of which may be relatively old. If a significant amount of microbial degradation of "old" organic matter occurs along the infiltration pathway, this can lead to erroneously old ages for the waters. It is also possible that dissolution of soil carbonates or carbonate rocks in the subsurface can add significant amounts of old carbon to the DIC in the waters, leading to the determination of anomalously old ages. Finally, significant inputs of magmatic carbon (for example, in active volcanic or geothermal areas) can also yield old ages. It is possible to recognize inputs of old carbon from some of these sources by measuring the stable carbon isotope ratios ($\delta^{13}C$ values) of the DIC or associated soil gas CO_2 (see below), but care should be taken in interpreting this data (Landmeyer *et al.* 1995; Johnson and DePaolo 1996).

^{36}Cl has the longest half-life (301,000 years) of these three tracers. During bomb testing in the mid-1950s, its atmospheric concentration increased to more than 1,000 times natural levels. Because of the long half-life, these concentrations do not change significantly over the short term (unlike tritium). Natural atmospheric concentrations of ^{36}Cl can vary considerably, based on latitude and proximity to coastal areas. Also, a variety of subsurface processes (including evapotranspiration, *in situ* production, and radioactive decay) can affect both the concentration of ^{36}Cl and the $^{36}Cl/Cl$ ratio.

^{36}Cl concentrations, when combined with chloride, ^{14}C, and ^{3}H concentrations, provide a powerful way of for quantifying infiltration rates through the vadose zone, origin of surface waters, erosion processes, and other hydrologic processes (Scanlon *et al.* 1990; Scanlon 1992; Cook *et al.* 1994; Murphy *et al.* 1996). In soil profiles of arid regions, high $^{36}Cl/Cl$ ratios (exceeding 9×10^{-12}) often appear at depths of 0.5-2 m from the surface (Philips 1994) resulting from slow migration of ^{36}Cl released from bomb pulse in the 1950s. Note that the background ratios of $^{36}Cl/Cl$ range from 0.5 to 1.5×10^{-12} in the western USA (Davis *et al.* 1998). The presence of bomb-pulse ^{36}Cl in deep waters at Yucca Mountain (the proposed high-level nuclear waste repository in Nevada), has significantly changed perceptions of the hydrology of the site, underscoring the potential role of fast pathways for infiltration of water through the vadose zone (Fabryka-Martin *et al.* 1993; Levy *et al.* 1997).

Reactive Isotope Tracers

Reactive isotopic tracers are those that can interact with the matrix to acquire its isotopic signature. They can be particularly useful for determining the nature of interaction between infiltrating fluid and specific subsurface features. Generally, oxygen and hydrogen isotopes are not considered reactive isotopes at low temperatures, but at higher temperatures (for example, in geothermal systems) they can be significantly shifted by interaction with the host rocks. The degree of shift can be used to estimate the minimum water to rock ratio in the system (for example, Taylor 1974).

At lower temperatures, the most sensitive isotopic tracers are those with low solubility in the water relative to their concentration in the matrix. Two examples of reactive isotope tracers are the $\delta^{13}C$ values of pore-water DIC and the $^{87}Sr/^{86}Sr$ ratios of dissolved Sr. There are a number of other isotopic tracers that have been or could potentially be used to study infiltration and transport in the vadose zone (for example, the isotopic ratios of metals and radionuclides to track their movement through the vadose zone; McCarthy *et al.* 1998). However, we will focus here on carbon and strontium isotopes.

The stable carbon isotope ratios of DIC in vadose zone pore-waters are directly related to the $\delta^{13}C$ values of associated CO_2. Their isotope compositions are the functions of factors such as the pH of the pore-fluids pH and the temperature (Wigley *et al.* 1978). Further, since CO_2 is a gas phase and is relatively mobile, it can be affected by exchange with the atmosphere (Cerling 1984). As discussed above, the general assumption is that pore-water DIC attains its carbon isotope composition in the relatively high CO_2 region within the root zone of surface soils. However, there are a number of other subsurface processes that can significantly affect the carbon isotope compositions of DIC/CO_2 in the unsaturated zone and are direct indications that these processes are occurring in the subsurface. Examples include microbial degradation of hydrocarbon contaminants (Conrad *et al.* 1997) and interaction with subsurface carbonate units (Johnson and DePaolo 1996).

The strontium isotopc ratio of dissolved strontium in water is very sensitive to interaction with the rock matrix, even at the relatively low temperatures characteristic of most groundwater systems. The concentrations of strontium in water are typically 1,000 times less than the concentrations in the rock. After relatively minor degrees of interaction with the matrix, the $^{87}Sr/^{86}Sr$ ratios of the water will quickly become dominated by the $^{87}Sr/^{86}Sr$ ratios of the rock. The length-scale of this change can be used to calculate factors such as the flow velocity of the fluids and/or the relative amounts of water and rock in the system (Johnson and DePaolo 1997a, b).

Characterization and Monitoring for Bioremediation*

Characterization and monitoring of bioremediation can be as simple as maintaining a fermentor for above-ground processes like prepared

*This section was contributed by Terry Hazen.

beds, land farming, bioslurry reactors, composting, and bioreactors. The terrestrial subsurface is much more difficult because of sampling problems, poorly defined interfaces, and spatial heterogeneity. For any type of bioremediation, careful consideration and planning must be given to the remediation objectives, sampling, the types of samples, frequency, cost, priority, and background literature for method verification. The microbiology and chemistry may be of less overall importance to the remediation of the site than the hydrology, geology, meteorology, toxicology, and engineering requirements. All of these things must be integrated into the plan for characterization and monitoring of any site. For examples of test plans for bioremediation, see Hazen *et al.* (1991), Lombard and Hazen (1994), and Nelson *et al.* (1994).

The type of sample used for monitoring and characterization of sediment or groundwater can have a significant impact on a bioremediation project. Fortunately, most bioremediation applications are shallow and eutrophic, owing to the nature of the waste mix usually deposited. Enzien *et al.* (1994) further underscored the need for careful sampling when they showed significant anaerobic reductive dechlorination processes occurring in an aquifer whose bulk groundwater was aerobic (greater than 2 mg/L O_2).

It is extremely difficult to determine the rate and amount of contaminant that is bioremediated in any environment. Many of the problems and measurements discussed above for mass balance also apply here.

In recent years, bioremediation studies have focused on the measurement of biodegradation products rather than organisms, due to the difficulty in measuring organisms. Soil and groundwater measurements of microorganisms often require long incubations or long preparation times, and the measurements are usually not specific to contaminant-degraders. Several methods have been used to determine the rate and amount of biodegradation: monitoring of conservative tracers, measurement of byproducts of anaerobic activity, intermediary metabolite formation, electron acceptor concentration, stable isotopic ratios of carbon, and the ratio of non-degradable to degradable substances. Helium has been used at a number of sites as a conservative tracer since it is non-reactive and non-biodegradable, and moves like oxygen (National Research Council 1993). By simultaneously injecting He with O_2 at known concentrations and comparing the subsurface ratios over time, the rates of respiration can be calculated. This technique has also been used to measure rates of injected methane consumption (Hazen 1991). Bromide has been successfully used as a conservative tracer for liquid-injection comparisons with nitrate, sulfate, and dissolved oxygen (National Research Council 1993). By-products of anaerobic biotransformation in the environment have been used to estimate the amount of biodegradation that has occurred in anaerobic environments, (for example, PCB-containing sediments). These byproducts include methane, sulfides, nitrogen gas, and reduced forms of iron and manganese (Harkness *et al.* 1993). Measurements of chloride changes have also proven useful in indicating the amount of chlorinated solvents that have been oxidized or reduced (Hazen *et al.* 1994). Consumption of electron acceptors (O_2, NO_3, or SO_4) has been used for measuring rates of biodegradation and bioactivity at some bioremediation sites (National Research Council 1993; Smith *et al.* 1991). Bioventing remediations of petroleum-contaminated sites rely on stable isotopic ratios of carbon, carbon dioxide production, and oxygen consumption to quantify biodegradation rates in the field (Hinchee *et al.* 1991; Hoeppel *et al.* 1991). Mixtures of contaminants (for example, petroleum hydrocarbons) can have their own internal standards for biodegradation. By comparing concentrations of nonbiodegradable components of the contaminant source with concentrations of degradable components from both virgin and weathered sources, the amount of degraded contaminant can be calculated. These measurements have been used on the Exxon Valdez spill cleanup (Glasser 1994) and at a number of other petroleum-contaminated sites (Breedveld *et al.* 1995).

Microbial ecologists have continually struggled with methods to identify the organisms in the environment, and to measure how many organisms are present and how active they are. For bioremediation, we need to know what contaminant-degraders are present, how many are present, and how active they are. We may also need to know if there are other organisms in the environment that are important in the biogeochemistry, and what proportion of the total community the degraders represent.

Plate counts can only provide a measurement of what microbes are present in the sample that will grow on the media used, under the conditions incubated. Since the number of possible media and possible incubations are infinite, the number of possible interpretations is also infinite. Generally, heterotrophic plate counts have been used to show that bacteria densities in the sediment or groundwater increase in response to biostimulation (Litchfield 1993). Using contaminant-enrichment media and either plates or most probable number (MPN) extinction dilution techniques, the number of contaminant-degraders can be estimated (National Research Council 1993). However, serious fallacies appear in the underlying assumptions of many of these assays. For example, diesel-degraders are determined using minimal media with a diesel-soaked piece of cotton taped to the top of the petri dish. In such a situation, it is unclear whether the colonies that are observed are using the diesel, or whether they are merely tolerant of the volatile components of the diesel fuel. MPN assays have also been used to conservatively measure methanotroph densities in soil and groundwater at chlorinated solvent-contaminated sites. These measurements involve sealing each tube under an air/methane headspace and then scoring positive only those tubes that are turbid and have produced carbon dioxide and used methane (Fogel *et al.* 1986). The incubation time for plate count and MPN contaminant-degrader assays is 1 to 8 weeks, thus negating their use for real-time monitoring and control.

A number of direct-count assays have been tried on contaminant degraders including direct fluorescent antibody (DFA) staining, acridine orange direct counts (AODC), and fluorescien isothiocyanate (FITC) direct counts. The fluorochrome stains only indicate the total number of organisms present in the sample; they do not indicate the type of organism or its activity. However, these techniques have been used in bioremediation studies to determine changes in the total numbers of organisms (Litchfield 1993). Increases in total counts have been found when contaminated environments are biostimulated. DFA shows promise but requires an antibody that is specific to the contaminant-degraders in that environment. The environment must be checked for organisms that may cross-react with the antibody and for contaminant-degraders that do not react with the antibody. DFA will be most useful in monitoring specific organisms added for bioaugmentation, though it has been used in biostimulation applications (Fliermans *et al.* 1994). Since the assay time is only hours for these direct techniques, they have significant advantages for real-time monitoring and rapid characterization.

Biological activity at bioremediation sites has been determined in a number of ways: Iodophenyl-Nitrophenyl, Tetrazolium (INT) Chloride activity/dehydrogenase, fatty acid analyses, acetate incorporation into lipids, ^{3}H-thymidine incorporation into deoxyribonucleic acid (DNA), BIOLOG™, phosphatase, and acetylene reduction. The INT test has been used in combination with direct counts since INT-formazan crystals can be detected in the cell. Cells with crystals are assumed to be actively respiring since the reaction occurs at the electron transport system of the cell. The assay requires only a 30-minute incubation; however, it can only be used in groundwater samples since particles in sediment samples cause too much interference with interpretation of the intracellular crystals. Barbaro *et al.* (1994) used this technique to measure microbial biostimulation of the Borden Aquifer in Canada.

Phospholipid fatty acid (PLFA) analyses have been used for characterization and monitoring at a number of bioremediation sites. An organism's PLFAs (signature compounds) may be unique to its species or even its strain, or the PLFAs may be conserved across physiological groups, families, or even kingdoms. Certain groups of fatty acids (cis and trans isomers) may also change in response to the physiological status of the organism. PLFAs have been used at bioremediation sites to provide direct assays for physiological status (cis/trans ratio), total biomass estimates; and presence and abundance of particular contaminant-degraders and groups of organisms, such as methanotrophs, actinomycetes and anaerobes (Phelps *et al.* 1989; Heipieper *et al.* 1995; Ringelberg *et al.* 1994). PLFAs would seem to be a panacea for characterization and monitoring of bioremediation. Unfortunately, the assays require -70°C sample storage and long extraction times, they have a fairly high detection limit (10,000 cells), and they require expensive instrumentation. Nevertheless, this technique merits careful consideration since it is a direct assay technique and is so versatile.

Radiolabeled acetate and thymidine incorporation into lipids and DNA, respectively, have been used at bioremediation sites to provide measurements of total community metabolic and growth responsiveness (Fliermans *et al.* 1988; Palumbo *et al.* 1995). These techniques require incubation, extraction, purification, and radiolabeled substrates, making interpretation of results difficult.

The BIOLOG™ assay has also been adapted to determine the activity of bacteria (with regard to contaminants) in groundwater and soil samples. The assay consists of a 96-well titer plate with carbon sources and an electron transport system indicator. It can be used to identify isolates (pure cultures of one organism separated from the community) and to examine the overall activity of a soil or water sample of a particular substrate. Gorden *et al.* (1993) adapted the assay to determine activity of different contaminants by using both contaminants and the electron transport system (ETS) indicator alone and adding contaminants to the plates with substrates to determine co-metabolic activity. The assay provides more rapid screening than other viable count techniques, but it suffers from some of the same problems, such as incubation conditions and repeatability. It is also difficult to determine if the contaminants are being transformed or tolerated.

Phosphatase and dehydrogenase enzyme assays have also been used to access bioactivity in soil and groundwater during bioremediation of terrestrial subsurface sites. Acid and alkaline phosphatase have been linked to changes in ambient phosphate concentrations and bioactivity at contaminated sites caused by biostimulation (Lanza and Dougherty 1991). The incubation, extraction, and interference caused by pH differences in samples make results difficult to interpret. Acetylene reduction has been used to indicate nitrogenase activity in a few bioremediation studies; however, the importance of nitrogen fixation for most bioremediation is probably insignificant, unless the site is oligotrophic, e.g. very low in nutrients (Hazen *et al.* 1994).

Nucleic acid probes provide, at least theoretically, one of the best ways to characterize and monitor organisms in the environment (Hazen and Jiménez 1988; Brockman 1995). Since many contaminants, especially the more recalcitrant ones, are degraded by only a few enzymes, it is possible to produce DNA or even ribonucleic acid (RNA) probes that will indicate the amount of that gene in the environment. This quantity reveals whether the functional group that can degrade or transform the contaminant is present, and if so, indicates its relative abundance. Since probes have also been found for species, families, and even kingdoms, soil and groundwater communities can be monitored. Recently, conserved regions in ribosomes have also been found, allowing samples to be probed for the relative abundance of ribosomes and, hence, the bioactivity of the total community (Ruminy *et al.* 1994). Bowman *et al.*

(1993) demonstrated that probes for methanotrophs indicated their presence in soil at trichloroethylene (TCE)-contaminated sites in South Carolina and Tennessee. Brockman *et al.* (1995) also showed that methane/air injection at the South Carolina site increased the methanotroph probe signal in sediment near the injection point in the aquifer. The probe-signal increases for methanotrophs coincided with increases in the MPN counts for methanotrophs. Table 3-16 summarizes bioremediation characterization and monitoring parameters.

TABLE 3-16 Bioremediation Characterization and Monitoring Parameters

Measurements	Parameter
Biomass	
Viable Counts	Plate counts, Most Probable Number (MPN), enrichments, BIOLOG™
Direct Counts	Acridine Orange Direct Count (AODC), Fluorescien Isothiocyanate (FITC), Direct Fluorescent Antibody (DFA)
Signature Compounds	Phospholipid Fatty Acid (PLFA), DNA, RNA
Bioactivity and Bioremediation	
Daughter Products	Cl, CO_2, CH_4, stable isotopic C, reduced contaminants
Intermediary Metabolites	Epoxides, reduced contaminants
Signature Compounds	PLFA, ribosome probes, BIOLOG™, phosphatase, dehydrogenase, Iodophenyl-Nitrophenyl, Tetrazolium Chloride (INT), acetylene reduction, recalcitrant contaminants
Electron Acceptors	O_2, NO_3, SO_4, (microrespirometer)
Conservative Tracers	He, CH_4, Cl, Br
Radiolabeled Mineralization	^{14}C, ^{3}H -labeled contaminants, acetate, thymidine
Sediment	
Nutrients	PO_4 NO_3, NH_4, O_2,, total organics, SO_4
Physical/Chemical	Porosity, lithology, cationic exchange, redox potential, pH, temperature, moisture, heavy metals
Toxicity	Microtox™, Mutatox™

Thus, sediment can be directly extracted and probed with DNA and RNA for bioremediation characterization and monitoring. As more nucleic acid sequences are found and mapped, it will be possible for us to construct sequences that will detect the amount of a gene from any organism in that environment involved in the desired transformation of the contaminant. Clearly, this ability will allow bioremediation injection strategies to have better control of the process in terms of effecting the desired changes in the functional group responsible for the bioremediation process. Unfortunately, nucleic acid probe technology has some serious obstacles to overcome before it becomes practical:

- The direct detection of nucleic acids in soil and groundwater requires lysis, extraction, and purification
- Soil humics and groundwater pH interfere with the nucleic acid signal
- The detection limit for most probe assays is 10,000 cells.

The extraction and purification steps also significantly increase the cost and analysis time. These problems are not insurmountable, but will impede realistic use of nucleic acid probes for bioremediation. Certainly, research in this area needs to be encouraged, given the sound theoretical advantages that these techniques provide for bioremediation.

DETERMINATION OF UNSATURATED HYDRAULIC PROPERTIES OF VARIABLY SATURATED SOILS AND ROCKS*

The unsaturated hydraulic properties of porous media are the water-retention and unsaturated hydraulic-conductivity functions that are the constitutive relationships of the Richards' equation for unsaturated flow in the vadose zone:

$$\frac{\partial \theta}{\partial t} = \frac{\partial}{\partial z}\left[K(h)\left(\frac{\partial h}{\partial z}+1\right)\right] \tag{3.10}$$

*This section was contributed by F.J. Leij and B. Faybishenko.

where h is the soil-water pressure head, θ is the volumetric water content, K is the unsaturated hydraulic conductivity—which can be expressed as a function of either h or θ, t is time, and z is the vertical distance (coordinate) taken positive upward.

The soil-water retention curve, $\theta(h)$, quantifies the ability of a soil to retain water by relating the soil-water content to the soil-water pressure head. The latter expresses the soil-water energy status. The soil-water content can also be expressed in terms of the effective water saturation (S_e). Some theoretical concepts of the soil-water energy status given in terms of the matric potential, soil-water suction, and the soil-water matric are discussed in Chapter 1. Experimental methods to determine both water saturation and matric potential are discussed in the section "Field Vadose Zone Characterization and Monitoring Methods," above.

The unsaturated hydraulic-conductivity function quantifies the ability of an unsaturated soil to transmit water under different saturations, by relating the hydraulic conductivity to either the pressure head, $K(h)$, or the water saturation, $K(S_e)$. Because this function is nonlinear, it renders Richards' equation nonlinear. Soil hydraulic properties can be determined using a variety of experimental field and laboratory methods, and can also be estimated with indirect methods, using different empirical and semi-empirical approaches (Klute, 1986; van Genuchten *et al.* 1992; 1999).

FACTORS AND PROCESSES AFFECTING HYDRAULIC PARAMETERS

The water-retention and hydraulic-conductivity functions are strongly affected by both the texture and structure of soils as well as by the physical and chemical properties of the fluids (typically air and water) occupying the pore space. Numerous investigations have attempted to elucidate how hydraulic properties depend on particle- and pore-size distributions, clay content, shrinkage and swelling phenomena, entrapped air, temperature, solution composition, and other physical and chemical properties of soil and soil solution.

Figure 3-37 shows a schematic of soil water retention curves (SWRCs) exhibiting hysteresis for a typical coarse-textured soil (Luckner *et al.* 1989). Both conductivity and retention functions are hysteretic under wetting and drying conditions. Variations in hydraulic properties affected by hysteresis, soil heterogeneity, preferential flow, soil contam-

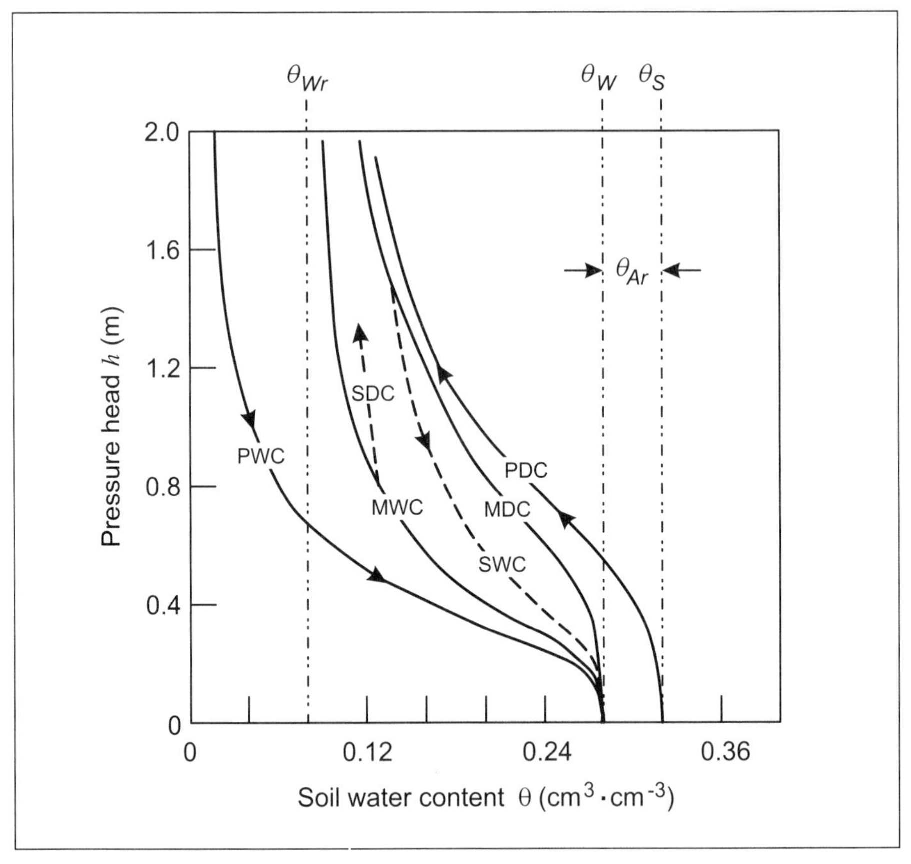

Figure 3-37. Hysteresis of the SWRC for a coarse-textured soil (Luckner *et al.* 1989). PDC is the primary drainage curve, PWC is the primary wetting curve, MDC is the main drainage curve, MWC is the main wetting curve, SWC is the scanning wetting curve and SCD is the scanning drainage curve.

ination, and air entrapment are of particular interest in predicting the effect of remediation activities.

Soil Heterogeneity and Preferential Flow

The natural spatial variability of soils physical properties causes hydraulic conductivity to vary by as much as several orders of magnitude over short distances. The heterogeneity of natural soil samples makes it difficult to evaluate the methods used to determine hydraulic properties because results are not accurate and/or reproducible. An additional complication is that different wetting and drying events during experiments create a multitude of spatial water-content distributions and preferential flow, which may result in nonuniqueness of hydraulic properties (Topp *et al.* 1980, Topp and Davis 1985; Smiles *et al.* 1971; Faybishenko 1986; Globus1987; Plagge *et al.* 1999).

Flow through macropores and cracks in structured soils creates preferential flow, especially under near-saturated and saturated conditions. Unfortunately, it is very difficult to accurately and reproducibly measure the hydraulic properties of a soil at low suctions (that is, close to saturation). Conventional methods for estimating hydraulic properties, which are suitable for capillary flow in the soil matrix, should be supplemented with techniques to quantify noncapillary flow in macropores. One approach assumes that the medium consists of several pore structures, and the hydraulic properties of the medium are obtained by summing the properties for individual structures (Ross and Smettem, 1993; Durner 1994). An example is provided by Mohanty and van Genuchten in the case study on the accompanying CD "Field and Modeling Studies of Preferential Flow in Macroporous Soils." Mohanty and van Genuchter determined the relative trend of soil-water retention and hydraulic- conductivity functions near saturation and established piecewise-continuous functions to take into account preferential flow through macropores. These hydraulic functions were then used to predict field-scale flow and transport in a flood-irrigated field at Las Nutrias, New Mexico.

Another type of preferential flow is fingering, which results from wetting-front instabilities. A well-known scenario for water fingering involves infiltration of water from a fine-textured soil horizon overlying a coarse-textured horizon (Hill and Parlange 1972; Parlange and Hill 1976). Initially, water in the smaller pores of the topsoil cannot enter the larger pores of the subsoil because of insufficient soil-water suction developed in the coarser material (the coarser subsoil has a lower air entry and a higher "water" entry value than the topsoil). Lateral flow will occur until the pressure is sufficient for water to enter the subsoil. Because the coarser subsoil can transmit a larger water flux than the topsoil, water flow in the coarser subsoil is confined to a number of fingers.

Other possible factors leading to water fingering involve air entrapment, hydrophobicity, changes in interfacial tension, increased water content with depth, continuous low-rate irrigation, and lenses of coarse soil acting as a capillary barriers (Raats 1973; Diment *et al.* 1982; Glass *et al.* 1989; Kung 1990a, b; Ritsema and Dekker 1995).

Entrapped Air

Air entrapment occurs when the air phase is discontinuous and is no longer connected to the atmosphere. Air remains as small, partly immobilized, disconnected bubbles, which can behave like solid spheres. Air can be entrapped in both dead-end and open pores. Entrapped air may reduce the soil's hydraulic conductivity by as much as two orders of magnitude (Faybishenko 1986, 1995, 1999a). Hysteresis of water retention curves may also be affected by air entrapment (Chahal 1965; Faybishenko 1983; Hopmans and Dane 1986; Kaluarachchi and Parker 1987; Stonestrom and Rubin 1989a, b).

Effect of Salinity and Contaminants

Chemical compounds dissolved in the aqueous phase can affect the hydraulic properties in several ways. Firstly, the solution chemistry may affect surface tensions (that is, the interfacial tension and contact angle, which are parameters in the Laplace-Young equation). Dissolved organic compounds tend to lower surface tension while ionic species, at high concentrations, may increase the surface tension. Sorption of chemical compounds by the solid may alter the contact angle, that is, the wettability of the solid. Demond *et al.* (1994) and Lord *et al.* (1997) determined changes in the capillary pressure in the presence of solutes. Demond *et al.* (1999) investigated the impact of speciation, sorption, and partitioning on the relationship between the primary drainage capillary pressure and saturation. For an aqueous system containing octanoic acid, lowering the pH below the pKa of 4.8 transformed the anionic into a neutral form of the acid. The corresponding decrease in surface tension resulted in a lowering of the soil-water pressure head. For a two-liquid medium containing o-xylene and water, lowering the pH led to preferential partitioning of the neutral compound into the o-xylene phase and a corresponding increase in capillary pressure. Furthermore, sorption of cetyltrimethyl ammonium caused the solid to become hydrophobic (increase in the contact angle beyond 90°) and lowering of capillary pressure. However, these processes are not well understood at this time.

Secondly, the pore geometry may be altered. Such changes may be attributed to changes in solution composition (that is, dispersion and flocculation), or changes in hydraulic regime (that is, shrinking and swelling). The detrimental effects of soil salinity on hydraulic conductivity of soils have been well-documented (Ghassemi *et al.* 1995). The pH is an important factor, affecting the charge distribution of clay particles. If the electrolyte level is low enough, depending on the type of cations, the repulsion of particles and the collapse of soil aggregates cause dispersion and changes in pore geometry. Furthermore, small particles may be dislodged and clog downstream pores. The inverse process of flocculation occurs when the electrolyte level is increased or if the cationic composition is changed (for example, during the displacement of sodium by calcium cations). Flocculation and aggregation are also promoted by a heterogeneous charge distribution across the soil particles. The negative faces of clay platelets and positively charged edges of other platelets attract each other to form a stable aggregate. However, the positive charge is highly variable, and an increase in pH or the addition of polyanions such as phosphates can lead to dispersion. Swelling and shrinking phenomena may occur during imbibition or release of water from the soil. Several scales of shrinking may be discerned, such as crystalline and osmotic swelling (Parker 1986). It is important that the swelling process will usually not be irreversible. Although the redistribution of soil colloids is not totally random, the complexity of the flow field and driving forces for colloids makes a physical description of the process rather difficult.

Numerous publications have been devoted to the influence of soil salinity on hydraulic properties. Shainberg and Levy (1992) and Lima *et al.* (1990), among others, reported an increase in water retention for relatively higher amounts of sodium. The behavior of the saturated conductivity as a function of salinity and pH were investigated by Rhoades and Ingvalson (1969) and Suarez *et al.* (1984), respectively. McNeal and Coleman (1966) demonstrated that the saturated hydraulic conductivity is reduced for lower total electrolyte concentrations and higher sodium levels. There are various mechanisms that contribute to this reduction in conductivity. Swelling will change the soil structure, soil pores may be partially or completely blocked (Quirk and Schofield 1955), and dispersion may result in movement of clay platelets, causing clogging of (smaller) soil pores, and consequently reducing the conductivity (cf. Shainberg and Levy 1992).

Temperature Effects

Soils near the land surface can experience diurnal and seasonal temperature fluctuations of as much as 50°C. Soil temperature affects fluid properties, including water density, viscosity, and surface tension as well as gas density, viscosity, and saturation vapor pressure. It also affects the contact angle, which, in turn, affects soil hydraulic properties. Ambient temperature affects evaporation from the surface, and the soil temperature fluctuations influence affect the intra-soil evaporation and condensation of water. Condensed water may become an important source of infiltration in the vadose zone in arid and semi-arid areas. Temperature gradients will act in different ways on water and vapor flow. Simulations of nonisothermal water flow and solute transport should consider the energy transfer and the effect of temperature on constitutive soil hydraulic parameters as well as solute-transport parameters.

It is important to consider the effect of temperature for several practical applications. At high-level radioactive waste disposal sites (for example, at Hanford), radioactive decay generates significant thermal energy. This, in turn, will change the soil and rock transport properties in and around the repository. Thermal extraction methods, often used to enhance petroleum recovery from reservoirs, are now being applied to improve the removal of nonaqueous phase liquid contaminants from the subsurface (Price *et al.* 1999; Udell 1998). Prediction of the performance of these remediation methods requires incorporation of the effects of temperature on the pertinent soil hydraulic properties (She and Sleep 1998).

To determine the effect of temperature on the capillary pressure, Grant and Salehzadeh (1996) found that the ratio $P_c/(dP_c/dT)$ was a linear function of temperature with a slope equal to 1. Faybishenko (1983) experimentally determined that the ratio dP_c/dT is equivalent to $0.008P_{10°C}$, where P_c is the matric potential in kPa, and T is the temperature (°C). A temperature change of 10°C causes a change in the pressure of 0.8 kPa. These data can be incorporated in the water-retention functions used in modeling (Grant and Salehzadeh 1996)

Instrumentation

The results of both field and laboratory determination of hydraulic properties depend on the type of instrumentation used. In particular, water-retention and unsaturated hydraulic-conductivity functions may depend on the flow geometry (axial, radial, or centrifugal) in laboratory cores as well as the magnitude and type of boundary conditions used to change the water content and pressure head during the experiment. Experimental methods are reviewed in detail by Klute (1986), Dirksen (1991), and Gee and Ward (1999).

Field Methods

Field methods are employed to observe flow and transport processes under natural conditions. *In situ* methods allow one to minimally disturb the soil profile and obtain realistic field-scale observations. However, these methods are time-consuming, difficult to implement, and frequently constrained to a narrow range of changes of saturation and water- pressure changes over the period of observations. Field results may be less accurate and reproducible than laboratory results, because of difficulty in controlling boundary conditions. A description of several field methods is presented below.

Instantaneous Profile Method

The instantaneous profile or unsteady drainage-flux method is used to determine the hydraulic conductivity by observing the water content and pressure head of a soil profile after the soil has been wetted (for example, by irrigation). Mass-balance calculations based on Richards' equation yield the hydraulic conductivity, whereas the observations of the water content and water pressure directly provide the water retention functions (Rose *et al.* 1965; Cassel 1974). The method is only applicable to well-drained soils with no significant lateral flow. There are several variations of the method (Green *et al.* 1986).

After the entire profile is wetted by ponding or irrigation, the soil can be covered to prevent evaporation. Arya *et al.* (1975) modified the method by using a zero-flux plane to allow evaporation. Pressure-head gradients are obtained by installing tensiometers at a sufficient number of depths. The water content profile is monitored using a neutron probe or TDR. The hydraulic conductivity at an arbitrary distance $z=L$ may be determined numerically by integrating the Richards equation:

$$\frac{\partial}{\partial t}\int_0^L \theta(z,t)dz = K(h)\left.\frac{\partial H(z,t)}{\partial z}\right|_{z=L} \tag{3.11}$$

The main problem in interpreting the results of this method is the discrepancy between the location and scale of observation for pressure heads and water contents (Flühler *et al.* 1976). The solution of equation (3.9) using the finite-difference method may yield inaccurate or even negative values for the unsaturated hydraulic conductivity (Jury *et al.* 1991). Flühler *et al.* (1976) determined that in wet soils, the relative errors of hydraulic conductivity are approximately 20 to 30 percent, and for dry soils the errors may exceed 100 percent. The errors are greater for low hydraulic gradients (less than 0.3) and at earlier stages of the infiltration experiment. Plagge *et al.* (1999) determined that higher hydraulic gradients cause preferential, accelerated flow and decrease the tortuosity effect, thereby increasing the calculated hydraulic conductivity.

Numerical inverse procedures are now available that allow greater flexibility of the initial and boundary conditions, as well as can take into account the locations of pressure-head and water-content measurements (cf. "Inverse Methods," below).

Gravity Drainage Experiments

The gravity drainage or simplified unsteady drainage-flux method is a modification of the instantaneous profile method. However, the method assumes that gravity is the dominant force for water flow and that a unit hydraulic gradient exists (Sisson 1987). Under such conditions, it is not necessary to monitor the soil-water pressure head, and the conductivity function, $K(\theta)$, can be readily obtained from water-content observations. The unit gradient assumption is a severe one, but reasonable for a deep vadose zone (Faybishenko 1986). Sisson (1987) and Sisson and van Genuchten (1991) proposed using the unit gradient water-flow models to fit the analytical functions for $dK/d\theta$ to the estimates for $dK/d\theta$ derived from field-measured water contents. Sisson and van Genuchten (1991) formulated the instantaneous profile data analysis in the form of a parameter optimization process. Application of the gravity-drainage analysis to the heterogeneous soils requires the use of a scaling procedure to transform the heterogeneous soil profile into an equivalent homogeneous soil profile (Shouse *et al.* 1991).

Ring Infiltrometry

Elrick and Reynolds (1992a) and Reynolds (1993) used analyses of three-dimensional, variably saturated flow to measure soil hydraulic properties, using a single-ring infiltrometer known as the Guelph Pressure Infiltrometer (GPI). Cylindrical infiltrometers have long been used to determine infiltration rate and saturated conductivity for essentially one-dimensional problems by using an inner and an outer ring (Bouwer 1986). An analytical solution equivalent to the expression by Wooding (1968) is fitted to the observed infiltration rate to determine the saturated conductivity. A very similar approach is followed to determine the conductivity with a well permeameter, where water is maintained at a constant level in a borehole, or by determining the sorptivity from the infiltration rate (Talsma 1969).

In the case of the GPI, a single ring (usually 0.10 m in diameter) is inserted approximately 0.02 to 0.05 m into the soil, and a Mariotte reservoir is used to supply water. The steady water flow rate through the ring is used to calculate the field-saturated hydraulic conductivity and matric flux potential. Parkin *et al.* (1999) reviewed some recent advances in the analysis of single-ring infiltrometer data, including the use of TDR results, which may improve experimental results in heterogeneous and low-permeability materials.

The steady-state constant-head/falling-head procedure is a refinement of the previous method (Elrick *et al.* 1992a, b). After establishing steady infiltration by maintaining a constant ponded head in the ring, the head is allowed to fall and is monitored as a function of time. An approximate analytical solution is then fitted to the observed head values to determine the field-saturated hydraulic conductivity and the matric flux potential of the soil. This method is less sensitive to soil heterogeneities because head values are used for many different times.

The early-time constant-head/falling-head analysis is similar, except that it is not necessary to attain steady infiltration, which can take a significant amount of time to achieve in low-permeability soils (Elrick *et al.* 1995). The analysis is based on the assumption that early-time constant head infiltration into low-permeability, capillary-dominated materials is linear with the square root of time. This method can be useful for clay soils, landfill caps and liners, and waste water impoundments (Fallow *et al.* 1993; Elrick *et al.* 1995).

Tension Infiltrometry

Tension infiltrometers allow the determination of unsaturated hydraulic parameters with minimal disturbance of the soil. A circular, porous plate is placed on the soil surface, with a good hydraulic contact between plate and soil. The plate is connected to a Mariotte reservoir to provide a water supply under constant suction to the soil. The experiment is repeated for different water suctions. The experimentally determined infiltration rate is used to optimize the hydraulic parameters. The tension infiltrometry can be used to determine the soil sorptivity from the infiltration rate and the soil water diffusivity from the advance of the water front (Clothier and White 1981). More recent analyses rely on parameterization of the hydraulic properties. The hydraulic conductivity may be estimated from formulae developed for the initial transient or the final steady phase of infiltration (Reynolds and Elrick 1991; White and Perroux 1989).

Inverse numerical methods based on using observed infiltration-rate data, soil-water-content, or pressure-head data as well as soil water content or pressure head can be used to estimate parameters for both water-retention and hydraulic-conductivity functions. Wang *et al.* (1998) combined tensiometer and TDR data with tension infiltrometry results for different disk sizes. It should be noted that tension infiltrometry allows one to obtain precise flow-rate measurement in soils near saturation, which is needed to assess the unsaturated hydraulic parameters for macroporous soils (Mohanty *et al.* 1994).

Crust Method

Using a crust method (Hillel and Gardner 1970; Bouma *et al.* 1971), a steady soil water flux is established by applying water at a relatively low rate, that is, below the saturated conductivity, to a soil pedestal through a crust such as a puddled soil material, hydraulic cement (Bouma *et al.* 1983), or gypsum or silica sand (Bouma and Denning 1972). Because of the low hydraulic conductivity of the crust, the soil is unsaturated below the crust. The soil's unsaturated hydraulic conductivity function, $K(h)$, is calculated using the flow rate and the water pressure head distribution measured in soils with tensiometers. Crusts of different hydraulic conductivity are used to obtain various points of the $K(h)$ curve. (It is recommended to start the experiment with the crust of a lowest hydraulic conductance.) For a layered soil profile, tensiometers may be required for each layer. The method is more laborious compared to tension disk infiltrometry, but it has the advantage of maintaining one-dimensional flow within the soil pedestal.

Large Columns

Large columns of at least 3 m in length may be designed in the field for *in situ* determination of unsaturated hydraulic properties. An example of such a column is shown in Figure 3-38. The column can be instrumented with tensiometers, TDR probes, piezometers, temperature and salinity sensors, suction cups, and other devices to allow a thorough investigation of water flow and chemical transport.

Single Tensiometer Experiments

The unsaturated hydraulic conductivity of soils can be determined by applying suction to the interior of the tensiometer and measuring the water flux into (or out of) the tensiometer from the soil, a process that will decrease the volume of air in the tensiometer and increase the air pressure. Using this method, Timlin and Pachepsky (1998) developed a way of calculating unsaturated conductivity using a two-dimensional finite-element model (2DSOIL), coupled with a Marquardt-Levenberg algorithm to fit the calculated fluxes to the measured fluxes.

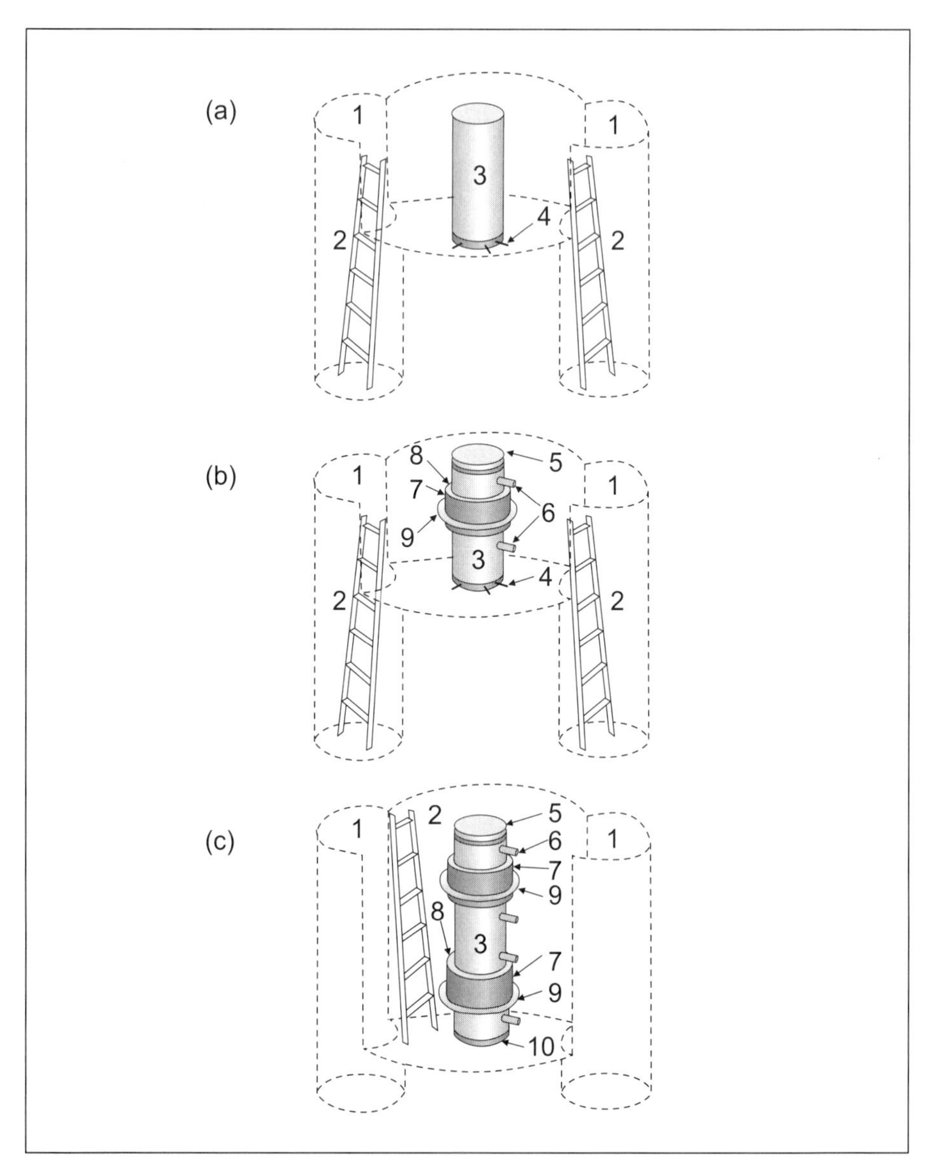

Figure 3-38. Sketch showing a field procedure of the preparation of a large vertical core between two boreholes (1) with ladders (2): (a) Upper segment (3) of uncovered core with cylindrical knife (4), (b) Upper segment is covered with top plate (5) and instrumented by probes (6) - tensiometers, piezometers and thermometers; the core is supported by backfill (7) placed between the core (3), cover (8), and platform (9), and (c) the sample has been cutoff, equipped with a bottom plate (10) (Faybishenko 1995)

Laboratory Methods

Laboratory determination of unsaturated hydraulic properties has many advantages over field methods, because boundary conditions and thermal regime can be carefully and easily controlled over a wider range of saturations. Laboratory results are also likely to be more accurate and reproducible. On the other hand, the laboratory soil-core volume tends to be smaller, and consequently a large number of soil cores are needed to adequately characterize a field site. Furthermore, sampling and handling of soil cores may inevitably affect the soil structure. Note that it is difficult to take soil samples of a loose structure and containing coarse materials such as pebbles and gravel. Because of the effect of instrumentation and expected changes in the soil structure, it is advisable to use several laboratory methods to determine hydraulic properties for a particular range of soil- water pressures.

Water Retention

The following contains a brief overview of methods to determine the water-retention function. Note that the previous section, "Field Vadose Zone Characterization and Monitoring," and publications by Klute (1986) and Gee and Ward (1999) include the techniques needed to measure the two variables used in assessing the retention-curve function: the soil-water content and matric potential.

The water-retention function for the matric pressure above approximately -80 kPa can be determined by regulating suction in a cell apparatus containing the soil sample. Figure 3-39 shows a schematic of a suction cell apparatus. A saturated soil sample is placed on a porous plate or membrane to which a hanging water column is attached for control of the soil matric pressure. By automatically monitoring the outflow, the method offers a convenient and accurate procedure to determine water retention near saturation (van den Elsen *et al.* 1999).

The application of this method is limited by the air-entry value of the porous plate and the length of the hanging water column. Suction tables, which use sand-silt packings as porous membranes, have been popular in the past to handle large numbers of core samples (Jamison, 1958). For pressures up to 20 bar (that is, where h = -200 m), the water-retention and unsaturated hydraulic-conductivity functions are determined using a pressure plate or a plate membrane apparatus.

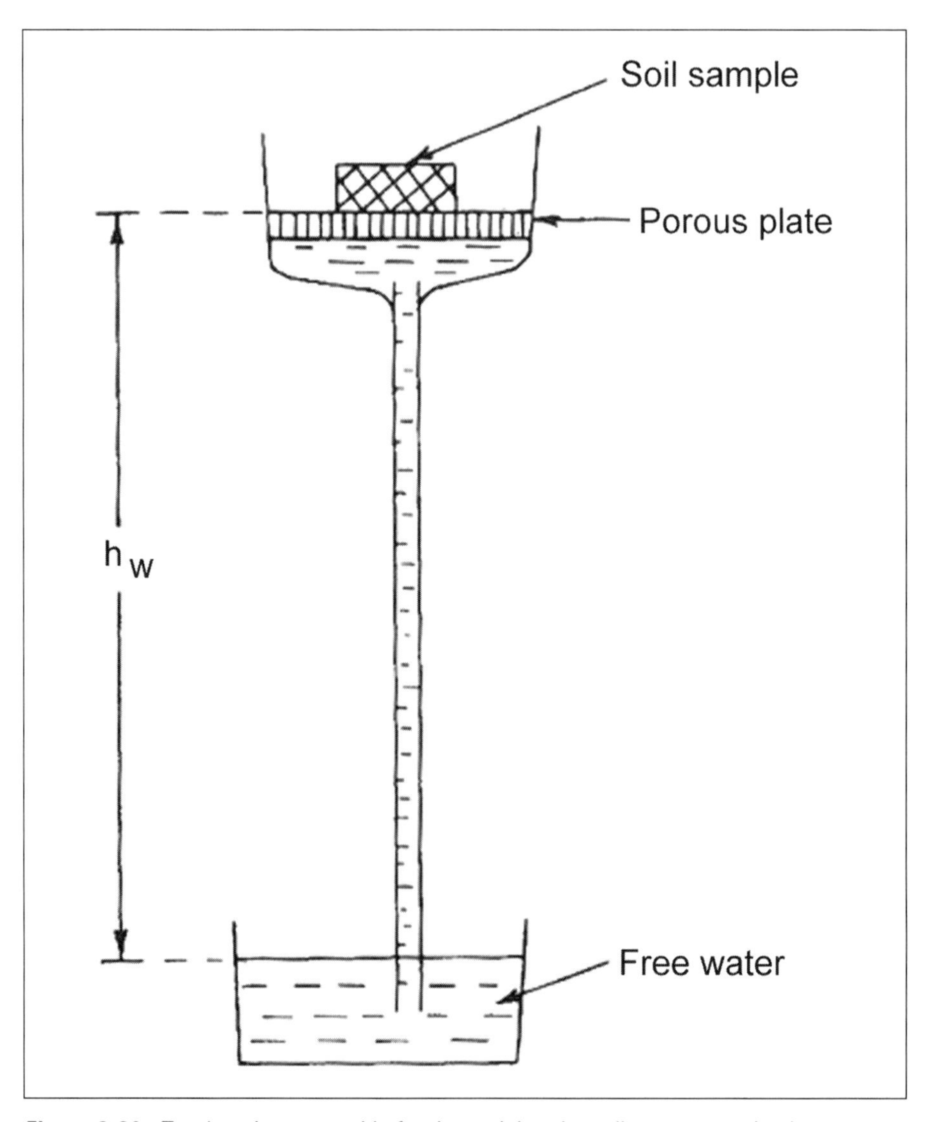

Figure 3-39. Tension plate assembly for determining the soil water retention by equilibrating a soil sample with a known matric suction value. This assembly is applicable for the range of matric suction of 0–0.8 bar (Hillel, 1981).

Figure 3-40 shows a generic setup of the pressure cell apparatus. Water is pushed out from the initially saturated sample by raising the gas-phase pressure in the cell above the plate while water and air below the plate are at atmospheric pressure. The soil sample is allowed to equilibrate for a sequence of air pressures used to obtain different soil matric pressures. The volume of water coming out of the soil sample as the pressure changes is used to calculate the changes in the soil saturation.

Different types of equipment and methods of changing the moisture content in a soil core are used, depending on the pressure range. For air pressures up to 1 bar (corresponding to h = -100 kPa), Tempe-pressure cells are frequently used. A soil core is held between two end caps, and air pressure is applied to the top cap while the bottom cap contains a ceramic plate attached to a burette measuring the outflow. The equili-

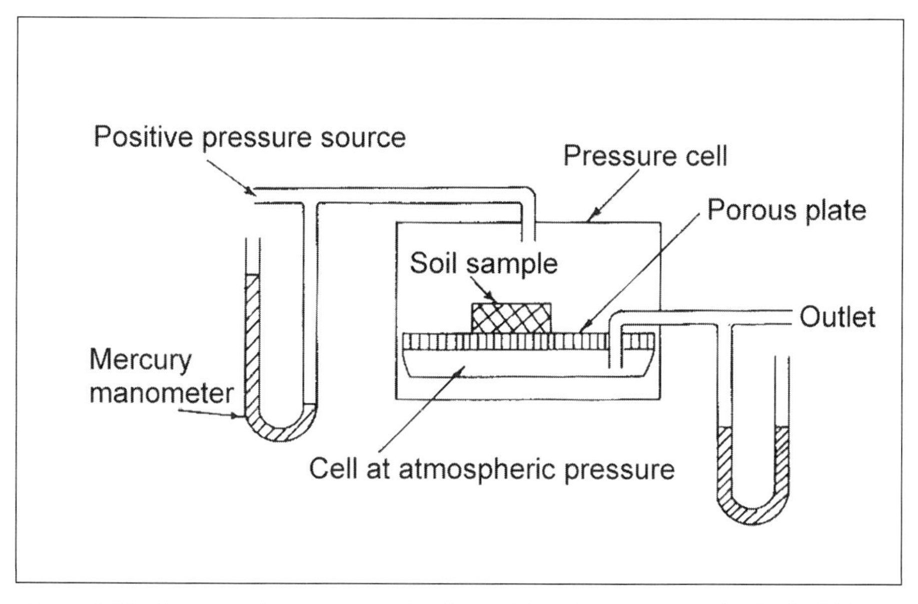

Figure 3-40. Pressure plate apparatus for determining the water retention in the high suction range using water extraction by applying positive air pressure above the soil sample. Note that the lower side of the porous plate is in contact with water at the atmospheric pressure (Hillel, 1981).

bration time depends on the plate bubbling pressure and the type of soils. The changes in the soil-water content are calculated from the outflow volume. If the outflow rate and the soil-water pressure head are monitored as a function of time, the unsaturated hydraulic conductivity function can be determined using inverse modeling (Eching *et al.* 1994). For higher pressures up to 20 bar (that is, h = -2 Mpa), several core samples or soil clods are placed on a ceramic plate in a pressure chamber. All equipment needed for this procedure is available commercially. In this case, the volumetric water content is obtained by weighing the samples. The bubbling pressure of the plate determines the range of pressures, typically 1 to 15 bar. Note that a disadvantage of the method is the long time to reach equilibrium.

Hydraulic Conductivity

The one-dimensional, head-controlled method flow experiments, has been the classical approach for steady-state determinations of the hydraulic conductivity (Klute and Dirksen 1986 and Dirksen 1991). The soil-water pressure head is controlled at both ends of the sample through a porous membrane, and tensiometers are installed in the vertical soil column to determine the hydraulic gradient needed to calculate the $K(h)$ function from Darcy's law. This method is effective for matric pressures of more than approximately -50 kPa. The disadvantage of the method is that the flux may vary over time because of changes in the porous-plate impedance and the plate-soil layer permeability of the plate-soil layer.

Figure 3-41 shows the experimental setup for the head-controlled method, which includes two porous plates to allow one to control the head at the top and bottom edges of the soil sample. Changes in soil-water content resulting from the pressure changes can be directly estimated from the outflow data. This setup can be used to determine both water-retention and hydraulic-conductivity functions using the steady and transient "one-step" and "multi-step" experiments. Using the one-step experiment, a vacuum of about -90 kPa is applied to drain water from the core, and the pressure is measured using a monitoring tensiometer installed in the core.

Multi-step flow experiments can be conducted using step-wise pressure changes in both the upper and lower porous plate. After the end of each transient step, the steady-state water-flow experiment is conducted.

The water-retention curve is determined from the outflow data during the transient regime of water flow. The unsaturated hydraulic-conductivity function is calculated from Darcy's law using the data on the steady-state flow rate and matric gradient in the core. Inverse modeling is used to analyze the transient flow data.

The flux-controlled method is an alternative for constant-head experiments. In this method, water is supplied at a prescribed rate to the soil column using porous plates connected to a controlled vacuum. Tensiometers are used to monitor the matric pressure in the soil core. Unit gradient conditions are established by adjusting flow rate or suction (van den Elsen 1999). The difficulty of this method lies in maintaining the uniform, steady-state water supply into the soil core, especially the very

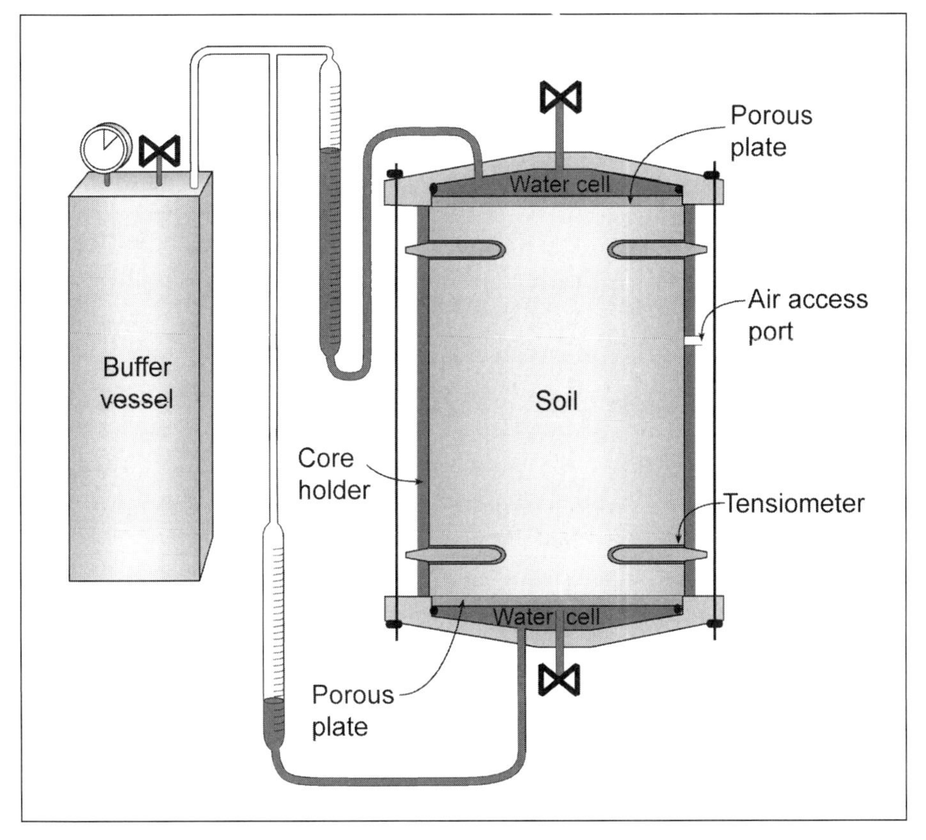

Figure 3-41. Principal scheme of core arrangement for soil drainage and saturation using axial flow experiments (Faybishenko, 1986)

small fluxes associated with lower conductivity values (cf. Dirksen 1999).

The evaporation method used by Wind (1969) constitutes a variation of the traditional flux-controlled method. An initially saturated soil sample is instrumented with tensiometers and placed on a balance. Water is allowed to evaporate from the top. The evaporation may be regulated by a fan. Hydraulic properties are determined in a similar manner for the instantaneous profile method. The method appears to have become more popular recently with the emergence of TDR, automated data collection and operation of equipment, and inverse modeling (cf. van Genuchten *et al.* 1999).

In a heterogeneous soil sample, the asymmetry of flow geometry with respect to the core axis and geometrical boundaries leads to a complex distribution of the moisture content and the flow field in the core (Finsterle and Faybishenko 1999b). Two-dimensional inverse modeling should be used to analyze the results of laboratory experiments for such samples.

Radial-flow analysis has been used for quite some time to determine hydraulic properties. An experiment with radial-flow geometry was proposed by Richards *et al.* (1937) and was further developed by Richards and Richards (1951), Gardner (1960), and Klute *et al.* (1964). Richards and Richards (1962) and Klute *et al.* (1964) developed analytical solutions for radial-flow experiments. Klute *et al.* (1964) discussed the advantages of using radial-flow geometry rather than the more common axial-flow geometry. They noted that soil shrinkage during drying is significantly reduced in a design with a central porous cylinder, thus preventing loss of contact between the sample and the boundary. Furthermore, the air trapped in the porous cylinder can be removed easily and with minimal disturbance to the boundary. Because of the reduced flow distance for radial- flow geometry, a larger sample volume can be tested in a shorter time (Gardner 1960). Timlin and Pachepsky (1998) recently reported on the measurement of the conductivity function by optimizing the inflow and outflow to a ceramic-cup tensiometer that was subjected to various suctions.

Figure 3-42 shows a schematic of a flow cell apparatus for radial, single-step and multi-step desaturation experiments on soil samples. Faybishenko (1986) and Dzekunov *et al.* (1987) used this setup for

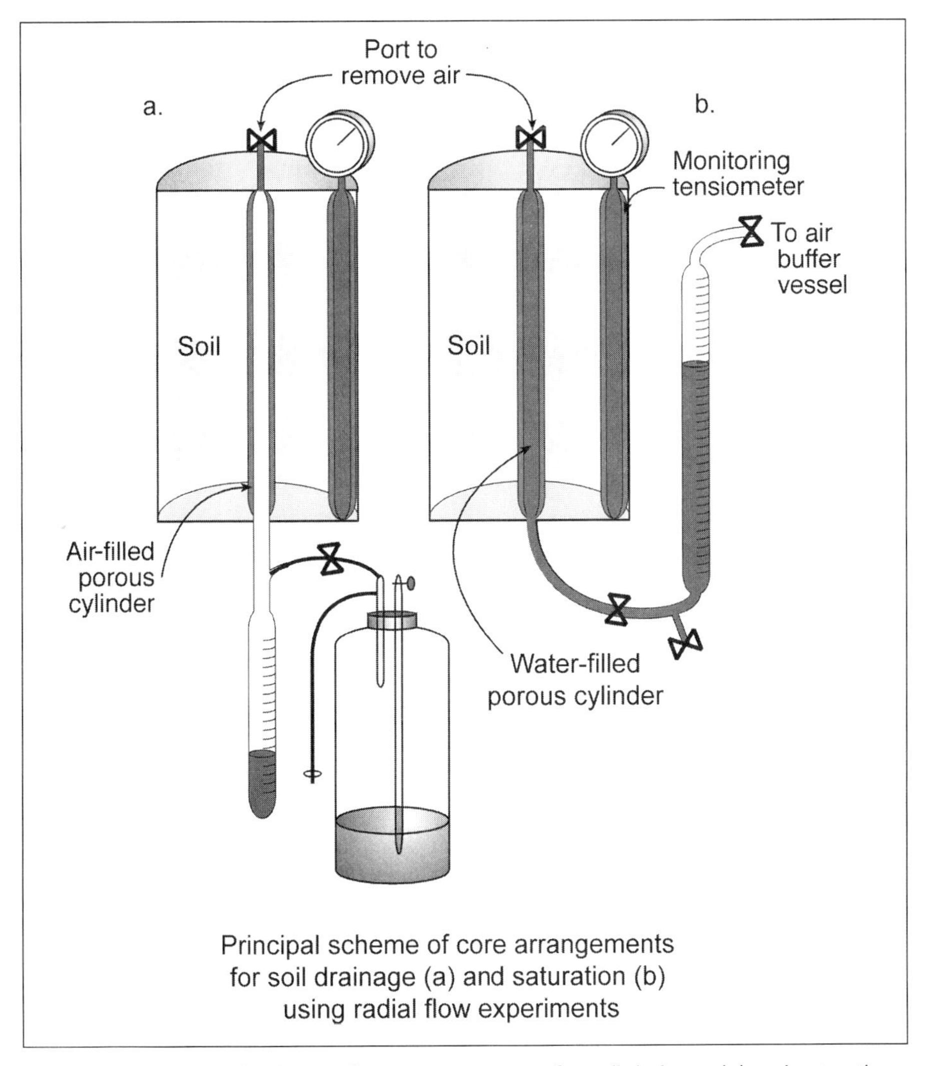

Figure 3-42. Principal scheme of core arrangements for soil drainage (a) and saturation (b) using radial flow experiments (Faybishenko, 1986)

radial-flow experiments with a central porous cylinder for both injection and extraction of water. A second porous cylinder was used as a monitoring tensiometer. Wetting and drying curves were obtained by applying one-step, multi-step, and continuously changing boundary pressures under isothermal and non-isothermal conditions. Soil cores (22 cm long and 15 to 18 cm in diameter) were conserved in a solid metal or plastic cylinder, and the annulus was filled with a paraffin-tar mixture. A ceramic cylinder with an air-entry pressure of about 1 bar was inserted along the axis of the core in the center of the soil core. This cylinder was attached to a vacuum-regulated burette to measure the cumulative water discharge. The outlet is located at the bottom to facilitate free movement of the extracted water into the measuring burette to inhibit air accumulation in the cylinder (Elrick and Bowman 1964; Klute *et al.* 1964). A tensiometer was inserted near the outer wall of the flow cell. As confirmed by numerical simulations (Finsterle and Faybishenko 1999a), a one-dimensional, radial model can accurately describe flow.

Determining hydraulic properties in deforming porous media is a challenging but important and pertinent problem. Angulo-Jaramillo *et al.* (1999) presented an example of one-dimensional infiltration into a free-swelling, undisturbed sample of compacted clay. The laboratory setup is shown in Figure 3-43. Water flow processes in a deformable, porous medium can be monitored using a ^{241}Am and ^{137}Cs dual gamma-ray system and tensiometers connected to a differential-pressure transducer.

Globus and Gee (1995) used a temperature gradient to determine soil-water diffusivity and hydraulic conductivity for moderately dry soils. A partially wetted sample at uniform water content is sealed and equilibrated under an applied thermal gradient. When equilibrium is reached, liquid-phase flow from the cool end is equal to vapor-phase flow from the warm end. The nonuniform water profile in the core is then used to determine the unsaturated hydraulic-conductivity and water-retention functions. The water-content profile can be determined by sampling or, possibly, nondestructive methods. This method is time consuming but allows determination of very low unsaturated conductivities ranging from 10^{-7} to 10^{-12} cm/s. Furthermore, the conductivity represents the combined effects of vapor and liquid flow, which may be important for many practical applications.

Ultracentrifuge methods* have been used fairly recently to determine unsaturated hydraulic properties (Conca and Wright 1990, 1992, 1998;

*The section on ultracentrifuge methods was written by James L. Conca and Judith Wright.

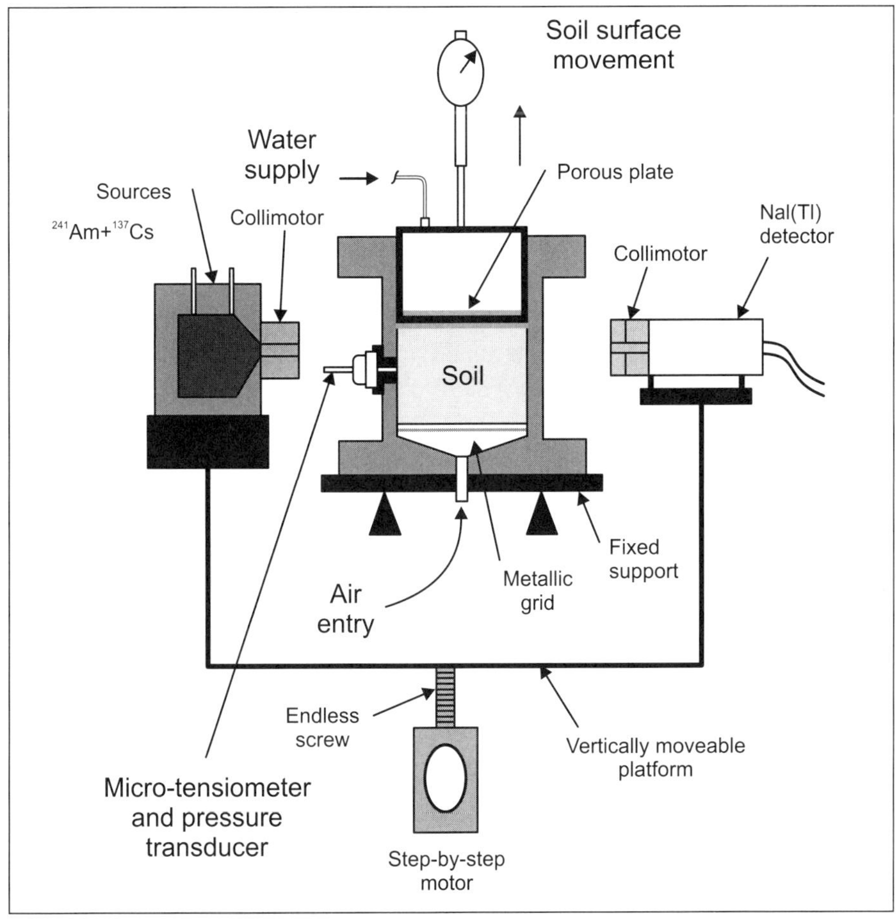

Figure 3-43. Rigid-wall permeameter for infiltration in unsaturated swelling soils with with a dual-energy gamma-ray scanner system and micro-tensiometer (Angulo-Jaramillo *et al.* 1999)

Conca *et al.* 1999). In this method, a saturated soil sample is placed in a high-speed centrifuge and spun around. The sample desaturates during the centrifugation until a certain soil matric head is reached, which corresponds to a specified centrifugal speed. Water-retention points can be determined by weighing the sample. However, it is also possible to determine the conductivity by supplying water to the soil sample during the centrifugation (Nimmo 1990; Nimmo *et al.* 1987). After steady flow conditions are established, the centrifugation is stopped and the sample is weighed to determine the water content. Retention and conductivity curves can be determined by repeating the process at increasingly higher speeds. Some disadvantages of the method are the limited sample size, compaction of finer-textured soils, and high equipment cost. The case study on the accompanying CD, "The UFA method for Characterization of Vadose Zone Behavior," by James L. Conca and Judith Wright, reports on the Unsaturated/Saturated Flow Apparatus (UFA) that determines hydraulic properties with the ultracentrifuge method. The device can be used to determine hydraulic conductivity, matric potential, electrical conductivity, vapor diffusivity, distribution coefficient, retardation factor, dispersivity, and thermal conductivity. The UFA instrument can achieve steady-state flow in hours using an ultracentrifuge with a constant, ultralow flow pump that provides fluid to the sample surface through a rotating seal assembly and microdispersal system. The ultracentrifuge can reach accelerations of up to 20,000 *g* (soils are generally run only to 1,000 *g* [3,000 rpm], an effective hydrostatic pressure of 2.5 bars). Constant flow rates can be reduced to 0.001 ml/h.

Multi-Liquid Systems

Hydraulic properties of porous media containing separate aqueous and nonaqueous liquid phases are of interest in petroleum and environmental engineering and remediation design. The pressure, saturation, and conductivity of two phases need to be considered. The consideration of parameters for two phases is in contrast with the hydraulic properties of air-water systems, where the air phase is usually of little interest and considered to be at atmospheric pressure. The terminology and methodology tends to vary somewhat for multi-fluid flow, but the principles are the same as for flow of water in an unsaturated soil (Dullien 1992; Corey 1994).

Retention

The retention curve, usually referred to as the capillary pressure–saturation curve, is frequently interpreted with the Laplace-Young equation for the pressure drop at the interface of a nonwetting and wetting fluid in a cylindrical tube:

$$P_c = P_n - P_w = \frac{2\sigma_{nw}}{r}\cos\phi_{snw} \tag{3.12}$$

where the subscripts n and w denote the nonwetting and wetting phases, ϕ_{saw} is the contact angle, σ is the tension at the air-water interface, r is the tube radius. In fractional or mixed wettability media, water and oil alternate as wetting fluids, depending on the saturation of the sample and the degree of hysteresis. The difference in oil and water pressure may therefore indicate a sign change (Bradford and Leij 1995). Lenhard and Parker (1988) measured the retention by controlling the saturation of the oil and water phases, rather than the pressures. Figure 3-44 shows a schematic of the experimental setup. The soil column contains hydrophobic and hydrophilic ring tensiometers, which are connected to fluid reservoirs to allow displacement of a known liquid volume into or from the column. The soil is allowed to equilibrate under atmospheric air pressure. The equilibrium liquid pressures are also determined with the tensiometers.

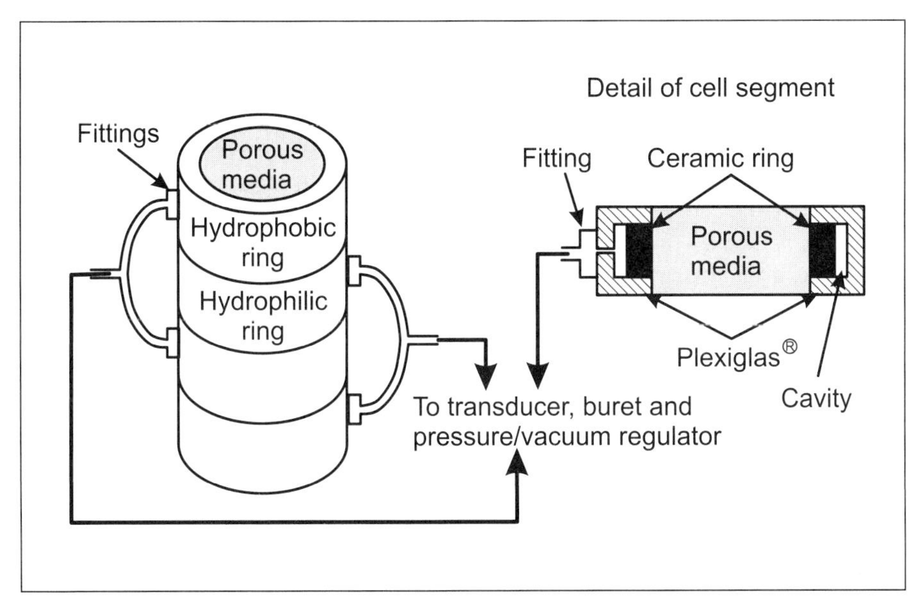

Figure 3-44. Schematic of an experimental apparatus consisting of hydrophilic and hydrophobic rings tensiometers for measuring retention in multi-liquid systems.

The experimental retention curve for a two-fluid system can be used to estimate the retention curve for a different pair of fluids in the same medium by the following scaling procedure:

$$P_{c2}(S) = \frac{\sigma_2}{\sigma_1}\frac{\cos\phi_2}{\cos\phi_1}P_{c1}(S) \tag{3.13}$$

where the subscripts 1 and 2 denote a two-fluid system with known and unknown retention curves. Frequently identical contact angles are assumed, in which case the scaling is based on the ratio of interfacial tensions. Furthermore, the retention curves for a three-fluid system, schematically shown in Figure 3-45, can be estimated from the corresponding curves for two-fluid systems using Leverett's assumption. Additional procedures are needed to account for surface tensions and wettability (Bradford and Leij 1996).

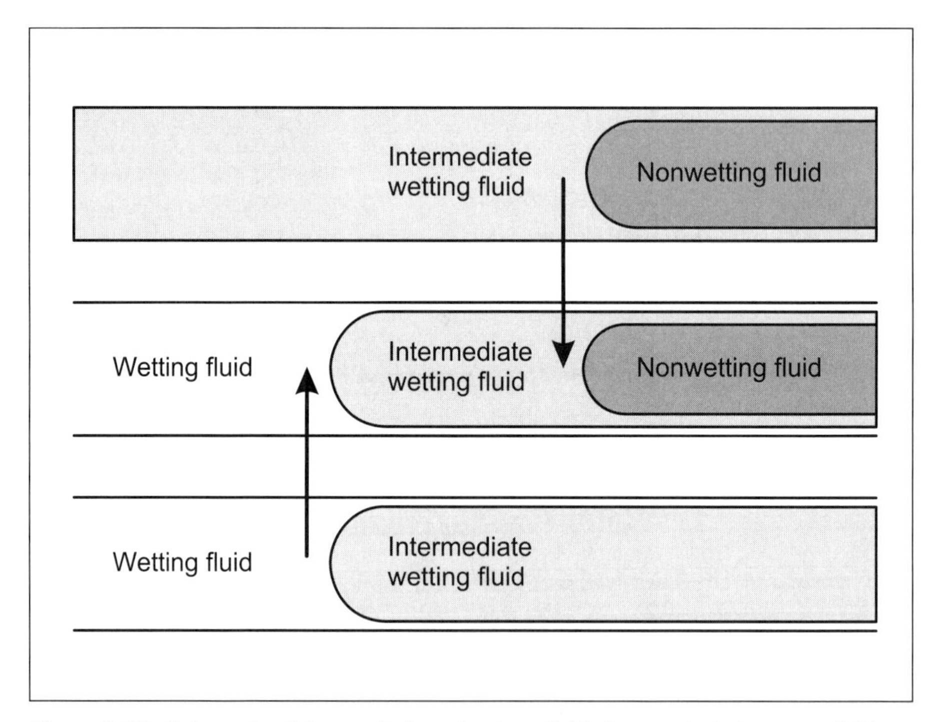

Figure 3-45. Schematic of the prediction of a three-fluid characteristic from two-fluid retention characteristic according to Leverett's principle.

Conductivity

The measurement of conductivities, usually referred to as relative permeability, is more cumbersome than the hydraulic conductivity for air-water systems, but the same principles of steady-state measurements are used (Demond and Roberts, 1993). A porous ceramic plate with Teflon inlays is placed on both sides of the soil core for separate application and collection of aqueous and nonaqueous liquids. Tensiometers are installed to determine liquid pressures. Ideally, the capillary pressure and liquid saturations should be constant throughout the sample. The approach can become quite time-consuming if the permeability needs to be characterized for three-fluid systems with mixed wettability (cf. Honarpour *et al.* 1986). Inverse modeling of transient flow processes can be used to determine hydraulic parameters for such multifluid systems.

INVERSE METHODS

The design of field and laboratory experiments is often restricted to simple steady-state boundary conditions because of limitations in the analytical methods used to determine the water unsaturated hydraulic functions. The determination of unsaturated hydraulic properties by inverse flow modeling has rapidly become popular as an alternative. Inverse solutions rely on numerical methods, which obviate most restrictive conditions and offer flexibility regarding the type of data that can be used, as well as the optimization and parameterization procedures employed to determine hydraulic parameters. Inverse modeling has been used to quantify hydraulic parameters of laboratory soil cores using one-step outflow (Toorman *et al.* 1992), multi-step outflow (Eching *et al.* 1994; Finsterle and Faybishenko 1999a), upward infiltration (Hudson *et al.* 1996), and evaporation (Simunek *et al.* 1998). Many applications to the field can be envisaged, such as the analysis of single- and multiple-disk infiltrometer data (Simunek and van Genuchten 1996, 1997; Zhang 1997). Specific implementation of inverse methods are summarized below and described in detail in Chapter 5.

A variety of optimization algorithms have been used to estimate several or all of the unknown hydraulic parameters from observed time-series data of the water flux, water content, and pressure head (Kool *et al.* 1987; Simunek and van Genuchten 1996; Abbaspour *et al.* 1997; Finsterle and Faybishenko 1999a). In the case of a multi-step outflow experiment where a tensiometer is inserted in the soil (Eching *et al.* 1994), the parameter vector $\boldsymbol{b}$ may be estimated by minimizing the following objective function (Hopmans and Simunek 1999):

$$O(\boldsymbol{b}) = W_Q \sum_{i=1}^{N} \{w_i[Q(t_i) - Q(t_i, \boldsymbol{b})]\}^2 +$$

$$W_h \sum_{j=1}^{M} \{w_j[h(t_j) - h(t_j, \boldsymbol{b})]\}^2 + W_\theta \sum_{l=1}^{L} \{w_l[\theta(t_l) - \theta(t_l, \boldsymbol{b})]\}^2 \quad (3.14)$$

where N, M, and L refer to the number of observations that were made over time (t) of cumulative outflow, Q, matric head, h, and volumetric water content, θ, respectively, while the corresponding simulated variables also include the parameter vector $\boldsymbol{b}$ as an independent variable. The objective function $O(\boldsymbol{b})$ is normalized with weighting factors W for different types of data; weighting of individual observation may be done with the factors w. The objective of this particular optimization procedure is to determine the parameter vector $\boldsymbol{b}$ that minimizes $O(\boldsymbol{b})$. As mentioned above, independently measured hydraulic data may also be included in the objective function; this will extend the optimized hydraulic parameters beyond the range of the outflow experiment.

The hydraulic properties are often parameterized according to van Genuchten (1980) leading to, for example, the parameter vector $\boldsymbol{b} = \{\theta_r, \theta_s, \alpha, n, K_s, L\}$. Several other parametric models are given in the subsection "Analytic Functions," below.

Knowledge of the initial values and mathematical and physical constraints on the parameters in question may improve the performance of the optimization algorithm. Frequently, the inverse problem is ill-posed (Yeh 1986). The mathematical model may not be convex, leading to different estimates for the vector $\boldsymbol{b}$ depending on the initial estimates. Furthermore, errors and uncertainty are associated with the physical model, observations, and parameterization. These errors can cause the vector $\boldsymbol{b}$ to be unstable, especially for higher dimensions of the parameter vector. For many applications, it may not even be possible to determine a "true" parameter vector.

INDIRECT METHODS

Because field and laboratory experiments for determining unsaturated hydraulic properties of soils are still relatively time consuming and expensive, and the results may not be accurate or reproducible (especially for heterogeneous soils), indirect methods of estimating the parameters can be used. Indirect methods encompass a wide array of procedures developed to estimate hydraulic properties with "surrogate" data that are easily measured and, hence, commonly available (for example, particle-size distribution, bulk density, organic carbon content, and porosity). The process of predicting *needed* soil data from *existing* soil data is known as the pedotransfer function approach and is based on empirical and quasi-empirical functions and algorithms that establish the relationships between surrogate input data and unsaturated hydraulic-parameter output data.

Analytical Functions

Analytical expressions to describe hydraulic properties are convenient for characterizing soils and for use in numerical models (van Genuchten *et al.* 1991; Marion *et al.* 1994). Many closed-form expressions have been proposed to describe hydraulic properties. A limited number of papers were published to evaluate their suitability (Alexander and Skaggs 1986; Globus 1987; Mualem 1976; van Genuchten and Nielsen 1985; Vereecken 1992; Leij *et al.* 1997). A case study on the accompanying CD, "Closed Form Expressions for Water Retention and Conductivity Data," by F.J. Leij, W.B. Russell, and S.M. Lesch, compares 14 water-retention and 11 unsaturated-conductivity functions. A few selected functions describing the hydraulic data fairly well are briefly presented below (see also Leij *et al.* 1997).

The Brooks-Corey water-retention function (Brooks and Corey, 1964) has long been used to describe water-retention data for relatively homogeneous soils with a narrow pore-size distribution:

$$S_e(h) = \frac{\theta - \theta_r}{\theta_s - \theta_r} = \begin{cases} 1 & \alpha h \leq 1 \\ (\alpha h)^{\lambda} & \alpha h > 1 \end{cases} \tag{3.15}$$

where S_e is the effective saturation, θ is the volumetric water content, and the subscripts r and s denote residual and saturated water contents. A disadvantage of the expression is the abrupt change of the $S_e(h)$ curve at $h = 1/\alpha$, denoting the pressure at which the largest pore drains.

The van Genuchten water-retention closed-form expression, which appears to be the most widely used since publication of the paper by van Genuchten (1980), is given by

$$S_e(h) = [1 + (\alpha h)^n]^{-m} \tag{3.16}$$

Frequently, the restriction $m \equiv 1\text{-}1/n$ is imposed in a pore-size distribution model to predict the unsaturated hydraulic conductivity.

Hutson and Cass (1987) proposed a two-part expression, including a parabolic equation close to saturation and a power function elsewhere:

$$S_e(h) = \begin{cases} 1 - (\alpha h)^2 \, S_i^{2/n}(1 - S_i) & h \le h_i \\ (\alpha h)^{-n} & h > h_i \end{cases} \tag{3.17}$$

where $h_i = 1/(\alpha\ S_i^{1/n})$ and $S_i = 2/(2 + n)$

This expression ensures continuity in both $S_e(h)$ and $dS_e(h)/dh$ at $h = h_i$, unlike the Brooks-Corey function.

The following unsaturated hydraulic conductivity function, similar to the one used by Gardner (1958), can be used to describe K(h) data:

$$K(h) = K_s[1 + (\alpha h)^n]^{-m} \tag{3.18}$$

where K_s is the saturated hydraulic conductivity.

The Brooks-Corey and van Genuchten water-retention functions were used to derive unsaturated hydraulic-conductivity functions given, correspondingly, by

$$K(S_e) = K_s\, S_e^{3+2/L} \tag{3.19}$$

$$K(S_e) = K_s S_e^{L}\,[1 - (1 - S_e^{1/m})^m]^2 \tag{3.20}$$

K_s is the saturated hydraulic conductivity (cm day^{-1}), and L is a lumped parameter that accounts for pore tortuosity and connectivity, and it is often assumed $L = 0.5$. Equation (3.20) is often called the Mualem-van Genuchten model, and the parameter m is given by $m = 1\text{-}1/n$; sometimes it is assumed that L=0.5.

Empirical Models of Pedotransfer Functions

Definition of Pedotransfer Functions

Empirical pedotransfer functions (PTFs) are used to determine soil hydraulic properties from soil texture and other readily available soil physical properties without relying on specific analytical models. The relationship between input and output parameters of a PTF is determined by a statistical-regression or neural-network analysis. These methods determine empirical coefficients for analytical functions by minimizing the difference between the predicted and observed results. Note that empirically determined parameters are mostly reliable only for conditions identical to those for which the PTFs were determined.

Whereas many PTFs exist for determining water-retention functions, only a few exist for saturated hydraulic conductivity, K_s, and unsaturated hydraulic conductivity. Initially, PTFs were used to determine specific points of the water-retention curves, such as the field capacity and wilting points (Jamison and Kroth 1958). Currently, PTFs are almost exclusively used to predict unsaturated hydraulic parameters, which are then used in numerical modeling.

Almost all PTFs rely on particle-size distribution data. With only a soil textural classification, simple "class" PTFs can be used to assess average hydraulic properties for each soil textural class (Carsel and Parrish 1988; Wösten *et al.* 1995). For a given particle-size distribution, a continuous PTF can be determined with textural percentages as independent variables. PTF predictions may be further improved by adding basic soil properties such as bulk density, porosity, or organic matter content (Rawls and Brakensiek 1985; Vereecken *et al.* 1989; Rawls *et al.* 1991). Additional improvements may be achieved by including one or more water-retention data points (Rawls *et al.* 1992; Williams *et al.* 1992). Ahuja *et al.* (1989) and Messing (1989) improved predictions of the saturated hydraulic conductivity, K_s, by using effective porosity data, which they defined as the total porosity minus the water content at -10 or -33 kPa pressure head.

The reliability and accuracy of PTFs are important issues since most PTFs are developed by sparse, noisy, and sometimes ambiguous data sets that may yield parameter estimates with large confidence intervals. The accuracy of PTFs should be assessed against independent data sets as it was done by Tietje and Tapkenhinrichs (1993), Kern (1995), and Tietje and Hennings (1996).

Regression Analysis

Brakensiek *et al.* (1984) and Rawls and Brakensiek (1985) presented PTFs that predict parameters from the Brooks-Corey equation (3.16) and saturated hydraulic conductivity, K_s as well as from porosity, ϕ, and sand and clay percentages (S and C, respectively). In these approaches, θ_s is set equal to the porosity while θ_r, $h_b = 1/\alpha$, λ, and K_s are related to S, C, and ϕ, employing a regression equation given by:

$$\begin{aligned} b_i = {} & a_1 + a_2 S + a_3 C + a_4 \phi + a_5 S^2 + a_6 C^2 + a_7 \phi^2 + a_8 S\phi + a_9 C\phi \\ & + a_{10} S^2 C + a_{11} S^2 \phi + a_{12} C^2 \phi + a_{13} SC^2 + a_{14} C\phi^2 + a_{15} S^2 \phi^2 + a_{16} C^2 \phi^2 \end{aligned} \tag{3.21}$$

where b_i is a component of the hydraulic parameter vector $\boldsymbol{b} = \{\theta_r, h_b, \lambda, K_s\}$, and a_i are the input coefficients. This model is applicable for sand percentages between 5% and 70% and clay percentages between 5% and 60% (Rawls *et al.* 1991, 1992). The coefficients a_i are summarized in Table 3-17.

Vereecken *et al.* (1989, 1990) provided expressions for water retention and unsaturated hydraulic conductivity for 182 Belgian soil horizons. The water retention was described by modifying equation (3.16) and setting $m = 1$. The unsaturated hydraulic conductivity was described according to Gardner (1958) (see also equation 3.18):

$$K(h) = \frac{K_s}{1 + (bh)^c} \tag{3.22}$$

where b and c are empirical parameters.

According to Vereecken *et al.* (1989, 1990), the expressions for θ_r, θ_s, α (1/cm), n, K_s (cm/day), b, and c are:

$$\begin{aligned} &\theta_r = 0.015 - 0.005C + 0.014OM \\ &\theta_s = 0.81 - 0.238BD + 0.014OM \\ &\ln(\alpha) = -2.486 + 0.025S - 0.023C - 2.617BD - .0351OM \\ &\ln(n) = 0.053 - 0.009S - 0.013C + 0.00015S^2 \\ &\ln(K_s) = 20.62 - 0.96\ln(C) - 0.66\ln(S) - 0.46\ln(OM) - 8.43BD \\ &\ln(b) = -0.73 - 0.01877S + 0.058C \\ &\ln(c) = -1.186 - 0.194\ln(C) - 0.0489\ln(Si) \end{aligned} \tag{3.23}$$

TABLE 3-17 **Coefficients of the PTFs Determined by Brakensiek et al. (1984) and Rawls and Brakensiek (1985) for Prediction of Brooks-Corey Parameters and K_s.**

Index a_i	Variables	θ_r cm³/cm³	Input Coefficients ln(h_b, cm)	ln(λ)	ln(K_s, cm/h)
1	Intercept	-0.01825	5.339674	-0.78428	-8.96847
2	S	0.000873	0	0.017754	0
3	C	0.005135	0.184504	0	-0.02821
4	ϕ	0.029393	-2.48395	-1.0625	19.52348
5	S^2	0	0	-5.3E-05	0.000181
6	C^2	-0.00015	-0.00214	-0.00273	-0.00941
7	ϕ^2	0	0	1.111349	-8.39522
8	$S\phi$	-0.00108	-0.04356	-0.03088	0.077718
9	$C\phi$	0	-0.61745	0	0
10	S^2C	0	-1.3E-05	-2.4E-06	1.73E-05
11	$S^2\phi$	0	-0.00072	0	0.001434
12	$C^2\phi$	0.000307	0.008954	0.007987	0.02733
13	SC^2	0	5.4E-06	0	-3.5E-06
14	$C\phi^2$	-0.00236	0.500281	-0.00674	0
15	$S^2\phi^2$	0	0.001436	0.000266	-0.00298
16	$C^2\phi^2$	-0.00018	-0.00855	-0.00611	-0.01949

where S, C, and Si are the percentages of sand, clay, and silt; BD signifies the bulk density (g/cm³); and OM stands for the organic matter percentage.

Neural-Network Analysis

Artificial neural networks (ANNs) have been used to estimate hydraulic parameters by Pachepsky *et al.* (1996), Schaap and Bouten (1996), and Tamari *et al.* (1996). Neural networks are sometimes described as "universal function approximators" because of their ability to "learn" how to approximate a continuous (nonlinear) function to a desired degree of accuracy (Hecht-Nielsen, 1990; Haykin, 1994). An advantage of PTFs derived with a neural-network method is that no *a priori* model is required. Instead, the optimal relations between input data (basic soil properties) and output data (hydraulic parameters) are determined during the analysis with an iterative calibration procedure. The relationship is established with weighting factors for nodes at input, hidden, and output layers of the network. The number of input and output nodes corresponds to the number of input and output variables of the PTF (Schaap and Bouten 1996).

Although artificial neural networks may yield more accurate PTFs than a regression analysis (Schaap *et al.* 1998, Tamari 1996), their disadvantages are that (1) except for trivially small networks, it is cumbersome to formulate explicit expressions for the PTF, (2) the network may be too complex, and (3) the calibration involves too many iterations (Schaap *et al.* 1999).

Water-Retention Function and Saturated Hydraulic Conductivity

With a database of 1,209 samples, Schaap *et al.* (1998) determined neural-network PTFs with the retention parameters given by equation (3.16), and with $m = 1\text{-}1/n$ and the saturated hydraulic conductivity, K_s. Schaap *et al.* used a feed-forward back-propagation method with input, hidden, and output layers calibrated with the bootstrap method (Efron and Tibshirani 1993). Schaap *et al.* (1998) calibrated neural-network models to predict retention parameters and K_s with five different data sets and then recalibrated the models for an enlarged database of 2,085 samples. A summary of these models is given in Table 3-18, which presents parameters based on measured data (unlike a similar PTF by Carsel and Parrish 1988, which is based on predictions made by another PTF developed by Rawls and Brakensiek 1985). In Table 3-18, Model 1 (TXT) uses the average of "observed" hydraulic parameters within textural classes of the USDA-SCS soil classification. Models 2 through 5 use progressively more detailed input data. Model 2 uses the sand, silt, and clay fractions (SSC). Model 3 additionally uses a bulk density value (BD). Models 4 and 5 also require one or two water-retention points (moisture content) at 33 and 1500 kPa suction (TH33 and TH1500). These particular points are routinely available in the large USDA/NRCS NSSC database (Soil Survey Staff 1995). It is evident from Table 3-18 that R^2 (coefficient of determination) increases and RMSE (the root mean square for error) decreases with respect to "observed" parameters when the number of input variables increases, which indicates that the accuracy of the water-retention-function calculations increases.

TABLE 3-18 **Coefficients of Correlation and RMSE between Observed and Predicted Water Retention Parameters and Saturated Hydraulic Conductivity**

		Water retention parameters					*Saturated Hydraulic Conductivity*	
		R^2				RMSE	R^2	RMSE
Model No.	*Input data*	θ_r	θ_s	α	n	cm³/cm³		Log (cm/day)
1	TXT	0.066	0.136	0.204	0.452	0.108	0.425	0.741
2	SSC	0.073	0.150	0.221	0.472	0.107	0.437	0.735
3	SSCBD	0.070	0.567	0.232	0.479	0.093	0.509	0.685
4	SSCBDTH33	0.090	0.589	0.380	0.569	0.066	0.609	0.611
5	SSCBDTH331500	0.339	0.585	0.550	0.743	0.063	0.613	0.610

Unsaturated Hydraulic Conductivity

Adopting similar techniques, Schaap and Leij (2000) developed neural-network PTFs to estimate parameters of the Mualem-van Genuchten model (equation 3.20). With a database of 235 soils, Schaap and Leij (2000) determined that from water-retention parameters θ_r, θ_s, α, and n, the unsaturated hydraulic conductivity can be predicted with an accuracy of 0.84 log(cm/day), that is, within one order of magnitude. Similar results were reported by Kosugi (1999) for a different water-retention function.

It is important to assess the uncertainty involved in determining unsaturated hydraulic parameters. The neural-network calibration with the bootstrap method made it possible to quantify the uncertainty associated with the PTF prediction for a given sample. Figure 2 of the case study "Estimation of Soil Hydraulic Properties" by Schaap *et al.* shows the water-retention and unsaturated hydraulic-conductivity curves with the 10% and 90% confidence intervals as predicted by Model 3 (Table 3-18) for a loamy sand and a clay sample. The entire probability distribution is given for the saturated hydraulic conductivity, K_S. The uncertainty in the predicted hydraulic properties for clay is higher than that for loamy sand. Note that the neural-network calibration for fine-textured soils was conducted with a relatively small number soil samples, and hence the PTF cannot predict hydraulic properties of the clay as accurately as for sand. The case study "Estimation of the Soil Hydraulic Properties," by Marcel G. Schaap, Feike J. Leij, and Martinus Th. Van Genuchten, U.S. Salinity Laboratory, USDA-ARS, describes several indirect methods of determining soil hydraulic properties, including neural network analysis (see page 501).

The five PTF models were implemented in a user-friendly Windows 95 program called Rosetta, which specifies input data to predict the hydraulic parameters θ_r, θ_s, α, n, K_s, K_0 and L of equations (3.16) and (3.23), where K_0 is used as matching point for the saturated conductivity in model (3.23) rather than the experimental value for K_s. The program also quantifies the standard deviation of the estimated parameters. Hydraulic parameters can then be used in simulation models while uncertainty estimates facilitate risk-based analyses of water and solute transport. Rosetta is available for download at http://www.ussl.ars.usda.gov/MODELS/rosetta/rosetta.HTM.

Finally, Table 3-19 includes the water-retention parameters θ_r, θ_s, α, n, and the saturated hydraulic-conductivity parameter, K_s, and the unsaturated conductivity parameters K_s, and L, obtained by textural averages (Model 1 in Table 3-18).

Methods Based on Particle-Size and Pore-Size Distributions

Particle-Size Distribution

For various textured soils, a remarkable similarity between cumulative particle-size distribution curves and water-retention curves was observed (Arya and Paris 1981; Haverkamp and Parlange 1986). The particle-size distribution (PSD) can determine a pore-size distribution function, which can then be used to estimate water-retention and unsaturated hydraulic-conductivity functions (Arya and Paris 1981; Arya *et al.* 1999a; Arya *et al.* 1999b; Hoffmann-Riem *et al.* 1999). This method by Arya is based on the hypothesis that the discrete soil particle domains are assembled together without mixing of various-size particles. The resulting assemblage has the same water-retention curves and unsaturated hydraulic-conductivity functions as its counterpart natural soil, in which natural particles are mixed.

TABLE 3-19 **Average Values of the Unsaturated Hydraulic Parameters for the Twelve USDA Textural Classes.** Standard deviations are given in parentheses.

Texture Class	N	θ_r cm^3/cm^3	θ_s cm^3/cm^3	Log(α) log(1/cm)	log(n)	K_s log(cm/day)	K_o log (cm/day)	L
Clay	84	0.098 (0.107)	0.459 (0.079)	-1.825 (0.68)	0.098 (0.07)	1.169 (0.92)	0.472 (0.26)	-1.561 (1.39)
C loam	140	0.079 (0.076)	0.442 (0.079)	-1.801 (0.69)	0.151 (0.12)	0.913 (1.09)	0.699 (0.23)	-0.763 (0.90)
Loam	242	0.061 (0.073)	0.399 (0.098)	-1.954 (0.73)	0.168 (0.13)	1.081 (0.92)	0.568 (0.21)	-0.371 (0.84)
L Sand	201	0.049 (0.042)	0.390 (0.070)	-1.459 (0.47)	0.242 (0.16)	2.022 (0.64)	1.386 (0.24)	-0.874 (0.59)
Sand	308	0.053 (0.029)	0.375 (0.055)	-1.453 (0.25)	0.502 (0.18)	2.808 (0.59)	1.389 (0.24)	-0.930 (0.49)
S Clay	11	0.117 (0.114)	0.385 (0.046)	-1.476 (0.57)	0.082 (0.06)	1.055 (0.89)	0.637 (0.34)	-3.665 (1.80)
S C L	87	0.063 (0.078)	0.384 (0.061)	-1.676 (0.71)	0.124 (0.12)	1.120 (0.85)	0.841 (0.24)	-1.280 (0.99)
S loam	476	0.039 (0.054)	0.387 (0.085)	-1.574 (0.56)	0.161 (0.11)	1.583 (0.66)	1.190 (0.21)	-0.861 (0.73)
Silt	6	0.050 (0.041)	0.489 (0.078)	-2.182 (0.30)	0.225 (0.13)	1.641 (0.27)	0.524 (0.32)	0.624 (1.57)
Si Clay	28	0.111 (0.119)	0.481 (0.080)	-1.790 (0.64)	0.121 (0.10)	0.983 (0.57)	0.501 (0.27)	-1.287 (1.23)
Si C L	172	0.090 (0.082)	0.482 (0.086)	-2.076 (0.59)	0.182 (0.13)	1.046 (0.76)	0.349 (0.26)	-0.156 (1.23)
Si Loam	330	0.065 (0.073)	0.439 (0.093)	-2.296 (0.57)	0.221 (0.14)	1.261 (0.74)	0.243 (0.26)	0.365 (1.42)

The pore-size distribution is estimated from the total soil-sample pore volume, which may be determined from particle and bulk density data, and particle-size distribution (PSD). The relationship between particle and pore radii is well defined for spherical particles arranged in a cubic, close-packed assemblage, but requires empirical corrections for natural-soil particles. By calculating pore radii, pressure heads are subsequently calculated. This conversion necessitates additional empirical corrections to account for water not held by capillary forces and unknown effects of fluid and solid properties on water retention.

Pore-Size Distribution

Analytical models for the conductivity function can be derived by representing the porous medium as an idealized medium, consisting of well-defined pores with a uniform pore-size or a known pore-size distribution. Usually, pore and solid geometry are simplified considerably. Since the pore system of real porous media is not as simple as these models assume, empirical parameters are included in the models to improve the fit between experimental and theoretical conductivity functions.

Two of the most popular pore-size distribution models for conductivity are those by Burdine (1953):

$$K(S_e) = K_s S_e^L [\int_0^{S_e} h^{-2}(x)dx \,/\, \int_0^1 h^{-2}(x)dx] \qquad (3.24)$$

and Mualem (1976):

$$K(S_e) = K_s S_e^L [\int_0^{S_e} h^{-1}(x)dx \,/\, \int_0^1 h^{-1}(x)dx]^2 \qquad (3.25)$$

where, as before, K_s is the saturated hydraulic conductivity, L is a pore-connectivity and tortuosity parameter, and x is a dummy integration variable.

Databases

Soil hydraulic databases can provide surrogate information when the direct measurement of soil hydraulic properties is not feasible because of cost, time, and uncertainty (Schaap and Leij 1998). Several collections of soil hydraulic data have been compiled (Mualem 1976; Rawls *et al.* 1985; Wösten and van Genuchten 1988; Leij *et al.* 1996; Lilly *et al.* 1999). The International Unsaturated Soil Hydraulic Database

(UNSODA) (Leij *et al.* 1996) contains data sets from around the world for a variety of porous media (see also the case study on the accompanying CD "The UNSODA Unsaturated Soil Hydraulic Database," by F.J. Leij, W.J. Alves, M. Th. van Genuchten, A. Nemes, and M.G. Schaap). The UNSODA database currently consists of 791 entries of field- and laboratory-measured water-retention, saturated and unsaturated hydraulic-conductivity, and particle-size-distribution data and bulk density from many international sources. Leij *et al.* (1996) documents UNSODA 1, intended for computers with a Disk Operating System (DOS). The database UNSODA 2 was developed for computers with the Windows 95 operating system using Microsoft Access. UNSODA 2 facilitates broad and user-friendly search and report procedures whereas UNSODA 1 has data entry and optimization routines. UNSODA 2 and supporting documentation may be obtained electronically (http://www.ussl.ars.usda.gov).

Several other databases of soil information and hydraulic data exist. The Natural Resources Conservation Service Soil Survey Laboratory in Lincoln, Nebraska, has compiled a database with soil characterization and profile description data. This database is distributed on a CD-ROM (http://www.statlab.iastate.edu/soils/ssl/cdinfo.html). Many other sites contain useful general information, such as those for Canadian (http://res.agr.ca/CANSIS/NSDB/) and Australian (http://www.cbr.clw.csiro.au/aclep) soils. The FAO and the International Soil Reference and Information Centre have compiled information for soils worldwide (FAO 1993, 1995). The Hydraulic Properties of European Soils (HYPRES) database will be maintained by the European Soil Bureau (Lilly *et al.* 1999). The Grizzly database contains hydraulic properties as well as structural and textural information on soils from different countries. Free copies of the database may be obtained electronically (ftp://lthe712c.hmg.inpg.fr/pub/Grizzly). Finally, commercial software is also available to estimate soil mechanical and hydraulic properties (http://www.soilvision.com/home.html).

REFERENCES

Abbaspour, K.C., M.T. van Genuchten, R. Schulin, and W. Schläppi. "A Sequential Uncertainty Domain Inverse Procedure for Estimating Subsurface Flow and Transport Parameters." *Water Resour. Res.*, 33(8) (1997): 1879–1892.

Acworth, R.I. "Investigation of Dryland Salinity using the Electrical Image Method." *Aust. J. Soil Res.*, 37 (1999): 623–636.

Ahuja, L.R., D.K. Cassel, R.R. Bruce, and B.B. Barnes. "Evaluation of Spatial Distribution of Hydraulic Conductivity Using Effective Porosity Data." *Soil Sci.*, 148 (1989): 404–411.

Alessi, R.S., and L. Prunty. "Soil–Water Determination using Fiber Optics." *Soil Sci. Soc. Am. J.*, 50 (1985): 860–863.

Alexander, L. and R.W. Skaggs. "Predicting Unsaturated Hydraulic Conductivity from the Soil Water Characteristic." *Trans. ASAE*, 29 (1986): 176–184.

Alexander, M. "Biodegradation: Problems of Molecular Recalcitrance and Microbial Fallibility." *Adv. Appl. Microbial.*, 7 (1965): 35–80.

Alfoldi, L. "Groundwater Microbiology: Problems and Biological Treatment-State-of-the-Art Report." *Wat. Sci. Tech.*, 20 (1988): 1–31.

Allison, G.B. "A Review of Some of the Physical and Chemical and Isotopic Techniques Available for Estimating Groundwater Recharge." *Estimation of Natural Groundwater Recharge*, I. Simmers (Ed.) (1988): 49–72.

Allison, G.B., G.W. Gee, and S.W. Tyler. "Vadose-Zone Techniques for Estimating Groundwater Recharge in Arid and Semiarid Regions." *Soil Science Society Of America Journal*, 58 (1994): 6–14.

Anderson, L.D., "Problems Interpreting Samples Taken with Large-Volume, Falling Suction Soil-Water Samplers." *Ground Water* (1986).

Anderson, T.A., E.A. Guthrie, and B.T. Walton. "Phytoremediation." *Environ. Sci. Tech.*, 27 (1993): 2630–2636.

Anderson, S.P., W.E. Dietrich, D.R. Montgomery, R. Torres, M.E. Conrad, and K. Loague. "Subsurface Flow Paths in a Steep Unchanneled Catchment." *Water Resources Research*, 33 (1997): 2637–2653.

Andraski, B.J. "Soil-Water Movement under Natural-Site and Waste-Site Conditions: A Multiple-Year Field Study in the Mojave Desert, Nevada." *Water Resour. Res.*, 33 (1997): 1901–1916.

Angulo-Jaramillo, R., M. Vauclin, R. Haverkamp, and P. Gérard-Marchant. "Dual Gamma-Ray Scanner and Instantaneous Profile Method for Swelling Unsaturated Materials." in *Proceedings of the Int. Workshop on the Characterization and Measurement of the Hydraulic Properties of Unsaturated Porous Media,* M.Th. van Genuchten, F.J. Leij and L. Wu (Eds.), Proc. Int. Workshop "Characterization and Measurement of the Hydraulic Properties of Unsaturated Porous Media." University of California, Riverside, CA (1999).

Arya, L.M., D.A. Farrell, and G.R. Blake. "A Field Study of Soil Water Depletion Patterns in the Presence of Growing Soybean Roots: 1. Determination of Hydraulic Properties of the Soil." *Soil Sci. Soc. Am. Proc.*, 39 (1975): 424–436.

Arya, L.M., and J.F. Paris. "A Physicoempirical Model to Predict Soil Moisture Characteristics from Particle-Size Distribution and Bulk Density Data." *Soil Sci. Soc. Am. J.*, 45 (1981): 1023–1030.

Arya, L.M., F.J. Leij, M.Th. van Genuchten, and P. J. Shouse. "Scaling Parameter to Predict the Soil Water Characteristic from Particle-Size Distribution Data." *Soil Sci. Soc. Am. J.*, 63 (1999a): 510–519.

Arya, L.M., F.J. Leij, P.J. Shouse, and M.Th. van Genuchten "Relationship Between the Hydraulic Conductivity Function and the Particle-Size Distribution." *Soil Sci. Soc. Am. J.*, (in press) (1999b).

ASTM 1998 - D 3404 – 91 "Guide for Measuring Matric Potential in the Vadose Zone Using Tensiometers." (1998).

Baehr, A.L. and M.F. Hult. "Determination of the Air-Phase Permeability Tensor of an Unsaturated Zone at the Bemidji, Minnesota Research Site." in *Proceedings of Technical Meeting, U.S. Geological Survey Toxic Substances Hydrology Program*, G.E. Mallard and S.E. Ragone (Eds.), Reston, VA (1989): 55–62.

Baehr, A.L., and M.F. Hult. "Evaluation of Unsaturated Zone Air Permeability through Pneumatic Tests." *Water Resour. Res.*, 27 (1991): 2605–2617.

Baier, D.C., F.K. Aljibury, J.K. Meyer, J.K., and A.K. Wolfenden. "Vadose Zone Monitoring is Effective for Evaluating the Integrity of Hazardous Waste Pond Liners." Baier Agronomy, Inc., Woodland, CA, November (1983).

Baker, K.H., and D.S. Herson. "*In Situ* Bioremediation of Contaminated Aquifers and Subsurface Soils." *Geomicrobiol. J.*, 8 (1990): 133–146.

Ball, J. and D.M. Coley. "A Comparison of Vadose Monitoring Procedures." in *Proceedings of the Sixth National Symposium and Exposition on Aquifer Restoration and Ground Water Monitoring*, NWWA/EPA (1986): 52–61.

Barbaro, S.E., H.J. Albrechtsen, B.K. Jensen, C.I. Mayfield, and J.F. Barker. "Relationships Between Aquifer Properties And Microbial-Populations in the Borden Aquifer." *Geomicrobiol. J.*, 12 (1994): 203–219.

Barentsen, P. "Short Description of a Field Testing Method with Cone-Shaped Sounding Apparatus." in *Proceedings of the 1st International Conference on Soil Mechanics and Foundations Engineering*, Cambridge, MA (1936): 1, B/3, 6–10.

Barnes, C.J., and G.B. Allison. "Water Movement in the Unsaturated Zone Using Stable Isotopes of Hydrogen and Oxygen." *Journal of Hydrology,* 100 (1998): 143–176.

Barker, J.A. "A Generalized Radial Flow Model for Hydraulic Tests in Fractured Rock." *Water Resour. Res.*, 24(10) (1988): 1796–1804.

Baumgartner, N., G.W. Parkin, and D.E. Elrick. "Soil Water Content and Potential Measured by Hollow Time Domain Reflectometry Probe." *Soil Sci. Soc. Am. J.*, 58 (1994): 315–318.

Bear, J. Dynamics of Fluids in Porous Media. NY American Elsevier Pub. Co. (1972).

Bekins, B.A., E.M. Godsy, and D.F. Goerlitz. "Modeling Steady-State Methanogenic Degradation of Phenols in Groundwater." *J Contam. Hydrol.*, 14 (1993): 279–294.

Benito, P.H., P.J. Cook, B. Faybishenko, B. Freifeld, and C. Doughty. "Cross-Well Air-Injection Packer Tests for the Assessment of Pneumatic Connectivity In Fractured, Unsaturated Basalt." *Rock Mechanics for Industry*, Amadei, Kranz, Scott & Smeallie (Eds.), Balkema, Rotterdam, Netherlands (1999): 843–851.

Bianchi, W.C. "Measuring Soil Moisture Tension Changes." *Agricultural Engineering*, 43 (1962): 398–404.

Bilskie, J. "Reducing Measurement Errors of Selected Soil Water Sensors." in *Proceedings of the int. workshop Characterization and Measurement of the Hydraulic Properties of Unsaturated Porous Media,* van Genuchten, M.Th., F. J. Leij and L. Wu (Eds.), Proc. Int. Workshop, "Characterization and Measurement of the Hydraulic Properties of Unsaturated Porous Media." University of California, Riverside, CA (1999).

Bonazountas, M., and J.M. Wagner. "SESOIL, A Seasonal Soil Compartment Model." United States Environmental Protection Agency, Report No. C-85875, published by the National Technical Information Service (NTIS), Washington, D.C. (1984).

Bond, W.R., and J.V. Rouse. "Lysimeters Allow Quicker Monitoring of Heap Leaching and Tailing Sites." Mining Engineering April (1985): 314–319.

Borden, R.C., and P.B. Bedient. "Transport of Dissolved Hydrocarbons Influenced by Reaeration and Oxygen Limited Biodegradation: 1. Theoretical Development." *Water Resources Research*, 22 (1986): 1973–1982.

Boulding, J.R., "Practical Handbook of Soil, Vadose Zone, and Ground-Water Contamination: Assessment, Prevention, and Remediation." Lewis Publishers, Chelsea, MI (1995).

Bouma, J., D.I. Hillel, F.D. Hole, and C.R. Amerman. "Field Measurements of Hydraulic Conductivity by Infiltration Through Artificial Crusts." *Soil Sci. Soc. Proc.*, 35 (1971): 362–364.

Bouma, J. and J.L. Denning, "Field Measurement of Unsaturated Hydraulic Conductivity by Infiltration Through Gypsum Crusts." *Soil Sci. Soc. Proc.*, 36 (1972): 846–847.

Bouma, J., A. Jongerius, and D. Schoondebeek. "Calculation of Hydraulic Conductivity of Some Saturated Clay Soils Using Micromorpho-Meteric Data." *Soil Science Society of America Journal,* 43 (1979): 261–265.

Bouma, J., C. Belmans, L.W. Dekker and W.J. Jeurissen. "Assessing the Suitability of Soils with Macropores for Subsurface Liquid Waste Disposal." *J. Environ. Qual.*, 12 (1983): 305–311.

Bouwer, H. "Intake Rate: Cylinder Infiltrometer." *Methods of Soil Analysis Part 1. Soil Science Society of America*, A. Klute (Ed.), Madsion, WI (1986): 825–844.

Bowman, J.P., L. Jiménez, I. Rosario, T.C. Hazen, and G.S. Sayler. "Characterization of the Methanotrophic Bacterial Community Present in a Trichloroethylene-Contaminated Subsurface Groundwater Site." *Appl. Environ. Microbiol.*, 59 (1993): 2380–2387.

Boynton, D., and W. Reuther. "A Way of Sampling Soil Gases in Dense Subsoils, and Some of Its Advantages and Limitations." in *Proceedings of Soil Science Society of America*, 3 (1938): 37–42.

Bradford, S.A., and F.J. Leij. "Wettability Effects on Scaling Two- and Three-Fluid Capillary Pressure-Saturation Relations." *Env. Sci. and Technol.*, 29 (1995): 1446–1455.

Bradford, S.A., and F.J. Leij. "Predicting Two-and Three-Fluid Capillary Pressure-Saturation Relations in Fractional Wettability Media." *Water Resour. Res.*, 32 (1996): 251–259.

Brakensiek, D.L., W.J. Rawls, and G.R. Stephenson. "Modifying SCS Hydrologic Soil Groups and Curve Numbers for Rangeland Soils." from *ASAE Paper No. PNR-84-203*, St. Joseph, MI (1984).

Breedveld, G.D., G. Olstad, T. Briseid, and A. Hauge. "Nutrient Demand in Bioventing of Fuel Oil Pollution." in *Proceedings of In Situ Aeration: Air Sparging, Bioventing, and Related Remediation Processes*, R. E. Hinchee, R. N. Miller and P. C. Johnson (Eds.), Battelle Press, Columbus, OH (1995): 391–399.

Bresler, E., and G. Dagan. "Solute Dispersion in Unsaturated Heterogeneous Soil at Field Scale, 2, Applications." *Soil Sci. Soc. Am. J.*, 43 (1979): 467–472.

Bristow, K.L., G.S. Campbell, and K. Calissendorff. "Test of a Heat-Pulse Probe for Measuring Changes in Soil Water Content." *Soil Sci. Soc. Am. J.*, 57 (1993): 930–934.

Bristow, K.L. "Measurement of Thermal Properties and Water Content of Unsaturated Sandy Soil Using Dual-Probe Heat-Pulse Probes." *Agric. Forest Meteor.*, 89 (1998): 75–84.

Brockman, F.J. "Nucleic-Acid-Based Methods for Monitoring the Performance of *In Situ* Bioremediation." *Molecular Ecology.*, 4 (1995): 567–578.

Brockman, F.J., W. Payne, D.J. Workman, A. Soong, S. Manley, and T.C. Hazen. "Effect of Gaseous Nitrogen and Phosphorus Injection on *In Situ* Bioremediation of a Trichloroethylene-Contaminated Site." *J Haz. Mat.*, 41 (1995): 287–298.

Brooks, R.H., and A.T. Corey. "Hydraulic Properties of Porous Media." *Hydrology Papers, No. 3*, Colorado State University, Fort Collins, CO (1964).

Burdine, N.T. "Relative Permeability Calculations from Pore-Size Distribution Data." *Petrol. Trans., Am. Inst. Min. Eng.*, 198 (1953): 71–77.

Burger, H.R. "Exploration Geophysics of the Shallow Subsurface." Prentice Hall, NJ (1992).

Butler, D.K. "Tutorial - Engineering and Environmental Applications of Microgravity." in *Proceedings of the Symposium on the Application of Geophysics to Engineering and Environmental Problems*, Knoxville, TN (1991): 139–177.

Campbell, G.S., and G.W. Gee. "Water Potential: Miscellaneous Methods." in *Methods of Soil Analysis. Part 1. of 2nd ed. Agron. Monogr. 9*, A, Klute (Ed.), ASA and SSSA, Madison, WI (1986): 619–633.

Campbell, G.S., C. Calissendorff, and J.H. Williams. "Probe for Measuring Soil Specific Heat Using a Heat-Pulse Method." *Soil Sci. Soc. Am. J.*, 55 (1991): 291–293.

Carsel, R.F. and R.S. Parrish. "Developing Joint Probability Distributions of Soil Water Retention Characteristics." *Water Resour. Res.*, 24 (1988): 755–769.

Cary, J.W., G.W. Gee, and C.S. Simmons. "Using an Electro-Optical Switch to Measure Soil Water Suction." *Soil Sci. Soc. Am. J.*, 55 (1991): 1798–1800.

Cassel, D.K. "*In Situ* Unsaturated Soil Hydraulic Conductivities for Selected North Dakota Soils." *Agric. Explt. St. North Dakota Bull*, 494 (1974).

Caster, A., and R. Timmons. "Pourous Teflont: Its Application in Groundwater Sampling." Timco Mfg. Inc., Prairie Du Sac, WI (1988).

Cerling, T.E. "The Stable Isotopic Composition of Modern Soil Carbonate and Its Relationship to Climate." *Earth and Planetary Science Letters,* 71 (1984): 229–240.

Chahal, R.S., "Effect of Temperature and Trapped Air on Matric Suction." *Soil Science*, 100 (4) (1965): 262–266.

Chan, D.B., and E.A. Ford. "*In Situ* Oil Biodegradation." *Military Engineer*, 509 (1986): 447–737.

Chapman, J.B., N.L. Ingraham, and J.W. Hess. "Isotopic Investigation of Infiltration and Unsaturated Zone Flow Processes at Carlsbad Cavern, New-Mexico." *Journal of Hydrology,* 133 (1992): 343–363.

Chen, G., M. Taniguchi, and S.P. Neuman. "An Overview of Instability and Fingering During Immiscible Fluid Flow in Porous and Fractured Media." NUREG/CR-6308. U.S. Nuclear Regulatory Commission, Washington, D.C. April (1995).

Chilakapati, A., T. Ginn, and J. Szecsody. "An Analysis of Complex Reaction Networks in Groundwater Modeling." *Water Resour. Res.,* 34(7) (1998): 1767–1780.

Cho, J.S. and D.C. DiGiulio. "Pneumatic Pumping Test for Soil Vapor Extraction." *Environmental Progress,* 11(3) (1992): 228–233.

Choubey, V.M., and R.C. Ramola. "Correlation between Geology and Radon Levels in Groundwater, Soil and Indoor Air in Bhilangana Valley, Garhwal Himalaya, India." *Environmental Geology*, 32(4) (1997): 258–262.

Clothier, B.E. and I. White. "Measurement of Sorptivity and Soil Water Diffusivity in the Field." *Soil Sci. Soc. Am. J.*, 45 (1981): 241–245.

Cohen, R.M., J.W. Mercer, and J.Matthews. "DNAPL Site Evaluation." S.K. Smoley (Ed.), Boca Raton, FL (1993).

Conca, J.L., and J.V. Wright. "Diffusion and Flow in Gravel, Soil, and Whole Rock." *Applied Hydrogeology,* 1 (1992): 5–24.

Conca, J.L., and J.V. Wright. "Diffusion Coefficients in Gravel Under Unsaturated Conditions." *Water Resources Research*, 26 (1990): 1055–1066.

Conca, J.L. and J.V. Wright. "The UFA Method for Rapid, Direct Measurements of Unsaturated Soil Transport Properties." *Australian J. of Soil Research,* 36 (1998): 291–315.

Conca, J.L., D.G. Levitt, P.R. Heller, T.J. Mockler, and M.J. Sully. "Direct UFA Measurements of Unsaturated Hydraulic Conductivity, Comparisons to van Genuchten/Mualem Estimations, and Applications to Recharge Mapping in Arid Regions." in M.Th. van Genuchten (1999).

Conrad, M.E., P.F. Daley, M.F. Fischer, B.B. Buchanan, T. Leighton, and M. Kashgarian. "Combined ^{14}C and $d^{13}C$ Monitoring of *In Situ* Biodegradation of Petroleum Hydrocarbons." *Environmental Science & Technology*, 31 (1997): 1463–1469.

Conrad, M.E., D.J. DePaolo, D.L. Song, and E. Neher. "Isotopic Evidence for Groundwater Flow and Biodegradation of Organic Solvents at the Test Area North Site, Idaho National Engineering and Environmental Laboratory." in *Ninth Annual V.M. Goldschmidt Conference*, Lunar and Planetary Institute, Houston, LPI Contribution 971 (1999): 58–59.

Cook, P. "*In Situ* Pneumatic Testing at Yucca Mountain." paper submitted to the *International Journal of Rock Mechanics* (1999).

Cook, P.G., and G.R. Walker. "Depth Profiles of the Electrical Conductivity from Linear Combinations of Electromagnetic Induction Measurements." *Soil Sci. Soc. Am. J.*, 56 (1992): 1015–1022.

Cook, P.G., I.D. Jolly, F.W. Leaney, G.R. Walker, G.L. Allan, L.K. Fifeld, and G.B. Allison. "Unsaturated Zone Tritium and Chlorine-36 Profiles from Southern Australia - Their Use as Tracers of Soil Water Movement." *Water Resources Research,* 30 (1994): 1709–1719.

Corey, A.T. "Mechanics of Immiscible Fluids in Porous Media." *Water Resources Publications* (1994): 252.

Corwin, D.L. "GIS Applications of Deterministic Solute Transport Models for Regional-Scale Assessment of Non-Point Source Pollutants in the Vadose Zone." in *Applications of GIS to the Modeling of Non-Point Source Pollutants in the Vadose Zone*, D.L. Corwin and K. Loague (Eds.), SSSA Special Publication No. 48, Soil Science Society of America, Madison, WI (1996): 69–100.

Craig, H. "Isotopic Variations in Meteoric Waters." Science, 133 (1961): 1702–1703.

Criss, R.E., and M.L. Davisson. “Isotopic Imaging of Surface Water/Groundwater Interactions, Sacramento valley, California.” *Journal of Hydrology*, 178 (1996): 205–222.

Crowe, R.D., and W.D. Wittekind. “Ferrocyanide Safety Program: In-tank Application of Electromagnetic Induction (EMI) Moisture Measurements.” in the FY 1995 Report, WHC-SD-WM-ER-520, Westinghouse Hanford Company, Richland, WA (1995).

Cusack, F., S. Singh, C. McCarthy, J. Grieco, M. Derocco, D. Nguyen, H. Lappinscott, and J.W. Costerton. “Enhanced Oil-Recovery - 3-Dimensional Sandpack Simulation of Ultramicrobacteria Resuscitation in Reservoir Formation.” *J. Gen. Microbiol.*, 138 (1992): 647–655.

Cushman, J.H., “On Measurement, Scale, and Scaling.” *Water Resour. Res.*, 22(2) (1986): 129–134.

Dalton, F.N. and M.Th. Van Genuchten. “The Time-Domain Reflectometry Method for Measuring Soil Water Content and Salinity.” *Geoderma*, 38 (1986): 237–250.

Davis, J.B. “Microbiology in Petroleum Exploration.” in *Unconventional Methods in Exploration for Petroleum and Natural Gas*, W.B. Heroy, (Ed.), Southern Methodist University, Institute for the Study of Earth and Man, SMU Press (1969): 139–157.

Davis, J. and A. Annan. “Ground-Penetrating Radar for High-Resolution Mapping of Soil and Rock Stratigraphy.” *Geophysical Prospecting,* 37 (1989): 531–551.

Davis, S.N., D. Cecil, M. Zreda, and P. Sharma. “Chlorine-36 and the Initial Value Problem.” *Hydrogeology Journal*, 6 (1998): 104–114.

Davisson, M.L., D.K. Smith, J. Kenneally, and T.P. Rose. “Isotope Hydrology of Southern Nevada Groundwater: Stable Isotopes and Radiocarbon.” *Water Resources Research*, 35 (1999): 279–294.

Dean, T.J., J.P. Bell, and A.J. Baty. “Soil Moisture Measurement by an Improved Capacitance Technique, Part I. Sensor Design and Performance.” *J. Hydrol.*, 93 (1987): 67–78.

Debyle, N.V., R.W. Hennes, and G.E. Hart. “Evaluation of Ceramic Cups for Determining Soil Solution Chemistry.” *Soil Science*, 146 (1988): 30–36.

DeFlaun, M.F., B.M. Marshall, E.P. Kulle, and S.B. Levy. “Tn5 Insertion Mutants of Pseudomonas Fluorescens Defective in Adhesion to Soil and Seeds.” *Appl. Environ. Microbiol.*, 60 (1994): 2637–2642.

Deka, R.N., M. Wairiu, P.W. Mtakwa, C.E. Mullins, E.M., Veenendaal, and J. Townend. "Use and Accuracy of the Filter-Paper Technique for Measurement of Soil Matric Potential." *European J. Soil Sci.*, 46 (1995): 233–238.

Delleur, J.W. (Ed.), The Handbook of Groundwater Engineering, CRC Press, Boca Raton, FL (1999).

Demond, A.H., and P.V. Roberts. "Estimation of Two-Phase Relative Permeability Relationships for Organic Liquid Contaminants." *Water Resour. Res.*, 29(4) (1993): 1081–1090.

Demond, A.H., F.N. Desai, and K.F. Hayes. "Effect of Cationic Surfactants on Organic Liquid-Water Capillary Pressure-Saturation Relationships." *Water Resour. Res.*, 30 (1994): 333–342.

Demond, A.H., K.F. Hayes, D.L. Lord, F. Desai, and A. Salehzadeh. "Impact of Organic Compound Chemistry on Capillary Pressure Relationships of Sands.", in *Proceedings of the Int. Workshop on Characterization and Measurement of the Hydraulic Properties of Unsaturated Porous Media,* van Genuchten, M.Th., F.J. Leij and L. Wu (Eds.), University of California, Riverside, CA (1999).

DeRyck, S.M., J.D. Redman, and A.P. Annan. "Geophysical Monitoring of a Controlled Kerosene Spills." in *Proceedings of the Symposium on the Application of Geophysics to Engineering and Environmental Problems*, San Diego, CA (1993): 5–20.

Devitt, D.A., R.B. Evans, W.A. Jury, T. Starks, B. Eklund, A. Gnolson, and J. van Eyk "Soil Gas Sensing for the Detection and Mapping of Volatile Organics." National Water Well Association, Dublin, OH (1987).

Diem, D., H.B. Kerfoot, and B.E. Ross. "Field Evaluation of a Soil-Gas Analysis Method for Detection of Subsurface Diesel Fuel Contamination." in *Proceedings of the Second National Outdoor Action Conference on Aquifer Restoration, Ground Water Monitoring and Geophysical Methods*, National Water Well Association, Dublin, OH (1987).

Diment, G.A., K.K. Watson, and P.J. Blennerhasset. "Stability Analysis of Water Movement in Unsaturated Porous Materials: 1. Theoretical Considerations." *Water Resour. Res.*, 18 (1982): 1248–1254.

Dirksen, C. "Unsaturated Hydraulic Conductivity." in *Soil analysis, Physical methods*, Marcel Dekker, K.A. Smith and C.E. Mullins (Eds.), NY (1991).

Dirksen, C. "Direct Hydraulic Conductivity Measurements for Evaluating Approximate and Indirect Determinations." in the *Proceedings of the Int. Workshop Characterization and measurement of the hydraulic properties of unsaturated porous media*, M.Th. Van Genuchten, F.J. Leij, and L. Wu (Eds.), Univ. of California, Riverside, CA (1999): 271–278.

Dirksen, C.E., and S. Dasberg. "Four Component Mixing Model for Improved Calibration of TDR Soil Water Content Measurements." *Soil Sci. Soc. Amer. J.*, 57 (1993): 660–667.

Doodge, J.C.I., and Q.J. Wang. "Comment on 'An Investigation of the Relation between Ponded And Constant Flux Rainfall Infiltration' by A. Poulovassilis *et al.*" *Water Res. Resour.*, 29(4) (1993): 1335–1337.

DOE, "Groundwater/Vadose Zone Integration Project Management Plan," DOE/RL-98-56, Rev. O, U.S. Department of Energy, Richland Operations Office, Richland, WA (1999).

Douglas, B.J. and R.S. Olsen. "Soil Classification using Electric Cone Penetrometer. Cone Penetration Testing and Experience." in *Proceedings of the ASCE National Convention*, American Society of Civil Engineers, St. Louis, MO (1981): 209–227.

Dragun, J. "Microbial Degradation of Petroleum Products in Soil." in *Proceedings of the Conference on the Environmental and Public Health Effects of Soils Contaminated with Petroleum Products*, Amherst, MA October 30–31 (1985).

Driscoll, F.G. "Groundwater and Wells." in the *2nd ed. Johnson Filtration Systems, Inc.*, St. Paul, MN (1986).

Dullien, F.A.L. "Porous Media. Fluid Transport and Pore Structure." Academic Press, San Diego, CA (1992).

Durner, W. "Hydraulic Conductivity Estimation for Soils with Heterogeneous Pore Structure." *Water Resour. Res.*, 30 (1994): 211–233.

Dzekunov, N.E., I.E. Zhernov, and B.A. Faybishenko. "Thermo-Dinamicheskie Metody Izucheniya Vodnogo Rezhima Zony Aeratsii (Thermodynamic Methods of Investigating the Water Regime In The Vadose Zone)." Nedra, Moscow (1987).

Eagleson. "Climate, Soil, and Vegetation, –3, A Simplified Model of Soil Moisture Movement in the Liquid Phase." *Water Res. Resour.*, 14(5) (1978): 722–730.

Eaton, R.R., C.K. Ho, R.J. Glass, M.J. Nicholl, and B.W. Arnold. "Three-Dimensional Modeling of Flow through Fractured Tuff at Fran Ridge." SAND95-1896, MOL.19970203.0139, SNSAND95189600.000, Sandia National Laboratories, Albuquerque, NM (1996).

Eching, S.O., J.W. Hopmans, and O. Wendroth. "Unsaturated Hydraulic Conductivity from Transient Multistep Outflow and Soil Water Pressure Data." *Soil Sci. Soc. Am. J.*, 58 (1994): 687–695.

Edwards, K.B., and L.C. Jones. "Air Permeability from Pneumatic Tests in Oxidized Till." *Journal of Environmental Engineering*, 120(2) (1994): 329–347.

Efron, B., and R. J. Tibshirani. "An Introduction to the Bootstrap–Monographs on Statistics and Applied Probability." Chapman and Hall, NY (1993).

Eklund, B. "Detection of Hydrocarbons in Groundwater by Analysis of Shallow Soil Gas/Vapor." API Publication No. 4394, Washington, D.C. (1985).

Eldredge, E.P., C.C. Shock, and T.D Stieber. "Calibration of Granular Matrix Sensors for Irrigation Management." *Agron. J.*, 85 (1993): 1228–1232.

Elrick, D.E., and D.H. Bowman. "Note on an Improved Apparatus for Soil Moisture Flow Measurements." *Soil Sci. Soc. Am. Proc.*, 28 (1964): 450–451.

Elrick, D.E., and W.D. Reynolds. "Infiltration from Constant-Head Well Permeameters and Infiltrometers." in *Advances in Measurement of Soil Physical Properties: Bringing Theory Into Practice,* SSSA Spec. Publ. 30, American Society of Agronomy, G.C. Topp, W.D. Reynolds, and R.E. Green (Eds.), Madison, WI (1992a): 1–24.

Elrick, D.E., and W.D. Reynolds. "Methods for Analyzing Constant-Head Well Permeameter Data." *Soil Sci. Soc. Am. J.*, 56 (1992b): 309–312 (1992b).

Elrick, D.E., G.W. Parkin, W.D. Reynolds, and D.J. Fallow. "Analysis of Early-Time and Steady State Single-Ring Infiltration under Falling Head Conditions." *Water Resour. Res.*, 31 (1995): 1883–1893.

Elrick, D.E., J.H. Scandrett, and E.E. Miller. "Tests of Capillary Flow Scaling." *Soil Sci. Am. Proc.*, 23 (1959): 329–332.

EML Procedures Manual. USDOE Report HASL-300. 28th Edition, Vol. 1, Sect. 2.2.3.7, Available at http://www.eml.doe.gov/publications/procman (1997).

Enfield, C.G., and S.R. Yates. "Organic Chemical Transport to Groundwater." in *Pesticides in the Soil Environment - SSSA Book Series*, 2 (1990): 271–302.

Enzien, M.V., F. Picardal, T.C. Hazen, R.G. Arnold, and C.B. Fliermans. "Reductive Dechlorination of Trichloroethylene and Tetrachloroethylene under Aerobic Conditions in a Sediment Column." *Appl. Environ. Microbiol.*, 60 (1994): 2200–2205.

Epstein, S., and T. Mayeda. "Variations of O^{18} Content of Waters from Natural Sources." *Geochimica et Cosmochimica Acta*, 4 (1953): 213–224.

Everett, L.G. "Monitoring in the Vadose Zone." *Ground Water Monitoring Review*, Summer (1981): 44–51.

Everett, L. G., L. G. Wilson, and L. G. McMillion. "Vadose Zone Monitoring Concepts for Hazardous Waste Sites." *Ground Water*, 20 (1982): 312–324.

Everett, L.G., L.G. Wilson, and E.W. Hoylman. "Vadose Zone Monitoring Concepts for Hazardous Waste Sites." Noyes Data Corporation, NJ (1984a).

Everett, L.G., E.W. Hoylman, L.G. Wilson, and L.G. McMillion. "Constraints and Categories of Vadose Zone Monitoring Devices." *Ground Water Monitoring Review*, Winter (1984b): 26–32.

Evett, R.S., and J.L. Steiner. "Precision of Neutron Scattering and Capacitance Type Soil Water Content Gauges from Field Calibration." *Soil Sci. Soc. Am. J.*, 59 (1995): 961–968.

Ewing, R.P., and B. Berkowitz. "A Generalized Growth Model for Simulating Initial Migration of Dense Non-Aqueous Phase Liquids." *Water Res. Resour*, 34(4) (1998): 611–622.

Fabryka-Martin, J., H. Bentley, D. Elmore and P.L. Airey. "Natural Iodine-129 as an Environmental Tracer." *Geochimica et Cosmochimica Acta*, 49 (1985): 337–347.

Fabryka-Martin, J., A.V. Wolfsberg, P.R. Dixon, S.Levy, J. Musgrave, and H.J. Turin. "Summary Report of Chlorine-36 Studies: Sampling, Analysis and Simulation of Chlorine-36 in the Exploratory Facility Los Alamos National Laboratory Milestone Report 3783M," Los Alamos National Laboratory, Los Alamos, NM (1996).

Fallow, D.J., D.E. Elrick, W.D. Reynolds, N. Baumgartner, and G.W. Parkin. "Field Measurement of Hydraulic Conductivity in Slowly Permeable Materials Using Early-Time Infiltration Measurements in Unsaturated Media." in *Hydraulic Conductivity and Waste Contaminant Transport in Soils,* D.E. Daniel and S.J. Treautwein (Eds.), ASTM Spec. Tech. Publ. 1142, American Society of Testing and Materials, PA (1993): 375–389.

Falta, R.W. "Analysis of Soil Gas Pump Tests." in *Proceedings of the ER '93 Environmental Remediation Conference*, Augusta, GA October 24–28, 1 (1993): 441–447.

Falta, R.W. "A Program for Analyzing Transient and Steady-State Soil Gas Pump Tests." *Ground Water*, 34(4) (1996): 750–755.

FAO "Global and National Soils and Terrain Digital Database (SOTER) ." Procedures manual. *FAO World Soil Resour.Rep.,* 74 (1993).

FAO "Digital Soil Map of the World and Derived Soil Properties." Version 3.5. FAO, Rome, Italy (1995).

Faybishenko, B.A. "Vliyanie Temperatury na Vlazhnost, Entropiu i Vsasyvaushchee Davlenie Vlagi v Suglinkakh." (Impact of temperature on moisture, entropy, and water potential in loams) *Pochvovedenie (Soil Science)*, 12 (1983): 43–38.

Faybishenko, B.A. "Water-Salt Regime of Soils under Irrigation." (title translated from Russian), *Agropromisdat*, Moscow (1986): 304.

Faybishenko, B. "Hydraulic Behavior of Quasi-Saturated Soils in the Presence of Entrapped Air: Laboratory Experiments." *Water Resour. Res*., 31(10) (1995): 2421–2435.

Faybishenko, B. "Comparison of Laboratory and Field Methods for Determination of Unsaturated Hydraulic Conductivity of Soils." *in Proceedings of the International Workshop Characterization and Measurement of the Hydraulic Properties of Unsaturated Porous Media* (1999a): 279–292.

Faybishenko, B. "Tensiometer for Shallow or Deep Measurements Including Vadose Zone and Aquifers." U.S. Patent 5,941,121 (1999b).

Faybishenko, B., B.J. Sisson, K. Dooley, W.E. McCabe, and H.W. McCabe. "Method for Borehole Instrumentation and Grouting in Rocks and Soils, Disclosure and Record of Invention." submitted to the LBNL Patent Department on 9/11/98 (IB-1443) (1998).

Faybishenko, B., C. Doughty, M. Steiger, J.C.S. Long, T. Wood, J. Jacobsen, J. Lore, and P. Zawislanski. "Conceptual Model of the Geometry and Physics of Water Flow in a Fractured Basalt Vadose Zone: Box Canyon Site, Idaho." LBLN Report 42925. Paper submitted to *Water Resour. Res.* (1999a).

Faybishenko, B., P.A. Witherspoon, C. Doughty, T.R. Wood, R.K. Podgorney, and J.T. Geller. "Multi-Scale Investigations of Liquid Flow in the Vadose Zone of Fractured Basalt." paper submitted to the AGU Monograph *Flow and Transport in Fractured Rocks*, LBNL Report No. 42910 (1999b).

Ferre, P.A., D.L. Rudolph, and R.G. Kachanoski. "A Multilevel Waveguide for Profiling Water Content using Time Domain Reflectometry." in *Proceedings of the Symposium and workshop on Time Domain Reflectometry in Environment, Infrastructure and Mining Applications,* Special Pub. SP-19-94 Northwestern University, Evanston IL (1994).

Ferre, P.A., D.L. Rudolph, and R.G. Kachanoski. "Water Content of a Profiling Time Domain Reflectometry Probe." *Soil Sci. Soc. Am. J.*, 62 (1998): 865–873.

Fewson, C.A. "Biodegradation of Xenobiotic and Other Persistent Compounds: The Causes of Recalcitrance." *Trends in BIOTECH*, 6 (1988): 148–153.

Finsterle, S., and B. Faybishenko. "Design and Analysis of an Experiment to Determine Hydraulic Parameters of Variably Saturated Porous Media." *Advances in Water Resources*, 22(1) (1999a): 431–444.

Finsterle, S., and B. Faybishenko. "What Does a Tensiometer Measure in Fractured Rocks?" in *Proceedings of the International Conference Characterization and Measurement of the Hydraulic Properties of Unsaturated Porous Media*, LBNL Report-41454 (1999b)

Fliermans, C.B., T.J. Phelps, D. Ringelberg, A.T. Mikell, and D.C. White. "Mineralization of Trichloroethylene by Heterotrophic Enrichment Cultures." *Appl. Environ. Microbiol.*, 54 (1988): 1709–1714.

Fliermans, C.B., J.M. Dougherty, M.M. Franck, P.C. McKinsey and T.C. Hazen. "Immunological Techniques as Tools to Characterize the Subsurface Microbial Community at a Trichloroethylene Contaminated Site." in *Applied Biotechnology for Site Remediation*, R.E. Hinchee, D.B. Anderson, and F.B. Metting Jr, (Eds.), San Diego, CA (1994).

Flint L.E. and A.L. Flint. "Shallow Infiltration Processes at Yucca Mountain—Neutron Logging Data, 1984–93." U.S. Geological Survey Open-file Report 95-4035 (1995).

Flühler, H., M.S. Ardakani, and L.H. Stolzy. "Error Propagation in Determining Hydraulic Conductivities from Successive Water Content and Pressure Head Profiles." *Soil Sci. Soc. Am. J.*, 40 (1976): 830–836.

Flury, M. "Experimental Evidence of Transport of Pesticides through Field Soils - A Review." *J. Environ. Qual.*, 25 (1996): 25–45.

Fogel, M.M., A.R. Taddeo, and S. Fogel. "Biodegradation of Chlorinated Ethenes by a Methane-Utilizing Mixed Culture." *Appl. Environ. Microbiol.*, 51 (1986): 720–724.

Fredrickson, J.K., R.J. Hicks, S.W. Li, and F.J. Brockman. "Plasmid Incidence in Bacteria from Deep Subsurface Sediments." *Appl. Environ. Microbiol.*, 54 (1988): 2916–2923.

Friedman, I. "Deuterium Content of Natural Water and Other Substances." *Geochimica et Cosmochimica Acta*, 4 (1953): 89–103.

Frischknecht, F.C., V.F. Labson, B.R. Speis, W.L. Anderson. "Profiling Methods using Small Sources." in *Electromagnetic Methods in Applied Geophysics, Vol. 2, S.E.G. Investigations in Geophysics 3*, M. Nabighian (Ed.) (1991): 105–270.

Gardner, W.R. "Measurement of Capillary Conductivity and Diffusivity with a Tensiometer." in the *Trans. 7th International Congress of Soil Science*, Madison WI, Elsevier, Amsterdam, 1 (1960): 300–305.

Gardner, W.R. "Some Steady-State Solutions of the Unsaturated Moisture Flow Equation with Application to Evaporation from a Water Table." *Soil Sc.,* 85 (1958): 228–232.

Gat, J.R., E. Mazor, and Y. Tzur. "The Stable Isotope Composition of Mineral Waters in the Jordan Rift Valley." *Journal of Hydrology*, 7 (1969): 334–352.

Gaye, C.B., and W.M. Edmunds. "Groundwater Recharge Estimation Using Chloride, Stable Isotopes and Tritium Profiles in the Sands of Northwestern Senegal." *Environmental Geology*, 27 (1996): 246–251.

Gee, G.W., M.D. Campbell, G.S. Campbell, and J.H. Campbell. " Rapid Measurement of Low Soil Water Potential Using a Water Activity Meter." *Soil Sci. Soc. Am. J.*, 56 (1992): 1068–1070.

Gee, G.W., and A.L. Ward. "Innovations in Two-Phase Measurements of Soil Hydraulic Properties." in *Proceedings of the Int. Workshop Characterization and Measurement of the hydraulic properties of unsaturated porous media*, M.Th. Van Genuchten, F.J. Leij, and L. Wu (Eds.), Univ. of California, Riverside, CA (1999): 241–269.

Germann, P.F., and K. Beven. "Kinematic Wave Approximation to Infiltration into Soils with Sorbing Macropores." *Water Resour. Res.*, 21(7) (1985): 990–996.

Ghassemi, F., A.J. Jakeman, and H.A. Nix. "Salinisation of Land and Water Resources." CABI, Canberra, Australia (1995).

Ghodrati, M. "Point Measurements of Solute Transport in Soil Using Fiber Sensors." *Soil Sci. Soc. of Am. J.*, 63 (1999): 471–479.

Ghodrati, M., and W.A. Jury. "A Field Study Using Dyes to Characterize Preferential Flow of Water." *Soil Sci. Soc. Am. J.*, 54 (1990): 1558–1563.

Glass, R.J., J.Y. Parlange, and T.S. Steenhuis. "Wetting Front Instability: 1. Theoretical Discussion and Dimensional Analysis." *Water Resour. Res.*, 25 (1989): 1187–1194, "2. Experimental Determination of Relationship between System Parameters and Two-Dimensional Unstable Flow Field Behavior in Initially Dry Porous Media.", *Water Resour. Res.,* 25 (1989): 1195–1207.

Glass, R.J., T.S. Steenhuis, and J.Y. Parlange. "Immiscible Displacement in Porous Media: Stability Analysis of Three-Dimensional, Axisymmetric Disturbances with Application to Gravity-Driven Wetting Front Instability." *Water Resources Research,* 27(8) (1991): 1947–1956.

Glass, R.J., and M.J. Nicholl. "Physics of Gravity Fingering of Immiscible Fluids within Porous Media: an Overview of Current Understanding and Selected Complicating Factors." *Geoderma*, 70(2–4) (1996): 133–166.

Glasser, J.A. "Engineering Approaches using Bioremediation to Treat Crude Oil-Contaminated Shoreline Following the Exxon Valdez Aaccident in Alaska." in *Bioremediation Field Experience*, P.E. Flathman, D.E. Jerger, and J.H. Exner, (Eds.), Lewis Publishers, Boca Raton, FL (1994): 81–106.

Globus, A.M. "Soil Hydrophysical Description of Agroecological Mathematical Models." (in Russian). *Gidrometeoizdat*, St. Petersburg, Russia (1987).

Globus, A.M., and G.W. Gee. "A Method to Estimate Moisture Diffusivity and Hydraulic Conductivity of Moderately Dry Soil." *Soil Sci. Soc. Am. J.*, 59 (1995): 684–689.

Goldstein, N.E., "Expedited Site Characterization Geophysics: Geophysical Methods and Tools for Site Characterization." *Lawrence Berkeley Laboratory, LBL-35384* (1994).

Gonfiantini, R., T. Dincer and A.M. Derekoy. "Environmental Isotope Hydrology in the Honda Region, Algeria." in *Isotope Techniques in Groundwater Hydrology*, International Atomic Energy Association, Vienna, 1 (1974): 293–316.

Gorden, D.S. and P.L.M. Veneman. "Soil Water Pressure Measurements in Subzero Air Temperatures." *Soil Science Society of America Journal*, 59(5) (1995): 1242–1243.

Gorden, R.W., T.C. Hazen, and C.B. Fliermans. "Rapid Screening for Bacteria Capable of Degrading Toxic Organic Compounds." *J. Microbiol. Meth.*, 18 (1993): 339–347.

Granovsky, A.V. and E.L. McCoy. "Air Flow Measurements to Describe Field Variation in Porosity and Permeability of Soil Macropores." *Soil Sci. Soc. Am. J.*, 61(6) (1997): 1569–1576.

Grant, S.A. and A. Salehzadeh. "Calculation of Temperature Effects on Wetting Coefficients of Porous Solids and their Capillary Pressure Functions." *Water Resour. Res.*, 32(2) (1996): 261–270.

Green, R.E., L.R. Ahuja, and S.K. Chong. "Hydraulic Conductivity, Diffusivity, and Sorptivity of Unsaturated Soils: Field Methods" in *Methods of Soil Analysis Part 1. Soil Science Society of America*, A. Klute (Ed.), Madsion, WI (1986): 771–798.

Griffin, J.W. and K.B. Olsen. "A Review of Fiber Optic and Related Technologies for Environmental Sensing Applications." in *ASTM Special Technical Publication*, D.M. Nielsen *et al.* (Eds.), 1118 (1992): 311–328.

Guzman, A.G., S.P. Neuman, C. Lohrstorfer, and R. Bassett. "Validation Studies for Assessing Unsaturated Flow and Transport Through Fractured Rock." in Bassett, R.L., S.P. Neuman, T.C. Rasmussen, A.G. Guzman, G.R. Davidson, and C.L. Lohrstorfer, NUREG/CR–6203 (1994): 4–1 through 4–58.

Guzman, A.G. and S.P. Neuman. "Field Air Injection Experiments." in Apache Leap Tuff INTRAVAL Experiments: Results and Lessons Learned, Rasmussen, T.C., S.C. Rhodes, A. Guzman, and S.P. Neuman, NUREG/CR–6096 (1996): 52–94.

Hantush, M.S. "Hydraulics of Wells," in Advances in Hydroscience, Academic Press, NY 1 (1964): 281–432.

Hansen, Edward A., and A.F. Harris. "Validity of Soil–Water Samples Collected with Porous Ceramic Cups." *Soil Science Society of America Proceedings*, 39 (1975): 528–536.

Harkness, M.R., J.B. McDermott, D.A. Abramowicz, J.J. Salvo, W.P. Flanagan, M.L. Stephens, F.J. Mondello, R.J. May, J.H. Lobos, K.M. Carroll, M.J. Brennan, A.A. Bracco, K.M. Fish, G.L. Warner, P.R. Wilson, D.K. Dietrich, D.T. Lin, C.B. Margan, and W.L. Gately. "*In Situ* Stimulation of Aerobic PCB Biodegradation in Hudson River Sediments." Science, 259 (1993): 503–507.

Haverkamp, R., and J.Y. Parlange. "Predicting the Water–Retention Curve from Particle Size Distribution: 1. Sandy Soils without Organic Matter." *Soil Sci.*, 142 (1986): 325–339.

Haykin, S. "Neural Networks, a Comprehensive Foundation." 1st ed. Macmillan College Publishing Company, NY (1994).

Hazen, T.C. "Test Plan for *In Situ* Bioremediation Demonstration of the Savannah River Integrated Demonstration Project." *DOE/OTD TTP No.: SR 0566–01.* WSRC–RD–91–23. WSRC Information Services, Aiken, SC (1991): 82.

Hazen, T.C. "Bioremediation." in *Microbiology of the Terrestrial Subsurface*, P. Amy and D. Haldeman (Eds.), CRC Press, Boca Raton, FL (1997): 247–266.

Hazen, T.C., and L. Jimenez. "Enumeration and Identification of Bacteria from Environmental Samples UUsing Nucleic Acid Probes." *Microbiol. Sci.*, 5 (1988): 340–343.

Hazen, T.C., L. Jimenez, G. López De Victoria, and C.B. Fliermans. "Comparison of Bacteria from Deep Subsurface Sediment and Adjacent Groundwater." *Microb. Ecol.*, 22 (1991): 293–304.

Hazen, T.C., K.H. Lombard, B.B. Looney, M.V. Enzien, J.M. Dougherty, C.B. Fliermans, J. Wear, and C.A. Eddy–Dilek. "Summary of *In Situ* Bioremediation Demonstration (Methane Biostimulation) Via Horizontal Wells at the Savannah River Site Integrated Demonstration Project." in *Proceedings of the Thirty–Third Hanford Symposium on Health and the Environment: In–Situ Remediation: Scientific Basis for Current and Future Technologies*, G.W. Gee and N.R. Wing (Eds.), Battelle Press, Columbus, OH (1994): 135–150.

Healy, R.W., C.A. Peters, M.R. DeVrines, P.C. Mitts, and D.L. Moffeti. "Study of the Unsaturated Zone at a Low-Level Radioactive-Waste Disposal Site near Sheffield, Ill." in Proceedings of the National Water Well Association Conference on Characterization and Monitoring of the Vadose Zone, Las Vegas, NV (1983): 820–31.

Healy, R.W., M.P. deVries, and R.G. Striegl. "Concepts and Data-Collection Techniques Used in a Study of the Unsaturated Zone at a Low-Level Radioactive-Waste Disposal Site Near Sheffield, Illinois." U.S. Geological Survey Water-Resources Investigations Report 85–4228 (1986).

Hearn, P.P., W.C. Steinkampf, D.G. Horton, G.C. Solomon, L.D. White, and J.R. Evans. "Oxygen-Isotope Composition of Ground Water and Secondary Minerals in Columbia Plateau Basalts: Implications for the Paleohydrology of the Pasco Basin." *Geology*, 17 (1989): 606–610.

Hearst, J.R. and P.H. Nelson. "Well Logging for Physical Properties." McGraw-Hill, NY (1985).

Hecht-Nielsen, R. "Neurocomputing." 1st ed. Addison-Wesley publishing company, Reading, MA (1990).

Heimovaara, T.J., W. Bouten, and J.M. Verstraten. "Frequency Domain Analysis of Time Domain Reflectometry Waveforms 2. A Four Component Complex Dielectric Mixing Model for Soils." *Water Resour. Res.*, 30 (1994): 201–209.

Heipieper, H.J., B. Loffeld, H. Keweloh, and J.A.M. Debont. "The cis/trans Isomerization of Unsaturated Fatty-Acids in *Pseudomonas P. putida* s12 - An Indicator for Environmental-Stress Due to Organic-Compounds." *Chemosphere*, 30 (1995): 1041–1051.

Herkelrath, W.N. and S.P. Murphy. "Automatic, Real-Time Monitoring of Soil Moisture in a Remote Field Area With Time Domain Reflectometry." *Water Resour. Res.*, 27(5) (1991).

Hewett, T.A., and R.A. Behrens. "Considerations Affecting the Scaling of Displacements in Heterogeneous Permeability Distributions." *SPE Formation Evaluation* (1993): 258–266.

Hilhorst, M.A., and C. Dirksen. "Dielectric Water Content Sensors: Time Domain Versus Frequency." in *Time Domain Reflectometry in Environmental, Infrastructure, and Mining Applications*, Special Publication *SP*19-94, U. S. Department of Interior and Bureau of Mines, (1994): 23–33.

Hill, D.E., and Parlange, J.Y. "Wetting Front Instability in Layered Soils." *Soil Sci. Soc. Am. Proc.*, 36 (1972): 697–702.

Hillel, D. *Fundamentals of Soil Physics.* Academic Press, New York (1980).

Hillel, D. and W.R. Gardner. "Measurement of Unsaturated Conductivity and Diffusivity by Infiltration through an Impeding Layer." *Soil Sci.*, 109 (1970): 149–153.

Hills, R.G., Wierenga, P.J., D.B. Hudson, and M.R. Kirkland. "The Las Cruces Trench Experiment: Experimental Results and Two-Dimensional Flow Predictions." *Water Resources Res.*, 27(10) (1991): 2707–2718.

Hinchee, R.E., D.C. Downey, R.R. Dupont, P.K. Aggarwal, and R.N. Miller. "Enhancing Biodegradation oOf Petroleum-Hydrocarbons through Soil Venting." *J. Haz. Mat.*, 27 (1991): 315–325.

Hinze, W.J. "The Role of Gravity and Magnetic Methods in Engineering and Environmental Studies." in *Geotechnical and Environmental Geophysics Vol 1: Review and Tutorial, SEG Investigations in Geophysics No. 5,* Stanley Ward (Ed.) (1990): 75–126.

Hoekstra, P. and M. Blohm. "Case Histories of Time-Domain Electromagnetic Soundings in Environmental Geophysics." in *Geotechnical and Environmental Geophysics Vol 2: Environmental and Groundwater, S.E.G. Investigations in Geophysics 5*, S Ward (Ed.) (1990): 1–17.

Hoeppel, R., R.E. Hinchee, and M.F. Arthur. "Bioventing Soils Contaminated with Petroleum Hydrocarbons." *J. Indust. Microbiol.*, 8 (1991): 141–146.

Hoffmann-Riem, H., M.Th. Van Genuchten, and H. Flühler. "General Model for the Hydraulic Conductivity of Unsaturated Soils." in *Proceedings of the Int Workshop on the Characterization and Measurement of the Hydraulic Properties of Unsaturated Porous Media*, M.Th. Van Genuchten, F.J. Leij, and L. Wu (Eds.), Univ. of California, Riverside, CA (1999): 31–42.

Honarpour, M., L. Koederitz, and A.H. Harvey. "Relative Permeability of Petroleum Reservoirs." CRC Press, Inc., Boca Ration, FL (1986).

Hook, W.R., N.J. Livingston, Z.J. Sun, and P.B. Hook. "Remote Diode Shorting Improves Measurement of Soil Water by Time Domain Reflectometry." *Soil Sci. Soc. Am. J.*, 56 (1992): 1384–1391.

Hopmans, J.W., and J. Simunek. "Review of Inverse Estimation of Soil Hydraulic Properties," in *Proceedings of the Int. Workshop on the Characterization and Measurement of the Hydraulic Properties of Unsaturated Porous Media,* van Genuchten, M.Th., F.J. Leij, and L. Wu (Eds.), University of California, Riverside, CA (in press) (1999).

Hopmans, J.W., and J.H. Dane. "Temperature Dependence of Soil Hydraulic Properties." *Soil Sci. Soc. Am.*, 50 (1) (1986): 4–9.

Horvath, R.S. "Microbial Co-Metabolism and the Degradation of Organic Compounds in Nature." *Bacteriol. Rev.*, 36 (1972): 146–155.

Horvitz, L. "Hydrocarbon Geochemical Prospecting After Thirty Years." in *Unconventional Methods in Exploration for Petroleum and Natural Gas*, W. B. Heroy, (Ed.), Southern Methodist University, Institute for the Study of Earth and Man, SMU Press (1969): 205–218.

Houghton, R.L., M.E. Berger, M. Curl. "Effects of Well-Casing Composition and Sampling Methods on Apparent Quality of Ground Water." in *Proceedings of the Fourth National Symposium on Aquifer Restoration and Ground Water Monitoring*, D.M. Nielsen, (Ed.), Natl. Water Well, Worthington, OH (1984).

Hubbell, J.M., and J.B. Sisson. "Advanced Tensiometer for Shallow or Deep Soil Water Potential Measurements." *Soil Sci.*, 163(4) (1998): 271–277.

Hudson, D.B., P.J. Wierenga, and R.G. Hills. “Unsaturated Hydraulic Properties from Upward Flow in Soil Cores.” *Soil Sci. Soc. Am. J.*, 60 (1996): 388–396.

Hutson, J.L. and A. Cass. “A Retentivity Function for Use in Soil-Water Simulation Models.” *J. Soil Sci.*, 38 (1987): 105–113.

Hutter, A.R. “A Method for Determining Soil Gas ^{220}Rn (Tthoron) Concentrations.” *Health Phys.*, 68 (1995): 835–839.

Hutter, A.R. “Spatial and Temporal Variations of Soil Gas 220Rn and 222Rn at Two Sites in New Jersey.” *Environmental International*, 22 (1996): 455–469.

Hutter, A.R. and E.O. Knutson. “An International Intercomparison of Soil Gas Radon and Radon Exhalation Measurements.” *Health Physics*, 74(1) (1998): 108–114.

Huang, F.C. and C.H. Dowding. “Telemetric and Multiplexing Enchancement of Time Domain Reflectometry Measurements.” in *the Proceedings, Symposium and Workshop on Time Domain Reflectometry in Environmental, Infrastructure, and Mining Applications,* Northwestern University, U.S. Bureau of Mines Special Publication SP-19–94 (1994): 34–45.

Illman W.A., D.L. Thompson, V.V. Vesselinov, G.Chen, and S.P. Neuman. “Single- and Cross-Hole Pneumatic Tests in Unsaturated Fractured Tuffs at the Apache Leap Research Site: Phenomenology, Spatial Variability, Connectivity and Scale.” NUREG/CR-5559 (1998).

Ingraham N.L., R.J. Johnson, and R. Broadbent. “Facility-Altered Stable Isotopic Ratios of Power Generation Cooling Wastewater - Opportunity for Tracing Leakages.” *Environmental Science & Technology*, 11 (1994): 1983–1986.

Jackson, D.R., F.S. Brinkley, and E.A. Bondietti. “Extraction of Soil Water Using Cellulose-Acetate Hollow Fibers.” *Soil Science Society of America Journal*, 40 (1976): 327–329.

Jamison, V.C. “Sand-Silt Suction Column for Determination of Moisture Retention.” *Soil Sci. Soc. Am. Proc.*, 22 (1958): 82–83.

Jamison, V.C., and M.E. Kroth. “Available Moisture Storage Capacity in Relation to Textural Composition and Organic Matter Content of Several Missouri Soils.” *Soil Sci. Soc. Am. Proc.*, 22(3) (1958): 189–192.

Jensen, B., Arvin, E., and A.T. Gundersen. “The Degradation of Aromatic Hydrocarbons with Bacteria from Oil Contaminated Aquifers.” in *Proceedings of the NWWA/API Conference on Petroleum Hydrocar-bons and Organic Chemicals in Groundwater*, Houston, TX, November 13–15 (1985): 421–435.

Johnson, T.M., and K. Cartwright. "Monitoring of Leachate Migration in the Unsaturated Zone in the Vicinity of Sanitary Landfills." State Geological Survey Circular 514, Urbana, IL (1980).

Johnson, T.M., and D.J. Depaolo. "Reaction-Transport Models for Radiocarbon in Groundwater - The Effects of Longitudinal Dispersion and the Use of Sr Isotope Ratios to Correct for Water-Rock Interaction." *Water Resources Research*, 32 (1996): 2203–2212.

Johnson, T.M., and D.J. Depaolo. "Rapid Exchange Effects on Isotope Ratios in Groundwater Systems, 1. Development of a Transport-Dissolution-Exchange Model." *Water Resources Research,* 33 (1997a): 187–195.

Johnson, T.M., and D.J. Depaolo. "Rapid Exchange Effects on Isotope Ratios in Groundwater Systems, 2. Flow Investigation Using Sr Isotope Ratios." *Water Resources Research*, 33 (1997b): 197–205.

Jones, V.T., and H.W. Thune. "Surface Detection of Retort Gases from an Underground Coal Gasification Reactor in Steeply Dipping Beds Near Rawlins, Wyoming." SPE Paper 11050, 57th Annual Fall Technical and Exhibition of the Society of Petroleum Engineers of *AIM*E, New Orleans, LA, September 26–29 (1982).

Jury, W.A., H. Elabd, and M. Resketo. "Field Study of Napropamide Movement through Unsaturated Soil." *Water Res. Resour.*, 22 (1986): 749–755.

Jury, W.A., W.R. Gardner, and W.H. Gardner. *Soil Physics,* J. Wiley, NY (1991).

Jury, W.A. and K. Roth. "Transfer Functions and Solute Movement through Soil: Theory and Applications." *Birkhäuser Verlag*, Basel, Switzerland (1990).

Kabala, Z.J., and P.C.D. Milly. "Sensitivity Analysis of Infiltration, Exfiltration, and Drainage in Unsaturated Miller-Similar Porous Media." *Water Res. Resour.*, 27(10) (1991): 2655–2666.

Kachanoski, R.G., E.G. Gregorich, and I.J. van Wesenbeeck. "Estimating Spatial Variations of Soil Water Content Using Noncontacting Electromagnetic Inductive Methods." *Can. J. Soil Sci.*, 68 (1988): 715–722.

Kaluarachchi, J.J., and J.C. Parker. "Effects of Hysteresis with Air Entrapment on Water Flow in the Unsaturated Zone." *Water Resour. Res.*, 23(10) (1987): 1967–1976.

Karably, L.S., and K.B. Babcock. "The Effects of Environmental Variables on Soil Gas Surveys." *Hazardous Materials Control*, January/February (1989): 36–43.

Keet, B.A. "Bioslurping State of the Art." in *Applied Bioremediation of Petroleum Hydrocarbons,* R.E. Hinchee, J.A. Kittel, and H. J. Reisinger (Eds.), Battelle Press, Columbus, OH (1995): 329–334.

Keller, C. "So, What is the Practical Value of Seamist?" in *Proceedings of the Fifth National Outdoor Action Conference on the Aquifer Restoration Ground Water Monitoring and Geophysical Methods*, Las Vegas, NV, May (1991).

Kerfoot, H.B. "Shallow-Probe Soil-Gas Sampling for Indication of Ground Water Contamination by Chloroform." *International Journal of Environmental and Analytical Chemistry*, 30 (1987): 167–181.

Kerfoot, W. B., and W. Sanford, "Four-Dimensional Perspective of an Underground Fuel Oil Tank Leakage." *In Proceedings of the NWWA/API Conference on Petroleum Hydrocarbons and Organic Chemicals in Ground Water, Houston, Texas*, National Water Well Association, Dublin, OH, November 12-14 (1986): 383-403.

Kern, J.S. "Evaluation of Soil Water Retention Models Based on Basic Soil Physical Properties." *Soil Sci. Soc. Am. J.*, 59 (1995): 1134–1141.

Keys, S.W. "Borehole Geophysics Applied to Ground-Water Investigations." National Water Well Association, Dublin, OH (1989).

Klinkenberg, L.J. "The Permeability of Porous Media to Liquids and Gases." Am. Pet. Inst., Drilling and Production Practice (1941): 200–213.

Klute, A., and C. Dirksen. "Hydraulic Conductivity and Diffusivity: Laboratory Methods." in *Methods of Soil Analysis, Part 1, Physical and Mineralogical Methods,* Second Edition, American Society of Agronomy, Inc., and Soil Science Society of America, Inc., A. Klute (Ed.), Madison, WI (1986): 687–734.

Klute, A., and Peters, D.B., "A Recording Tensiometer with a Short Response Time." *Proceedings*, Soil Science Society of America, Vol. 26 (1962): 87–88.

Klute, A. (Ed.), "Methods of Soil Analysis, Part I. Physical and Mineralogical Methods." American Society of Agronomy, Madison, WI (1986).

Klute, A., F.D. Whisler, and E.J. Scott. "Soil Water Diffusivity and Hysteresis Data from Radial Flow Pressure Cells." *Soil Sci. Am. Proc.*, 28 (1964): 160–163.

Knighton, M. Dean, and Dwight E. Streblow. "A More Versatile Soil Water Sampler." *Soil Science Society of America Journal*, 45 (1981): 158–159.

Komor, S.C., and D.G. Emerson. "Movements of Water, Solutes, and Stable Isotopes in the Unsaturated Zones of Two Sand Plains in the Upper Midwest." *Water Resources Research*, 30 (1994): 253–267.

Kool, J.B., J.C. Parker, and M.Th. van Genuchten. "Parameter Estimation for Unsaturated Flow and Transport Models - A Review." *J. Hydrol.*, 91 (1987): 255–293.

Kornev, V.G. "Vsasyvayushchaya Ssila Ppochvy." (Suction force of soils*), Journal of Agronomy*, 22(1) (1924).

Kosugi, K. "General Model for Unsaturated Hydraulic Conductivity for Soils with Lognormal Pore-Size Distribution." *Soil Sci. Soc. Am. J.*, 63 (1999): 270–277.

Kriukov, P.A. and F.T. Manheim. "Extraction and Investigative Techniques for Study of Interstitial Waters of Unconsolidated Sediments; A Review." in *Dynamic Environment of the Ocean Floor*, Fanning, Kent A. (Ed.), D.C. Heath and Co., Lexington, MA (1982): 3–26.

Kueper, B.H., W. Abbott, and G. Farquhar. "Experimental Observations of Multiphase Flow in Heterogeneous Porous Media." *J. of Contam. Hydrology*, 5 (1989): 83–95.

Kung, K.J.S. "Preferential Flow in a Sandy Vadose Zone: 1. Field Observation." *Geoderma*, 46 (1990a): 51–58.

Kung, K.J.S. "Preferential Flow in a Sandy Vadose Zone: 2. Mechanism and Implications." *Geoderma*, Vol. 46 (1990b): 59–71.

Kutilek, M., and D.R. Nielsen. "Soil Hydrology, Cremlingen-Destedt." *Catena-Verl.* (1994).

Lahti M.L., L.K. Killoran, R.F. Holub, G.M. Reimer. "New Rapid Method to Determine Radon and Thoron from a Single Counting Sequence Using a Field Portable Alpha Particle Scintillometer." *J. Radioanal. Nucl. Chem.*, 236 (1998): 253– 256.

Landmeyer, J.E., and P.A. Stone "Radiocarbon and Delta-C-13 Values Related to Ground-Water Recharge and Mixing." *Ground Water*, 33 (1995): 227–234.

Lankston, R.W. "High-Resolution Refraction Seismic Data Acquisition and Interpretation." in *Geotechnical and Environmental Geophysics Vol 1: Environmental and Groundwater,* S.E.G. Investigations in Geophysics 5, Stanley Ward (Ed.) (1990): 45–73.

Lanza, G. R., and J.M. Dougherty. "Microbial Enzyme-Activity and Biomass Relationships in Soil Ecotoxicology." *Environmental Toxicology and Water Quality*, 6 (1991): 165–176.

Lehmann, B.E., S.N. Davis, and J.T. Fabryka-Martin. "Atmospheric and Subsurface Sources of Stable and Radioactive Nuclides Used for Groundwater Dating." *Water Resources Research*, 29 (1993), 2027–2040.

Leij, F.J., W.J. Alves, M. Th van Genuchten, and J.R. Williams. "The UNSODA Unsaturated Soil Hydraulic Database, Version 1.0," *EPA report EPA/600/R-96/095*, EPA National Risk Management Laboratory, G-72, Cincinnati, OH http://www.ussl.ars.usda.gov/models/unsoda.htm (1996).

Leij, F.J., W.B. Russell, and S.M. Lesch. "Closed-Form Expressions for Water Retention and Conductivity Data." *Ground Water*, 35(5) (1997): 848–858.

Lenhard, R J., and J.C. Parker. "Experimental Validation of the Theory of Extending Two-Phase Saturation-Pressure Relations to Three-Fluid Phase Systems for Monotonic Drainage Paths." *Water Resour. Res.*, 24(3) (1988): 373–380.

Levin, M.J., and D.R. Jackson. "A Comparison of *In Situ* Extractors for Sampling Soil Water." *Soil Science Society of America Journal*, 41 (1977): 535–536.

Levy, S.S., D.S. Sweetkind, J.T. Fabryka-Martin, P.R. Dixon, J.L. Roach, L.E. Wolfsberg, D. Elmore, and P. Sharma. "Investigations of Structural Controls and Mineralogic Associations of Chlorine-36 Fast Pathways." in the ESF, Milestone Report SP2301M4, Los Alamos National Laboratory, Los Alamos, NM (1997).

Libby, W.F. "History of Tritium." in A.A. Moghissi and M.W. Carter (Eds.), Tritium, Messenger Graphics/Las Vegas Publishers, Las Vegas, NV (1971): 3–11.

Lichtler, W.F., D.I. Stannard, and E. Kouma. "Investigation of Artificial Recharge of Aquifers in Nebraska." in *U.S. Geological Survey Water-Resources Investigations Report (*1980): 80–93

Lieberman, S.H., S.M. Inamn, G.A. Therialt, S.S. Cooper, P.G. Malone, Y. Shimizu, and P.W. Lurk. "Fiber Optic-Based Chemical Sensors for *In Situ* Measurement of Metals and Aromatic Organic Compounds in Seawater and Soil Systems." *Proc. SPIE,* 1269 (1990): 175–184.

Lieberman, S.H., G.A. Theriault, S.S. Cooper, P.G. Malone, R.S. Olsen, and P.W. Lurk. "Rapid, Subsurface, *In Situ* Field Screening of Petroleum Hydrocarbon Contamination using Laser Induced Fluorescence over Optical Fibers" in *the Proceedings of the Second International Symposium on Field Screening Methods for Hazardous Wastes and Toxic Chemicals*, U.S. Environmental Protection Agency, Las Vegas, NV (1991): 57–63.

Lilly, A., J.H.M. Wösten, A. Nemes, and C. Le Bas. "The Development and Use of the HYPRES Database in Europe." in *Proceedings of the Int Workshop on the Characterization and Measurement of the Hydraulic Properties of Unsaturated Porous Media*, M.Th. Van Genuchten, F.J. Leij, and L. Wu (Eds.), Univ. of California, Riverside, CA (1999): 1283–1294.

Lima, L.A., M.E. Grismer, and D.R. Nielsen. "Salinity Effects on Yolo Loam Hydraulic Properties." *Soil Sci.*, 150 (1990): 451–458.

Litaor, M.I. "Review of Soil Solution Samplers." *Water Resources Research*, 24 (1988): 727–733.

Litchfield, C.D. "*In Situ* Bioremediation: Basis and Practices." in *Biotreatment of Industrial and Hazardous Waste*, M.A. Levin and M.A. Gealt (Eds.), McGraw-Hill, Inc. NY (1993): 167–196.

Litschmann, T. "Virrib®: A Soil Moisture Sensor and its Application in Agriculture." Commun. in *Soil Sci. Plant Anal.*, 22(5&6) (1991): 409–418.

Liu, B., F. Phillips, S. Hoines, A.R. Campbell, and P. Sharma. "Water Movement in Desert Soil Traced by Hydrogen and Oxygen Isotopes, Chloride, and Chlorine-36, Southern Arizona." *Journal of Hydrology*, 168 (1995): 91–110.

Loague, K., and D.L. Corwin. "Uncertainty in Regional-Scale Assessments of Non-Point Source Pollutants." in Applications of GIS to the Modeling of Non-Point Source Pollutants in the Vadose Zone, D.L. Corwin and K. Loague (Eds.), SSSA Special Publication No. 48, Soil Science Society of America, Madison, WI (1996): 131–152.

Loaiciga, H.A., S. Renehan, and S. Weeks. "Survey of Current Practice of Environmental Characterization and Monitoring Techniques: Selected DOE Sites." Santa Barbara, CA, Contract Bechtel, NV PO 13440, September 1 (1997).

Lombard, K.H., and T.C. Hazen. "Test Plan for the Soils Facility Demonstration - Petroleum Contaminated Soil Bioremediation Facility (U)." WSRC-TR-94-0179. Westinghouse Savannah River Company, Aiken, SC (1994).

Long, J.C.S., (Ed.) "Rock Fractures and Fluid Flow: Contemporary Understanding and Applications." Committee on Fracture Characterization and Fluid Flow, National Academy Press, Washington, D.C. (1996).

Long, F.L. "A Glass Filter Soil Solution Sampler." *Soil Science Society of America Journal*, 42 (1978): 834–835.

Loosli, H.H. "A Dating Method with ^{39}Ar." *Earth and Planetary Sciences,* 63 (1983): 51–62.

Lord, D.L., K.F. Hayes, A.H. Demond, and A. Salehzadeh. "Influence of Organic Acid Solution Chemistry on Subsurface Transport Properties, I. Surface and Interfacial Tension." *Environ. Sci. Technol.*, 31(7) (1997): 2045–2051.

Luckner, L., M.Th. van Genuchten, and D.R. Nielsen. "A Consistent Set of Parametric Models for the Two–Phase Flow of Immiscible Fluids in the Subsurface." *Water Resour. Res.*, 25 (1989): 2187–2193.

Lunne, T., P.K. Robertson, and J.J.M. Powell. *Cone Penetration Testing in Geotechnical Practice,* Blackie Academic & Professional, London (1997).

Major, D., E. Cox, E. Edwards, and P.W. Hare. "The Complete Dechlorination of Trichloroethene to Ethene under Natural Conditions in a Shallow Bedrock Aquifer Located in New York State." in *Proceedings of the EPA Symposium on Intrinsic Bioremediation of Ground Water*, U.S. Environmental Protection Agency, EPA/540/R-94/515, Denver, CO, August 30-September 1, (1994).

Malicki, M.A., and W.M. Skierucha. "A Manually Controlled TDR Soil Moisture Meter Operating with 300 Psi Rise-Time Needle Pulse." *Irrig. Sci.*, 10 (1989): 153–163.

Mallants, D., M. Vanclooster, N. Toride, J. Vanderborght, M.Th. van Genuchten, and J. Feyen. "Comparison of Three Methods to Calibrate TDR for Monitoring Solute Movement in Undisturbed Soil." *Soil Sci. Soc. Am. J.*, 60(3) (1996): 747–754.

Manos, C.G. Jr., K.R. Williams, W.D. Balfour, and S.J. Williamson. "Effects of Clay Mineral-Organic Matter Complexes on Gaseous Hydrocarbon Emissions from Soils." in *Proceedings of the NWWA/API Conference on Petroleum Hydrocarbons and Organic Chemicals in Ground Water-Prevention, Detection and Restoration*, Houston, TX, No-vember 13–15 (1985).

Marion, J.M., D. or D.E. Rolston, M.L. Kavas, and J.W. Biggar. "Evaluation of Methods for Determining Soil-Water Retentivity and Unsaturated Hydraulic Conductivity." *Soil Sci.*, 158 (1994): 1–13.

Marthaler, H.R, W. Vogelsanger, F. Richard, and P.J. Wierenga. "A Pressure Transducer for Field Tensiometers." *Soil Science Society of America Journal*, 47 (1983): 624–627.

Martin, M. and T.E. Imbrigiotta. "Contamination of Ground Water with Trichloroethylene at the Building 24 Site at Picatinny Arsenal, New Jersey." in *Proceedings of the EPA Symposium on Intrinsic Bioremediation of Ground Water,* August 30-September 1, Denver, CO. U.S. Environmental Protection Agency, EPA/540/R-94/515 (1994).

Massmann, J.W. "Applying Groundwater Flow Models in Vapor Extraction System Design." *Journal of Environmental Engineering*, 115(1) (1989).

Massmann, J.W. and M. Madden. "Estimating Air Conductivity and Porosity from Vadose-Zone Pumping Tests." *Journal of Environmental Engineering*, 120(2) (1994): 313–328.

Mayer, C.L. "Draft Interim Guidance Document for Soil-Gas." *Surveying*, U.S. EPA Environmental Monitoring Systems Laboratory, Office of Research and Development, Contract No. 68-03-3245, September (1989): 124.

McCarthy, J.H. Jr., and Reimer, G.M. "Advances in Soil Gas Geochemical Exploration for Natural Resources: Some Current Examples and Practices." *Journal of Geophysical Research*, 91(B12) (November 1986): 327–338.

McCarthy, J.F., W.E. Sanford, and P.L. Stafford. "Lanthanide Field Tracers Demonstrate Enhanced Transport of Transuranic Radionuclides by Natural Organic Matter." *Environmental Science & Technology*, 32 (1998): 3901–3906.

McKim, H.L., J.E. Walsh, and D.N. Arion. "Review of Techniques for Measuring Soil Moisture *In Situ*." U.S. Army Corps of Engineers, Cold Regions Research and Engineering Lab Special Report (1980): 80–31.

McMahon, P.B., and K.F. Dennehy. "Water Movement in the Vadose Zone at Two Experimental Waste-Burial Trenches in South Carolina." in *Proceedings of the National Water Well Association Conference on Characterization and Monitoring of the Unsaturated Vadose Zone*, Denver, CO (1985): 34–54.

McNeal, B.L., and N.T. Coleman. "Effect of Solution Composition on Soil Hydraulic Conductivity." *Soil Sci. Soc. Am. Proc.*, 20 (1966): 308–312.

McNeill, J.D. "Use of Electromagnetic Methods for Groundwater Studies." in *Geotechnical and Environmental Geophysics Vol 1: Review and Tutorial, SEG Investigations in Geophysics No. 5,* Stanley Ward (Ed.) (1990): 191–218.

McWhorter, D.B. "Unsteady Radial Flow of Gas in the Vadose Zone." *J. of Contam. Hydrol.*, 5 (1990): 297–314.

Merry, W.M., and C.M. Palmer. "Installation and Performance of a Vadose Monitoring System." in *Proceedings of the NWWA Conference on Characterization and Monitoring of the Vadose Zon*e, NWWA (1986): 107–125.

Messing, I. "Estimation of the Saturated Hydraulic Conductivity in Clay Soils from Soil Moisture Retention Data." *Soil Sci. Soc. Am. J.*, 53 (1989): 665–668.

Miller, E.E., and A. Salehzadeh. "Stripper for Bubble-Free Tensiometry." *Soil Sci. Soc. Am. J.*, 57 (1993): 1470–1473.

Miller, R.V., and J.S. Poindexter. "Strategies and Mechanisms for Field Research in Environmental Bioremediation." *American Academy of Microbiology* (1994).

Miller, W., *et al.* "Natural Analogue Studies in the Geological Disposal of Radioactive Wastes." Amsterdam, New York, *Elsevier, Studies in Environmental Science* 57 (1994).

Milly, P.C.D. "An Event-Based Simulation Model of Moisture and Energy Fluxes at a Bare Soil Surface." *Water Res. Resour.*, 22(12) (1986): 1680–1692.

Milly, P.C.D., and P.S. Eagleson. "Effects of Special Variability in Water Budget Modeling." *Water Res. Resour.*, 23(11) (1987): 2135–2143.

Mohanty, B.P., M.D. Ankeny, R. Horton, and R.S. Kanwar. "Spatial Analysis of Hydraulic Conductivity Measured Using Disc Infiltrometers." *Water Resour. Res.*, 30 (1994): 2489–2498.

Morrison, R.D., *Ground Water Monitoring Technology*, Timco Mfg., Inc., Prairie Du Sac, WI (1983).

Morrison, R., and J. Szecsody. "Sleeve and Casing Lysimeters for Soil Pore Water Sampling." *Soil Science*, 139 (1985): 446–451.

Morrison, R.D., and J.E. Szecsody. "A Tensiometer and Pore Water Sampler for Vadose Zone Monitoring." *Soil Science*, 144 (1987): 367–372.

Mott Metallurgical Corp., Sales Division, *Catalog of Products*, Farmington, CT (1988).

Mualem, Y. "A New Model For Predicting the Hydraulic Conductivity of Unsaturated Porous Media." *Water Resour. Res.*, 12 (1976): 513–522.

Mualem, Y. "A Catalogue of the Hydraulic Properties of Unsaturated Soils." *Research Project Report No. 442.* Technion, Israel Inst. of Technol., Haifa (1976b).

Murdoch, L.C., W.W. Slack, W. Harrar, R.L. Siegrist. "Embedded Sidewall Samplers and Sensors to Monitor Subsurface Conditions." *Ground Water* (1999) (in press).

Murphy, E.M., T.R. Ginn, and J.L. Phillips. "Geochemical Estimates of Paleorecharge in the Pasco Basin: Evaluation of the Chloride Mass Balance Technique." *Water Resources Research*, 32 (1996): 2853–2868.

Nadeau, R.J., T.S. Stone, and G.S. Clinger. "Sampling Soil Vapors to Detect Subsurface Contamination: A Technique and Case Study." in *Proceedings of the, NWWA Conference on Characterization and Monitoring of the Vadose (Unsaturated) Zone*, November 19–21, 1985, Denver, Colorado, National Water Well Association, Dublin, Ohio, November 19–21 (1985): 215–226.

Nagpal, N.K. "Comparison Among and Evaluation of Ceramic Porous Cup Soil Water Samplers for Nutrient Transport Studies." *Canadian Journal of Soil Science*, 62 (1982): 685–694.

National Research Council. "*In Situ* Bioremediation: When Does it Work?" *National Academy Press*, Washington D.C. (1993).

Nativ, R., E. Adar, O. Dahan, and M. Geyh. "Water Recharge and Solute Transport Through the Vadose Zone of Fractured Chalk Under Desert Conditions." *Water Resour. Res.*, 31(1995): 253–261.

Neary, A.J., and F.Tomassini. "Preparation of Alundum/Ceramic Plate Tension Lysimeters for Soil Water Collection." *Canadian Journal of Soil Science*, 65 (1985): 169–177.

Nelson, J.K., G. Compeau, T. Maziarz, and W.R. Mahaffey. "Laboratory Treatability Testing for Assessment of Field Applicability." in: Bioremediation Field Experience (1994).

Neuman, S.P. "Stochastic Continuum Representation of Fractured Rock Permeability as an Alternative to the REV and Fracture Network Concepts." in Rock Mechanics, Proc. Of the 28th U.S. Symposium, I.W. Farmer, J.J.K. Dalmen, C.S. Desai, C.E. Glass and S.P. Neuman (Eds.), Balkema, Rotterdam, Netherlands (1987): 533–561.

Newman, B.D, A.R. Campbell, and B.P.Wilcox. "Tracer-Based Studies of Soil Water Movement in Semi-Arid Forests of New Mexico." *Journal of Hydrology*, 196 (1997): 251–270.

Nicholl, M.J., R.J. Glass, and H.A. Nguyen. "Wetting Front Instability in an Initially Wet Unsaturated Fracture." in *Proceedings of the Fourth High Level Radioactive Waste Management International Conference*, Las Vegas, NV (1993).

Nielsen, D.M. and R. Schalla. "Design and Installation of Ground-Water Monitoring Wells." *Practical Handbook of Ground-WaterMonitoring,* D.M. Nielsen (Ed.), Lewis Publishers, Chelsea, MI (1991): 239–331.

Nimmo, J.R., J. Rubin, and D.P. Hammermeister. "Unsaturated Flow in a Centrifugal Field: Measurement of Hydraulic Conductivity and Testing of Darcy's Law." *Water Resour. Res.*, 23(1) (1987): 124–134.

Nimmo, J.R., "Experimental Testing of Unsaturated Flow Theory at Low Water Contents in a Centrifugal Field." *Water Resources Research*, 26 (1990): 1951–1960.

Nissen, H.H., P. Moldrup, T. Olesen, and P. Raskmark. "Printed Circuit Board Time Domain Reflectometry Probe: Measurements of Soil Water Content." *Soil Science*, 164 (1999): 454–466.

Nkedi-Kizza, P., P.S.C. Rao, and A.G. Hornsby. "Influence of Organic Cosolvents on Sorption of Hydrophobic Organic Chemicals by Soil." *Environ. Sci. Technol.,* 19 (1985): 975–979.

Noborio, K., K.J. McInnes, and J.L. Heilman. "Measurements of Soil Water Content, Heat Capacity, and Thermal Conductivity with a Single Tdr Probe." *Soil Science.*, 161 (1996): 22–28.

O'Connor, K.M. and L.V. Wade. "Applications of Time Domain Reflectometry in the Mining Industry." in *Proceedings of the Symposium and Workshop on Time Domain Reflectometry in Environmental, Infrastructure, and Mining Applications* held at Northwestern University, Evanston, Illinois, September 17–19, 1994 USBM special publication SP 19–94 (1994): 494–506.

Owens, J.W., S.P. Wasik, and H. Devoe. "Aqueous Solubilities and Enthalpies of Solution of n-Alkylbenzenes." *J. Chem. Eng. Data*, 31(47) (1986): 47–51.

Pachepsky, Ya.A., D. Timlin, and G. Varallyay. "Artificial Neural Networks to Estimate Soil Water Retention from Easily Measurable Data." *Soil Sci. Soc. Am. J.*, 60 (1996): 727–733.

Paltineanu, I.C., and J.L. Starr. "Real-Time Soil Water Dynamics Using Multisensor Capacitance Probes: Laboratory Calibration." *Soil Sci. Soc. Am. Proc.*, 61 (1997): 1576–1585.

Palumbo, A.V., S.P. Scarborough, S.M. Pfiffner, and T.J. Phelps. "Influence of Nitrogen and Phosphorus on the *In Situ* Bioremediation of Trichloroethylene." *Appl. Biochem. Biotech.*, 51 (1995): 635–647.

Parizek, R.R., and Lane. "Soil-Water Sampling Using Pan and Deep Pressure-Vacuum Lysimeters." *Journal of Hydrology*, 11 (1970): 1–21.

Parker, J.C. "Hydrostatics of Water in Porous Media." in *Soil Physical Chemistry*, D.L. Sparks (Ed.), CRC Press, Boca Raton, FL (1986).

Parkin G.W., D.E. Elrick, and W.D. Reynolds. "Recent Advances in Using Ring Infiltrometers and TDR to Measure Hydraulic Properties of Unsaturated Soils." in *Proceedings of the Int. Workshop Characterization and Measurement of the Hydraulic Properties of Unsaturated Porous Media*, van Genuchten, M.Th., F.J. Leij and L. Wu (Eds.), University of California, Riverside, CA (1999).

Parlange, J.Y., and Hill. "Theoretical Aanalysis of Wetting Front Instability in Soils." *Soil Science*, 122 (1976): 236–239.

Paul, E.A., and F.G. Clark. "Soil Microbiology and Biochemistry." San Diego: Academic Press (1989).

Pedersen, T.A. and J.T. Curtis. "Soil Vapor Extraction Technology Reference Book." PB91-168476, EPA/540/2-91/003 (1991).

Peterson, J.E., B.N. Paulsson, and T.V. McEvilly. "Applications of Aalgebraic Reconstruction Techniques to Crosshole Seismic Data." *Geophysics*, 50 (1985): 1566–1580.

Phelps, T.J., D. Ringelberg, D. Hedrick, J. Davis, C.B. Fliermans, and D.C. White. "Microbial Biomass and Activities Associated with Subsurface Environments Contaminated with Chlorinated Hydrocarbons." *Geomicrobiol. J.*, 6 (1989): 157–170.

Phene, C.J., D.A. Clark, G.E. Cardon, and R.M. Mead. "Soil Matric Potential Sensor Research and Applications." in *Advances in Measurement of Soil Physical Properties: Bringing Theory into Practice*, SSSA Spec. Publ., No. 30, SSSA, Madison, WI (1992): 263–280.

Phillips, F.M. "Environmental Tracers for Water Movement in Desert Soils of the American Southwest." *Soil Science Society of America Journal*, 58 (1994): 15.

Phillips, F.M. "The Use of Isotopes and Environmental Tracers in Subsurface Hydrology." *Reviews of Geophysics*, 33 (1995): 1029–1033.

Plagge, R., P. Häupl, and M. Renger. "Transient Effects on the Hydraulic Properties of Porous Media." in *Proceedings of the Int. Workshop Characterization and Measurement of the Hydraulic Properties of Unsaturated Porous Media*, van Genuchten, M.Th., F.J. Leij and L. Wu (Eds.), University of California, Riverside, CA (1999).

Price, L.C. "Aqueous Solubility of Petroleum as Applied to Its Origin and Primary Migration." *Am. Assoc. Petrol. Bull*, 60(2) (1976): 213–244.

Price, S.L., R.S. Kasevich, M.A. Johnson, D. Wiberg, and M.C. Marley. "Radio Frequency Heating for Soil Remediation." J. Air Waste Manag. Assoc., 49(2) (1999): 136.

Quin, B. F., and Forsythe, L. J., "All-Plastic Suction Lysimeters for the Rapid Sampling of Percolating Soil Water." *New Zealand Journal of Science*, 19 (1976): 145–148.

Quirk, J. P., and R. K. Schofield. "The Effect of Electrolyte Concentration on Soil Permeability." *J. Soil Sci.*, 6 (1955): 163–178.

Raats, P.A.C. "Unstable Wetting Fronts in Uniform and Non-Uniform Soils." *Soil Sci. Soc. Am. Proc.,* 37 (1973): 681–685.

Rahman, S. S. and G.V. Chilingarian. "Casing Design: Theory and Practice." Series title: *Developments in Petroleum Science*, 42 (1995): 373.

Rasmussen, T.C., S. C. Rhodes, A. Guzman, and S. P. Neuman. "Apache Leap Tuff INTRAVAL Experiments: Results and Lessons Learned." *Rep. NUREG/CR-6096*, prepared for U.S. Nuclear Regulatory Commission, Washington, D.C. 1995.

Rawlins, S. L., and G. S. Campbell. "Water Potential: Thermocouple Psychrometry," in *Methods of soil analysis. Part 1.*, 2nd ed., A. Klute (Ed.), Agron. Monogr. 9, ASA and SSSA, Madison, WI 1986: 597–618.

Rawls, W.J. and D.L. Brakensiek. "Prediction of Soil Water Properties for Hydrologic Modeling." in *Watershed Management in the Eighties*. Jones, E.B. and T.J. Ward (Ed.), *Proc. Irrig. Drain. Div.*, ASCE, Denver, CO April 30 - May 1, 1985: 293–299.

Rawls, W.J., T.J. Gish, and D.L. Brakensiek. "Estimating Soil Water Retention from Soil Physical Properties and Characteristics." in *Advances in Soil Science*, Stewart, B.A., Springer-Verlag, NY (1991).

Rawls, W.J., L.R. Ahuja and D.L. Brakensiek. "Estimating Soil Hydraulic Properties from Soils Data." in *Proceedings* of the Int. Worksh. *Indirect Methods For for Estimating the Hydraulic Properties of Unsaturated Soils*. M.Th. van Genuchten, M.Th., F.J. Leij, and L.J. Lund (Eds.), University of California, Riverside, CA (1992): 329–340.

Redman, J.D., and S.M. DeRyck. "Monitoring NAPLs in the Subsurface with Multilevel Tdr Probes." in *Proceedings of the Symposium and Workshop on Time Domain Reflectometry in Environment, Infrastructure and Mining Applications*, Special Pub. SP-19–94, Northwestern University, Evanston, IL (1994).

Reece, C. F. "Evaluation of a Line Heat Dissipation Sensor for Measuring Soil Matric Potential." *Soil Sci. Soc. Am. J.*, 60 (1996): 1022–1028.

Reeves, M., and E.E. Miller. "Estimating Iinfiltration for Erratic Rainfall." *Water Res. Resour.*, 11(1) (1975): 102–110.

Reimer, G.M. "Application of Reconnaissance Techniques for Determining Soil-Gas Radon Concentrations-An Example from Prince Georges County, Maryland." *Geophys. Res. Lett.*, 17 (1990): 809–812.

Remenda, V.H and G. van der Kamp. "Contamination from Sand-Bentonite Seal in Monitoring Wells Installed in Aquitards." *Ground Water*, 35(1) (1997): 39–46.

Ren, T., K. Noborio, and R. Horton. 1999a. "Measuring Soil Water Content, Electrical Conductivity and Thermal Properties with a Thermo-Tdr Probe." *Soil Sci. Soc. Am. J.*, (in press) (1999).

Reynolds, W. D. "Saturated Hydraulic Conductivity: Field Measurement." in *Soil Sampling and Methods of Analysis*, M. R. Carter (Ed.), Canadian Society of Soil Science, Lewis Publishers, Boca Raton, LA (1993): 599–613.

Reynolds, W.D., and D.E. Elrick. "Determination of Hydraulic Conductivity using a Tension Infiltrometer." *Soil Sci. Soc. Am. J.*, 55 (1991): 633–639.

Rhoades, J.D., and R.D. Ingvalson. "Macroscopic Swelling and Hydraulic Conductivity Properties of Four Vermiculite Soils." *Soil Sci. Soc. Am. J.*, 53 (1969): 1215–1219.

Rhoades, J.D., and J.D. Oster. "Solute Content." in *Methods of Soil Analysis, Part I,* Agronomy Series Monograph No 9, Second Edition, American Society of Agronomy, Inc. Madison, WI (1986): 985–1006.

Rhoades, J.D., N.A. Manteghi, P.J. Shouse, and W.J. Alves. "Soil Electrical Conductivity and Soil Salinity: New Formulations and Calibrations." *Soil Science Society of America Journal*, 53 (1989): 433–439.

Richards, L.A. "Capillary Conduction of Fluids Through Porous Mediums." *Physics*, 1 (1931): 318–333.

Richards, L.A., and W. Gardner. "Tensiometers for Measuring the Capillary Tension of Soil Water." *Journal of American Society of Agronomy*, 28 (1936): 352–358.

Richards, L.A., M.B. Russell, and O.R. Neal. "Further Developments on Apparatus for Field Moisture Studies." *Proceedings, Soil Science Society of America*, 2 (1938): 55–64.

Richards, L.A., P.L. Richards. "Radial-Flow Cell for Soil-Water Measurements." *Soil Sci. Soc. Am. Proc.*, 26(6) (1962): 515–518.

Richards, S.J., L.S.Willardson, S. Davis, and J.R. Spencer. "Tensiometer Use in Shallow Ground-Water Studies." *Proceedings, American Society of Civil Engineers* 99, (IR4) (1973): 457–464.

Ringelberg, D.B., G.T. Townsend, K.A. Deweerd, J.M. Suflita, and D.C. White. "Detection of the Anaerobic Dechlorinating Microorganism Desulfomonile Ttiedjei in Environmental Matrices by its Signature Lipopolysaccharide Branched-Long-Chain Hydroxy Fatty-Acids." *Fems Microbiology Ecology*, 14 (1994): 9–18.

Ripp, J.A., and J.F. Villaume. "A Vadose Zone Monitoring System for a Flyash Landfill." in *Proceedings* of the *National Water Well Association Conference on Characterization and Monitoring of the Unsaturated (Vadose) Zone*, Nov., Denver, CO (1985): 73–96.

Ritsema, C.J., and L.W. Dekker. "Distribution Flow: A General Process in the Top Layer of Water Repellent Soils." *Water Resour. Res.*, 31(5) (1995): 1187–1200.

Robbins, G.A., and M.M. Gemmell. "Factors Requiring Resolution in Installing Vadose Zone Monitoring Systems." *Ground Water Monitoring Review*, Summer (1985): 75–80.

Robinson, E.S. and C. Coruh. *Basic Exploration Geophysics*, John Wiley and Sons (1988).

Roffman, H.K., M.D. Neptune, J.W. Harris, A. Carter, and T. Thomas. "Field Screening for Organic Contaminants in Samples From Hazardous Waste Sites." in *Proceedings of the NWWA/API Conference on Petroleum Hydrocarbons and Organic Chemicals in Ground Water*, Houston, TX, November 13–15 (1985).

Rogers, K.P. and E.J. Poziomek. "Fiber Optic Sensors for Environmental Monitoring." *Chemospere*, 33 (1996): 1151–1174.

Rolston, D.E. "Gas flux." in *Methods of Soil Analysis, Part 1, Physical and Mineralogical Methods,* Second Edition, A. Klute, (Ed.), American Society of Agronomy, Inc., and Soil Science Society of America, Inc., Madison, WI (1986).

Rose, A.W., A.R. Hutter, and J.W. Washington. "Sampling Variability of Radon in Soil Gases." *Journal of Geochemical Exploration*, 38 (1990): 173–191.

Rose, C.W., W.R. Stern, and J.E. Drummond. "Determination of Hydraulic Conductivity as a Function of Depth and Water Content for Soil *In Situ*." *Water Resour. Res.*, 3(1) (1965): 1–9.

Ross, P.J., and K.J.R. Smettem. "Describing Soil Hydraulic Properties with Sums of Simple Functions." *Soil Sci. Soc. Am. J.*, 57 (1993): 26–29.

Rossabi, J., T.R Jarosch, B.D. Riha, B.B. Looney, D.G. Jackson, C.A. Eddy-Dilek, R.S. Van Pelt, and B.E. Pemberton. "Determining Contaminant Distribution and Migration by Integrating Data from Multiple Cone Penetrometer-Based Tools." in *Proceedings of ISC '98*, Atlanta, GA, Balkema Press (1998).

Roth, K., R. Schulin, H. Fluhler, and W. Attinger. "Calibration of Time Domain Reflectometry for Water Content Measurement Using a Composite Dielectric Approach." *Water Resour. Res.*, 26 (1990): 2267–2273.

Rubin, Y., S. Hubbard, A. Wilson, and M. Cushey. "Aquifer Characterization." Chapter 10 in *The Handbook of Groundwater Engineering,* J. Delleur (Ed.), CRC Press, NY (1998).

Rubin, Y., E. Majer, and S. Hubbard. "Use of Geophysical Data for Hydrogeological Site Characterization." presented at the *D.O.E. Subsurface Contaminants Focus Area Annual Review*, Augusta, GA, April (1999).

Ruimy, R., V. Breittmayer, V. Boivin, and R. Christen. "Assessment of the State of Activity of Individual Bacterial-Cells by Hybridization with a Ribosomal-RNA-Targeted Fluorescently Labeled Oligonucleotidic Probe." Fems Microbiology Ecology, 15 (1994): 207–213.

Ryan, B.J., M.P. DeVries, G. Garklavs, J.R. Gray, R.W. Healy, P.C. Mills, C.A. Peters, and R.G. Striegl. "Results of Hydrologic Research at a Low-Level Radioactive-Waste Disposal Site Near Sheffield, Illinois." *U.S. Geological Survey Water-Supply Paper 2367* (1991).

Salvucci, G.D., and D. Entekhabi. "Equivalent Steady Soil Moisture Profile and the Time Compression Approximation in Water Balance Modeling." *Water Res. Resour.*,30(10) (1994): 2737–2749.

Scanlon, B.R., P.W. Kubik, P. Sharma, B.C. Richter, and H.E. Gove. "Bomb Chlorine-36 Analysis in the Characterization of Unsaturated Flow at a Proposed Radioactive Waste Facility, Chihuahuan Desert, Texas." *Nuclear Instruments and Methods in Physics Research*, B52 (1990): 489–492.

Scanlon, B.R. "Evaluation of Liquid and Vapor Water Flow in Desert Soils Based on Chlorine-36 and Tritium Tracers and Nonisothermal Flow Simulations." *Water Resources Research*, 28 (1992): 285–297.

Schaap, M.G. and W. Bouten. "Modeling Water Retention Curves of Sandy Soils Using Neural Networks." *Water Resour. Res.*, 32 (1996): 3033–3040.

Schaap, M.G., and F.J. Leij. "Database Related Accuracy and Uncertainty of Pedotransfer Functions." *Soil Science*, 163 (1998): 765–779.

Schaap, M.G., Leij F.J. and van Genuchten M.Th. "Neural Network Analysis for Hierarchical Prediction of Soil Water Retention and Saturated Hydraulic Conductivity." *Soil Sci. Soc. Am. J.*, 62 (1998): 847–855.

Schaap, M.G., F.J. Leij, and M.Th. Van Genuchten. "A Bootstrap-Neural Network Approach to Predict Soil Hydraulic Parameters." in *Proceedings of the Int. Workshop on the Characterization and Measurement of the Hydraulic Properties of Unsaturated Porous Media*, M.Th. Van Genuchten, F.J. Leij, and L. Wu (Eds.), Univ. of California, Riverside, CA (1999): 1237–1250.

Schaap, M.G., and F.J. Leij. "Improved Prediction of Unsaturated Hydraulic Conductivity with the Mualem-van Genuchten Model." *Soil Sci. Soc. Am. J.*, (in press) (2000).

Scheinfeld, R.A., and T.G. Schwendeman. "The Monitoring of Underground Storage Tanks, Current Technology." in *Proceedings of the NWWA/API Conference on Petroleum Hydrocarbons and Organic Chemicals in Ground Water*, Houston, TX, November 13–15 (1985): 244–264.

Schnoor, J.L., L.A. Licht, S.C. McCutcheon, N.L. Wolfe, and L.H. Carreira. "Phytoremediation of Organic and Nutrient Contaminants." *Environ. Sci. Tech.*, 29 (1995): 318A–323A.

Selker, J.S., C.K. Keller, and J.T. McCord. "Vadose Zone Processes." Boca Raton, FL, Lewis Publishers (1999).

Shainberg, I., and G.J. Levy. "Physico-Chemical Effects of Salt upon Infiltration and Water Movement in Soils." in *Interacting Processes in Soil Science, Advances in Soil Science*, R.J. Wagenet, P. Baveye, and B.A. Stewart (Eds.), Lewis Publishers, Boca Raton, FL (1992): 38–93.

Shan, C. "Analytical Solutions for Determining Vertical Air Permeability in Unsaturated Soils." Water Resources Research, 31(9) (1995): 2193–2200.

Shan, C., R.W. Falta, and I. Javandel. "Analytical Solutions for Steady State Gas Flow to a Soil Vapor Extraction Well." Water Resour. Res., 28(4) (1992): 1105–1120.

She, H.Y. and B. Sleep. "The Effect of Temperature on Capillary Pressure-Saturation Relationships for Air-Water and Perchloroethylene-Water Systems." Water Resour. Res., 34 (1998): 2587–2597.

Sheets, K.R., and J.M. Hendrickx. "Non-Invasive Soil Water Content Measurement Using Electromagnetic Induction." *Water Resour. Res.*, 31 (1995): 2401–2410.

Shock, C.C., J. Barnum, and M. Seddigh. "Calibration of Watermark Soil Moisture Sensors for Irrigation Management." Irrigation Association. Proc. Int. Irrigation Show, San Diego, CA (1998): 139–146.

Shouse, P.J., M.Th. van Genuchten, and J.B. Sisson. "A Gravity-Drainage/Scaling Method for Estimating the Hydraulic Properties of Heterogeneous Soils." in *Proceedings of the Vienna Symposium, Hydrological Interactions Between Atmosphere, Soil and Vegetation Hydrological Interactions Between Atmosphere, Soil and Vegetation,* G. Kienitz, P.C.D. Milly, M.Th. van Genuchten, D. Rosbjerg, and W.J. Shuttleworth (Eds.), August 1991, IAHS Publ. No. 204 (1991): 281–291.

Si, B.C., R.G. Kachanoski, Z.F. Zhang, G.W. Parkin, and D.E. Elrick. "Estimation of Hydraulic Parameters Under Constant Flux Condition Using Multipurpose TDR Probes." in *Proceedings of the Int. Workshop Characterization and Measurement of the Hydraulic Properties of Unsaturated Porous Media,* van Genuchten, M.Th., F.J. Leij and L. Wu (Eds.), University of California, Riverside, CA (1999).

Silkworth, D.R., and D.F. Grigal. "Field Comparison of Soil Solution Samplers." *Soil Science Society of America Journal*, 45 (1981): 440–442.

Simunek, J., and M.Th. van Genuchten. "Numerical Model for Simulating Multiple Solute Transport in Variably-Saturated Soils." in *Proceedings of Water Pollution III: Modeling, Measurements, and Prediction*, L.C. Wrobel and P. Latinopoulos, Computation Mechanics Publication, Ashurst Lodge, Ashurst, Southampton, UK (1995): 21–30.

Simunek, J. and M.Th. van Genuchten. "Estimating Unsaturated Soil Hydraulic Properties from Tension Disc Infiltrometer Data by Numerical Inversion." *Water. Resour. Res*, 32 (1996): 2683–2696.

Simunek, J., and M.Th. van Genuchten. "Parameter Estimation of Soil Hydraulic Properties from Multiple Tension Disc Infiltrometer Data." *Soil Sci.*, 162 (1997): 383–398.

Simunek, J., O. Wendroth, M.Th. van Genuchten. "Parameter Estimation Analysis of the Evaporation Method for Determining Soil Hydraulic Properties." Soil Sci. Soc. Am. J., 62 (1998): 894–904.

Sisson, J.B. "Drainage from Layered Field Soils: Fixed Gradient Models." *Water Resour. Res.*, 23 (1987): 2071–2075.

Sisson, J.B., and M.Th. Van Genuchten. "An Improved Analysis of Gravity Drainage Experiments for Estimating the Unsaturated Soil Hydraulic Functions." *Water Resour. Res.*, 27 (1991): 569–575.

Smiles, D.E., G. Vachaud, and M. Vaucin. "A Test of Uniqueness of the Soil Moisture Characteristic During Transient, Nonhysteresis Flow of Water in a Rigid Soil." *Soil Sci. Soc. Amer. Proc.*, 35 (1971): 534–539.

Smith, C.N., and R.F. Carsel. "A Stainless-Steel Soil Solution Sampler for Monitoring Pesticides in the Vadose Zone." *Soil Science Society of America Journal*, 50 (1986): 263–265.

Smith, R.L., B.L. Howes, and S.P. Garabedian. "*In Situ* Measurement of Methane Oxidation in Groundwater by Using Natural-Gradient Tracer Tests." *Appl. Environ. Microbiol.*, 57 (1991): 1997–2004.

Soil Survey Staff. "Soil Survey Laboratory Information Manual." *Soil Surv. Invest. Rep. no 45,* Natl. Soil Surv. Center, Lincoln, NE (1995).

SoilMoisture Equipment Corp., Sales Division, *Catalog of Products*, Santa Barbara, CA (1988).

Song, Y., J.M. Ham, M.B. Kirkham, and G.J. Kluitenberg. "Measuring Soil Water Content under Turfgrass Using the Dual-Probe Heat-Pulse Technique." *J. Am. Soc. Hort. Sci.*, 123 (1998): 937–941.

Song, Y., M.B. Kirkham, J.M. Ham, and G.J. Kluitenberg. "Dual Probe Heat Pulse Technique for Measuring Soil Water Content Sunflower Water Uptake." *Soil and Tillage Res.* (in press) (1999).

Sophocleous, M. and C.A. Perry. "Experimental Studies in Natural Groundwater Recharge Dynamics: Analysis of Observed Recharge Events." *Journal of Hydrology*, 81 (1985): 297–332.

Spittler, T.M., and W.S. Clifford, "A New Method for Detection of Organic Vapors in the Vadose Zone." in *Proceedings of the NWWA Conference on Characterization and Monitoring of the Vadose (Unsaturated) Zone*, November 19–21, 1985, Denver, CO, National Water Well Association, November 19–21, Dublin, OH, 1985: 236–246.

Sposito, G., *The Thermodynamics of Soil Solutions*, Oxford University Press, NY (1981).

Sposito, G. and W.A. Jury. "Miller Similitude and Generalized Scaling Analysis." in Scaling in Soil Physics: Principles and Applications, D. Hillel and D.E. Elrick, (Eds.), SSSA Special Publication No. 25, Soil Science Society of America, Madison, WI (1990): 13–22.

Stahl, W., H. Aust, and A. Dounas. "Origin of Artesian and Thermal Water Determined by Oxygen, Hydrogen and Carbon Isotope Analyses of Water Samples from the Sperkios Valley, Greece." *Isotope Techniques in Groundwater Hydrology*, International Atomic Energy Association, Vienna, 1 (1974): 317–339.

Stannard, D.I. "Theory, Construction, and Operation of Simple Tensiometers." *Ground Water Monitoring Review,* 6(3), 1986: 70–78.

Starr, J.L., and I.C. Paltineanu. "Soil Water Dynamics Using Multisensor Capacitance Probes in Nontraffic Interrows of Corn." *Soil Sci. Soc. Am. J.*, 62 (1998): 114–122.

Starr, M.R. "Variation in the Quality of Tension Lysimeter Soil Water Samples from a Finnish Forest Soil." *Soil Science*, 140 December (1985): 453–461.

Swensen, B. "Unsaturated Flow in a Layered, Glacial-Contact Delta Deposit Measured by the Use of O-18, Cl- and Br- as Tracers." *Soil Science*, 162 (1997): 242–253.

Stephens, D.B. *Vadose Zone Hydrology*, Lewis Publishers: Boca Raton, FL (1996).

Stevenson, C.D. "Simple Apparatus for Monitoring Land Disposal Systems by Sampling Percolating Soil Waters." *Environmental Science and Technology*, 12 (1978): 329–331.

Stonestrom, D.A., and J. Rubin. "Air Permeability and Trapped-Air Content in Two Soils." *Water Resources Research*, 25(9) (1989a): 1959–1969.

Stonestrom, D.A., and J. Rubin. "Water Content Dependence of Trapped-Air in Two Soils." *Water Resour. Res.*, 25 (9) (1989b): 1947–1958.

Suarez, D.L., J.D. Rhoadcs, R. Lavado, and C.M. Grieve. "Effect of pH on Saturated Hydraulic Conductivity and Soil Dispersion." *Soil Sci. Soc. Amer. J.*, 48 (1984): 50–55.

Talsma, T. "*In Situ* Measurement of Sorptivity." *Aust. J. Soil Res.*, 7 (1969): 269–277.

Tamari, S., J.H.M. Wösten, and J.C. Ruiz-Suárezz. "Testing an Artificial Neural Network for Predicting Soil Hydraulic Conductivity." *Soil Sci. Soc. Am. J.*, 60 (1996): 1732–1741.

Tanner, A.B. "A Tentative Protocol for Measurement of Radon Availability from the Ground." *Rad. Prot. Dosim.*, 24(1/4) (1988): 79–83.

Tarara, J.M., and J.M. Ham. "Measuring Soil Water Content in the Laboratory and Field with Dual-Probe Heat-Capacity Sensors." *Agron. J.*, 89 (1997): 535–542.

Taylor, H.P., Jr. "Application of Oxygen and Hydrogen Isotopes to Problems of Hydrothermal Alteration and Ore Deposition." *Economic Geology*, 69 (1974): 843–883.

Taylor, S.A., and G.L. Ashcroft. *Physical Edaphology, The Physics of Irrigated and Non Irrigated Soils*, W.H. Freeman and Company, San Francisco, CA (1972).

Telford, W.M., Geldart, L.P. and Sheriff, R.E., *Applied Geophysics Second Edition*, Cambridge University Press (1990).

Thomas, A. "*In Situ* Measurement of Moisture in Soil and Similar Substances by 'Fringe' Capacitance." *J. Sci. Instrum.*, 43 (1966): 21–27.

Thomas, J.M., and C.H. Ward. "*In Situ* Biorestoration of Organic Contaminants in the Subsurface." *Environ. Sci. Technol.*, 23 (1989): 760–766.

Thomas, J.M., and C.H. Ward. "Subsurface Microbial Ecology and Bioremediation." *J. Haz. Mat.*, 32 (1992): 179–194.

Tietje, O. and M. Tapkenhinrichs. "Evaluation of Pedotransfer Functions." *Soil Sci. Soc. Am. J.,* 57 (1993): 1088–1095.

Tietje, O., and V. Hennings. "Accuracy of the Saturated Hydraulic Conductivity Prediction by Pedo-Transfer Functions Compared to the Variability within FAO Textural Classes." *Geoderma*, 69 (1996): 71–84.

Timlin, D. and Y. Pachepsky. "Measurement of Unsaturated Soil Hydraulic Conductivities Using a Ceramic Cup Tensiometer." *Soil Science*, 163 (1998): 625–635.

Tokunaga, T. "The Pressure Response of the Soil Water Sampler and Possibilities for Simultaneous Soil Solution Sampling and Tensiometry." *Soil Science*, 54(3) (1992): 171–183.

Toorman, A.F., P.J. Wierenga, and R.G. Hills. "Parameter Estimation of Hydraulic Properties Form One-Step Outflow Data." *Water Resour. Res.*, 28 (1992): 3021–3028.

Topp, G.C., J.L. Davis, and A.P. Annan. "Electromagnetic Determination of Soil Water Content: Measurement in Coaxial Transmission Lines." *Water Resour. Res.*, 16 (1980): 574–582.

Topp, G.C., and J.L. Davis. "Measurement of Soil Water Content Using Time-Domain Reflectometry (Tdr): A Field Evaluation." *Soil Sci. Soc. Am. J.*, 49 (1985): 19-24.

Tremblay, D., D. Tulis, P. Kostecki, and K. Ewald. "Innovation Skyrockets at 50,000 LUST Sites, EPA Study Reveal Technology Use at LUST Sites." *Soil & Groundwater Cleanup* December (1995): 6–13.

US EPA. *RCRA Ground-Water Monitoring Technical Enforcement Guidance Documen*t, Office of Waste Programs Enforcement, Office of Solid Waste and Emergency Response, OSWER-9950.1 (1986).

US EPA. *Environmental Response Team, Standard Operating Procedure 2051: Charcoal Tube Sampling,* November 7 (1988a).

US EPA. *Environmental Response Team, Standard Operating Procedure 2052: Tenax Tube Sampling*, November 8 (1988b).

US EPA. *Bioremediation of Hazardous Waste Sites Workshop*, CERI-89-11. Washington, DC (1989).

Udd, E., *Fiber Optic Sensors: Aan Introduction for Engineers and Scientists,* NY, Wiley & Sons (1991): 476.

Udell, K.S. "Reactive Transport/Enhanced Remediation - Application of *In Situ* Thermal Remediation Technologies for DNAPL removal." IAHS publication, 250 (1998): 367.

Ullom, W.L. "Ethylene and Propylene in Soil Gas: Occurrence, Sources and Impact on Interpretation of Exploration Geochemical Data." *Bulletin*, Association of Petroleum Geochemical Explorationists, 4(1) (December 1988): 62–81.

van den Elsen, E., J. Stolte, and G. Veerman. "Three Automated Laboratory Systems for Determining the Hydraulic Properties of Soils." in *Proceedings of the Int. Workshop on the Characterization and Measurement of the Hydraulic Properties of Unsaturated Porous Media*, M.Th. Van Genuchten, F.J. Leij, and L. Wu (Eds.), Univ. of California, Riverside, CA (1999): 329–340.

van der Kamp, G., and R. Schmidt. "Monitoring of Total Soil Moisture on a Scale of Hectares Using Groundwater Piezometers." *Geophys. Res. Letters*, 24 (6) (1997): 719–722.

van der Ploeg, R.R., and Beese, F. "Model Calculations for the Extraction of Soil Water by Ceramic Cups and Plates." *Soil Science Society of America Journal*, 41 (1977): 466–470.

van Genuchten, M.Th. "A Closed-Form Equation for Predicting the Hydraulic Conductivity of Unsaturated Soils." *Soil Sci. Soc. Am. J.*, 44 (1980): 892–898.

van Genuchten, M.Th. and D.R. Nielsen. "On Describing and Predicting the Hydraulic Conductivity of Unsaturated Soils." *Annales Geophysicae*, 3 (1985): 615–628.

van Genuchten, M.Th., F.J. Leij, and S.R. Yates. 1991. *The RETC Code for Quantifying the Hydraulic Functions of Unsaturated Soils*, EPA/600/2-91/065 (1991).

van Genuchten, M.Th. and F.J. Leij. "On Estimating the Hydraulic Properties of Unsaturated Soils." in *Proceedings of the International Workshop on Indirect Methods for Estimating the Hydraulic Properties of Unsaturated Soils*. Univ. of California, Riverside, CA (1992): 1–14.

van Genuchten, M.Th., F.J. Leij, and L.J. Lund (Eds.) "Proc. Int. Workshop Indirect Methods for Estimating the Hydraulic Properties of Unsaturated Soils." University of California, Riverside, CA (1992).

van Genuchten, M.Th., F.J. Leij, and L. Wu (Eds.) "Proc. Int Workshop Characterization and Measurement of the Hydraulic Properties of Unsaturated Porous Media." Univ. of California, Riverside, CA (1999).

van Golf-Racht, T.D. "Fundamental of Fractured Reservoir Engineering." *Development of Petroleum Science,* 12, Elsevier, Amsterdam-Oxford, NY (1982).

van Nostrand, R.G., and K.L. Cook. "Interpretation of Resistivity Data." *U.S.G.S. Prof. Paper No. 499* (1966).

Vereecken, H. "Derivation and Validation of Pedotransfer Functions for Soil Hydraulic Properties." in *Proceedings of the International Workshop on Indirect Methods for Estimating the Hydraulic Properties of Unsaturated Soils.* Univ. of California, Riverside, CA (1992): 473–488.

Vereecken, H., J. Maes, J. Feyen, and P. Darius. "Estimating the Soil Moisture Retention Characteristic from Texture, Bulk Density, and Carbon Content." *Soil Sci.*, 148 (1989): 389–403.

Vereecken, H., J. Maes and J. Feyen. "Estimating Unsaturated Hydraulic Conductivity from Easily Measured Soil Properties." *Soil Sci.*, 149 (1990): 1–12.

Vesselinov, V.V., and S.P., Neuman. "Numerical Inverse Interpretation Of Multistep Transient Single-Hole Pneumatic Tests in Unsaturated Fractured Tuffs at the Apache Leap Research Site." in Theory, Modeling and Field Investigation in Hydrogeology: A Special Volume in Honor of Shlomo P. Neuman's 60th Birthday, Geological Society of America, in press (1999).

Villa Nova, N.A., K. Reichardt, P.L. Libardi, and S.O. Moraes. "Direct Reading 'Air-Pocket' Tensiometer." *Soil Technology*, 2 (1989): 403–407.

von Hippel, A.R. (Ed.). *Dielectric Materials and Applications*. MIT Press, Cambridge, MA (1954): 1–304.

Wagemann, R., and B. Graham. "Membrane and Glass Fibre Filter Contamination in Chemical Analysis of Fresh Water." *Water Research*, 8 (1974): 407–412.

Wagenet, R.J., J. Bouma, and J.L. Hutson. "Modeling Water and Chemical Fluxes as Driving Forces in Pedogenesis." in *Qualitative Modeling of Soil Forming Processes*, R.B. Bryant, R.W. Arnold, and M.R. Hoosbeek eds., SSSA Special Publ. 39, ASA/CSSA/SSSA, Madison, WI (1994): 17–35.

Wang, D., S.R. Yates, and F.F. Ernst. "Determining Soil Hydraulic Properties using Tension Infiltrometers, Time Domain Reflectometry, and Tensiometers." *Soil Sci. Soc. Am. J.*, 62 (1998): 318–325.

Ward, A.L. "Dielectric Measurements in the Presence of High Ionic Conductivity Using Time Domain Reflectometry." in *1998 Annual Meeting Abstracts,* p. 183, Soil Sci. Soc. Am., Madison WI (1998): 183.

Ward, A.L., A.P. von Bertoldi, and R.G. Kachanoski. "An Improved TDR Probe for the Measurement of Soil Water Content at High Salinity Levels." in *1992 Annual Report*, Department of Land Resource Science, University of Guelph, Guelph, ON (1992): 38–39.

Ward, A.L., R.G. Kachanoski and D.E. Elrick. "Laboratory Measurement of Solute Transport Using Time Domain Reflectometry." *Soil Sci. Soc. Am J.*, 58 (1994): 1031–1039.

Ward, A.L., J.M. Leather, D.S. Knowles, and S.H. Lieberman. "Development and Testing of New Sensors for Rapid *In Situ* Moisture Logging and Pore Space Visualization by Cone Penetrometry." *PNNL*-11744, Pacific Northwest National Laboratory, Richland, WA (1996).

Ward, A.L., G.W. Gee, and M.D. White. "A Comprehensive Analysis of Contaminant Transport in the Vadose Zone Beneath Tank SX-109." PNNL-11463, UC-702. Pacific Northwest National Laboratory. Richland, WA, February (1997).

Ward, A.L., M.J. Fayer, J.C. Ritter, and R. E. Clayton. "Automated Measurement of the Hydraulic Properties Vadose Zone Core Samples." in *1998 Annual Meeting Abstracts*, Soil Sci. Soc. Am., Madison WI (1998): 186.

Ward, S.H., "Resistivity and Induced Polarization Methods." in *Geotechnical and Environmental Geophysics, S.E.G. Investigations in Geophysics,* Stanley Ward (Ed.), 1(5) (1990): 147–189.

Warrick, A.W., and Amoozegar-Fard. A. "Soil Water Regimes Near Porous Cup Water Samplers." *Water Resour. Res.*, 13(2) (1977): 203–207.

Warrick A.W., and A. Amoozegar-Fard. "Infiltration and Drainage Calculations Using Spatially Scaled Hydraulic Properties." *Water Res. Resour.*, 15 (1979): 1116–1120.

Watson, K.K. "A Recording Field Tensiometer with Rapid Response Characteristics." *Journal of Hydrology*, 5 (1967): 33–39.

Weeks, E.P. "Field Determination of Vertical Permeability to Air in the Unsaturated Zone." *Geol. Surv. Prof. Paper 1051*, U.S. Geological Survey, Washington, D.C. (1978).

Wesson, T.C., and Armstrong, F.E. "The Determination of C 1 –C 4 Hydrocarbons Adsorbed on Soils." *Bartlesville Energy Research Center Report of Investigations* BERC/RI-75/13, U.S. Energy Research and Development Administration, Office of Public Affairs, Technology Information Center, Bartlesville, OK, December (1975).

Whalley, W.R., P.B. Leeds-Harrsion, P. Joy, and P. Hoefsloot. "Time Domain Reflectometry and Tensiometry Combined in an Integrated Soil Water Monitoring System." *J. Agric. Engr. Res.*, 59 (1994): 141-144.

Wheatcraft, S.W., and J.H. Cushman. "Hierarchical Approaches to Transport in Porous Media." in *U.S. National Report to International Union of Geodesy and Geophysics,* Rev. Geophysics (supplement), AGU, Washington, D.C. (1991): 263–269.

White, K.D., J.T. Novak, C.D. Goldsmith, and S. Bevan. "Microbial Degradation Kinetics of Alcohols in Subsurface Systems," in *Proceedings of the NWWA/API Conference on Petroleum Hydrocarbons and Organic Chemicals in Ground Water-Prevention, Detection and Restoration*, Houston, TX, November 13–15 (1985).

White, I., and K.M. Perroux. "Estimation of Unsaturated Hydraulic Conductivity from Field Sorptivity Measurements." *Soil Sci. Soc. Am. J.*, 53 (1989): 324–329.

White, I., M.J. Sully, and K.M. Perroux. "Measurement of Surface-Soil Hydraulic Properties: Disk Permeameters, Tension Infiltrometers, and Other Techniques." *Bringing Theory into Practice.* SSSA Spec. Publ., G. C. Topp *et al.* (Eds.), Soil Sci. Soc. Am., Madison, WI 30 (1992): 69–103.

White, I., S.J. Zegelin, G.C. Topp, A. Fish. "Effect of Bulk Electrical Conductivity on TDR Measurement of Water Content in Porous Media." *in Proceedings of the Symposium and Workshop on Time Domain Reflectometry in Environmental, Infrastructure, and Mining Applications*, Northwestern University, Evanston, IL, September 17–19, 1994: 294–308, and USBM special publication SP (1994): 19–94.

Wierenga, P.J., M. Young, A. Warrick, *et al.* "Maricopa Environmental Monitoring Site." *Material of the Unsaturated Zone Monitoring Technology Transfer Workshop*, Casa Grande, AZ, February 11–12 (1998).

Wierenga, P.J., R.G. Hills, and D.B. Hudson. "The Las Cruces Trench Site: Characterization, Experimental Results, and One-Dimensional Flow Predictions." *Water Resources Res.*, 27 (1991): 2695-2705.

Wigley, T.M.L., L.N. Plummer and F.J. Pearson. "Mass Transfer and Carbon Isotope Evolution in Natural Water Systems." *Geochimica et Cosmochimica Acta*, 42 (1978): 1117–1139.

Williams, B.G., and G.C. Baker. "An Electromagnetic Induction Technique for Reconnaissance Surveys of Soil Salinity Hazards." *Australian Journal of Soil Research*, 20 (1982): 107–118.

Williams, R.D., L.R. Ahuja, and J.W. Naney. "Comparison of Methods to Estimate Soil Water Characteristics from Limited Texture, Bulk Density, and Limited Data." *Soil Sci.*, 153 (1992): 172–184.

Wilson, J.T., J.W. Weaver, and D.H. Kampbel. "Intrinsic Bioremediation of Tce in Ground Water a an Npl Site in St. Joseph, Michigan." in *Proceedings of the EPA Symposium on Intrinsic Bioremediation of Ground Water*, August 30-September 1, Denver, CO. U.S. Environmental Protection Agency, EPA/540/R-94/515 (1994).

Wilson, L.G. "Monitoring in the Vadose Zone: A Review of Technical Elements and Methods." U.S. Environmental Protection Agency, EPA-600/ 7-80-134 (1980).

Wilson, L.G. "The Fate of Pollutants in the Vadose Zone, Monitoring Methods and Case Studies." *Thirteenth Biennial Conference on Ground Water*, September (1981).

Wilson, L.G. "Monitoring in the Vadose Zone: Part II." *Ground Water Monitoring Review*, Winter (1982): 31–42.

Wilson, L.G. "Monitoring in the Vadose Zone: Part III." *Ground Water Monitoring Review*, Winter (1983): 155–165.

Wilson, L.G. "Methods for Sampling Fluids in the Vadose Zone." *Ground Water and Vadose Zone Monitoring*, ASTM STP 1053, ASTM, 1990: 7–24.

Wilson, L.G., L.G. Everett, and S.J. Cullen (Eds.), "Handbook of Vadose Zone Characterization and Monitoring." Boca Raton, FL, Lewis Publishers (1995).

Wind, G.P. "Capillary Conductivity Data Estimated by a Simple Method." in *Water in the Unsaturated Zone, Proc. Wageningen Symposium,* P.E. Rijtema and H. Wassink (Eds.), IAHS, Gentbrugge, Unesco, Paris, 1 (1969): 181–191.

Wittmann, S.G., Quinn, K.J., and Lee, R.D. "Use of Soil Gas Sampling Techniques for Assessment of Ground Water Contamination." in *Proceedings NWWA/API Conference on Petroleum Hydro-carbons and Organic Chemicals in Ground Water-Prevention, Detec-tion and Restoration,* Houston, TX, November 13–15, 1985: 291–309.

Wolff, R.G. "Weathering Woodstock Granite, near Baltimore, Maryland." *American Journal of Science*, 265 (1967): 106–117.

Wood, T.R., and G.T. Norrell. "Integrated Large-Scale Aquifer Pumping and Infiltration Tests: Groundwater Pathways." *OU 7-06: Summary Report*, Idaho National Engineering Laboratory Report INEL-96/0256 (1996).

Wood, W.W. "A Technique Using Porous Cups for Water Sampling at Any Depth in the Unsaturated Zone." *Water Resour. Res.*, 9(4) (1973): 486–488.

Wooding, R.A. "Steady Infiltration from a Shallow Circular Pond." *Water Resour. Res.*, 4 (1968): 1259–1273.

Wösten, J.H.M., M.H. Bannink, and J. Beuving. "Water Retention and Hydraulic Conductivity Characteristics of Top- and Sub-Soils in the Netherlands: The Staring Series." Soil Survey Institute, Wageningen, The Netherlands. Report 1932 (1987).

Wösten, J.H.M., and M.Th. van Genuchten. "Using Texture and Other Soil Properties to Predict the Unsaturated Soil Hydraulic Functions." *Soil Sci. Soc. Am. J.*, 52 (1988): 1762–1770.

Wösten J.H.M., P.A. Finke and M.J.W. Jansen. "Comparison of Class and Continuous Pedotransfer Functions to Generate Soil Hydraulic Characteristics." *Geoderma*, 66 (1995): 227–237.

Yeh, W. W-G. "Review of Parameter Identification Procedures in Groundwater Hydrology: The Inverse Problem." *Water Resour. Res.* 22(2) (1986): 95–108.

Yilmaz, O. "Seismic Data Processing." *Society of Exploration Geophysics Series: Investigation in Geophysics*, 2 (1987).

Yokuda, E., and R. Smith. "A New Probe for *In Situ* TDR Moisture Measurement." *Paper* #93–106. Idaho, National Engineering Laboratory, Idaho Falls, ID (1993).

Young, M. "Use of Suction Lysimeters for Monitoring in the Landfill Linear Zone." in *Proceedings of the Monitoring Hazardous Waste Sites*, Geotechnical Engineering Division, American Society of Civil Engineers, Detroit, MI, October (1985).

Zawislanski, P.T., and B. Faybishenko. "New Casing and Backfill Design for Neutron Logging Access Boreholes." *Groundwater*, 30(1) (1999).

Zhang, R. "Determination of Soil Sorptivity and Hydraulic Conductivity from the Disk Infiltrometer." *Soil Sci. Soc. Am. J.*, 61 (1997): 1024–1030.

Zohdy, A.A., G.P. Eaton, and D.R. Mabey. "Application of Surface Geophysics to Ground-Water Investigations." *Techniques of Water-Resources Investigations of the United States Geological Survey, Book 2, Chapter D1*, (1974).

CASE STUDIES

LARGE-SCALE FIELD INVESTIGATIONS IN FRACTURED BASALT IN IDAHO: LESSONS LEARNED

Thomas Wood, INEEL, and **Boris Faybishenko,** LBNL

INTRODUCTION

The U.S. Department of Energy (DOE) is challenged with the responsibility for environmental restoration of large, contaminated waste sites. A fundamental component of this responsibility is the assessment of the relative degree of environmental risk posed by various facilities. At the Idaho National Engineering and Environmental Laboratory (INEEL), hazardous and radioactive materials have been handled, stored, and discharged at a number of site locations. Contaminants include organics, metals, radioactive fission products, and transuranic elements. It is a formidable task to characterize such sites and quantify the fate and transport of those contaminants in the complex hydrogeologic system of the Snake River Plain (SRP) subsurface. A reasonable characterization of fundamental transport properties is necessary to provide adequate information for determining long-term environmental risk and for evaluating remedial alternatives. Parameters required for estimating contaminant transport rates in the vadose zone and in groundwater typically have been derived from limited laboratory and small-scale field experiments, EPA publications, textbook tables and, to the extent available, field data collected at the INEEL's Radioactive Waste Management Complex (RWMC). When these parameters are unknown or large uncertainty exists, conservative estimates are used because of the importance of the results for making decisions that may affect the health of nearby populations.

The RWMC includes one of the largest subsurface waste disposal facilities in the DOE complex. Past disposal of low-level, mixed, and transuranic radioactive wastes occurred by direct discharge or burial in shallow, unlined pits and trenches within the surficial sediments. Environmental Assessments of the RWMC are being conducted to evaluate risks to people and the environment from contaminants potentially migrating from this facility. Of particular concern and interest is the potential migration of contaminants through the vadose zone to the underlying SRP aquifer, which is the sole source of water for residents living on the ESRP. Flooding of the RWMC has occurred three times in the past, potentially increasing the mobility of some of the subsurface contaminants.

The INEEL embarked on a program of integrated field and laboratory investigations to significantly enhance our ability to responsibly predict contaminant fate and transport properties through the vadose zone and within the SRP aquifer. The overall program has principally consisted of a large-scale Aquifer Pumping and Infiltration Test (APIT), a series of supporting laboratory experiments, and numerical modeling. The APIT consisted of two major tests: a large-scale aquifer pumping test (LSPT) and a large-scale infiltration test (LSIT). The location for the LSPT was approximately 1.2 km south-southwest of the RWMC. The LSIT was located approximately 1.4 km south of the RWMC and east of the Big Lost River Spreading Areas (Figures 1 and 2). Wood and Norrell (1996) provide the most complete summary reference for the LSPT and LSIT.

The focus of the APIT was investigating hydrologic properties of the SRP basalt. These basalts occur beneath the INEEL in thick sequences of stacked basaltic lava flows, infrequently separated by sedimentary interbeds. Individual basalt flows are typically highly fractured, 3 to 12 m thick, of limited areal extent, and show a lobate distribution in plan view. The basalt flows show an extreme elongation in one direction, giving them a finger- or lenticular structure with a width typically ranging from 20 to 60 m. Fracturing is caused by thermal contraction as a basalt flow cools. The surfaces of the basalt flow, where cooling rates are large, are generally more densely fractured than the interior. Fracture density decreases toward the center of the flow, as the cooling rate is typically lower away from the basalt flow boundaries. For basalt flows that were laid down over recent still-warm basalt flows, cooling patterns and consequently fracture patterns in the upper and lower halves of the basalt flow may differ greatly, with the lower half containing few fractures at the lower boundary. Sedimentary interbeds may separate basalt flow units that were formed at widely separated times. These interbeds range from a few centimeters to as much as 15 m thick; some have great areal extent, while others are limited to the regions between adjacent basalt flows. At the LSIT site the first major interbed occurs at a depth of 70 m (Burgess, 1995). Geophysical and borehole evidence suggests total basalt thickness in the Snake River Plain may exceed three kilometers.

At the scale of the LSIT, or the macro-scale, the Snake River Plain Basalts is often considered an over-sized porous medium, with the grains represented by individual basalt flows interiors, and the pore space comprised of the rubble and clinker zones that form at basalt flow margins during cooling. Regional-scale groundwater flow through the Snake River Plain aquifer is commonly studied at the macro-scale where wells are often separated by thousands of meters and model domains encompass hundreds of square kilometers. The aquifer is highly transmissive, with an anisotropic permeability arising from the large aspect ratio of the individual basalt flow fingers, which determines the topology of the network of rubble zones. This anisotropy has implications for liquid infiltration through the vadose zone as well, suggesting that a relatively large amount of lateral spreading should accompany vertical infiltration. At the macro-scale, the sedimentary interbeds of low permeability, which often separate basalt flows, may act as confining units, and as potential locations for perched water zones, where contaminants may accumulate.

Figure 1. Large Scale Infiltration Test, 1994.

TEST DESIGN AND METHODS

The large-scale infiltration test, which was conducted at the RWMC of the INEEL, was probably the largest infiltration test performed to date in America. It was designed to assess flow and transport phenomena at the scale of several basalt flows, separated by rubble zones, between the surface and the sedimentary interbed at the depth of 70 m. An area of approximately 26,000 m^2 was flooded in 1994 for a month using water that was pumped from the underlying aquifer and then was piped into the 200 m diameter infiltration basin. The total volume of water supplied into the pond was 38.5 million gallons (Starr and Rohe 1995). After 6 days of flooding, several short-lived radioactive tracers, as well as a NaBr tracer, were added to

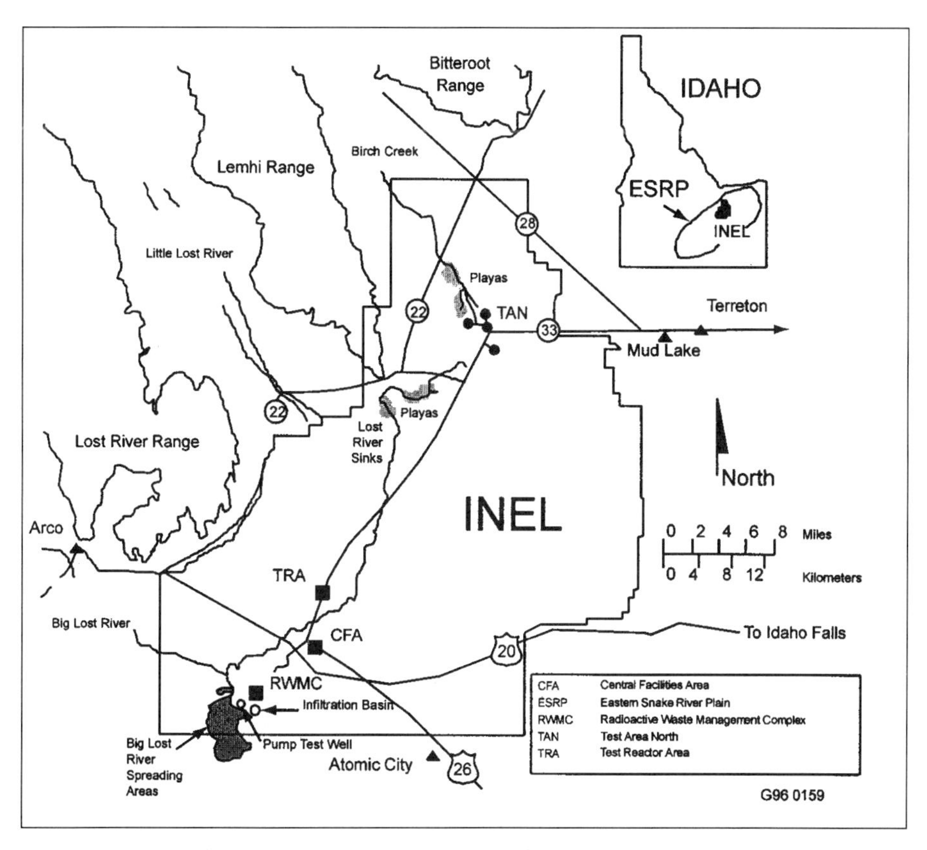

Figure 2. Location of the INEEL with respect to regional features, the state of Idaho, and the Eastern Snake River Plain.

the infiltration basin. The Snake River Plain aquifer water was used. After conservative and reactive tracers were added to the infiltration basin and mixed, no new water was added to the basin for 11 days while the tracer pulse infiltrated. Subsurface migration of infiltrating water and tracers was monitored by several techniques, primarily to the depth of the sedimentary interbed at 70 m. Thus, the LSIT test investigated the bulk hydrologic properties on a "block" of basalt 70 m thick and 200 m in diameter.

The monitoring well network consisted of 66 wells, located within and around the infiltration pond (Figure 3). It was anticipated that preferential flow and lateral spreading of water along rubble zones and the formation of perched water bodies above the sedimentary interbeds would occur (Norrell, *et al.* 1994). Monitoring for

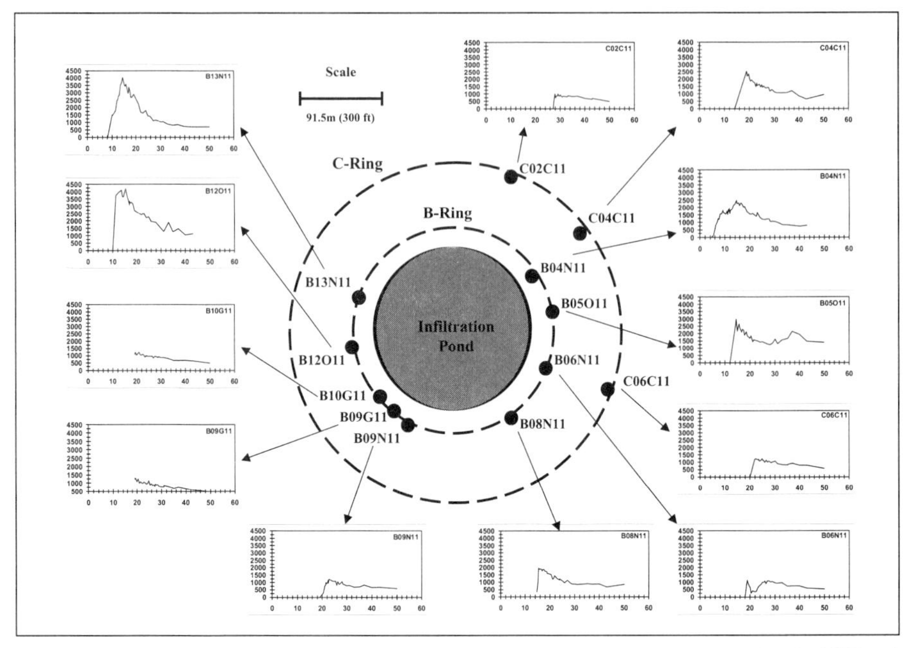

Figure 3. General layout of the infiltration pond and the borehole monitoring system during the LSIT, and breakthrough curves of Se-75 in pci/L determined in boreholes screened in a perched water zone.

subsurface water movement began immediately after the beginning of flooding using multiple monitoring systems. The following types of data were collected:

(a) Neutron logging to determine the changes in the moisture content of rocks in the near vicinity of the monitoring wells.

(b) Gamma spectroscopy data to track the movement of the gamma-emitting radioactive tracers.

(c) Water sampling to determine the water appearance and tracer breakthrough curves.

(d) DC electrical resistivity measurements to provide understanding about the spatial distribution of the moisture content over the entire area.

(e) Perched water level measurements.

TEST RESULTS

Despite the expectation of lateral flow, on which the monitoring network was designed, no water or tracer was observed beyond the footprint of the infiltration pond except for perched water that formed on top of the sedimentary interbed at a 70 m depth. The perched water moved down the slope of the interbed through a widespread rubble zone immediately above the sedimentary interbed. The fact that infiltrating water was not observed in partially saturated fractured basalt outside the basin suggests high permeability of vertical fractures and the horizontal rubble zone, and a very low permeability of the interbed. Thus, the LSIT showed that in addition to the effect of intra-basalt fracturing, which is studied at the intermediate scale, one has to take into account the geometry and permeability of rubble zones and sedimentary layers.

Figure 4 shows the arrival of water at different depths in the vadose zone, which was determined using neutron logging. This figure also includes water arrival into open boreholes screened above the sedimentary interbed at a depth of about 70 m, which is an indication of the formation of perched water. The left dashed line in Figure 4 shows the first water arrival at a different depth. Water travel times directly beneath the basin, as determined by Porro and Bishop (1995), was 5 m/day. The average wetting-front travel time, based on the CPN data, is a reasonable estimate at this scale of observation. Smaller scales of observation would have to consider local hydrogeologic features that may account for wide departures from the average wetting-front travel time presented here. Many observations made during the LSIT

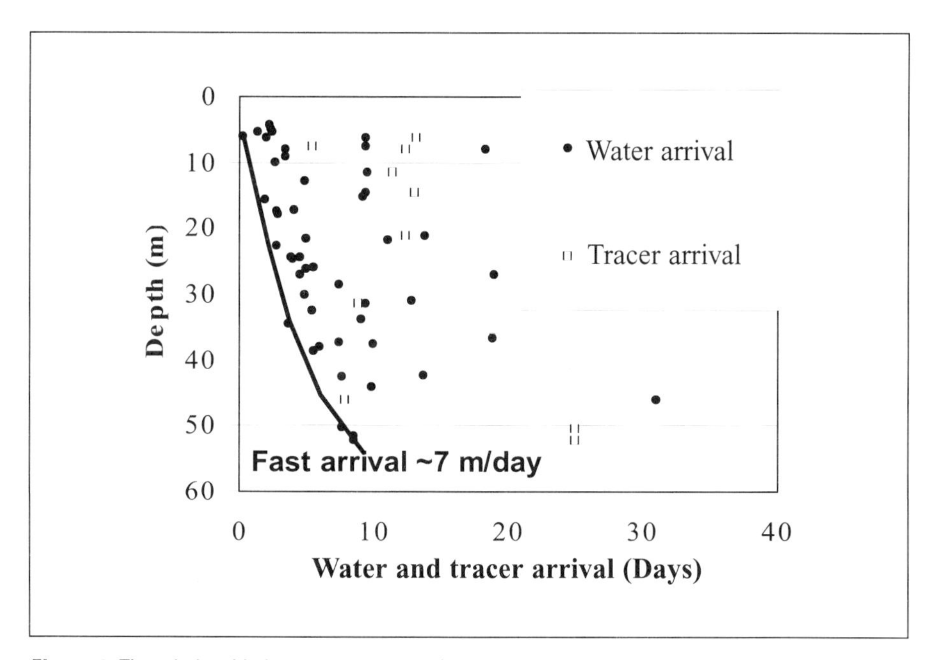

Figure 4. The relationship between the time of water arrival (determined from neutron logging) and the Se-75 tracer arrival (determined from water sampling) and depth.

confirm that small-scale features can and do control local moisture movement, however, at the scale of the test, these local features tended to average out; 5 m/d is a reasonable approximation of the wetting-front travel time under ponded conditions.

Despite the fact that clean water was added to the infiltration basin for 6 days prior to the addition of tracers, the first water observed in several wells contained tracer. Also, tracer-free water was observed in several lysimeters throughout the course of the test. Intervals containing insufficient water for the collection of samples were observed between water bearing zones. Therefore, taken sample point by sample point, Figure 4 may show little correlation between the depth and tracer arrival time. We learned that in this environment, depth is less important in many instances than the sampling method and the geometry of the fractured system. Therefore, interpretation of the data set must consider the sampling method and the fracture geometry with respect to the sampled point to adequately calculate standard transport parameters (Faybishenko *et al.* 1997). For instance, the interpretation of breakthrough curves determined using water sampling with suction lysimeters may have non-unique solutions because these devices interrogate different rock volumes depending on the moisture content and permeability of the surrounding fractures and rock matrix.

Figure 5 illustrates a variety of tracer breakthrough curves observed at different locations during the LSIT. One can see a conventional BTC near the surface, multi-modal curves as affected by migration from different fractures, and no tracer detected at some points. It is interesting to note that at some locations, which became saturated quickly after the beginning of flooding, the tracer did not arrive during the month-long duration of the LSIT. This can be attributed to changes of water flow pathways with time. Water may flow into dead-end, non-conductive fractures easily, but if it cannot continue flowing out of the fractures, no subsequent water (containing tracer) can flow into these fractures by advection. Tracer can enter these fractures only by diffusion or flow through the surrounding basalt, both of which are slow processes. Thus, a combination of the water and tracer experiments is needed to identify zones of preferential flow and locations of dead-end, non-conductive fractures.

DISCUSSION AND LESSONS LEARNED

Failure of the Conventional Approach to Data Analysis

For the LSIT, 101 locations were monitored for water and tracer (Newman and Dunnivant 1995). Of the 101 monitoring locations, water was recovered from 30 and the conservative tracer, Se-75, was found in only 26. The shapes of the breakthrough curves varied significantly between the different wells and lysimeters. Despite their best efforts, Dunnivant and Newman (1995) were able to model only 10 of 26 break-

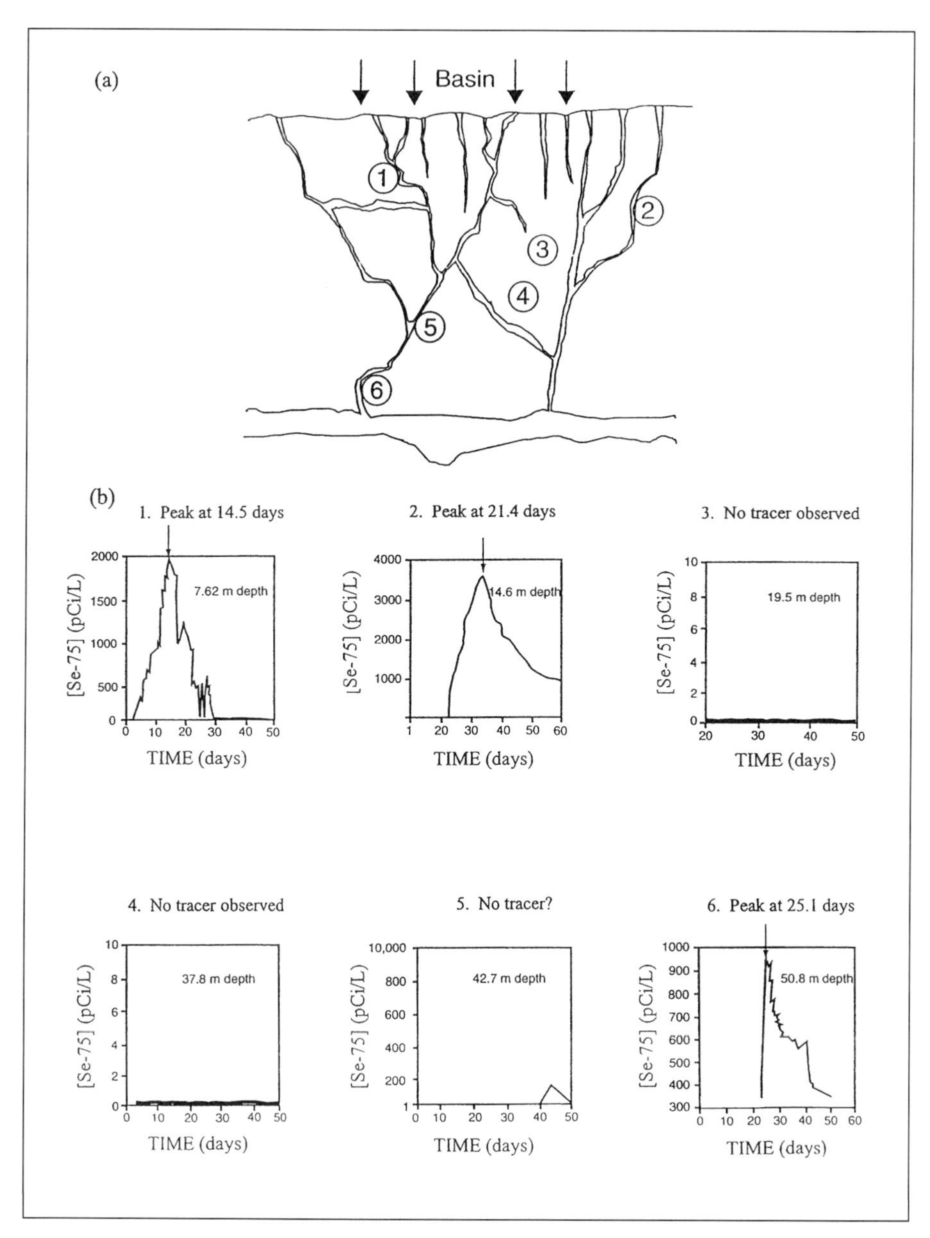

Figure 5. A conceptual model (a) of the fracture pattern and corresponding breakthrough curves (b) for Se-75 for several locations shown on the conceptual model in Figure (a).

through curves using a one-dimensional version of the transport equation with constant water velocity. Overall this amounts to a 90% failure rate of the conventional porous media model approach.

Reinterpretation of the breakthrough curves from the LSIT (Faybishenko et al. 1997) suggests that the geometric components of the fracture pattern and the location of the sampling point within the fracture system may control the arrival and concentration of tracer measured. Clearly, a one-dimensional transport model of flow in this system at this scale is inadequate for understanding the movement of water through a fractured basalt vadose zone.

Uncertainties of Measurements

We determined the following uncertainties and pitfalls for measuring/monitoring in a fractured rock vadose zone from our analysis of data from the LSIT:

- Not all components of the fracture system can be observed.
- Observable components can not be monitored simultaneously.
- Monitoring in different parts of the system is carried out with different accuracy and over different time scales.
- Each measurement represents an average over space and time, however, the volume affected can not be determined.
- During active monitoring, the water-tracer system is disturbed to a certain degree, and non-conductive fractures can be affected by water sampling.
- Two adjacent points may have different BTCs.
- The shape of the BTC depends not only on the exact point of sample collection, but also on flow paths above and below the point.

CONCLUSIONS

Flow through vertical flow paths was the major component of flow during the LSIT, significantly exceeding the lateral movement of water. However, this flow cannot be considered as simple one-dimensional steady-state flow, but represents flow through a network of flow paths, which may vary over time.

In the field, measurements are made on the scale of single fractures or perhaps multiple fractures; these individual measurements must be combined and used to develop an understanding of flow and transport at a larger scale, that is, to develop a macro-scale conceptual model of tracer transport.

Without an overall understanding of the geometry and physics of flow at the macroscale, we can describe what is seen, but cannot use this information for making further predictions, and therefore cannot make meaningful assessments of contaminant transport.

REFERENCES

Burgess, J. D. (1995), *Results of the Neutron and Natural Gamma Logging, Stratigraphy, and Perched Water Data Collected During a Large-Scale Infiltration Test*, Engineering Design File, ER-WAG7-60, INEL- 95/062.

Faybishenko, B., T. R. Wood, T. M. Stoops, C. Doughty and J. Jacobsen (1997), *A Conceptual model of Tracer Transport in Fractured Basalt: Large Scale Infiltration Test Revisted* Geological Society of America 1997 Annual Meeting.

Porro, I., and C. W. Bishop (1995), *Large Scale Infiltration Test CPN Data Analysis*, EDF No. WAG7-58, INEL-95/040.

Starr, R. C., and M. J. Rohe (1995), *Large-Scale Aquifer Stress Test and Infiltration Test: Water Management System Operation and Results*, Idaho National Engineering Laboratory, INEL-95/059.

Newman, M. E., and F. M. Dunnivant (1995), *Results from the Large-Scale Aquifer* Pumping *and Infiltration Test: Transport Through Fractured Media*, Engineering Design File ER-WAG7-77, INEL-95/146.

Norrell, G. T., I. Porro, F. M. Dunnivant, J. M. Hubbell, M. C. Pfeifer, R. C. Starr, M. E. Newman, and C. W. Bishop (1994), *Integrated Large-Scale Aquifer Pumping and Infiltration Test—Conceptual Design of the Large-Scale Infiltration Test*, EGG-ER-11363.

Wood, T. R. and G. T. Norrell (1996), *"Integrated Large-Scale Aquifer Pumping and Infiltration Tests, Ground Water Pathways OU 7-06, Summary Report"*, INEL-96/0256, Rev. 0, October 1996.

ACKNOWLEDGEMENTS

The authors thank Patti Kroupa and Kirk Dooley for their tireless support and leadership in conducting the field tests. Additionally, we thank the countless team members and participants that worked on collecting and interpreting the massive data sets for the LSIT and the LSPT. The LSIT was funded primarily by the Office of Environmental Management, Environmental Restoration Program, U.S. DOE Idaho Operations Office, under contract DE-AC07-94ID13223. Preparation of the paper was supported by the DOE EMSP Program.

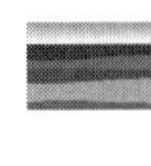

GROUNDWATER CONTAMINATION IN THE PERCHED AQUIFER AT THE DOE PANTEX PLANT: SUCCESSFUL EXPEDITED SITE CHARACTERIZATION

Caroline B. Purdy and **Jacqueline C. Burton**

INTRODUCTION

In 1993, the Office of Science and Technology (EM-50) within the DOE Office of Environmental Management (EM) was seeking a more efficient characterization process to overcome the extended schedules and costly characterization plans projected for its contaminated sites. A scientific team from Argonne National Laboratory developed an efficient characterization process for use on the Comprehensive Environmental Response, Compensation, and Liability Act (CERCLA) and the Resource Conservation and Recovery Act (RCRA) characterization projects for several federal facilities. That process is now known as the American Society for Testing and Materials (ASTM) Standard D 6235-98, Standard Practice for Expedited Site Characterization (ESC) of Vadose Zone and Groundwater Contamination at Hazardous Waste Contaminated Sites. The Characterization, Monitoring, and Sensor Technology Crosscutting Program in EM-50 chose to demonstrate the effectiveness of the ESC process on a DOE site with a complex groundwater contamination problem. The comprehensive and convincing results behind the final Conceptual Site Model (CSM) characterizing the contaminant sources and migration pathways at the chosen site, the DOE Pantex Plant near Amarillo, TX, are presented in this case study.

The ESC is structured to fit within the existing regulatory framework, and is suited to CERCLA remedial and RCRA facility investigations, as well as other voluntary contaminant cleanups. The ESC is intended to collect only the information required to meet the characterization objectives, and to terminate characterization activities once the objectives are met. The ESC determines contaminant sources, migration pathways, distribution, and fate, so that an accurate risk assessment and course of action can be taken (ASTM Standard D 6235-98; Benson *et al.* 1998; Burton 1994; Burton *et al.* 1993; Gelb 1998).

Key features of the ESC process include:

- Judgment-based sampling and measurements to characterize the vadose zone and groundwater contamination.
- An experienced, multidisciplinary team of specialists that integrates geophysical, geological, hydrological, and chemical data into a coherent and consistent CSM.
- Limitation of field deployments by mobilizing all field teams simultaneously (when possible) so that specialists can interpret the data together on a daily basis.

- A dynamic, flexible work plan that can be modified based on the changes made to the CSM, as it evolves with new information obtained daily.
- A field team leader who selects the type and location of measurements required to optimize data collection activities during field mobilization.
- Selection of field-based technologies (when reasonable) that provide quick data turnaround for daily decision-making.
- Non-intrusive or minimally invasive methods to increase the data density at minimum cost (like seismic and electromagnetic measurements).
- Multiple measurement techniques that corroborate one another.

The sequential process stressed in an ESC program begins with an exhaustive study of the existing data and an assessment of its usability in the construction of a preliminary CSM. A program is designed to identify the geological features controlling the migration of contaminants. Using this information, the appropriate sampling locations are selected to confirm the anticipated pathways. Gaining an understanding of the geology of the site is extremely cost-effective when the appropriate technologies are used, and such knowledge can be used to focus the contaminant sampling program in only the areas suspected to contain contamination.

In traditional characterization programs, expensive exploratory drilling combined with a statistically based sampling design is most often used to locate contaminants. This approach frequently determines where the contaminant is missing, rather than where it is present, thus resulting in multiple-phased drilling programs extending over months or years. Laboratory analysis that may last 1 to 3 months also contributes to the length of the characterization process.

The ESC approach has proven to be more advantageous than traditional characterization programs by (1) reducing the cost and schedule for characterization, (2) producing a more comprehensive CSM describing contaminant locations and migration pathways, and (3) gaining credibility with regulators because of the quality of characterization communicated through frequent (sometimes daily) updates.

In order to illustrate the application of the ESC, we will describe the results of the site characterization at the DOE Pantex plant. In 1993, for the Zone 12 Groundwater Operable unit at the DOE Pantex Plant, project managers had just completed the second phase of the drilling of 25 wells designed to locate contaminants in a perched aquifer 76 m (250 ft) below surface. It was suspected that the perched aquifer contributed contamination to the primary drinking source, the Ogallala Aquifer at 122m (400 ft). The goal of this case study is to illustrate the ESC program, developed and managed by Argonne National Laboratory, that was introduced at this point and replaced the existing planned activities which included another round of 13 additional wells. The Pantex Plant operations historically included R&D activities, fabrication, testing of chemical explosives, and assembly/disassembly of the nation's nuclear weapons.

GENERAL GEOLOGY, HYDROLOGY, AND HYDROGEOLOGY OF THE SITE

The Pantex Plant, located in the Texas panhandle (Figure 1), is a semiarid site situated at a 1,148 m (3,500 ft) elevation with 51 cm (20 in.) of precipitation per year. Ephemeral lakes, called playas, are the only topographical features that break the flatness of the region. The surface water runoff collects in playas where water evaporates or percolates into the subsurface. Playa 1 (Figure 1) is wet all year; all other playas within the Pantex area are dry, except after significant precipitation (Burton *et al.* 1995). At Pantex, groundwater occurs at two levels, directly above bedrock and in a shallower, perched zone. Although both water zones occur in the Tertiary Ogallala Formation, the convention at the Pantex site is to refer to the shallow perched zone as the "perched aquifer" and the deeper zone as the "Ogallala aquifer." The Ogallala aquifer is unconfined with a saturated thickness of 65.6 to 131.2 m (200 to 400 ft) in the vicinity of Pantex. To the northeast of the plant, the Amarillo City drinking wells are located in the Ogallala. Historically, groundwater flows east in the Ogallala, but it has been redirected to the northeast under Zone 12 by the pumping of city wells. The upper part of the Ogallala Formation (typically 46-76 m or 150-250 ft below the surface) contains a persistent zone of low-permeability, fine-grained material called the fine-grained zone (FGZ). This zone provides a relatively impermeable barrier and serves to perch water above the main Ogallala aquifer in the vicinity of the Pantex Plant. Although water from the perched aquifer is not used, it is expected to be a source of both organic and inorganic contaminants that migrate downward into the Ogallala aquifer. The Zone 12 Groundwater RFI investigated the primary groundwater contamination of explosives, chromium, and TCE.

PRE-ESC CSM

The generally accepted hydrological model for this site assumed that water penetrated the subsurface through the playa floors. Perched aquifer water table distribution confirmed radial flow away from the playas (Figure 2). Contaminants were expected to reach the perched aquifer via penetration through the playa regions (Figure 3). It was also suspected that some surface water penetrated through the trenches that direct wastes discharged from the Zone 12 area to Playa 1 (Figure 2).

Discontinuity of the perched aquifer was suggested in the areas where three dry wells were drilled (Figure 4). Leakage through the FGZ could occur if the FGZ were thinned or absent in those areas. However, whether the perched aquifer flow was continuous throughout the Zone 12 region or whether leakage occurred through the FGZ at the dry wells was not resolved by the well drilling program used in the original characterization plan.

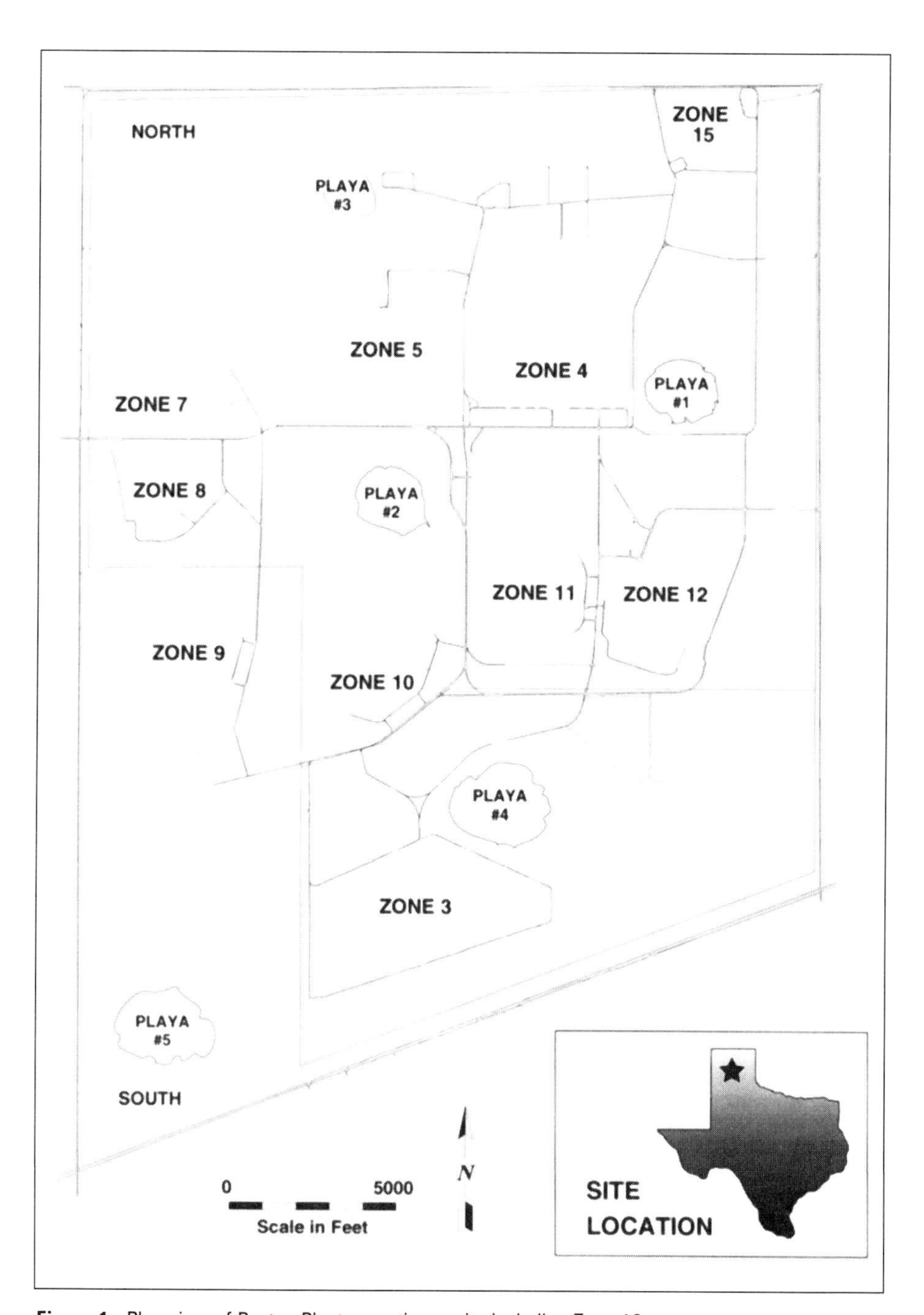

Figure 1. Plan view of Pantex Plant operations units including Zone 12.

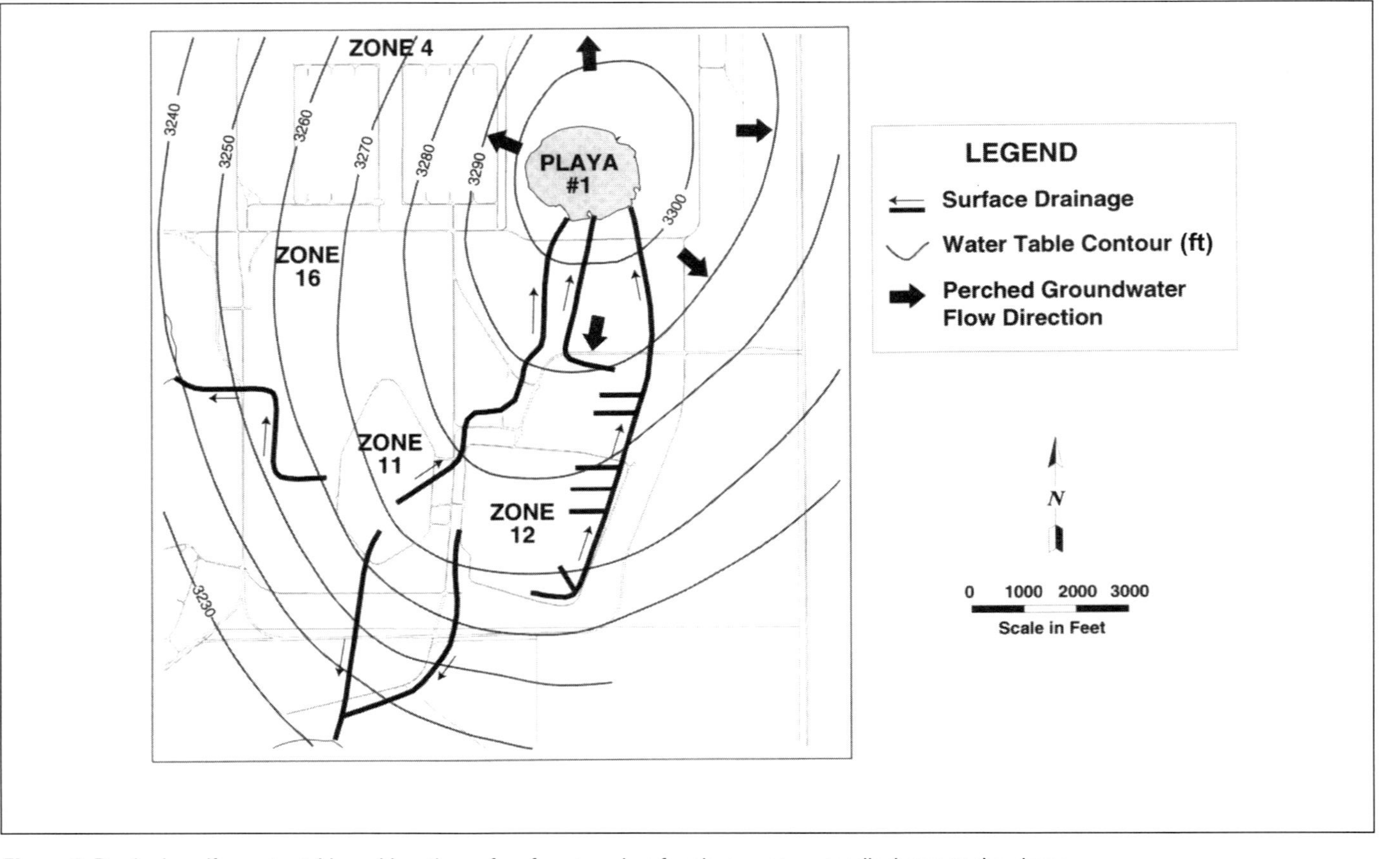

Figure 2. Perched aquifer water table and locations of surface trenches for plant wastewater discharge to the playas.

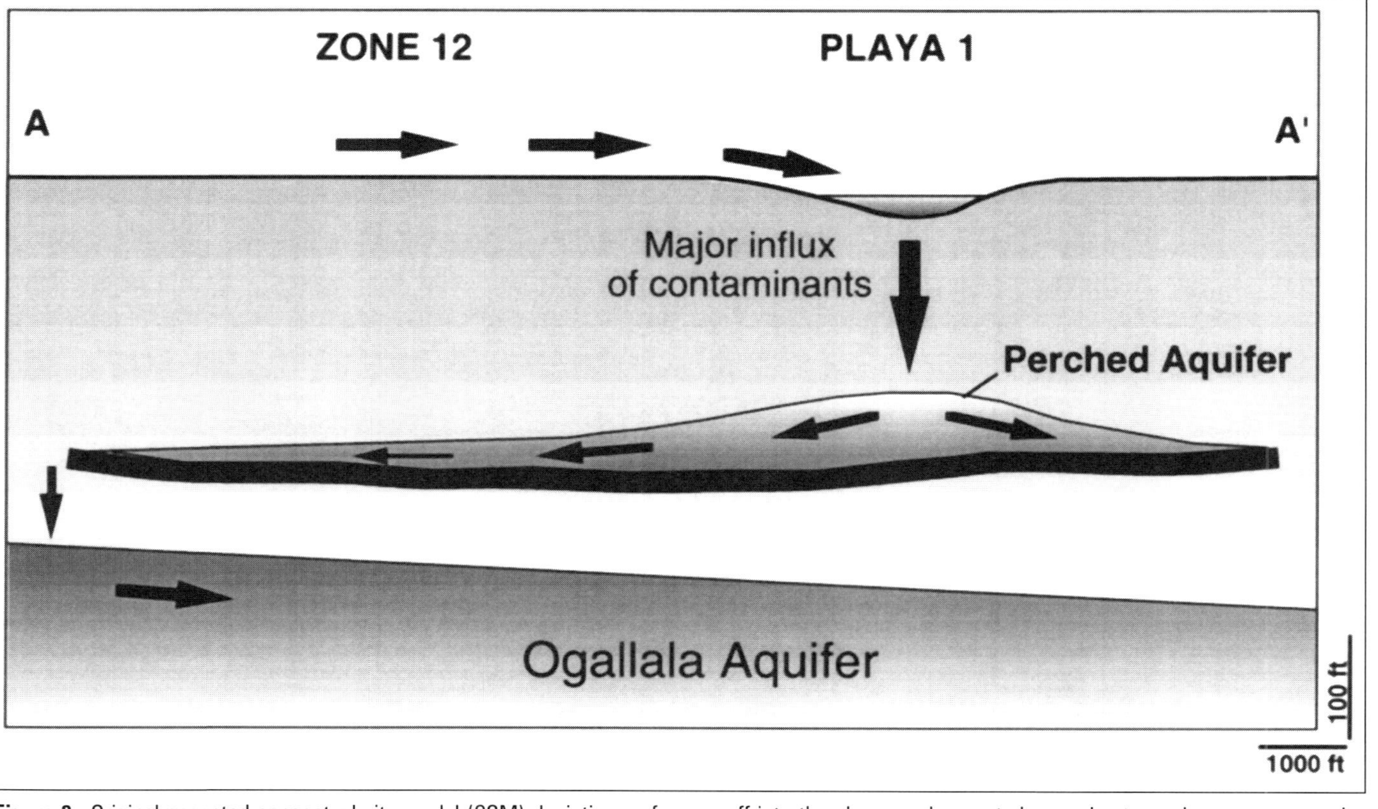

Figure 3. Original accepted conceptual site model (CSM) depicting surface runoff into the playas and expected groundwater recharge source to the perched aquifer.

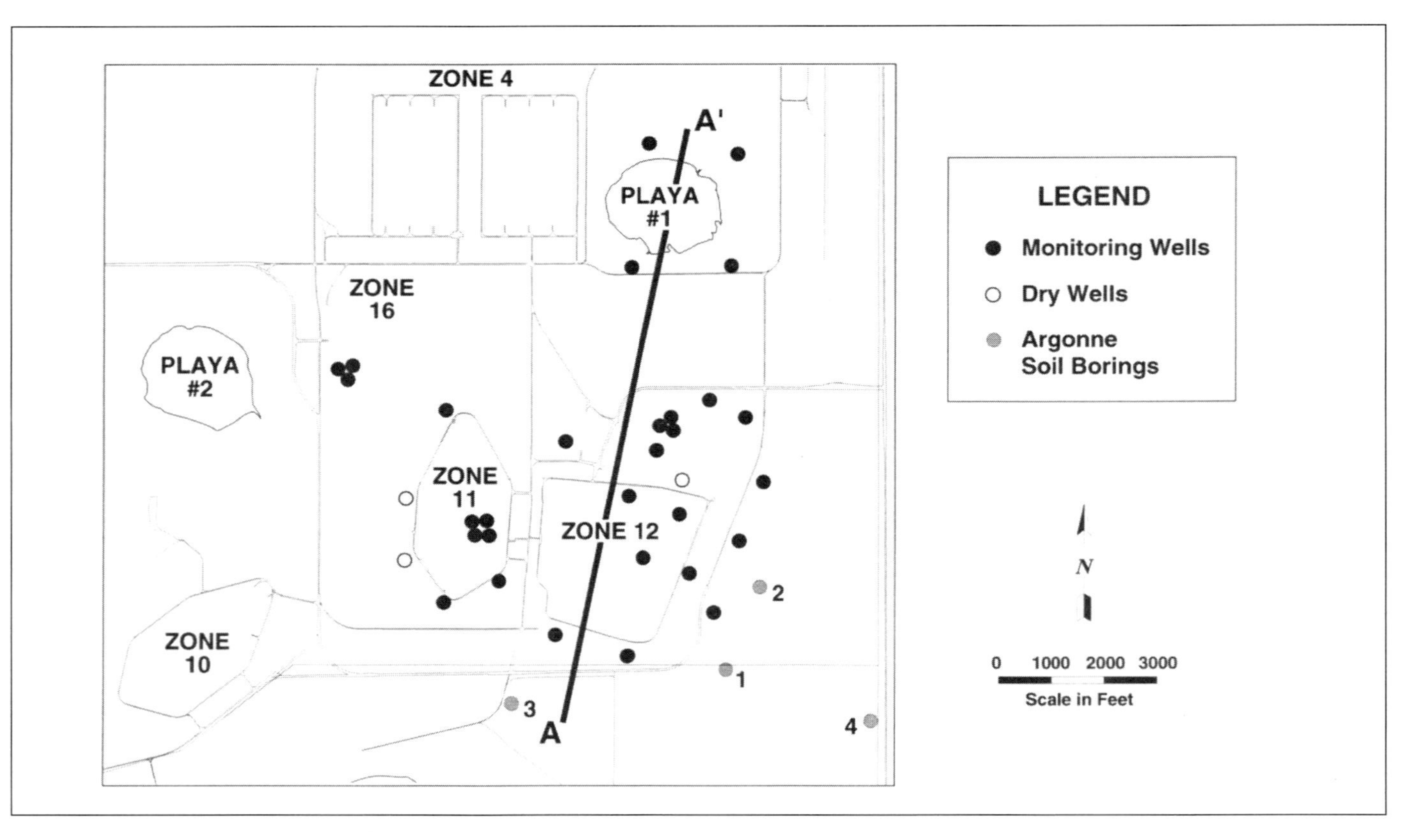

Figure 4. Plan view of Pantex site showing dry well locations, a cross section used in Figure 3 and Figure 7, and the location of four (Argonne) soil borings drilled to confirm the presence of a paleochannel directing contaminant flow.

EPA Region VI and Texas Natural Resources Conservation Commission (TNRCC) regulators identified six technical issues to be resolved by the site characterization.

1. How was the perched aquifer recharged?
2. What was the distribution of the perched aquifer and what was the flow pattern?
3. Was there discontinuity in the perched aquifer and was the perched aquifer leaking into the Ogallala?
4. What was the contamination distribution in the vadose zone?
5. What was the contamination distribution in the perched aquifer?
6. Was there contamination in the Ogallala aquifer and where would it be likely to occur?

ANALYSIS OF EXISTING DATA AND ESC

A complete description of the ESC program used at Pantex is covered in the publications of Burton and Meyer (1998), Burton *et al.* (1995), and Ferguson (1995). Based on a review of the data generated by the Pantex characterization program, Argonne set out to resolve the data inconsistencies using the ESC approach. First, the contaminant distribution in the perched groundwater was inconsistent with the accepted CSM. Concentrations did not gradually decrease in the perched aquifer tracking away from the playas, the suggested source of contamination. The highest concentrations occurred within the Zone 12 area (Figure 5a and b), suggesting significant penetration through the Zone 12 vadose zone, or the interplaya regions. The vertical penetration could have occurred below the trenches. Second, the 17 miles of seismic data collected throughout the Pantex site was uninterpretable. Reprocessing a portion of the data showed convincing reflections where the perched aquifer existed. Strong reflections attributed to perched water occurred under the trenches, again suggesting vertical water flow below the trenches. Previously, this was not expected to be a dominant groundwater pathway to the perched zone. Where the water reflections were absent, or the dry holes occurred, the perched aquifer had been previously suspected to be either thinned or absent (pre-ESC CSM). Argonne proposed an alternative model. There could be topographical features in the FGZ surface such that "highs" in the surface could rise above the perched groundwater, thus creating dry zones that disrupt the continuous flow. The FGZ highs would create isolated perched water regions. Third, the geochemistry of the perched groundwater sampled from monitoring wells showed regions of different types of water, confirming distinct and isolated perched water zones. The ^{18}O (Figure 6a), the tritium (^{3}H) (Figure 6b), as well as the Si and Cl (Figure 6c) data distinguished the different zones and suggested a barrier (or hydraulic discontinuity) between these zones. The dry holes coincided with the areas of water chemistry

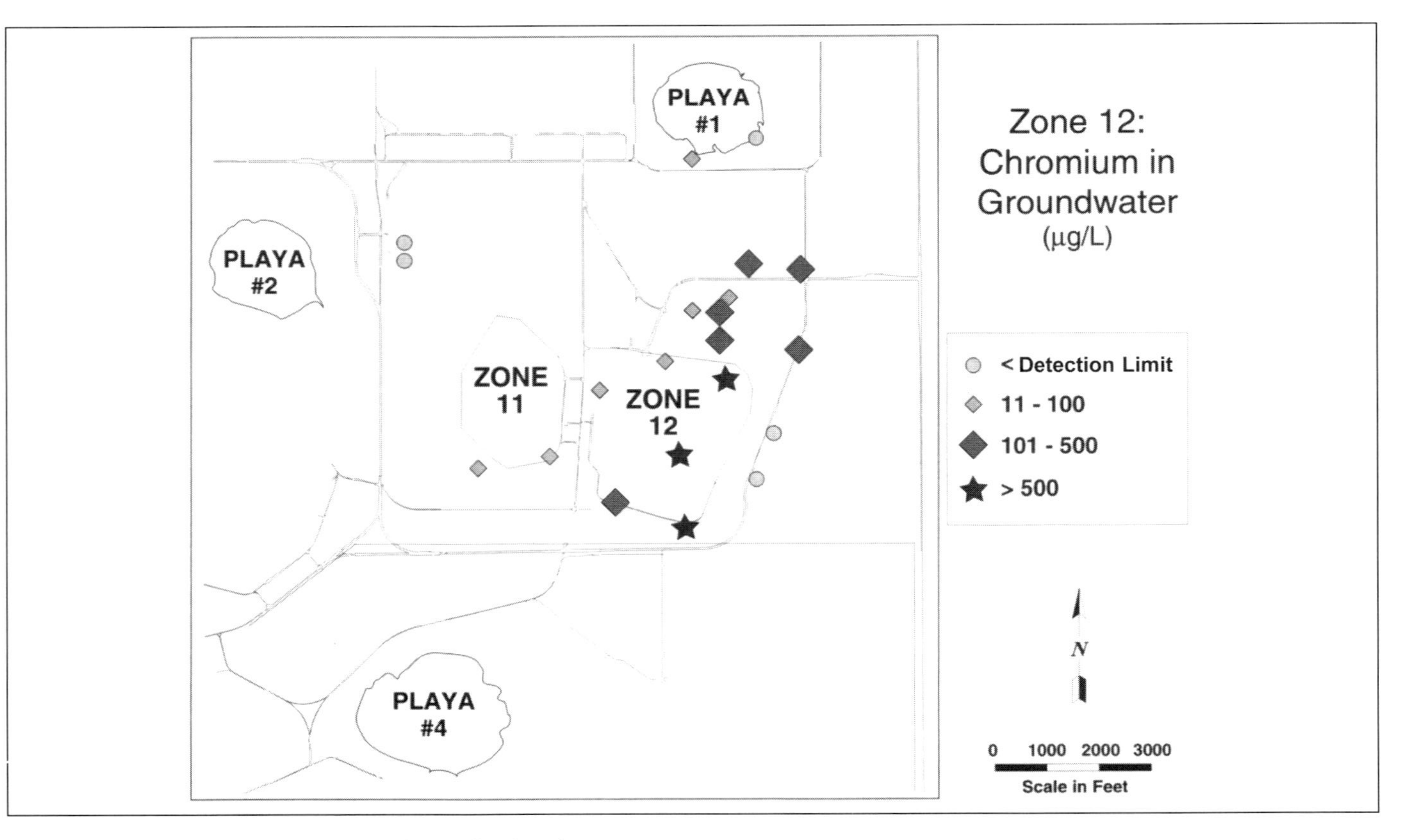

Figure 5a. Contaminant concentration distribution for chromium.

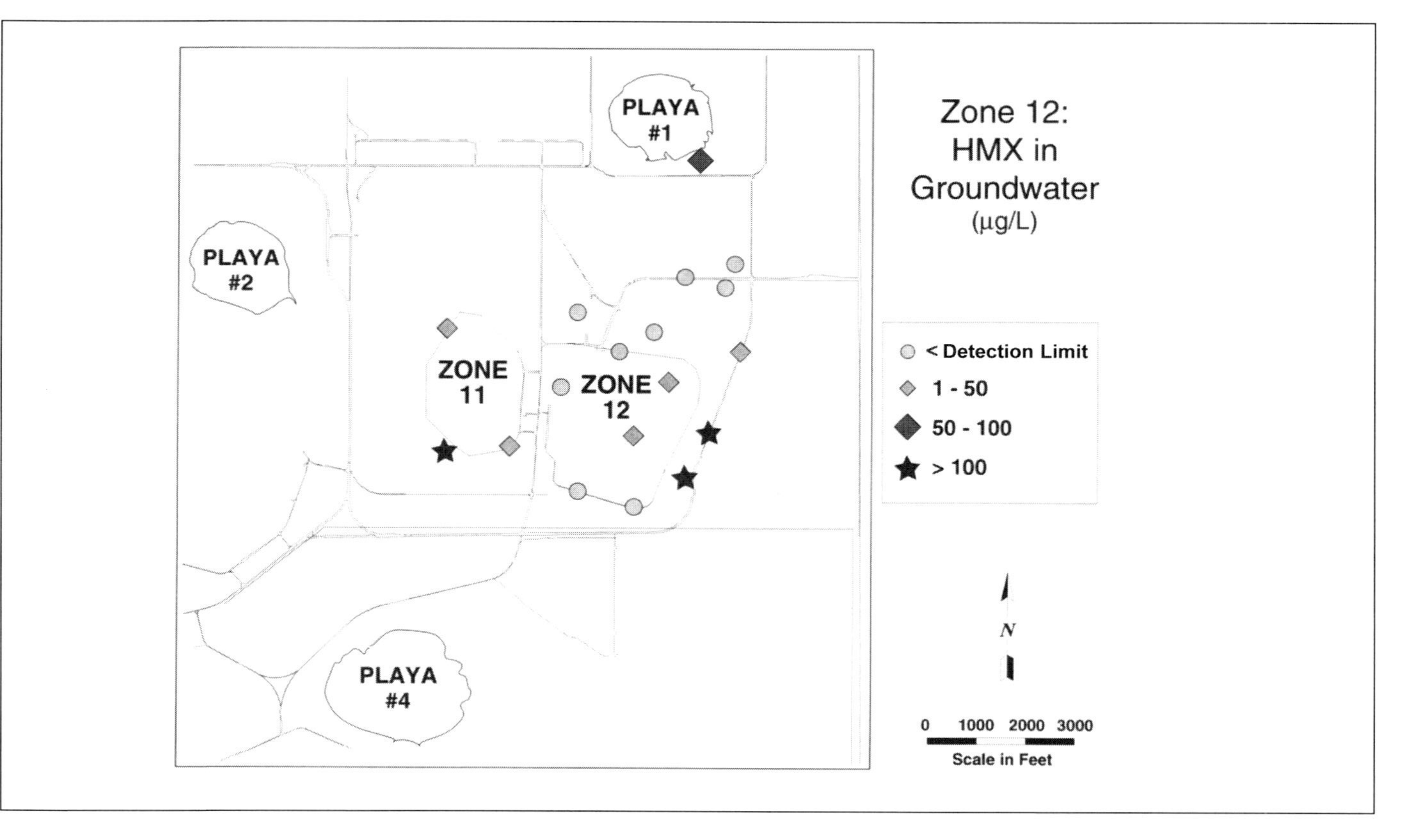

Figure 5b. Contaminant concentration distribution for HMX explosives.

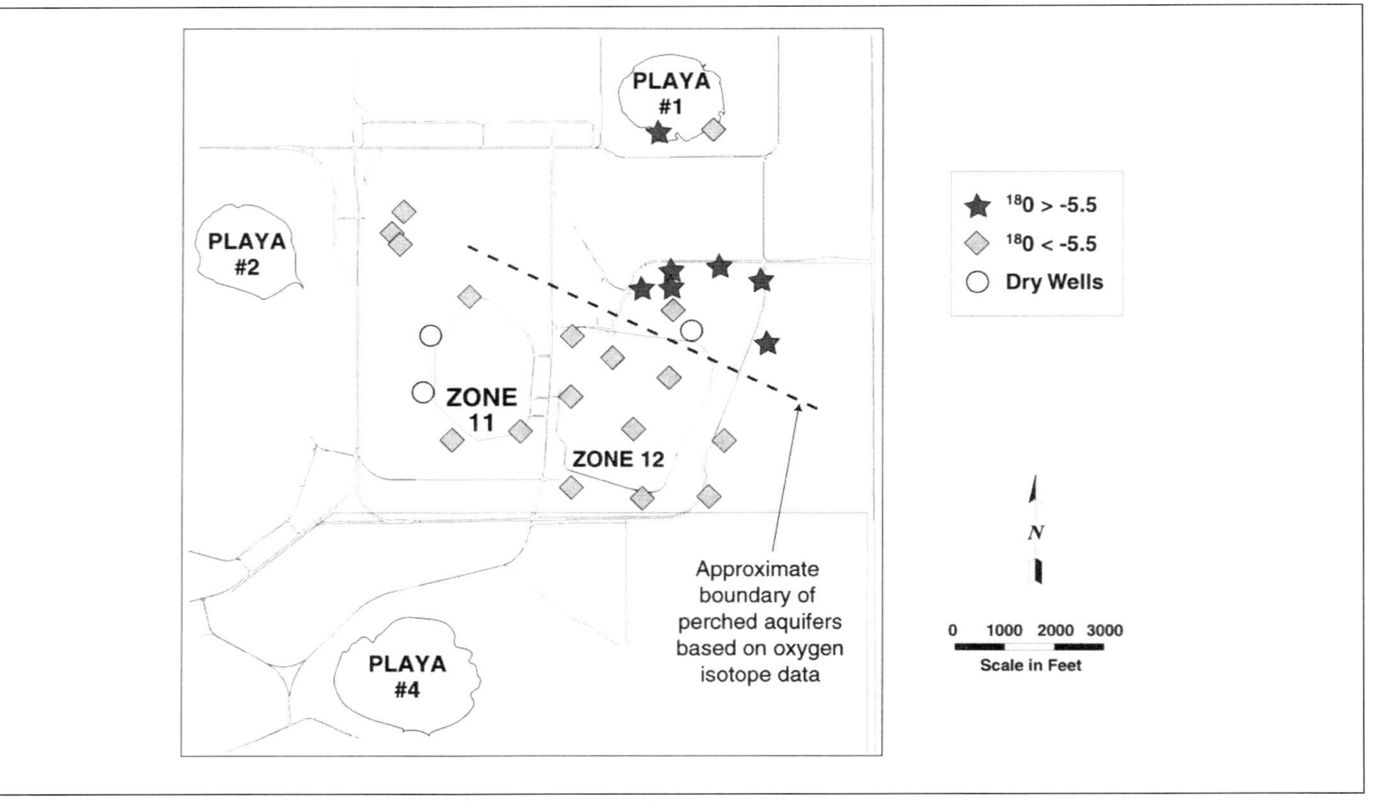

Figure 6a. Isotopic concentration distribution for $\delta^{18}O$ in per mil units ($^{0}/00$).

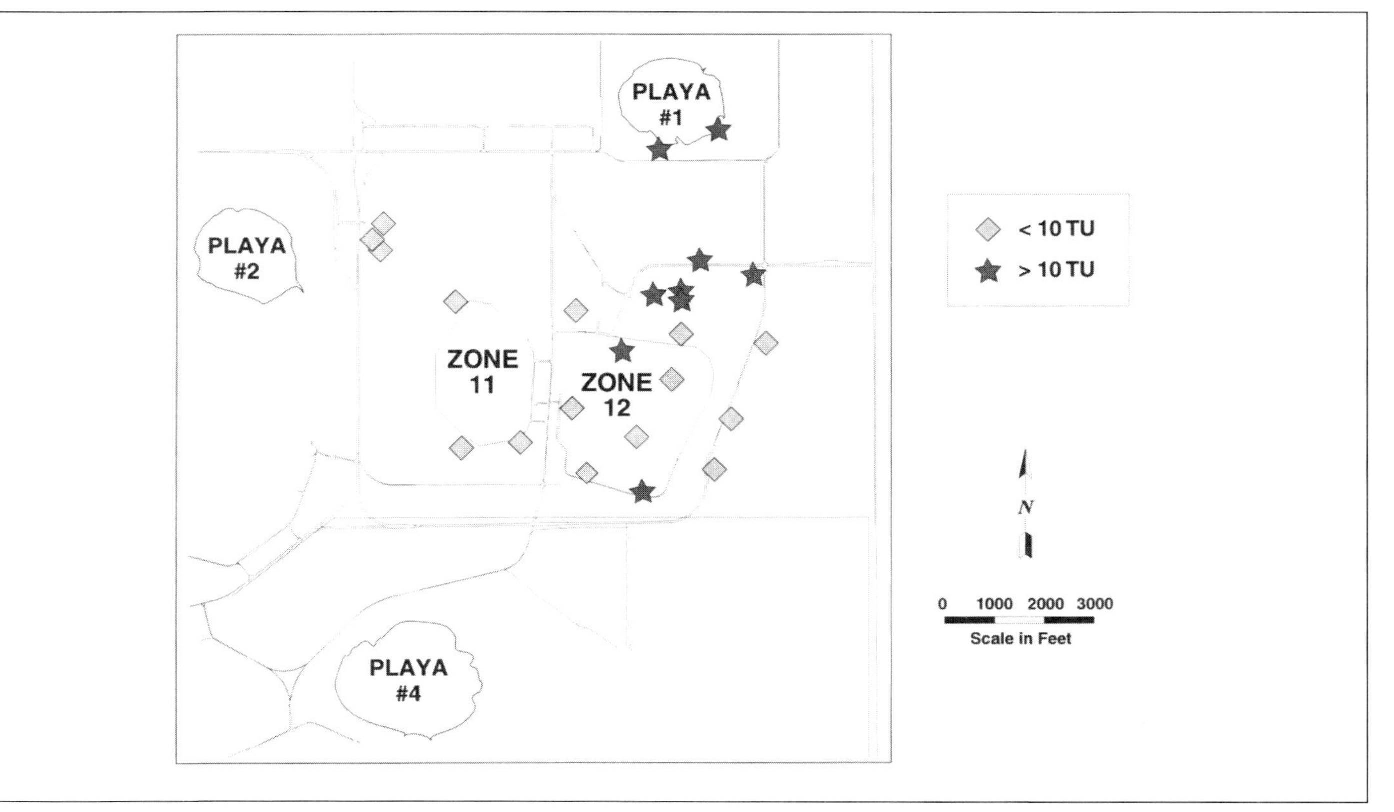

Figure 6b. Isotopic concentration distribution for tritium (^3H) in tritium units (TU).

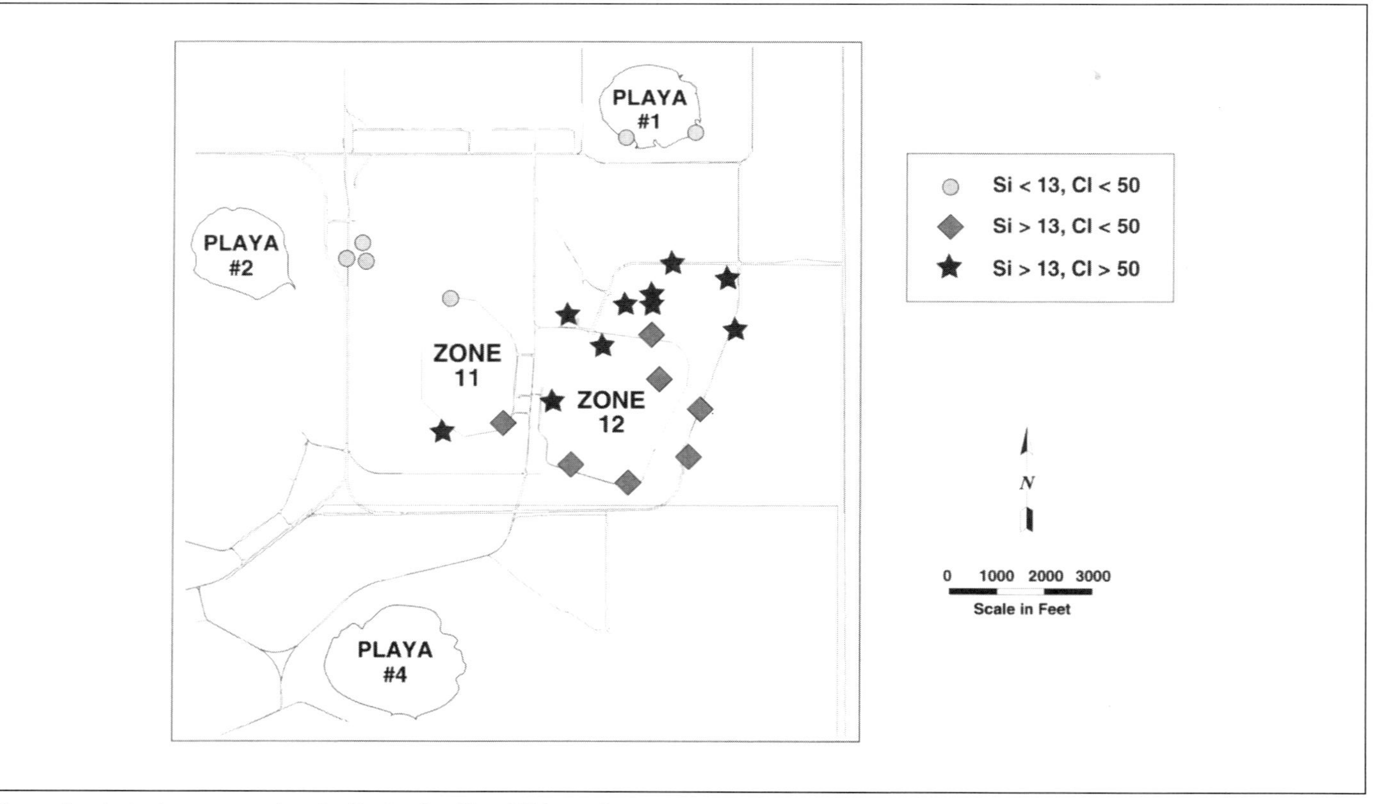

Figure 6c. Isotopic concentration distribution for Si and Cl in mg/L.

transition. Fourth, a gravel-filled paleochannel had been identified in the Zone 12 region using well logs. It was hypothesized that the ridges forming the channel could create the highs in the FGZ layer, and thus produced barriers isolating the perched aquifer zones (Figure 7).

Post-ESC CSM

The data suggested that contaminants primarily penetrated vertically through the Zone 12 vadose zone into the perched aquifer and then migrated along the paleochannel, where the axis and flow track from the northwest to the southeast (Figure 8). Thus, the projected contaminant migration pathway was directed to the southeast corner of the Pantex property where the channel was expected to continue off-site. To test this model, four soil borings (Figure 4) were drilled and cored to the FGZ: one each to confirm the southern ridge of the channel (Hole 1), the center of the channel (Hole 2), and outside the southern ridge of the channel (Hole 3). Hole 4, drilled on the southeast boundary of the site, was placed to confirm the expected southeast extension of the postulated channel. The drillings confirmed the existence of the channel, which explained the isolated perched aquifer zones, and predicted where the contaminant flow might extend off-site. The HydroPunch sampling at Hole 4 confirmed the existence of explosive contamination within the channel at the southeast boundary of the site.

CONCLUSION

The CSM that evolved during the ESC project resulted from the integration of primarily already-existing geochemical, geophysical, hydrological, and chemical data. A careful review of past well logs and a reprocessing of seismic data created a better understanding of the geology that was controlling the contaminant migration. Attention was then focused on where the soil borings should be placed to gain the maximum information. The next phase in the original characterization plan called for 13 additional wells to be placed around the boundaries of the plant. We now know that these wells would have shown no contamination, and would not have added any information to the CSM or to the understanding of the flow patterns. Confirmation of the highs in the FGZ surface, which equated to increased thickness, led to the conclusion that no leakage was occurring through this barrier. A careful lithological study of the soils in the FGZ supported this theory. Thus, no apparent pathway existed to the Ogallala aquifer under contaminated areas.

Chemical monitoring data had indicated higher contaminant levels within Zone 12. The contaminant distribution and the confirmation of saturated zones under the trenches provided evidence of vertical flow from the Zone 12 surface to the perched aquifer where the channel directed the flow off site. With all the data consistent with the new CSM, the characterization program ended and a remediation plan was

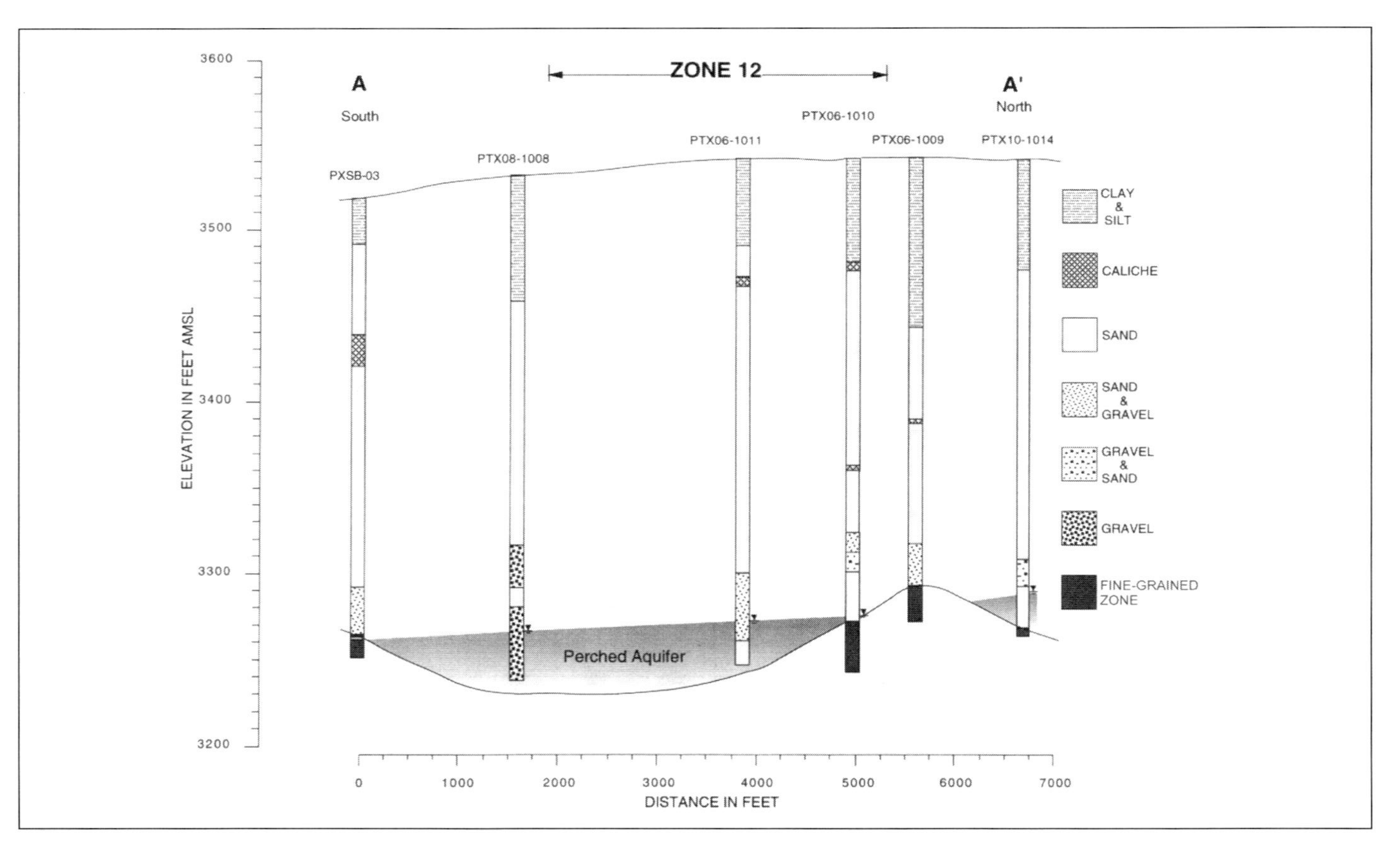

Figure 7. Cross section A-A′ (shown on Figure 4) depicting a gravel-filled channel below Zone 12, which creates isolated regions in the perched aquifer.

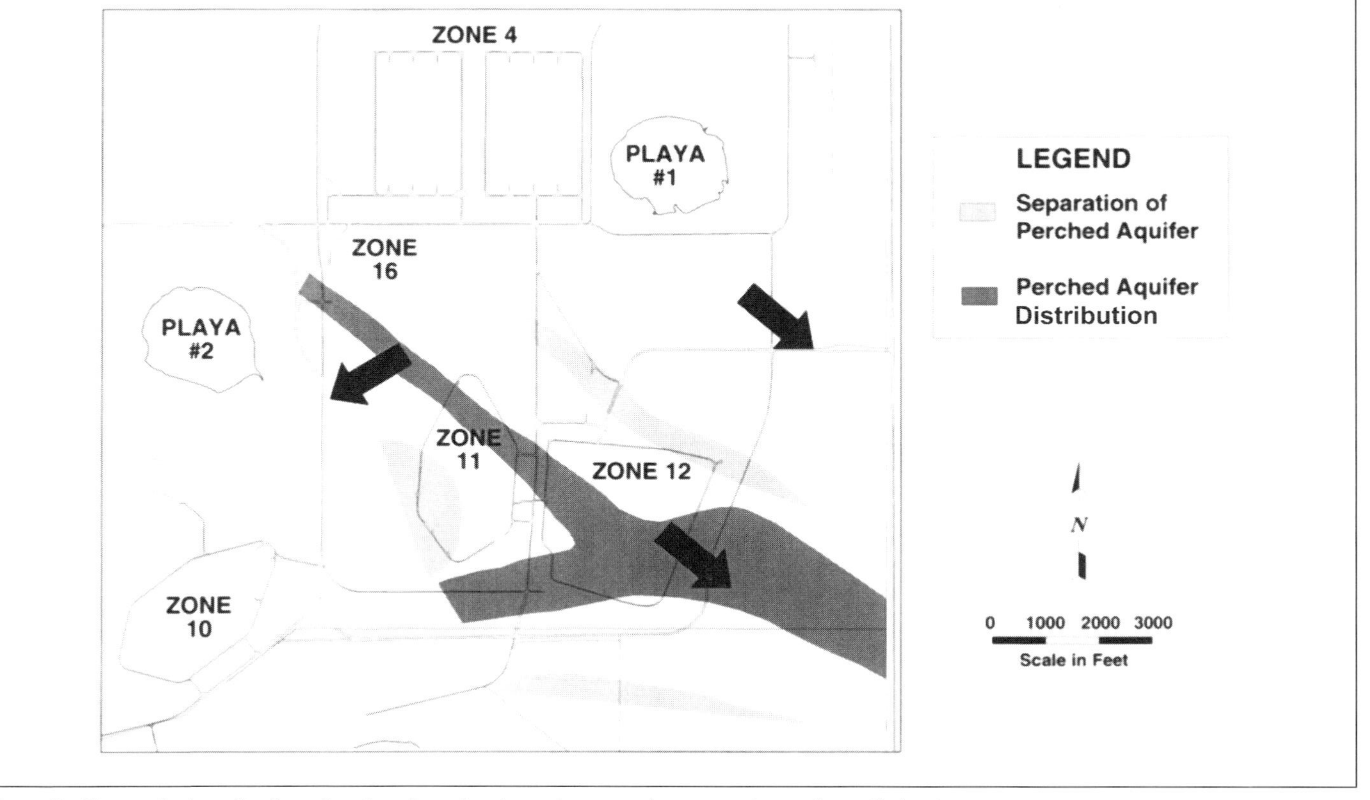

Figure 8. The perched aquifer flow direction along the channel transporting contaminants flow off-site.

developed to prevent migration farther off site via the paleochannel pathway. The original proposed remediation plan called for clay liners at the bottom of the playas, which we now know would not have prevented the primary contaminant flow. Finally, a cost study showed that the ESC program at Pantex reduced the schedule by 60 percent and the projected costs by 70 percent (Starke 1996).

REFERENCES

ASTM (1998), Standard practice for expedited site characterization of vadose zone and ground water contamination at hazardous waste contaminated sites, *D6235-98. West Conshohocken, Pennsylvania: American Society for Testing and Materials.*

Benson, R., A. Bevolo, and P. Beam (1998), *ESC: How It Differs From Current State of the Practice*, Proceedings of the Symposium on the Application of Geophysics to Environmental and Engineering Problems, 22-26 March, Chicago,IL, 531-540.

Burton, J.C. (1994), Expedited site characterization for remedial investigations at federal facilities. *FER III & WM II Conference and Exhibition*, 1407-1415.

Burton, J.C. and W.T. Meyer (1998), *QuickSite^{SM}, the Argonne expedited site characterization methodology,* Geotechnical Site Characterization, edited by Robertson and Mayne, Balkema, Rotterdam, 115-120.

Burton, J.C., J.L. Walker, T.V. Jennings, P.K. Aggarwal, B. Hastings, W.T. Meyer, C.M. Rose, and C.L. Rosignolo (1993), Expedited site characterization: A rapid cost-effective process for preremedial site characterization, *Superfund XIV*, 809-826.

Burton, J. C., Walker, J. L., Aggarwal, P. K., and Meyer, W. T. (1995), *Argonne's Expedited Site Characterization: An Integrated Approach to Cost-and Time-Effective Remedial Investigation*, Air & Waste Management Association 88th Annual Meeting, 18-23 June, San Antonio, TX, 27 pp.

Ferguson, D.J. (1995), *The successful application of Argonne's expedited site characterization to DOE Pantex groundwater investigations*, Oral presentation at Seventh National Technology Information Exchange (TIE) Workshop, Cincinnati, Ohio.

Gelb, S.B. and J.D. Wonder (1998), *ESC Demonstration: D-Area Oil Seepage Basin-Savannah River Site: A Case Study*, Proceedings of the Symposium on the Application of Geophysics to Environmental and Engineering Problems, 22-26 March, Chicago, IL 551-560.

Starke, T.P. (1996), *Cost effectiveness analysis of expedited site characterization at the DOE Pantex Plant*, LA-UR-96-2945, Los Alamos National Laboratory, Los Alamos, New Mexico.

COMPARISON OF VADOSE ZONE SOIL AND WATER ANALYTICAL DATA FOR CHARACTERIZATION OF EXPLOSIVES CONTAMINATION

Wilson S. Clayton, Ph.D., P.G., and **Peter Wirth, P.E.**

INTRODUCTION AND BACKGROUND

This case study presents data from a United States Department of Defense (DOD) Superfund site at which vadose zone characterization was critical to the determination of potential contaminant migration and the evaluation of remedial options. The data show the value of collecting pore water samples for vadose zone contaminant characterization. Conversely, the case study also illustrates the disutility of laboratory analysis of soil samples for vadose zone characterization. The data presented were collected as part of the Remedial Investigation (RI) at the Milan Army Ammunition Plant (MLAAP), Operable Unit Number 5, located in Milan, Tennessee.

Historical operations at the MLAAP resulted in the release of explosive compounds (nitrobodies) to the surface and subsurface at several areas of the site. One area, known as the Open Burning Ground (OBG), involved extensive soil and groundwater contamination resulting from disposal of ordnance waste materials. The OBG is approximately 144 acres of land that has been used continuously since 1942 for the destruction, burial, and disposal of reject munitions and explosive-contaminated wastes. Nitrobodies were released to the ground surface and shallow trenches as granular solids derived from the ordnance waste.

A widely used modern explosive that is commonly found at the OBG is Cyclotrimethylenetrinatramine (RDX). RDX is a moderately soluble (25 to 45 mg/L at 10 °C to 25 °C) nitroaromatic compound with an Environmental Protection Agency (EPA) drinking water Lifetime Health Advisory (HA) of 0.002 mg/L. RDX is present in groundwater above the HA of 0.002 mg/L over an area extending approximately 2 km down-gradient from the OBG. The distribution and transport mechanisms of RDX through the vadose zone from the ground surface to the groundwater are of great interest in order to characterize: (1) the locations of source areas, and (2) the potential pathways of contaminant migration, which are needed for the design of remedial alternatives for the vadose zone and groundwater.

SITE GEOLOGY AND HYDROGEOLOGY

Site geology and hydrogeology at the OBG are typical of the geographical area, which lies within the Mississippi embayment of the gulf coastal plain. The primary shallow hydrologic unit is the Tertiary Memphis Sand Aquifer, which consists primarily of sand with discontinuous silt and clay lenses. This unit extends from the ground surface to a depth of approximately 80 m. The Memphis Sand is underlain by the Flour Island formation, which is a clay and silt deposit approximately 15 to 20 m thick. Thin fluvial and alluvial deposits locally overlay the Memphis Sand. Infiltration rates at the site are approximately 0.000696 m/day (0.25 m/year). Depth to groundwater is approximately 30 m, and groundwater flows northwest at a gradient of approximately 0.015. The aquifer has a bulk hydraulic conductivity of about 30 m/day, and intervening finer-grained lenses have hydraulic conductivity as low as 0.01 m/day, or less.

SAMPLING AND ANALYSIS CONDUCTED

Contaminant sources within the OBG were defined based upon data gathered from aerial photographs, surface geophysics, interviews with current and former plant personnel, site inspection, and laboratory quantification of RDX concentrations in surface soil samples. Laboratory analysis of RDX concentrations in soil and water samples was accomplished using High Pressure Liquid Chromatography (HPLC) by EPA Method 8330. Within the OBG area, hundreds of soil samples were analyzed as part of a subsurface drilling program to define vadose zone and groundwater contamination. Water sampling was conducted from more than 40 groundwater monitoring wells and 20 vadose zone pressure-vacuum lysimeters.

COMPARISON OF SOIL SAMPLE AND SOIL-PORE WATER ANALYTICAL DATA

Table 1 shows dissolved RDX detected in vadose zone pore water samples collected from lysimeters, and total RDX detected in soil samples collected at the lysimeter location. Of the 20 soil samples analyzed, only three contained RDX above the method detection limit (MDL) of 0.587 mg/kg. However, 15 of the 20 lysimeter samples contained a detectable concentration of RDX, and 14 of these samples were above the HA of 0.002 mg/L.

Figure 1 shows that there is virtually no relationship between RDX concentrations in soil pore water collected using vacuum lysimeters, and paired soil samples taken during lysimeter installation. Only one set of paired soil-water analytical data is within the range of phase distribution depicted by the solid lines that are based on distribution coefficient (K_d) values of 0.1 and 1.0. The lack of correlation between paired soil and lysimeter data may be attributable to the method detection limit (MDL) of the RDX soil analysis (0.587 mg/kg), inconsistencies in analytical sub-sampling and extraction techniques, or heterogeneities, which may result in a different K_d value for each individual soil sample. Even in the absence of sorption ($K_d = 0$), the concentration of dissolved RDX in pore water in excess of 4 mg/L could result in detection of RDX above the MDL by soil analysis. However, for several sampling locations, RDX was not detected in soil samples, while lysimeter samples contained

TABLE 1 **Comparison of paired vadose zone soil and pore water RDX analytical results.**

mg/kg Soil Sample	mg/L RDX in in Soil-Pore Water Sample
<MDL	N/D
<MDL	0.00163
<MDL	0.00294
<MDL	N/D
2.04	0.74
<MDL	0.566
<MDL	0.421
<MDL	2.06
<MDL	1.54
<MDL	N/D
<MDL	N/D
<MDL	0.0167
<MDL	0.00336
<MDL	N/D
4.04	0.0145
<MDL	0.0228
<MDL	6.56
2.27	8.4
<MDL	7.39
<MDL	2.99

Note: MDL is 0.587 mg/kg
N/D denotes not detected

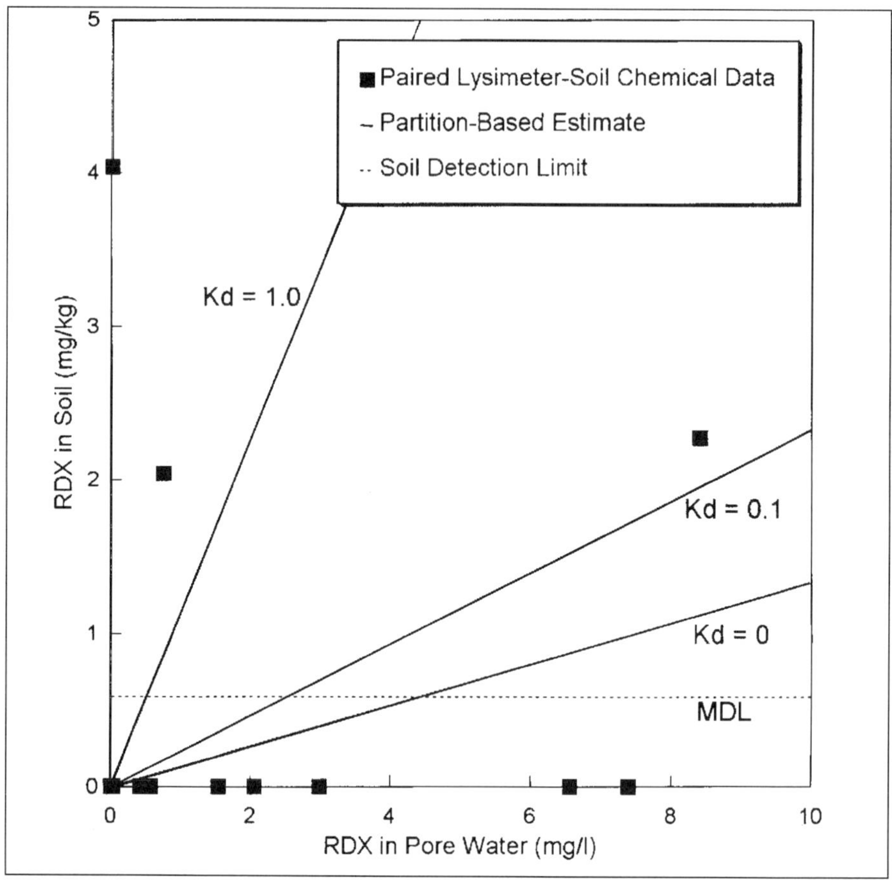

Figure 1. Comparison of analytical results from paired lysimeter and soil samples to relationships estimated based on equilibrium partitioning shows that RDX soil concentrations are a poor predictor of vadose zone pore water contamination.

elevated concentrations of dissolved RDX. This may reflect technical difficulties with RDX extraction from soil samples as part of the laboratory analytical procedures.

COMPARISON OF SOIL-PORE WATER DATA TO CONTAMINANT SOURCE AREAS AND GROUNDWATER IMPACTS

While the soil sample analytical data showed little correlation to the vacuum lysimeter data, the lysimeter data showed excellent correlation to contaminant source areas and groundwater impacts (Figure 2). In fact, the lysimeter data supplemented groundwater contamination data and historical data on disposal activities and soil sampling, which are not presented in this case study.

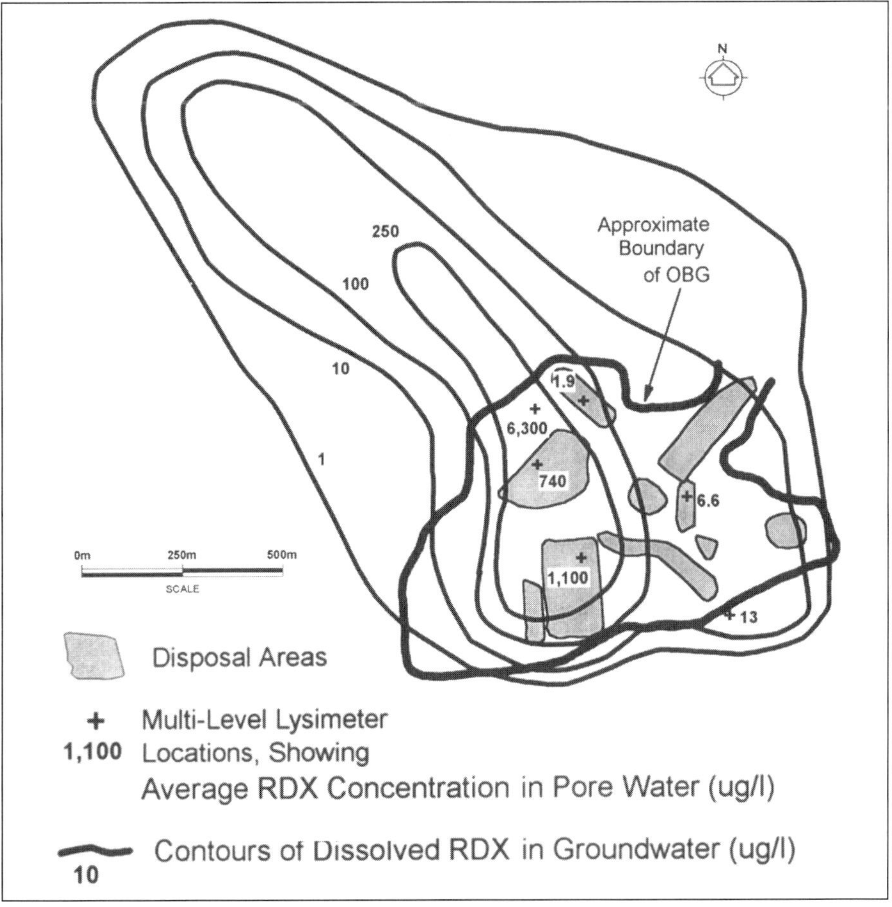

Figure 2. Generalized maps of Open Burning Ground (OBG), including known ordnance disposal areas, locations of multi-level lysimeters, average dissolved RDX in pore water samples collected from lysimeters, and contours of dissolved RDX in groundwater. Note the strong correlation between lysimeter results and the areal extent of groundwater contamination.

CONCLUSIONS

Soil analytical method detection limits are commonly evaluated against soil treatment standards in order to determine whether sufficient laboratory analytical detection is achievable. Since soil treatment standards are generally related to direct human exposure to soils, this is a useful approach to consider risks related to direct soil exposures. In the case study presented, analysis of vadose zone pore water samples collected using pressure-vacuum lysimeters was shown to be highly effective in characterizing the distribution and magnitude of vadose zone contamination, and in identifying and ranking source areas causing groundwater contamination. The lysimeter data were also used to calibrate a variably saturated flow and transport model, which was utilized to simulate vadose zone-groundwater interactions.

REFERENCES

Fluor Daniel, Inc. *Final Remedial Investigation Report, Milan Army Ammunition Plant, Southern Study Area (Operable Unit No. 5)*, Golden, Colorado: Fluor Daniel Inc.,1998.

CONE PENETROMETER (CPT)-BASED SOIL MOISTURE PROBES

Joe Rossabi

One of the most important characteristics of the vadose zone is the amount of water in the predominately two-fluid phase system. Figures 1 and 2 compare soil moisture data from two locations at Savannah River Site (SRS), separated by a distance of approximately 10 miles with baseline laboratory results. Two cone penetration data sets less than 10 feet apart are shown in each plot. Between the sets is a conventionally drilled hole with multiple Shelby tube samples. The excellent match of the CPT-based probe data to the standard methods is a clear indication of the validity of the new technique. Due to the higher vertical resolution, the additional structure in the CPT method is probably accurate and a representation of the degree of heterogeneity in the system. Figure 1 is a plot of data taken at the TNX area of SRS, and Figure 2 is a plot of data acquired in the M area (Integrated Demonstration Site) of SRS. Details of the test and a complete data set are provided in the excellent report compiled by personnel from Argonne National Laboratory, entitled "Evaluation Report: Study of Three Soil Moisture Probes with Laboratory Sample Results" (1997).

Two other CPT-based methods, the GeoVis video microscope for the CPT (US Navy SPAWAR Systems Center) and the Cone Permeameter (Science and Engineering Associates, Inc.), provide information that supports and complements the soil moisture data obtained by these probes. The Cone Permeameter gives a local measurement of the gas permeability of the sediments in the vadose zone. The GeoVis provides continuous video information at fine grain size resolution (10 micron). The degree of saturation of sediments in front of the window can be estimated from this visual record.

REFERENCES

Applied Geosciences and Environmental Management Section, Environmental Research Division. Evaluation Report: Study of Three Soil Moisture Probes with Laboratory Sample Results. Internal Document, Argonne National Laboratory, Argonne, IL (1997).

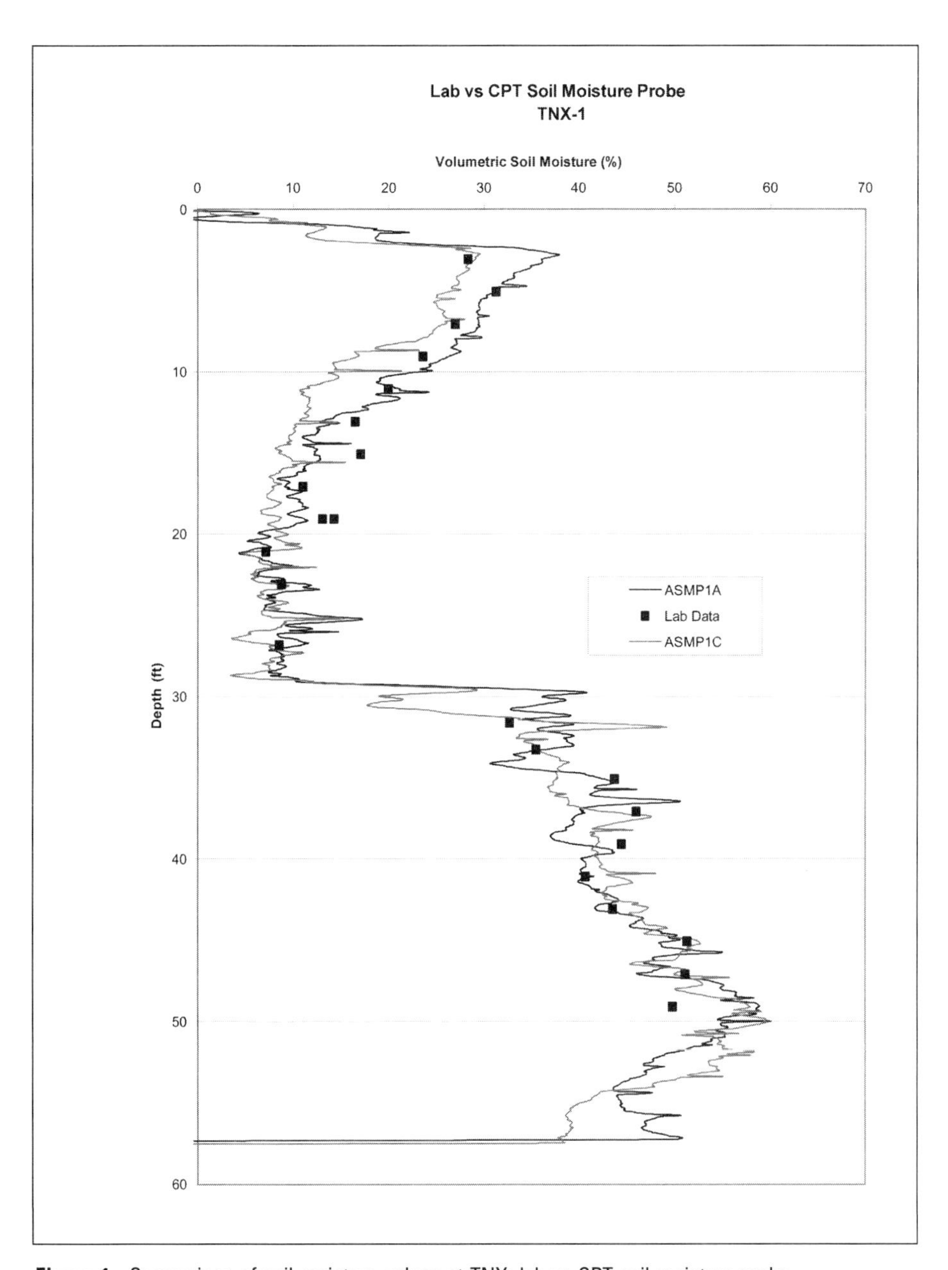

Figure 1. Comparison of soil moisture values at TNX: lab vs CPT soil moisture probe.

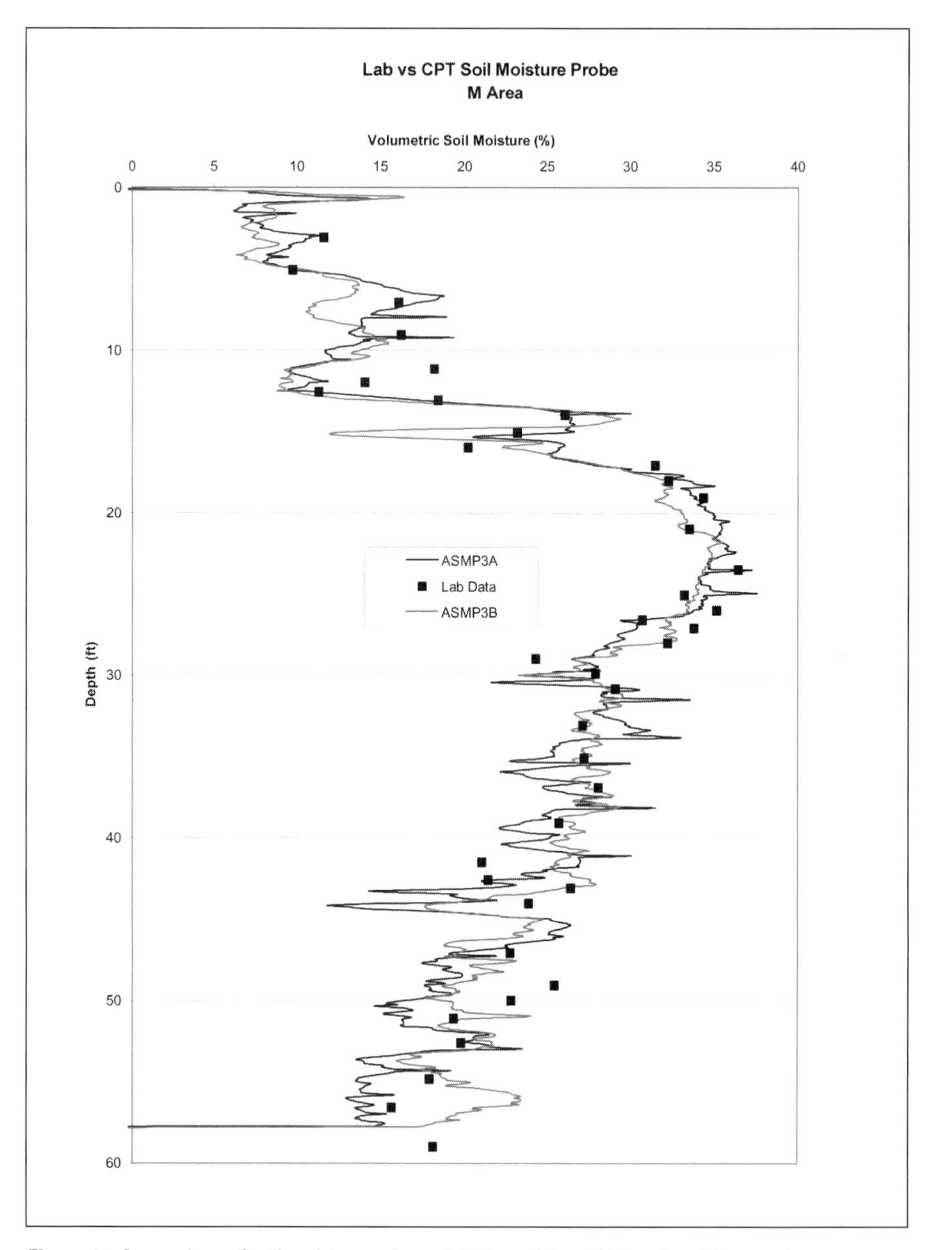

Figure 2. Comparison of soil moisture values at M Area: lab vs CPT soil moisture probe

CONE PENETROMETER-BASED RAMAN SPECTROSCOPY FOR NAPL CHARACTERIZATION IN THE VADOSE ZONE

J. Rossabi, B. D. Riha, J. Haas, C. A. Eddy-Dilek, A. Lustig, M. Carrabba, K. Hyde, and **J. Belo**

ABSTRACT

Cone penetrometer-based Raman spectroscopy was used to identify separate phase tetrachloroethylene (PCE) contamination in the vadose zone at two locations during field tests conducted at the Savannah River Site (SRS). Clear, characteristic Raman spectral peaks for PCE were observed and, because of the uniqueness of Raman spectrum for a given compound, are compelling evidence that this method is a viable dense nonaqueous phase liquid (DNAPL) characterization technique. In addition, the spectral data indications of DNAPL correlate with soil concentration data collected in the same zones. The Raman spectroscopic activities conducted in these tests represent the first *in situ* direct measurement of DNAPL in the subsurface.

Based on data from this field work, the Raman technique may require a threshold of DNAPL to provide an adequate optical cross-section for spectroscopic response. Similar to Cone Penetrometer Technology (CPT)-based, laser-induced fluorescence techniques, the probability of detecting DNAPL using this technique depends on the probability of droplets of DNAPL coming into contact with the optical window. It is likely that this technique requires a separate phase liquid rather than an aqueous solution for adequate response.

INTRODUCTION

RAMAN SPECTROSCOPY SENSORS

The Raman effect occurs when the interaction of light and matter results in the addition or subtraction of energy quanta in the scattered light (inelastic scattering). The energy shifts in the scattered light are correlated to the vibrational modes of the particular compound and constitute the Raman spectrum for that compound. The vibrational modes of the compound depend on the elemental constituents, energy state, and steric configuration of the molecule. The number of modes and associated energy of these modes is unique to the molecule, and therefore produces a unique Raman spectrum for the compound (Colthup *et al.* 1990).

Raman spectra have been used to both identify unknown compounds and to probe their molecular state (Carrabba *et al.* 1990). Because of the number of vibrational modes possible for a given compound, Raman is most easily observed when probed with a monochromatic light source such as a laser.

Raman is similar to fluorescence spectroscopy in that both techniques result in light that emanates from the compound in all directions and that is wavelength-shifted from the original source light. Fluorescence spectra, however, are a result of an electronic transition caused by the quantum absorption and subsequent release of energy (in the form of light) of the electron energy state of the compound. Raman spectroscopy does not involve an electronic energy transition and is inherently weaker in response than the fluorescent effect (Carrabba *et al.* 1992). Also, the fluorescence peaks are generally much broader with respect to wavelength than Raman peaks and can therefore obscure the Raman spectrum. As a result, it is important to minimize the fluorescence signal when performing Raman spectroscopy. Generally, this involves changing the wavelength of the monochromatic probe light (usually to a longer, lower-energy wavelength).

METHOD

A Raman spectroscopy system built by EIC, Inc., for subsurface cone penetrometer investigations, was used for this work (Carrabba *et al.* 1998). The system consists of an infrared laser source and Echelle spectrometer located within the Department of Energy's Site Characterization and Analysis Penetrometer System (SCAPS) truck. The source and detector are connected by fiber optics through the cone rods to an optics assembly located in the cone near the tip and sensor ends of the penetrometer. The optical assembly includes several lenses and filters designed to optimize light introduction to the formation and recovery of the Raman spectral signal. A sapphire window mounted to the cone rod is the interface between the formation and the sensor assembly. Raman spectral samples were generally taken every 3 ft as the cone penetrometer was pushed down. In sediments known from previous soil sampling and analysis to be likely to contain DNAPL, Raman spectral samples were taken every 0.5 ft to give detailed information about the depth and location of DNAPL. Generally, these zones correlated to depths of fine-grained sediment deposition in the vadose zone, which can be detected using the standard cone penetrometer sensor suite (tip pressure, sleeve friction, and pore pressure). The scans were integrated for 10 seconds as a rule, and longer (60 seconds or 120 seconds) only if there were suspicion of DNAPL presence due to spectral or geologic evidence. All spectra presented in this report were collected with a 10-second integration time unless otherwise marked on the figures. Background intensity varied slightly between pushes as a result of differences in geology as well as the examination of and the minor adjustments to the optical window, or train, between pushes. The spectral data are printed as relative units with respect to the abscissa, but correlate to the intensity (the number of photons received by the detector) at a specific wavelength. Specific DNAPL compounds such as tetrachloroethylene (PCE) are identified by their unique spectral signature. Because of the probe wavelength (infrared laser diode at 785 nm), background fluorescence was expected to be low unless fluorescent compounds were present. Pure PCE essentially does not fluoresce at the wavelength of investigation, but some compounds (synthetic oils or natural organics) that are soluble in DNAPL may fluoresce. It is possible that high spectral fluorescence is indicative of DNAPL-containing, dissolved, fluorescing compounds.

To assess the capability of the Raman technology in identifying DNAPL, soils and sediments were sampled in the same locations as the CPT/Raman pushes. The PCE and trichloroethylene (TCE) concentrations were determined using laboratory analyses from soil and sediment samples taken during drilling. In clay layers known to contain DNAPL in the vadose zone, samples were taken every 0.5 ft to give the same detailed information about the distribution of DNAPL concentration as acquired by the Raman probe.

RESULTS AND DISCUSSION

The first location used to evaluate the capability of Raman technology was adjacent to the A/M Area Seepage Basin. This was previously a waste disposal basin for caustic waste and spent machine oils, machine lubricants, and solvents. Therefore, background fluorescence was expected to reflect some of these impurities. Dense contaminants from the 20-ft deep basin accumulated below the lighter aqueous phase in the basin and eventually made their way through the vadose zone and down below the water table. The site geology determined the vertical and lateral gravity-driven contaminant transport. The cone penetrometer test probe provided data.

The sleeve friction, tip pressure, pore pressure, and conductivity are shown in Figure 1. These data can be used to evaluate the lithological profile of the vadose zone. From the surface to a depth of 24 ft, the data indicate a fine-grain or clayey zone. A clean sand (probably of medium to coarse-grain size, given the low conductivity and unchanging pore pressure response) follows from 24 ft to 40 ft. From 40 ft to 50 ft, some slightly finer-grain sediments are evident followed by 10 ft of coarse-grain sediment (50 to 60 ft). At 60 ft and 67 ft, two fine-grain zones (probably clays) appear. From the conductivity plots, only the 67-ft clay is relatively moist. These clay zones are expected to retain contaminant. Below the fine-grain zone at 67 ft, the profile indicates predominantly coarse-grained sand to a depth of 99 ft. There are a few thin zones of finer-grain sediments, most notably at 84 and 87 ft. At 99 ft, a 2-ft thick wet clay zone is evident that is likely to contain DNAPL. A multi-layered (16 in. to 3 ft each layer) clay and sand zone was encountered between the depth of 100 ft and the water table at a depth of 130 ft. The lithology inferred from this CPT plot is representative for all of the pushes adjacent to the M basin. Depths of the various zones shift slightly between CPT pushes, but the general patterns remain the same within the area around MSB-3D. These data inferring the geology of the area are critical in evaluating and focusing characterization efforts.

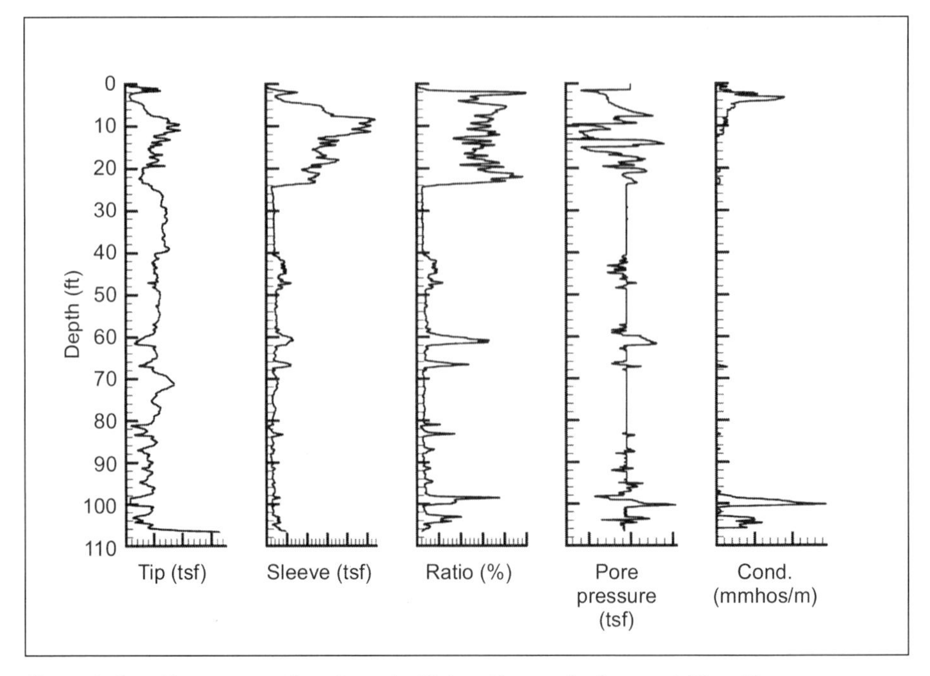

Figure 1. Cone Penetrometer Data From the Vadose Zone at the Savannah River Site.

Knowing the behavior of contaminants when encountering these different types of material allows the optimization of characterization resources. For example, fine-grain zones that are not completely water-saturated tend to absorb and retain contaminants by capillary forces. Therefore, the logical place to look for contaminants in the vadose zone is within these layers. Raman spectra should be acquired with fine vertical depth resolution in the vicinity of these fine-grain zones. Vertical data resolution in the coarse-grain zones can be sparser without loss of important contaminant information.

Initial spectra of spectral grade PCE and DNAPL collected from well MSB-3D were acquired by holding samples against the optical window of the probe before any work below the surface was performed. These spectra are provided in Figures 2 and 3, respectively, and were acquired using a laser excitation wavelength of 785 nm. The Raman spectral range shown is from 200 cm^{-1} to 600 cm^{-1} to capture several modes characteristic of PCE. Specifically, these peaks occur at 236 cm^{-1} and 448 cm^{-1}, with smaller peaks at 345 cm^{-1} and 513 cm^{-1}. There are also some characteristic peaks occurring further from the excitation line, but they are weaker than the two principal peaks. The spectrum acquired from the SRS DNAPL sample (well

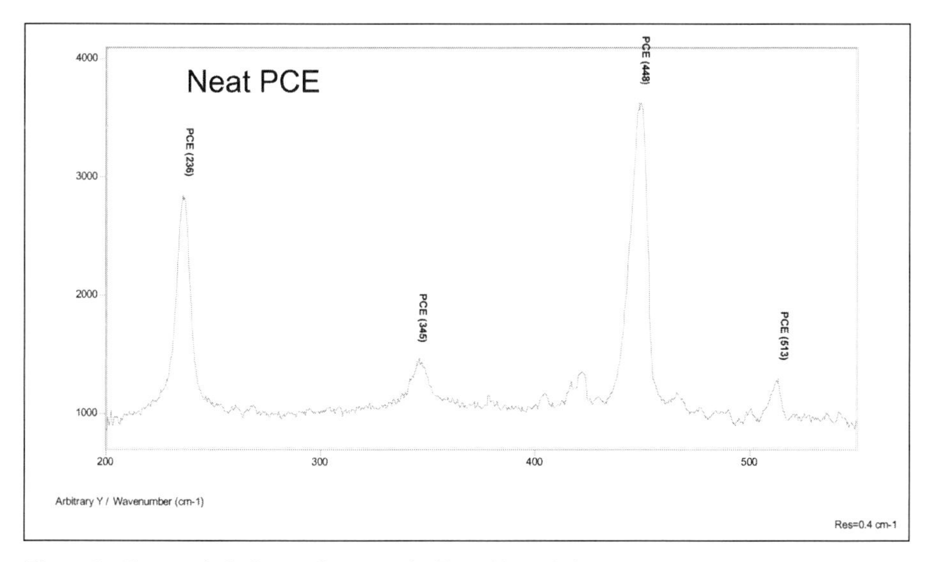

Figure 2. Characteristic Raman Spectrum for Tetrachloroethylene.

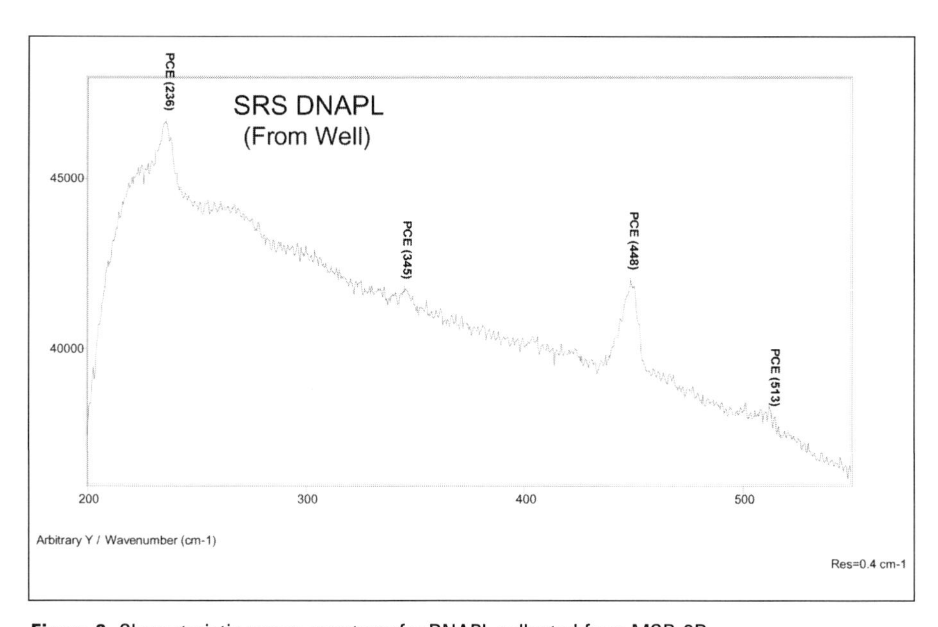

Figure 3. Characteristic raman spectrum for DNAPL collected from MSB-3D.

MSB-3D) does show the two strong peaks of PCE. However, it also indicates a high broad background signal characteristic of fluorescence emission from the sample. This is probably because of the presence of co-constituents of PCE in the DNAPL sample.

For all pushes, spectra were taken every 3 ft through the coarse layers, and every 0.5 ft through the clay layers. Raman spectral data collected from two consecutive pushes in February positively identified PCE located at depths of 100.2, 101.5, and 102.0 ft below ground surface, which corresponds to a clayey zone indicated by CPT geophysical data for that push. Fluorescent background levels were found to be much higher when DNAPL was present as compared to average background fluorescence levels for uncontaminated soils. Figure 4 illustrates the large difference in fluorescence background over small vertical distances, as well as the presence or absence of characteristic Raman peaks from the data collected during the first push (RMA-01). It is clear that PCE is present at depths of 101.5 ft and 102 ft from the Raman spectra, but the fluorescence data strongly infer the presence of solvent at 100.9 ft, as well. The fluorescence technique is inherently more sensitive than Raman as long as the contaminant fluoresces to a significant degree.

The effect of data collected during penetration or removal of the cone rods is shown in Figure 5. In this figure, the Raman spectra at the same depth indicate the presence of PCE upon penetration but not upon removal of the rods. In Figure 5, however, the background fluorescence of both spectra is comparable, implying that PCE

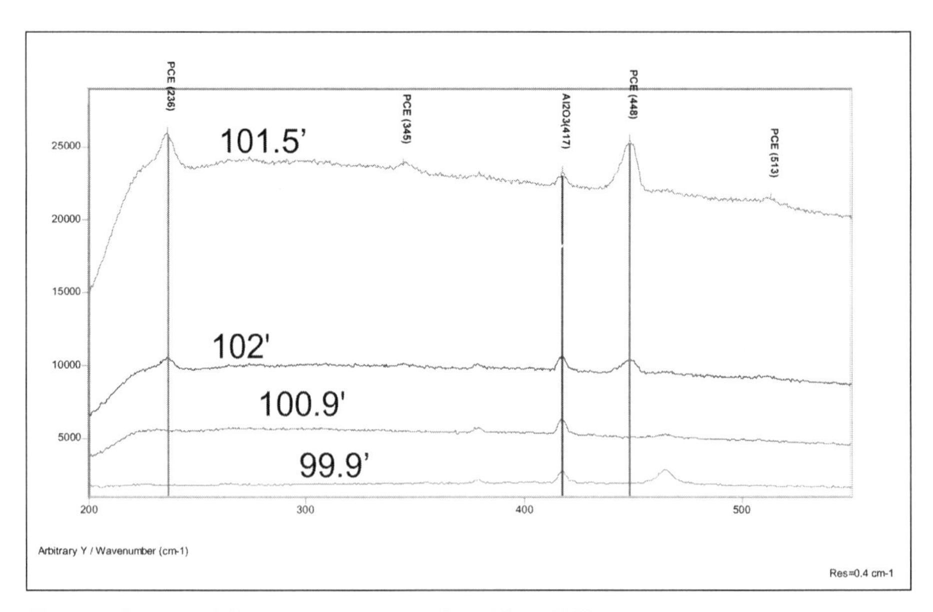

Figure 4. Raman and fluorescence spectra collected from RMA-01.

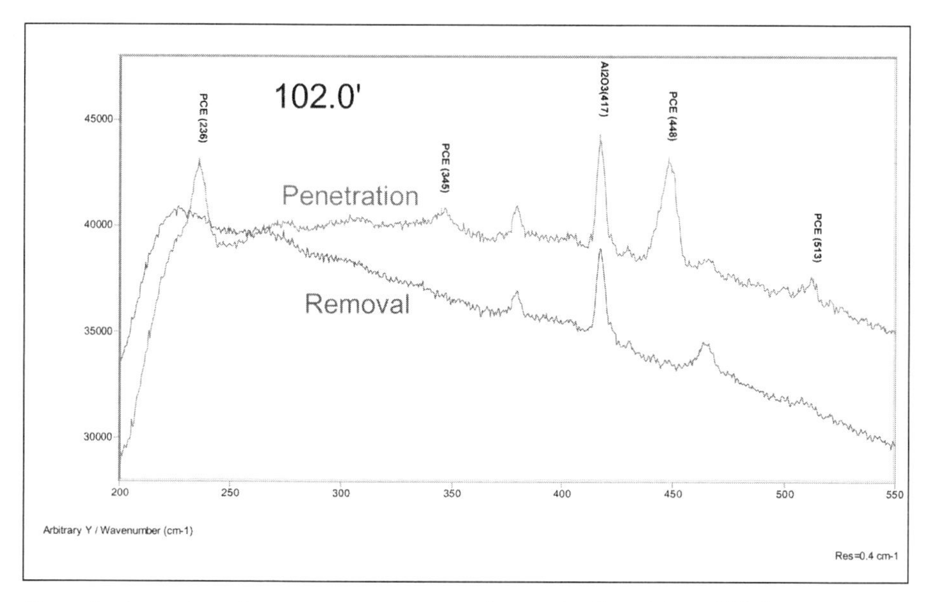

Figure 5. Comparison of spectra on penetration and removal collected from RMA-01.

is indeed present at the depth both during penetration and removal. These observations may be explained in several ways. Raman spectroscopy may require a threshold amount of DNAPL (either of a particular thickness or covering a certain percentage of the optical window) to produce the observable peaks characteristic of DNAPL. Fluorescence probably requires a much lower threshold to respond. On penetration, it is conceivable that a relatively substantial drop (still less than 1 cm in diameter) of DNAPL was squeezed out of the pore throats as the cone passed through and compressed the formation. On retrieval of the rods, the borehole at that depth might have been wallowed by wander in the penetration path of the rods, or the drop may have moved or volatilized below the Raman threshold of response. In either case, this observed behavior supports the conceptual model of small, dispersed blobs of DNAPL rather than pools at SRS.

Sediment samples taken from depths of 97 to 104 ft within a 5-ft radius of the CPT pushes showed high concentrations of both PCE and TCE. The sediment samples were collected by conventional hollow-stem auger drilling with split-spoon sampling. Figure 6 depicts the agreement between the two methods from the M Area Seepage Basin. PCE and TCE soil concentrations (μg/g) are plotted against depth using small, solid shapes. The Raman data are plotted with large open symbols at arbitrary values of 2,000 for PCE and 3,000 for TCE, which simply indicate presence or absence of PCE or TCE as indicated by spectral peaks. The cross symbols indicate

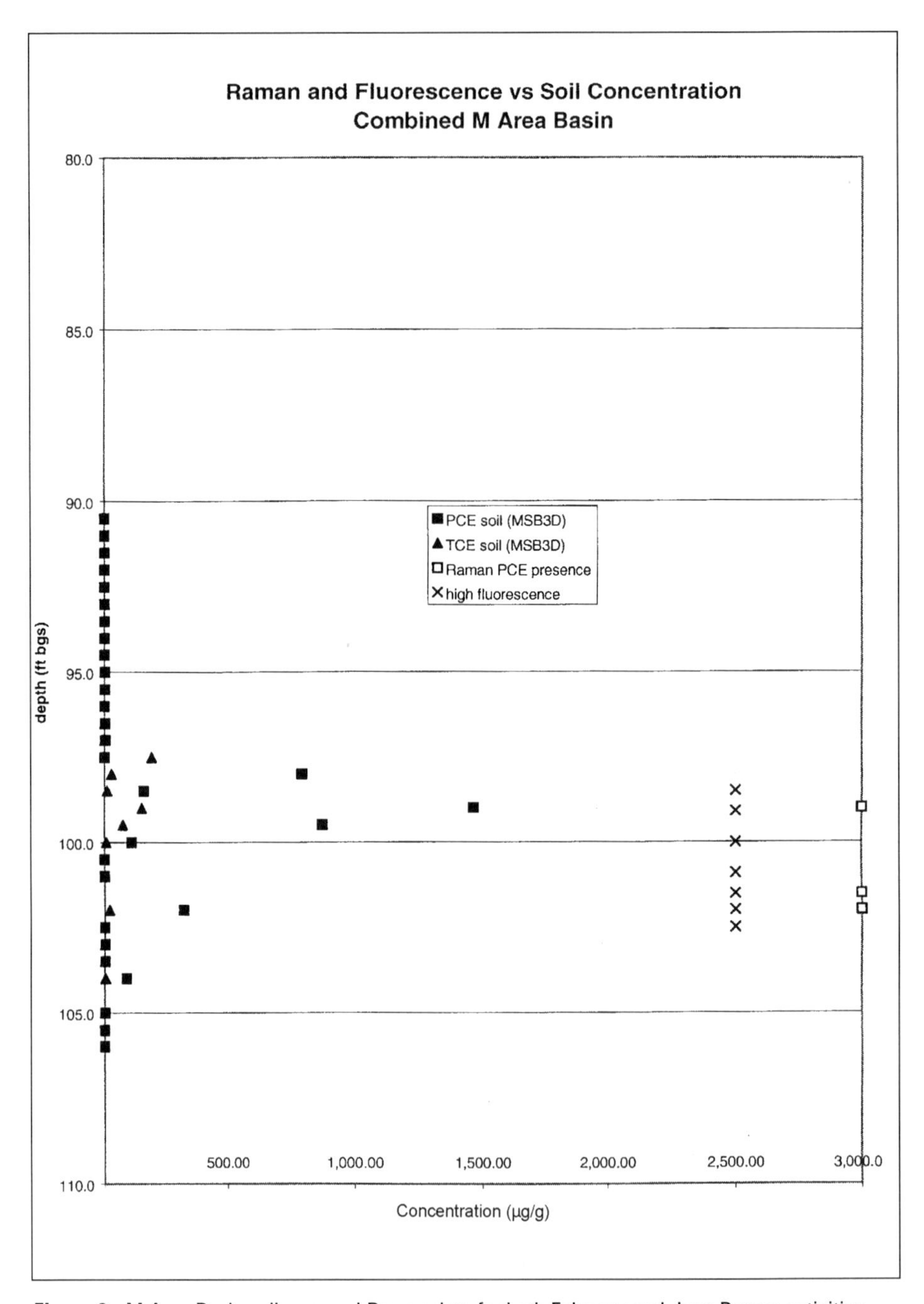

Figure 6. M Area Basin soil core and Raman data for both February and June Raman activities.

depths that were characterized by high fluorescence in the background of the Raman spectrum.

CPT-based optical spectroscopy was also performed at the 321 M Solvent Storage Tank. At this site, a 1,200 gallon spill of unused solvent occurred in 1985. Because this was a clean solvent spill, with no reported release of co-constituents, the fluorescence signal was expected to be weak at this site. Figure 7 shows some of the Raman spectra collected (RMA-5). In this figure, clear Raman peaks of PCE are evident at 25.6 ft and 26.6 ft. As expected, the background fluorescence is relatively low in these plots. The spectra at 24.6 ft and 27.6 ft show no PCE Raman peaks, and in fact, exhibit slightly higher fluorescence than the spectra with Raman hits. Upon examination of all of the spectra from this push, this slightly higher fluorescence does not appear to be significant at this time.

The 321 M CPT lithology plots are shown in Figure 8. All pushes conducted at the 321 M Solvent Storage Tank Area required hand-augering to a depth of 6 ft before pushing, to avoid cutting cables, pipes, or other underground interference. In general, the upper 30 ft at this site consist of clayey materials. Variations from this pattern occur near man-made structures where backfilled and compacted sand was used, and at occasional heterogeneities. For example, a clayey sand layer is seen in this figure between 20 and 23 ft. From 30 ft to 43 ft there is mostly clayey sand with small clean sand intervals at 33, 35, and 38 ft, and a 1-ft clay interval at 34 ft. From 43 ft to 46 ft is a layer of clean sand. Upon examination of the CPT lithology plots in

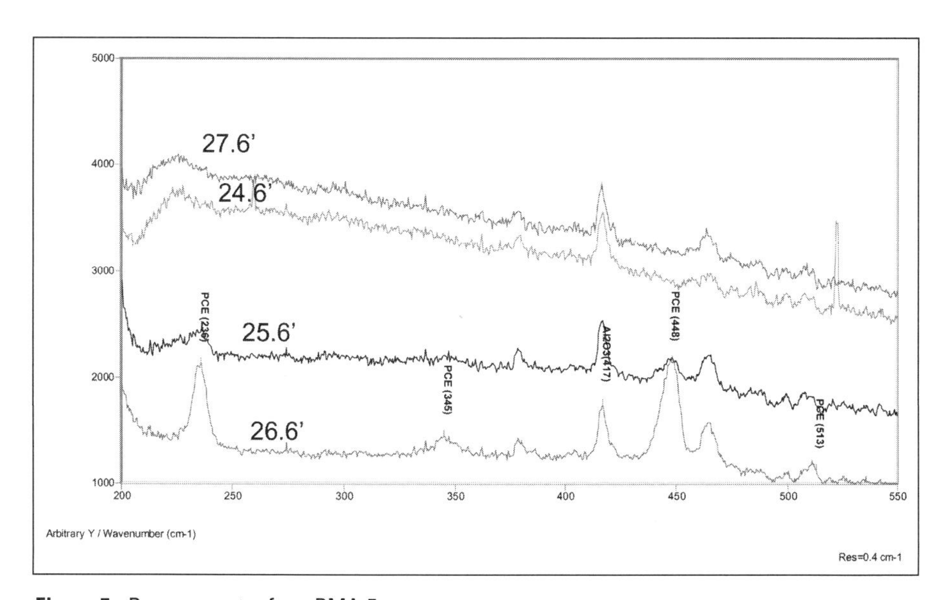

Figure 7. Raman spectra from RMA-5.

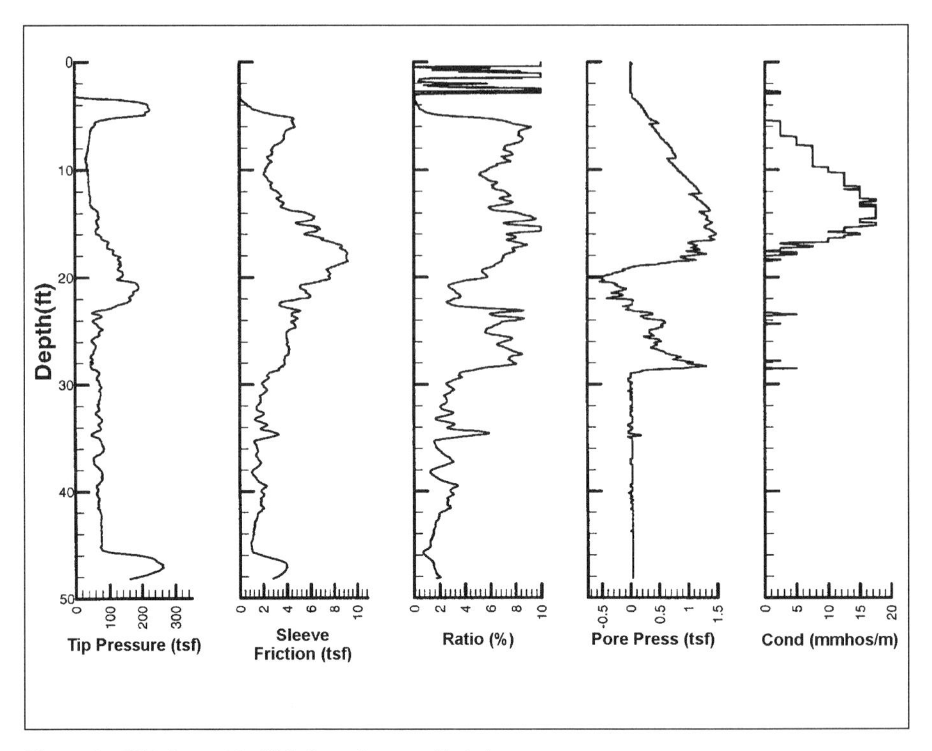

Figure 8. CPT plot at 321 M Solvent Storage Tank Area.

Figure 8 and the Raman plots in Figure 7, it is clear that although good portions of the solvents are found in the clays, the sands below the clays contain solvents as well. From this observation, we can hypothesize that the solvents are still descending.

The results from the 321 M Solvent Storage Area are shown in Figure 9. Two CPT/Raman pushes were done here within 20 ft of each other. The sediment samples were taken about 15 ft from the second Raman push. Spectra were taken at 3-ft intervals in coarse-grain sediments and at 0.5-ft intervals in finer-grained sediments until the probe was retrieved at a depth of 40 ft. The data collected in February indicated PCE at depths 25.6 and 26.6 ft, which corresponds to the geophysical logging of a finer-grained layer from 23 to 28 ft. The spectra with DNAPL peaks did not show higher background fluorescence, which can be explained by the lack of co-constituents in the DNAPL at this site.

Depicted in data from the first June push at the 321 M Area, the Raman spectra indicated the presence of PCE at depths of approximately 21 to 28 ft, which

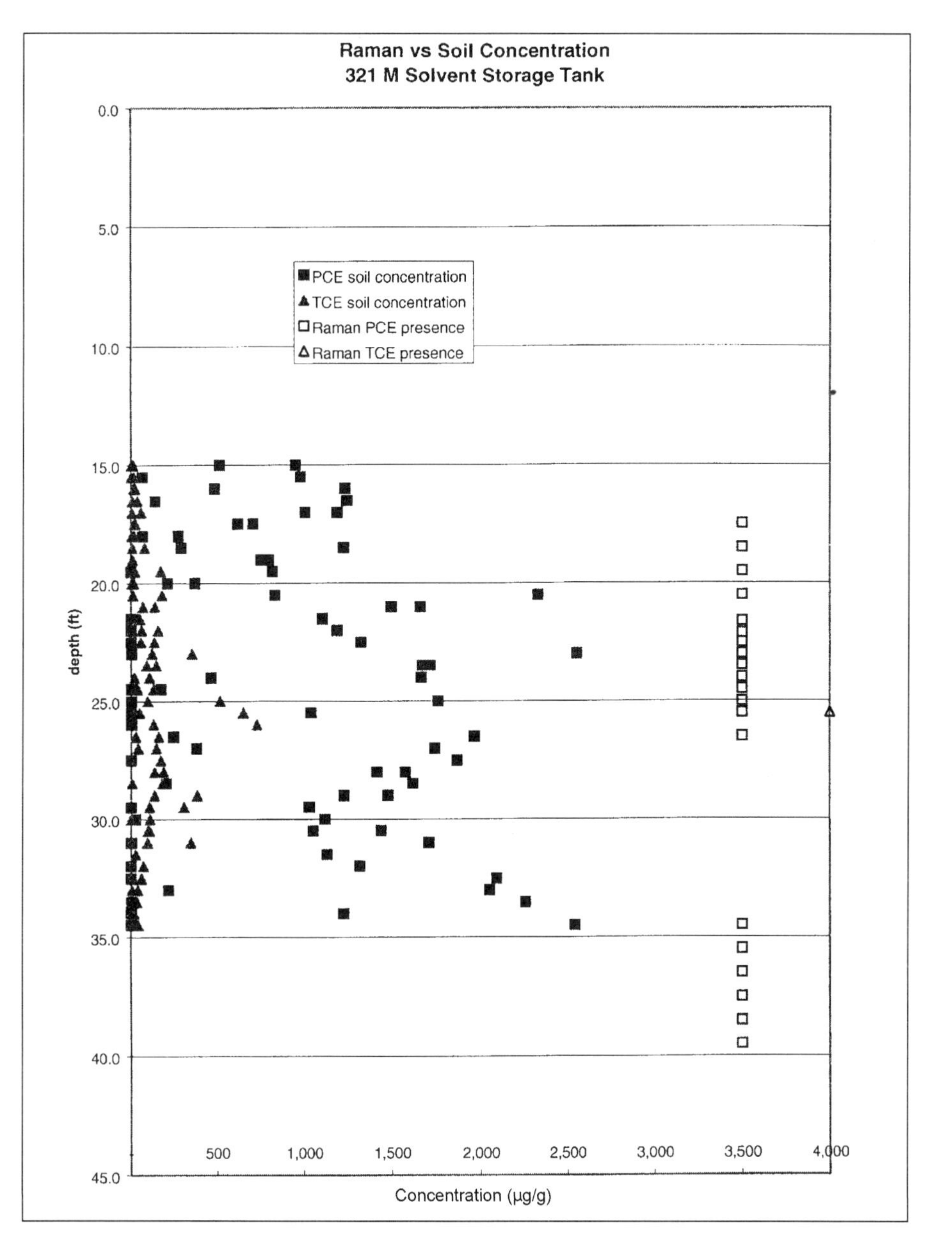

Figure 9. 321 M Solvent Storage Tank Area soil core and Raman data.

corresponds to a fine-grain layer, and 34 to 40 ft, which corresponds to a coarse-grain layer. The presence of TCE was detected at 27.5 ft, which is just below a fine-grain layer. The second push showed spectral evidence of PCE from 17 to 23 ft, which corresponds to a fine-grain layer from 17 to 20 ft and a coarse layer from 20 to 23 feet.

Figure 10 depicts a vertical sequence of Raman plots with varying degrees of PCE peak intensities. The trace, taken with a 10-second integration time at a depth of 17.5 ft, does not indicate the presence of PCE. When integrated for 60 seconds, however, PCE is clearly evident in the spectrum, as shown in Figure 11. Longer integration times produce longer and more expensive characterization pushes. Although for most characterizations a 1- or 2-minute pause for a measurement is acceptable, there must be a balance between desired sensitivity and characterization cost. Often the cost can be offset by intelligent selection of measurement intervals.

CONCLUSIONS

CPT-based Raman spectroscopy positively identified DNAPL at the M Area Seepage Basin and at the 321 M Area Solvent Storage Tanks. The very high signal-to-noise ratio of characteristic peaks for PCE are compelling evidence that the technique can identify contaminants in the subsurface. The spectral data correspond with the soil concentration data. Figure 6 from the M Area Seepage Basin, shows good agreement and successful detection of a thin layer of DNAPL-contaminated clay. In both the M Area Basin and the 321 M Solvent Storage Tank Areas, each DNAPL detection by the Raman probe was confirmed by laboratory analysis of sediment samples.

Some concern arises over the areas of contamination that were not detected by the Raman technology. In Figure 9, from the 321 M Area, there is some disagreement between the soil and Raman data from depths of 15 to 21 ft. DNAPL concentrations were high from 15 to 21 ft, but the spectral analyses did not show evidence of DNAPL peaks. One explanation for this inequality could be the heterogeneity of these sediments and very small droplets of DNAPL residing in clay lenses of varying depths and thickness'. The Raman technique may require a threshold of DNAPL to provide an adequate optical cross-section for spectroscopic response. It is also probable that the technique requires a separate phase liquid rather than an aqueous solution for adequate response.

Similar to CPT-based, laser-induced fluorescence techniques, the detection limit of the technique is strongly correlated to the probability of contaminant droplets appearing on the optical window. This is related to the amount of contaminant in the sediments, the types of sediments, and other factors such as moisture content and degree of heterogeneity of sediments. DNAPL droplets may be squeezed out of a formation and onto or away from the optical window. It is very important to synthesize

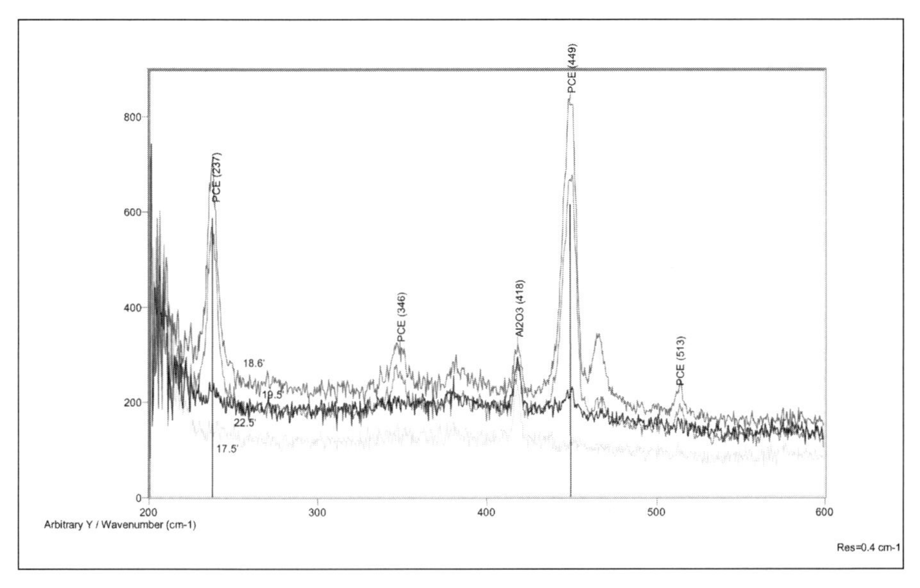

Figure 10. Raman spectra (RMA-5).

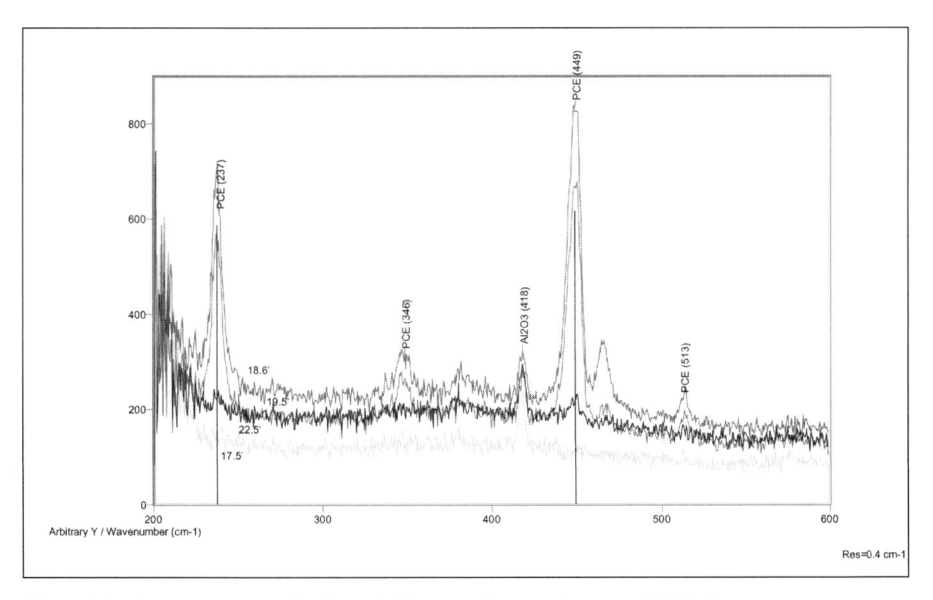

Figure 11. Raman spectra with 10- and 60-second integration times (RMA-5).

the spectroscopic characterization plan with an understanding of the subsurface geology obtained both prior to and concurrent with the spectroscopy.

Spectral integration times are important for determining the detection limits of the technique. Generally spectra were taken for 10 seconds. If peaks were seen or there was suspicion of DNAPL, a longer spectrum was taken. There were instances when the only suspicion of DNAPL came not from spectral evidence, but from prior experience and testing, and knowledge of the geology. In such cases, longer spectra were taken which often showed distinct DNAPL peaks. The danger of missing a DNAPL hit must be balanced with the expense of pausing to acquire a spectrum with a longer integration time and better signal-to-noise ratio or lower detection limit. Some experience with and refinement of the technique may suggest the appropriate integration time for adequate detection capabilities.

The Raman spectroscopic activities conducted in these tests represent the first *in situ* direct measurement of DNAPL in the subsurface. The Raman technique offers one of the very few viable DNAPL characterization techniques to date.

REFERENCES

Carrabba, M. M., J. W. Haas, K. M. Spencer, R. W. Forney, T. M. Johnston, and J. M. Sylvia (1998), Field Raman Spectrograph for Environmental Analysis, *Final Report under Contract DE-AC21-92MC29108, Federal Energy Technology Center*, Morgantown, WV.

Carrabba, M. M, K. M. Spencer, R. B. Edmonds, R. D. Rauh (1992), and J. W. Haas, Spectro-electrochemical technologies and instrumentation for environmental and process monitoring, *SPIE Proceedings vol 1637, Environmental and Process Monitoring Technologies*.

Carrabba, M. M, L. S. Robblee, and R. D. Rauh (1990), *The prospect of utilizing surface enhanced Raman spectroscopy (SERS) for bio- and biomedical sensing*, SPIE Proceedings vol 1201, Optical Fibers in Medicine V, 14-19 January, Los Angeles, CA.

Colthup, N. B., L. H. Daly, and S. E. Wiberley (1990), *Introduction to Infrared and Raman Spectroscopy*, 3rd ed., Academic Press, San Diego.

GAMMA BOREHOLE LOGGING FOR VADOSE ZONE CHARACTERIZATION AROUND THE HANFORD HIGH-LEVEL WASTE TANKS

David S. Shafer, *Desert Research Institute*
James F. Bertsch and **Carl J. Koizumi**, *MACTEC-ERS*
Edward A. Fredenburg, *Lockheed Martin Hanford Company*

INTRODUCTION

Shortly after World War II, the U.S. Atomic Energy Commission (AEC) began building large, carbon-steel, single-shell tanks (SSTs) at the Hanford Site in Washington State to store high level waste generated from chemical production of plutonium, and later, uranium, which were separated from irradiated fuel rods. Today, 149 SSTs, most ranging in capacity from 2.00 to 3.78 million L (0.53 to 1 million gal), store 132 of the 204 million L (35 of the 54 million gal) of high level nuclear waste at Hanford. The balance of the waste is stored in double-shelled tanks (DSTs) built after 1968. Treatment of the SST waste remains one of the most complex tasks facing the U.S. Department of Energy (DOE), successor to the AEC and manager of Hanford. Preliminary evaluation of the tank inventory showed that over 3.8 million L (1 million gal) of waste has leaked from or spilled around the SSTs (Gephart and Lundgren 1997). Although the tank farms have had various types of in-tank and vadose zone monitoring, recent spectral gamma logging in boreholes around the SSTs, along with analysis of historic gross gamma logging data, have raised concerns about the potential impacts of the tank waste on soils, sediments, and groundwater at Hanford. Because of the recent contaminants, DOE is planning to conduct new research in order to better understand water seepage and contaminant transport in the vadose zone around and beneath the tanks and other waste disposal sites at the Central Plateau of Hanford.

The SSTs and the vadose zone sediments impacted by contaminant releases from them are managed by the Office of River Protection (formerly the Tank Waste Remediation System Program) of the DOE Richland Operations Office (DOE-RL). The SST farms, as well as the newer DST farms, are regulated by the Washington State Department of Ecology (Ecology) under its Resource Conservation and Recovery Act (RCRA). Since the inception of the Spectral Gamma Logging System (SGLS) project, DOE-RL and Ecology have negotiated milestones for SST farm vadose zone characterization as part of the Tri-Party Agreement between DOE, Ecology, and the U.S. Environment Protection Agency that guides the clean-up of Hanford. SST vadose zone characterization is one part of a Hanford Groundwater/Vadose Zone Integration Project led by the DOE-RL Environmental Restoration Project. More information on it is available at http://www.bhi-erc.com/vadose. More information on the results of SGLS logging at Hanford is available at http://www.doegjpo.com/programs/hanf.

SETTING AND BACKGROUND

Hanford is situated in the Pasco Basin, a physical and structural depression in the Columbia Plateau created by tectonic activity and the folding of the Columbia River basalts (Reidel *et al.* 1994). The SSTs are in the Central Plateau, or 200 Area, of Hanford (Figure 1). The regional groundwater aquifer discharges to the Columbia River, which flows adjacent to the Hanford site. The river lies 61 to 91 m (200 to 300 ft) deep across the Central Plateau. Most of the vadose zone sediments overlying the basalts (Figure 2) beneath the SSTs are permeable sands and gravels, bedded but

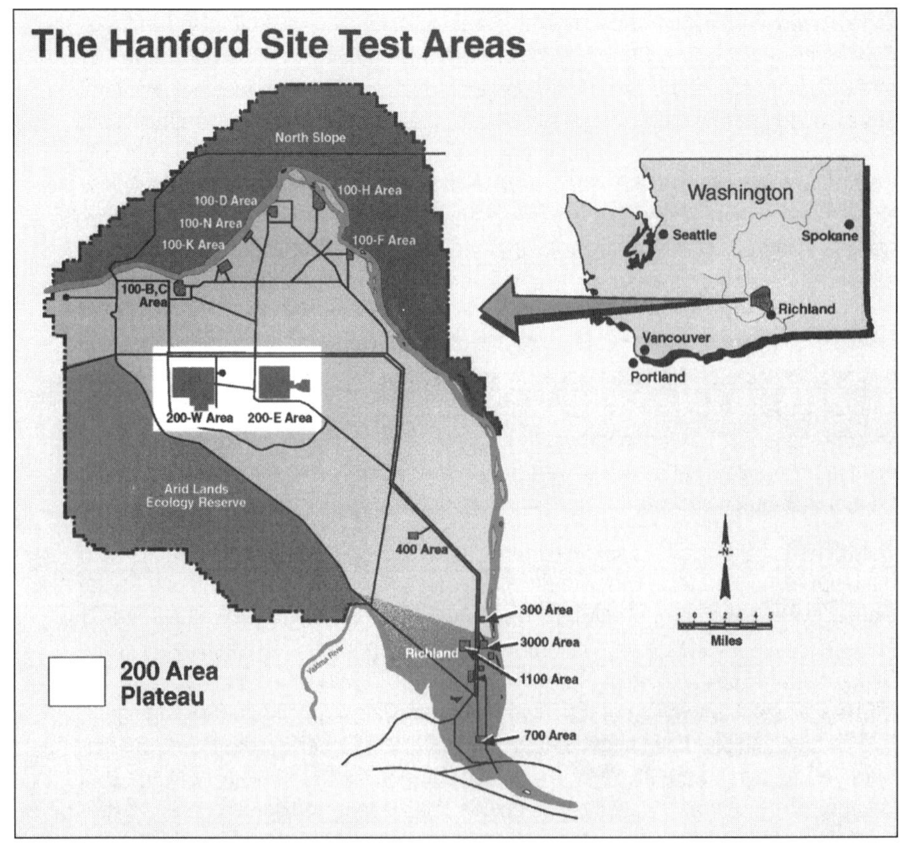

Figure 1. Location map of Hanford and the 200 West and 200 East areas of the Central Plateau where the high level waste tanks are located.

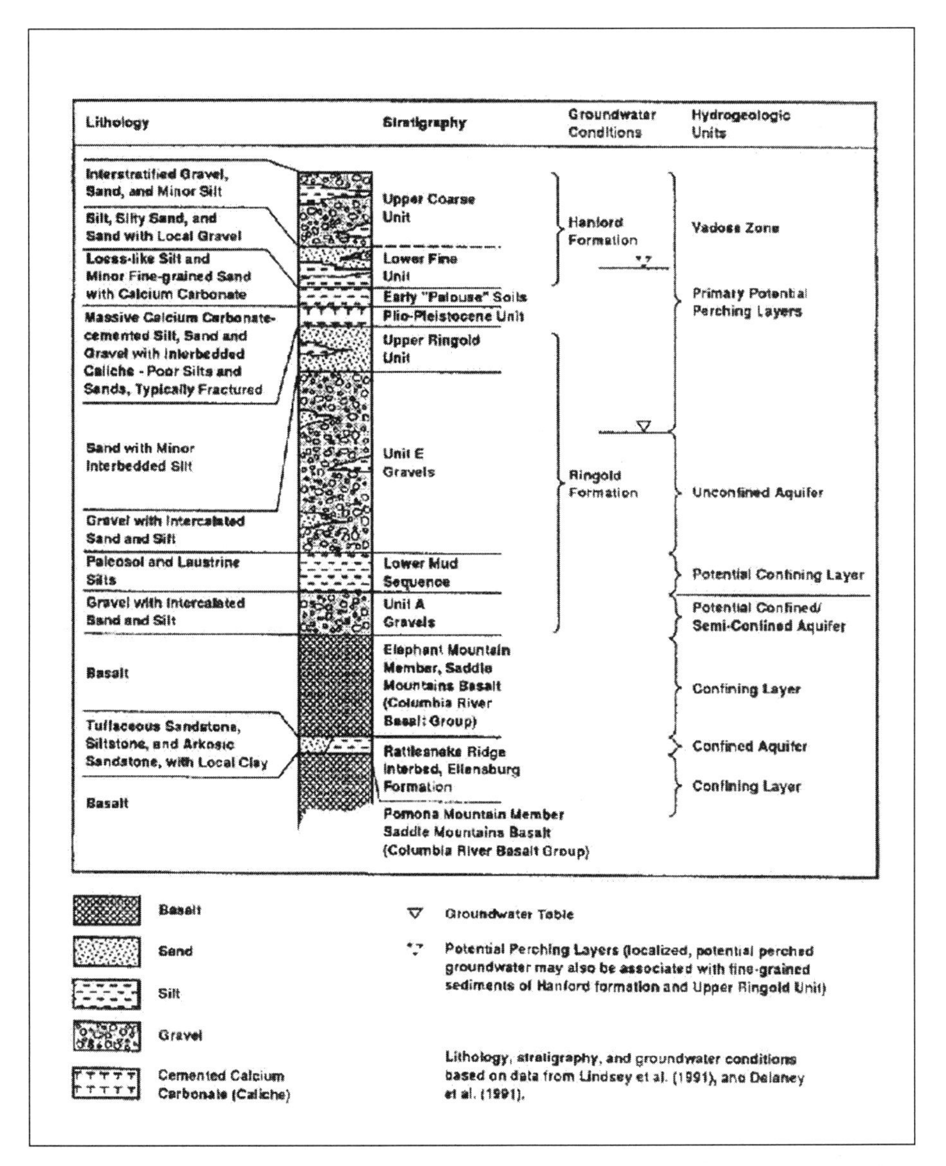

Figure 2. Conceptual hydrogeological column of the western side of the Central Plateau of Hanford from DOE (1992).

poorly consolidated. These sediments are of the Hanford Formation, deposited by glacial floodwaters from the Pleistocene Missoula or Spokane floods which swept repeatedly across eastern Washington, most recently between approximately 12,000 and 13,000 years ago. While bedding is present, the layers are highly discontinuous laterally, and there are frequent sharp contrasts in texture between adjacent layers. In general, however, the upper portions of the Hanford formation are more permeable than lower units (Kincaid *et al.* 1993). Beneath the Hanford Formation are the less permeable and locally cemented clays, sands, silts, and gravels of the Ringold Formation, deposited and formed by a variety of fluvial, paleosol, alluvial, and lacustrine processes (Caggiano 1996; Lindsey *et al.* 1994). The Ringold Formation is 10 to 100 times less permeable than the Hanford Formation (Wurstner and Devary 1993). Particularly in the western part of Central Plateau, a Plio-Pleistocene "caliche zone" separates the Hanford Formation from the Ringold Formation. This fine-grained, calcium carbonate-rich unit, with lenses of caliche, may be as thick as 2.5 m (8.2 ft). It can cause local perching and lateral spreading of water, and can slow the vertical contaminant transport. The caliche layer is frequently fractured. In several places this layer is covered by loess-like silt and fine-grained, calcareous sands of the Early "Palouse" Soil (Caggiano 1996; DOE 1992).

The SSTs are grouped in 12 tank farms. Each SST consists of a single carbon-steel shell surrounded by concrete and buried about 1.8 m (6 ft) deep. Of the total radioactivity of the tanks' contents, 24 percent is from ^{137}Cs and 75 percent from ^{90}Sr. However, the remaining 1 percent includes longer-lived radionuclides, such as the beta-emitting ^{99}Tc, that pose a long-term threat to groundwater (Gephart and Lundgren 1997). As early as 1956, it was suspected that some of the SSTs had leaked. Today, at least 67 tanks are known or suspected to have leaked, although no tanks are known to be actively leaking (DOE 1998d).

A long-standing conceptual model of contaminant transport at Hanford in the vicinity of the SSTs considered diffusive-advective solute transport through interstitial pores of the unconsolidated sediments of the vadose zone in which ^{137}Cs was believed to be highly retarded. It was assumed that ^{137}Cs did not move deeper than a few meters below the tank bases. However, recent spectral gamma logging conducted at boreholes indicates that, at least locally, ^{137}Cs has moved to considerably greater depths (at least 40 m [131.2 ft]) in the SX Tank Farm (DOE 1996).

It has recently been recognized at the Hanford site, that one of the most important factors leading to deep propagation of contaminants is preferential flow phenomena in heterogeneous soils and sediments (Ward *et al.* 1997). Preferential flow may occur along zones of increased permeability such as clastic dikes. Another mechanism of preferential flow at Hanford is the density-driven flow as solute density reaches 1.4 or higher, and pH is as high as 13-14. Preferential flow can also occur within the annular space of ungrouted boreholes. Gravel has been placed as a surface cover to reduce worker exposure to contamination, and to decrease the resuspension of contaminated surface soils by wind (Gee *et al.* 1992, 1998; Rogers *et al.* 1998). Infiltration from the surface increases up to 10 times that of under natural conditions.

As part of leak detection efforts in the SST farms, carbon-steel cased boreholes (called drywells) were driven into sediments surrounding the tanks. Many of the wells were emplaced in the 1950's to a depth of approximately 23 m (75 ft), and deepened to about 38 m (125 ft) in the 1970's. In general, boreholes were completed above the Plio-Pleistocene unit to avoid creating potential pathways for contaminants through the annulus of boreholes (DOE 1997a). In all, 760 boreholes were constructed around 132 tanks. Most were drilled using the cable-tool (percussion) method.

Gross gamma logging in boreholes began in the 1960's with Geiger-Mueller detectors that were retrieved manually. In the mid-1970's, the gross gamma program was upgraded to automated systems with downhole detector probes that were withdrawn from the boreholes at a set rate. Three downhole probes were used for monitoring: a sodium-iodide detector, a lower efficiency tool containing three Geiger-Mueller tubes, and a shielded Geiger-Mueller probe. Probes with different detector efficiencies allowed measurements over a great range of gamma ray flux intensity. The data were analyzed to determine intervals of elevated count rates that could indicate leaks from an adjacent SST, but were not analyzed for characterizing contaminant distribution (DOE 1998a).

Advances in logging tools have been driven by many factors in several industries. Among the most important drivers was logging for uranium exploration in the United States and Canada. The significance of this activity is reflected today in the passive spectral gamma tools used at Hanford. The AEC and its successor agencies sponsored research for the National Uranium Resource Evaluation (NURE) program, which began in 1974. Among the needs addressed was distinguishing the spectral signals of uranium from radioactive isotopes of potassium and thorium, a process frequently referred to as "KUT" (potassium-uranium-thorium) logging (Evans *et al.* 1979). Additional advances included methods for the direct detection of ^{238}U when it became apparent for certain ore bodies that ^{226}Ra was not always in decay equilibrium with ^{238}U (Conaway *et al.* 1980). By 1990, a system was available at Hanford that used log measurements based on KUT calibration standards at the DOE Grand Junction Office (GJO). The standards were developed for the NURE program to derive calibrations for man-made radionuclides such as ^{137}Cs and ^{60}Co. This Radionuclide Logging System (RLS) used a single sonde with an 18 percent efficient high-purity germanium (HPGe) detector (Koizumi *et al.* 1991).

The SGLS system, the results of which are a major focus of this case study, represents a refinement of the RLS, with a sonde with a 35 percent efficient HPGe detector and corrections developed for borehole casing and water-filled boreholes (Koizumi 1994). Results and interpretations of the SGLS logging in the SST farms, along with reassessment of historic gross gamma logging records, have raised a host of issues about the vadose zone at Hanford, many of which are still outstanding. However, one clear outcome is that the logging became a major impetus for an ongoing reassessment of the conceptual model for contaminant transport in the vadose zone at Hanford.

METHODS AND RESULTS

SGLS data has been collected at 15 cm (6 in) intervals in the boreholes with the sondes held stationary for 100 seconds at each counting point. Annual calibration of the equipment is traceable to national gamma ray counting standards, and daily pre- and post-logging spectrum tests have been performed to validate data. The current equipment becomes saturated when contaminant levels exceed 8,000-10,000 pCi/g of ^{137}Cs. Most of the signal that reaches the detector comes from contaminants within a 50 cm radius of the borehole. The primary contaminant detected is ^{137}Cs. Other elements detected are ^{60}Co, ^{235}U, ^{238}U, ^{125}Sb, ^{152}Eu, and ^{154}Eu. A new SGLS baseline log for all of the SST boreholes will be provided in 1999 (DOE 1998c). Periodic relogging and comparison of results against the baseline were used to determine whether gamma contaminants from past leaks were still migrating and if new leaks have occurred.

In general, the gross gamma logs do show contamination, as indicated by the SGLS system at the SX Farm, to a depth of at least 40 m (130 ft), except in places where the greater sensitivity (as low as 1 pCi/g for ^{137}Cs) of the SGLS system prevails (DOE 1996). The most significant issues to emerge with the SGLS were the interpretation of the distribution of the contaminants in the sediments, and hypotheses on how the contaminants have moved through the vadose zone.

Two different hypotheses for the spatial ^{137}Cs distribution have been debated. The first is that tank leaks and spills generated the contaminant plumes (Figure 3). The second model is that the boreholes themselves, with their unsealed annular space, created preferential pathways during leaks or spills. Under the latter model, the deepest detection of gamma-emitting contaminants would not necessarily represent the leading edge of a broad contaminant plume. In addition, when the casing was driven through sediments already contaminated by leaks, spills, or releases, "dragdown" contributed to contamination around the borehole.

In 1996, DOE-RL formed an independent panel, the "SX Expert Panel", to review evidence for these two hypotheses (DOE 1997b). Implementation of some of their recommendations for enhancing the SGLS, as well other types of characterization, suggest that contaminants have been transported under both scenarios, but that other transport processes occur as well.

One of the SX Expert Panel recommendations was to apply shape factor analysis (SFA) to provide more insight on the distribution of the gamma contaminants being

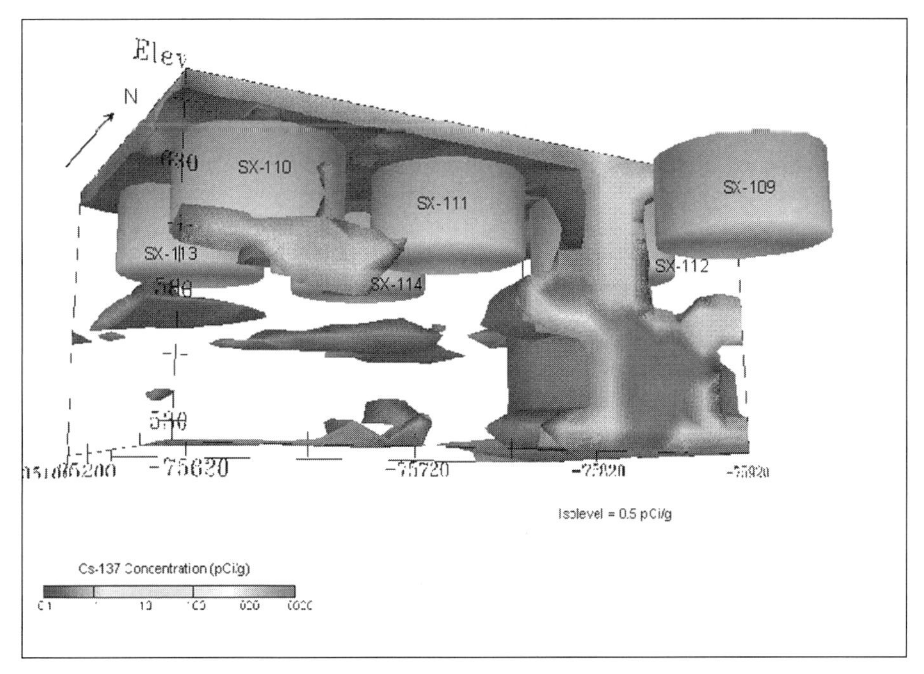

Figure 3. Three dimensional interpretation, unmodified by shape factor analysis, of possible contaminant distribution based on the kreiging of spectral gamma logging data of ^{137}Cs from boreholes around SSTs in the SX Farm. Such an interpretation is central to a conceptual model of broad-based contaminant plumes from SST releases. Units are in pCi/g. From DOE (1996).

detected by logging instruments. Wilson (1997) used variations in gamma ray source distribution to create distinct spectral profiles that are the result of Compton scattering caused by interaction of gamma rays with matter that is located between the gamma ray source and the detector. These interactions cause some higher-energy gamma rays to be converted to ones of lower energy. SFA ratios of counts from various parts of the spectrum were calculated to suggest how contaminants are distributed in the vadose zone at different depths. Beginning with the S Tank Farm (DOE 1998b), SFA ratios were calculated for ^{137}Cs and ^{60}Co in the continuum window from 60 to 650 keV to the spectral peak for each radionuclide. Additionally, the ratios between total gamma counts in the lower-energy portion of the continuum (60 to 350 keV) to the high-energy spectrum (350 to 650 keV) were determined. A high percentage of lower-energy counts would suggest that some contamination is more distant from the source than would be the case if the contamination were remote or uniformly distributed in the sediments. In contrast, a small percentage of lower-energy particles, indicating most photons were traveling a short distance to the detector and not interacting with the media, could indicate that the contamination was largely adjacent to the borehole. It is theorized that SFA can also be used to detect the presence of ^{90}Sr from the presence of brehmsstrahlung radiation, X-rays produced by the deceleration of beta particles emitted by ^{90}Sr and detected in the lower-energy portion of the gamma spectra. A preponderance of lower-energy radiation in a region of high total gamma ray activity may indicate the presence of ^{90}Sr (DOE 1998a). Composition of tank wastes suggests that ^{90}Sr is almost undoubtably a contaminant from some tank leaks (Gephart and Lundgren 1997). However, ^{90}Sr has not been detected in significant amounts with the SGLS (DOE 1998c).

SFA has been used to refine the SGLS 3-D visualizations by eliminating zones in which ratios suggest that contamination is only local to the borehole (Figure 4).

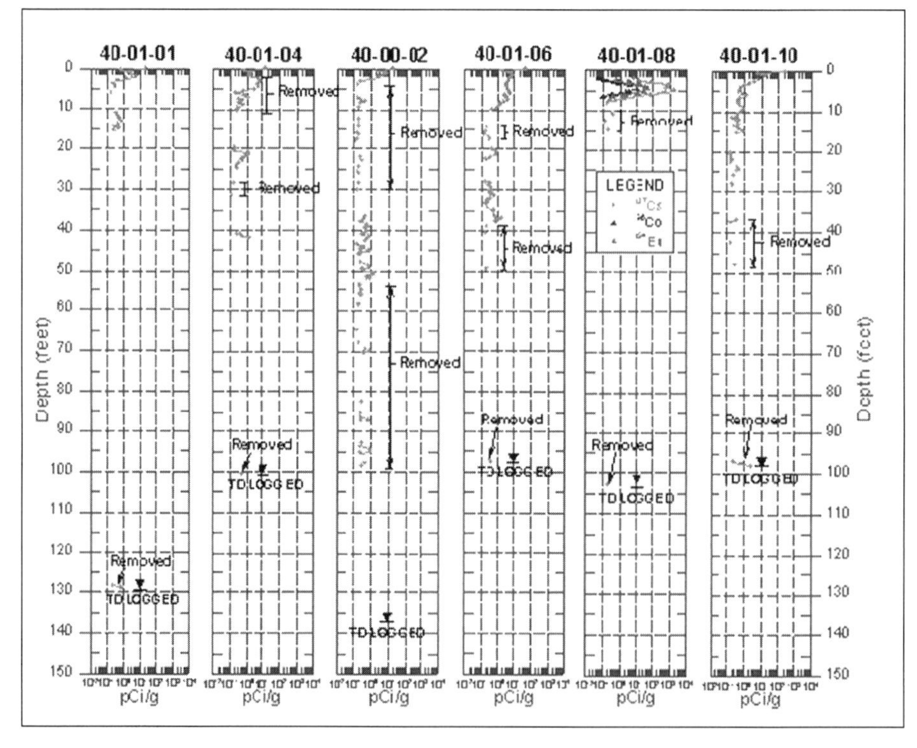

Figure 4. Correlation plots of ^{137}Cs, ^{60}Co, and ^{154}Eu concentrations in six boreholes surrounding SST ^{101}S. Zones labeled "removed" represent intervals of gamma contamination detected that, based on shape factor analysis, are believed to represent contamination only distributed inside, at the bottom, or immediately adjacent to the borehole. From DOE (1998b).

Such localized pollution might develop if contaminants were pushed or dragged during casing emplacement (or, without dragdown, simply because the casing was advanced through sediments already contaminated), or if contaminated liquids migrated in the annular space. Developing the 3-D visualizations also eliminated low gamma counts detected at the base of boreholes when no counts were detected in the sediment for a significant distance above the borehole bottoms. In many such cases, it has been concluded that these counts were probably the result of contaminated soil particles falling or washing down to the base of the boreholes (DOE 1998b).

The SX Expert Panel also recommended continuation of the analysis of gross gamma records (DOE 1997b). Randall and Price (1998) analyzed data collected between 1975 and 1994 for leak detection monitoring in the SX Farm. Of 98 boreholes analyzed, 53 exhibited vertical zones of contamination. Forty-eight boreholes showed contamination at depths greater than 15 m (50 ft), that is, an approximate depth of the tank bases. Of the contaminant zones, nine showed the increase of gross gamma count rates over time, suggesting the lateral movement of gamma-emitting contaminants long after the leak or spilled occurred. The unstable zones in the SX Farm are at depths of 14.6 to 25.9 m (48 to 85 ft), near or slightly below the base of the tanks. No decreases in gamma counts were observed except as a function of radioactive decay. The value of the gross gamma records is significantly enhanced when analyzed and interpreted in conjunction with the SGLS logs. Because the shape of the log profiles from the two data sets are similar, the SGLS allows what were previously gross gamma peaks to be resolved into concentration profiles of specific gamma-ray-emitting contaminants.

The advantage of the gross gamma logs is that they provide count rates for some zones in which the SGLS equipment is saturated. In addition, in some cases, records (albeit of varying quality) extend back more than 20 years. Consequently, in conjunction with the SGLS data, the gross gamma logs may provide sufficient information to determine rates of contaminant migration for gamma contaminants during and after past tank leak and spill events. However, spectral gamma relogging for comparison against the SGLS logs around some tanks may be necessary in order to answer questions about the response of gamma-emitters to possible future events such as a new tank leak or spill (that might remobilize pre-existing contaminants) or recharge events (for example, rapid snowmelts or water line breaks). The latter is suspected of contributing to ^{99}Tc, nitrates, and other contaminants reaching groundwater from leaks and spills in over half the SST farms (DOE 1998d).

DISCUSSION

An examination of the gross gamma records and comparison with tank waste inventories could have led to conclusions that elevated gross gamma readings at depth in some boreholes indicated that ^{137}Cs had migrated to depths greater than expected. No other high-energy gamma-emitter is present in tank waste that would have been detected by gross gamma equipment (in some cases) decades after particular tank leak events. In essence, changes to the conceptual models probably should have been made well before the use of the SGLS. Under such a scenario, another role of the SGLS would have been to confirm and quantify some elements of the conceptual model, rather than to act as a major impetus for revising it. In addition, work such as that by Randall and Price (1998) with the gross gamma records, in combination with conclusions about probable contaminant identity, could have led to conclusions that in many cases ^{137}Cs has not migrated significantly deeper after the initial tank leak event. With a more comprehensive examination of historic data from the beginning, an alternative strategy in sequencing the use of logging tools might have been to utilize a high-rate logging tool to shed light on the mass balance distribution of ^{137}Cs and other gamma-emitting contaminants. A high-rate logging sonde based on planar high-purity germanium technology (Hall and Sengstock 1991) is being assembled and calibrated, and is scheduled to be deployed in late 1999 in the SST farms. The detector will be coupled with a resistive feedback preamplifier configured for high-count rates which will have a positive bias supply input. Energy resolution should range from 2.0-2.5 keV for the 662 keV energy peak for ^{137}Cs at high-count rate. It is anticipated that the system will provide spectral gamma data in environments of dose rates up to 30 R/hr (J. Bertsch, personal communication, 1999). Data from the high-rate logging could be used to test the hypothesis that the preponderance of ^{137}Cs, even if it has locally migrated deeper in the vadose zone, remains within 10 to 15 m (33 to 50 ft) of the bases of the SSTs.

Two other lessons learned deserve mention. First, much of the rhetoric surrounding the first SGLS results led to a detrimental debate that all gamma contaminant movement was either in the form of broad-based plumes, or that it was all the result of movement along borehole annular space. This debate of end extremes probably slowed both scientific acceptance and public understanding that a variety of contaminant transport mechanisms, including natural preferential flow phenomena, could and probably do affect contaminant distribution in the Central Plateau at Hanford. The second lesson deals with the effect of the emplacement of the SST boreholes on the sediments around them (for example, contaminant dragdown). It points to the need, as characterization progresses, to carefully understand how the technique (especially any invasive method, such drilling and subsurface sampling) impacts the vadose zone.

CONCLUSIONS

The most significant limitation of passive spectral gamma logging is that it can detect and identify only the presence of gamma emitters, and, even with enhancements such as SFA, it is limited to direct measurements in a relatively small region surrounding the boreholes. Consequently, gamma logging can provide insight, but it is insufficient to address all issues of contaminant migration in the vadose zone. Nevertheless, the SGLS has met the need for an effective scoping characterization tool that uses existing boreholes for the vadose zone around the SSTs. Also, the logging has provided an estimate of the minimum depth of penetration of contaminants into the vadose zone from tank leaks and spills. In some cases, high count rates detected to the base of boreholes suggest that in all likelihood the gamma contaminants extend deeper (see, for example, DOE 1998a). In addition, by helping to identify the gamma emitters that contribute to the gross gamma count anomalies, it has allowed greater use of historic data to assess the behavior of gamma contaminants.

The logging data could also indirectly indicate the expected distribution of other contaminants in the vadose zone, such as non-gamma-emitting radionuclides and RCRA-regulated constituents such as nitrates and chromium. However, based on chemistry alone, this is doubtful, because the gamma-emitters exist largely as cations, while the radionuclides projected to contribute most to long-term risk from groundwater (for example, ^{99}Tc and ^{129}I) are beta-emitters and probably exist as anions. Consequently, while spectral gamma logging has proved to be of significant value, other characterization methods will need to be, and are being, deployed.

REFERENCES

Caggiano, J.A. (1996), Groundwater Water Quality Assessment Monitoring Plan for Single-Shell Tank Waste Management Area S-SX, WHC-SD-EN-AP-191, Rev. 0, Westinghouse Hanford Company, Richland, Washington.

Conaway, J., Q. Bristow, and P. Killeen (1980), Optimization of gamma-ray logging techniques for uranium, *Geophysics, 45*.

DOE (1992), S Plant Aggregate Area Management Study Report, DOE/RL-91-60, Rev. 0, Westinghouse Hanford Company, Richland, Washington.

DOE (1996), Hanford Tank Farms Vadose Zone, SX Tank Farm Report, prepared by MACTEC - ERS for the U.S. Department of Energy Grand Junction Office, Grand Junction, Colorado.

DOE (1997a), Hanford Tanks Initiative Retrieval Performance Evaluation Criteria Assessment Work Plan, prepared by Jacobs Engineering Group for the U.S. Department of Energy, Richland, Washington.

DOE (1997b), TWRS Vadose Zone Contamination Issue Expert Panel Status Report, DOE/RL-97-49, Rev. 0, U.S. Department of Energy, Richland Operations Office, Richland, Washington.

DOE (1998a), Hanford Tank Farms Vadose Zone, BX Tank Farm Report, prepared by MACTEC-ERS for the U.S. Department of Energy Grand Junction Office, Grand Junction, Colorado.

DOE (1998b), Hanford Tank Farms Vadose Zone, S Tank Farm Report, prepared by MACTEC-ERS for the U.S. Department of Energy Grand Junction Office, Grand Junction, Colorado.

DOE (1998c), Hanford Tank Farms Spectral Gamma Logging Status Report, prepared by MACTEC-ERS for the U.S. Department of Energy Grand Junction Office, Grand Junction, Colorado.

DOE (1998d), Tank Waste Remediation System Vadose Zone Program Plan DOE/RL-98-49, Rev.0, prepared by Lockheed Martin Hanford Company for the U.S. Department of Energy, Richland Operations Office, Richland, Washington.

Evans, H., D. George, J. Allen, B. Key, D. Ward, and M. Mathews (1979), *A borehole gamma ray spectrometer for uranium exploration*, 20th Annual Logging Symposium Transactions, Society of Professional Well Log Analysts, Tulsa, Oklahoma.

Gee, G.W., M.L. Rockhold, M.J. Fayers, and M.D. Campbell (1992), Variations in recharge at the Hanford Site, *Northwest Science, 66 (4).*

Gee, G.W., A.L. Ward, and K.M. Thompson (1998), Effect of sediment layering on preferential flow under high level radioactive waste tanks, American Geophysical Union *EOS* April Supplemental.

Gephart, R. E., and R. E. Lundgren (1997), Hanford Tank Clean Up: A Guide to Understanding the Technical Issues, Pacific Northwest National Laboratory PNL-10773, Richland, Washington.

Hall, D., and G.E. Sengstock (1991), Introduction to high-count-rate germanium gamma-ray spectroscopy, *Radioactivity and Radiochemistry, 2 (2).*

Kincaid, C.T., J.W. Shade, G.A. Whyatt, M.G. Piepho, K. Rhoades, J.A. Voogd, J.H. Westsik, Jr., M.D. Freshley, K.A. Blanchard, and B.G. Lauzon (1993), Performance Assessment of Grouted Double-Shell Tank Waste Disposal at Hanford, WHC-SD-WM-EE-004, Rev. 0. Westinghouse Hanford Company, Richland, Washington.

Koizumi, C.J. (1994), Vadose Zone Monitoring Project at the Hanford Tank Farms, Calibration Plan for Spectral Gamma-Ray Logging Systems, U.S. Department of Energy Report P-GJPO-1778, Rust Geotech, Grand Junction, Colorado.

Koizumi, C.J, W. Ulbricht, and J. Brodeur (1991), *Intrinsic germanium gamma-ray data from the New American Petroleum Institute spectral gamma-ray calibration models*, 4th International Symposium on Borehole Geophysics for Minerals, Geotechnical, and Groundwater Applications, Minerals and Geotechnical Logging Society, Toronto, Ontario, Canada.

Lindsey, K.A., S.P. Reidel, K.R. Fecht, J.L. Slate, A.G. Law, and A.M. Tallman (1994), *Geohydrologic setting of the Hanford site, south-central Washington*, in Geological Field Trips of the Pacific Northwest: 1994 Geological Society of America Annual Meeting, edited by D.A.

Swanson and R.A. Haugerud, Department of Geological Sciences, University of Washington, Seattle, Washington.

Randall, R., and R. Price (1998), *Vadose zone monitoring results for Hanford's SX tank farm using existing gross gamma data collected as secondary leak detection*, Department of Energy, Nevada Operations Office Vadose Zone Monitoring Workshop Proceedings, Las Vegas, Nevada.

Reidel, S.P., N.P. Campbell, K.R. Fecht, and K.A. Lindsey (1994), Late Cenozoic structure and stratigraphy of south-central Washington, *Washington Division of Geology and Earth Resources Bulletin, 80.*

Rogers, P.M., R.W. Lober, G. Wroblicky, and D.S. Shafer (1998), Potential preferential flow paths in the vadose zone at Hanford and their importance to remediation, American Geophysical Union *EOS* April Supplemental.

Wilson, R.D. (1997), Hanford Tank Farms Vadose Zone, Spectral Shape-Analysis Techniques Applied to the Hanford Tank Farms Spectral Gamma Logs, GJO-HAN-7, prepared by MACTEC-ERS for the U.S. Department of Energy Grand Junction Office, Grand Junction, Colorado.

Wurstner, S.K., and J.L. Devary (1993), Hanford Site Ground-Water Model: Geographic Information System Linkages and Model Enhancements, FY 1993, PNL-8991, Pacific Northwest National Laboratory, Richland, Washington.

NEAR SURFACE INFILTRATION MONITORING USING NEUTRON MOISTURE LOGGING, YUCCA MOUNTAIN, NEVADA

Alan L. Flint and **Lorraine E. Flint**, *U.S. Geological Survey*

INTRODUCTION

Net infiltration, that component of precipitation that percolates below the zone of influence of evapotranspiration (ET) to recharge the groundwater system, is an important boundary condition for hydrologic flow models used to calculate flux through the thick unsaturated zone at Yucca Mountain (Winograd and Thordarson, 1975). A data set has been developed consisting of volumetric water-content readings in 98 neutron access boreholes (neutron holes) monitored from 1984 (or when each hole was drilled) through 1995. Neutron logging is considered a reliable and dependable field technique for measuring subsurface moisture (Chanasyk and Naeth, 1996; Evett and Steiner, 1995; Tyler, 1988), and has been used in water balance studies in arid and semi-arid environments (McElroy, 1993; Sharma *et al.* 1991; Nixon and Lawless, 1960) and in bedrock (Sternberg *et al.* 1996). The purpose of this case study is to provide readers with an understanding of a long-term monitoring program, including description of the environment, methodologies under which the data were collected, and various approaches to analyzing and interpreting the data. The results of the analysis of this data set has contributed to an understanding of the mechanisms and processes by which precipitation becomes net infiltration for various topographic positions and surficial materials at Yucca Mountain.

SITE DESCRIPTION

Yucca Mountain is in the northern Mojave Desert in southern Nevada (Figure 1). The mountain is a faulted ridge consisting of a series of layered, nonwelded and welded, variably fractured and saturated pyroclastic rocks. The topography of the mountain has been defined by erosional processes on the eastern sloping ridge and along faults and fault scarps that have created a series of washes, which are down cut, to varying degrees, into different bedrock layers. The washes are primarily east-west and northwest trending, with gentle to steep slopes, and have available energy loads that vary greatly for much of the year. The bedrock exposed at the surface, or directly underlying the alluvial cover, ranges from nonwelded, unfractured rock that has 40-percent porosity, to densely welded, highly fractured rock that has less than 10-percent porosity. The alluvial/colluvial surficial deposits have varying degrees of soil development and thickness and have a gravelly texture; rock fragments (greater than 2 mm) constitute between 20 and 80 percent of the total volume. More stable surfaces, generally the flat upland ridges, have higher clay contents. Deeper soils in the center of many washes have developed cemented calcium carbonate layers.

Average annual precipitation at the site is about 170 mm (Hevesi *et al.* 1991). Precipitation occurs as localized thunderstorms in the summer when the ET demand

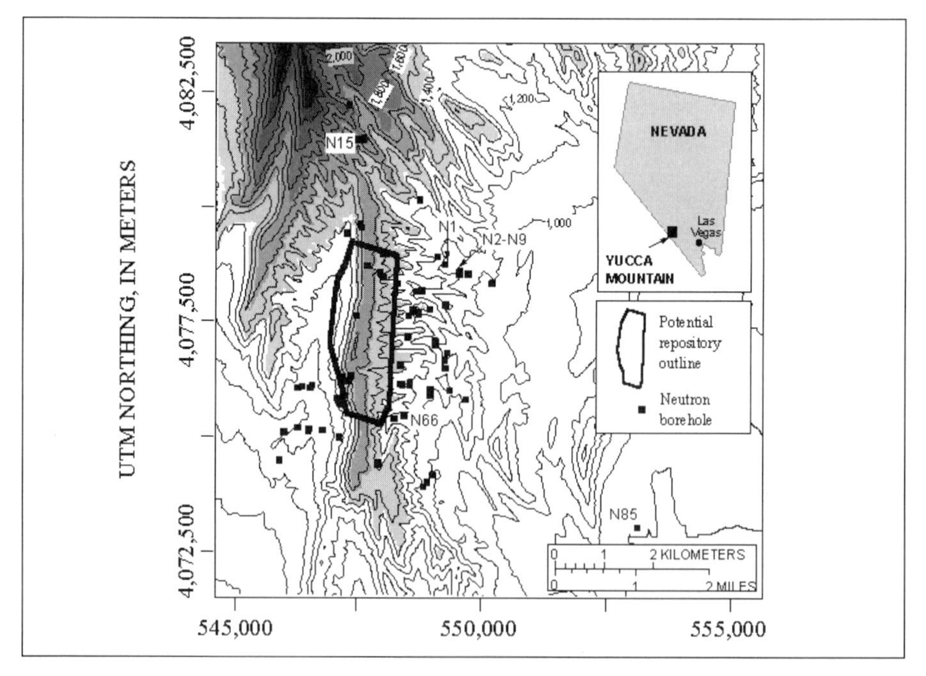

Figure 1. Study area, potential repository location, and location of neutron boreholes [contour interval is 100 m and ranges from 900 m elevation (white) to 2,200 m (dark)]

is very high. Most of the water is lost to ET processes within several days, unless there is enough rainfall to produce runoff or subsequent storms provide additional water for deeper penetration. Thunderstorms in the area can create runoff in one wash while an adjacent wash receives no rain at all. Precipitation in the winter occurs in large stratiform storm patterns as snow or rain. ET is low at this time of year, and lower precipitation rates or slowly melting snow may penetrate deeper into the soil profile.

In this study, infiltration processes were examined for four geomorphologic elements—ridgetop, sideslope, terrace, and channel (Figure 2). Over the area indicated by Figure 1, the ridgetop locations encompass about 14 percent of the total area, the sideslopes 62 percent, the terraces 22 percent, and the active channels 2 percent. The ridgetop, which is flat to gently sloping, is higher in elevation and has thin soils mostly developed in place with high clay content and high water-holding capacity that reduces rapid evaporation. Although the ridgetop locations have relatively thin to no surficial deposits, they are relatively stable morphologically. Existing soils are fairly well developed, and thin calcium carbonate layers are common. Some perennial channels have somewhat thicker soils, and some concentrated surface runoff occurs. Bedrock at ridgetop locations is moderately to densely welded (5- to 25-percent porosity) and moderately to highly fractured. Vegetation consists of well-established, fairly shallow-rooted blackbrush/desert thorn associations (O'Farrell and

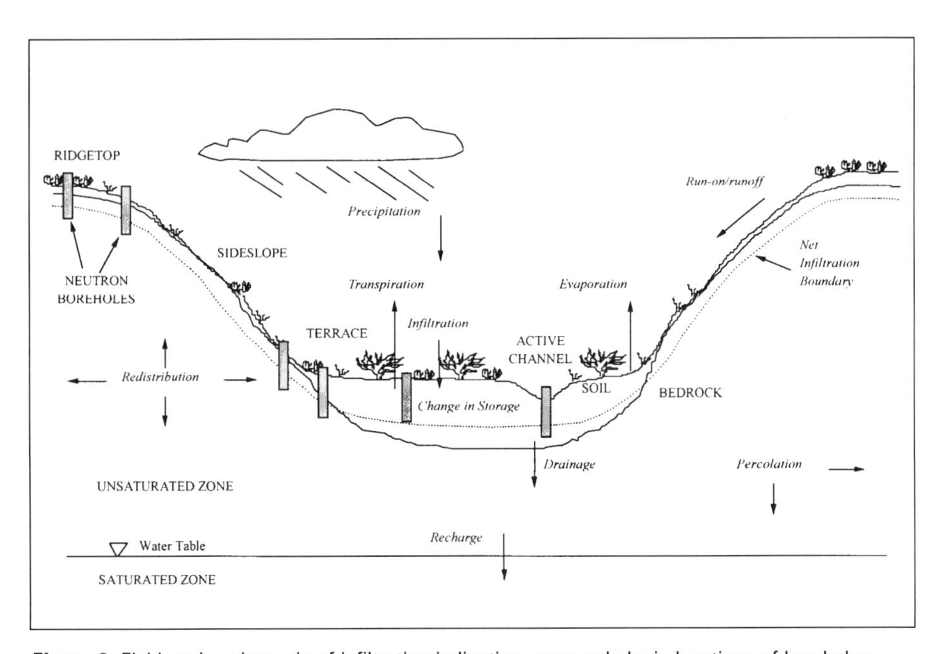

Figure 2. Field scale schematic of infiltration indicating geomorphologic locations of boreholes

Collins, 1983). The slope and elevation in this position tend to retain snowfall in the winter for several weeks at a time.

Because of the difficulty of drilling boreholes at steep sideslope locations, field data are limited to sites on the lower sideslopes of washes. The thickness of soils and the slopes at sideslope locations is different from the terrace and channel locations. Soil cover for sideslope positions is thin to nonexistent and, in most locations, bedrock is densely welded and highly fractured. The sideslopes are approximately north or south facing and, and therefore have different seasonal solar radiation loads. In some locations, side channels concentrate runoff water. In general, sideslopes tend to be more sparsely vegetated than the other geomorphologic elements.

Terraces and channels are located at lower elevations in the main washes, and have thin soil cover in the upper washes and thick soils in the lower washes. A very small percentage of exposed bedrock exists in the washes, almost all of which is non-welded, highly porous tuff. The soil has varying degrees of calcium carbonate cementation that is commonly quite extensive. The porosity of the soil ranges from 15 to 50 percent. The terrace surface is relatively flat and dissected by old alluvial and active channels. Contrary to the terraces, periodic runoff occurs in active channels during extreme precipitation conditions. The terraces generally are well vegetated, having deeply rooted creosote and other smaller plants. The active channels occupy a very small surface area of the wash and are more sparsely vegetated than the terraces.

DATA COLLECTION METHODS

Boreholes were drilled using dry-drilling technology and the ODEX 115 drilling and casing system (Hammermeister *et al.* 1985). This method uses simultaneous advancement of the casing with the deepening of the borehole to minimize the effects of air drilling on the borehole walls. The steel casing used for the Yucca Mountain boreholes has a 130-mm inside diameter. Formation volumetric water content is minimally disturbed, and some of the cuttings produced from drilling fill the small annular spaces between the casing and the formation, minimizing, but not eliminating, the void space. This type of drilling enabled continuous core sampling. These samples were used to measure field volumetric water contents in different lithologies in order to develop field calibration equations.

Volumetric watercontent profiles were measured in the boreholes using neutron moisture meters (Model 503, Campbell Pacific Nuclear, Pacheco, Calif.), with an ^{241}Am-Be, 50 mCi source and ^{3}He detector. This meter detects an area surrounding the probe of approximately 0.7 m vertically and 0.18 m radius horizontally (Klenke and Flint, 1991). Despite the unconventional casing for neutron logging, which commonly uses 50-mm inside diameter casing, neutron logging has been shown to be effective in large diameter boreholes (Tyler, 1988). These neutron probes were successfully calibrated using field samples collected during the drilling of several boreholes. The calibration holes were drilled through alluvium, welded tuff, and non-welded tuff, and samples were collected every 0.8 m and preserved on-site to maintain field moisture conditions. Samples were then processed in a laboratory to determine volumetric water content. Linear calibration equations were developed relating the neutron counts from the boreholes to the volumetric water content of the samples (Figure 3).

Five tanks were constructed to maintain known water conditions. They are steel barrels, 1.5 m in diameter, and provide a continuous measurement medium (Klenke and Flint, 1991). These tanks were used to assess long-term drift of the meters so that corrections of neutron decay could be made over time, rather than use count ratios which have more long-term errors. They also were used to develop transfer equations for meters that were not used to measure field moisture conditions at the time the samples were collected for field calibration.

Boreholes were located to emphasize representative geomorphologic locations and spatial coverage (Figures 1 and 2). Seventy-four boreholes were drilled between 1984 and 1986 and range from 4.6 to 36.6 m in depth. Primarily located in washes, these boreholes penetrated the entire alluvial thickness and extended about 1 m into

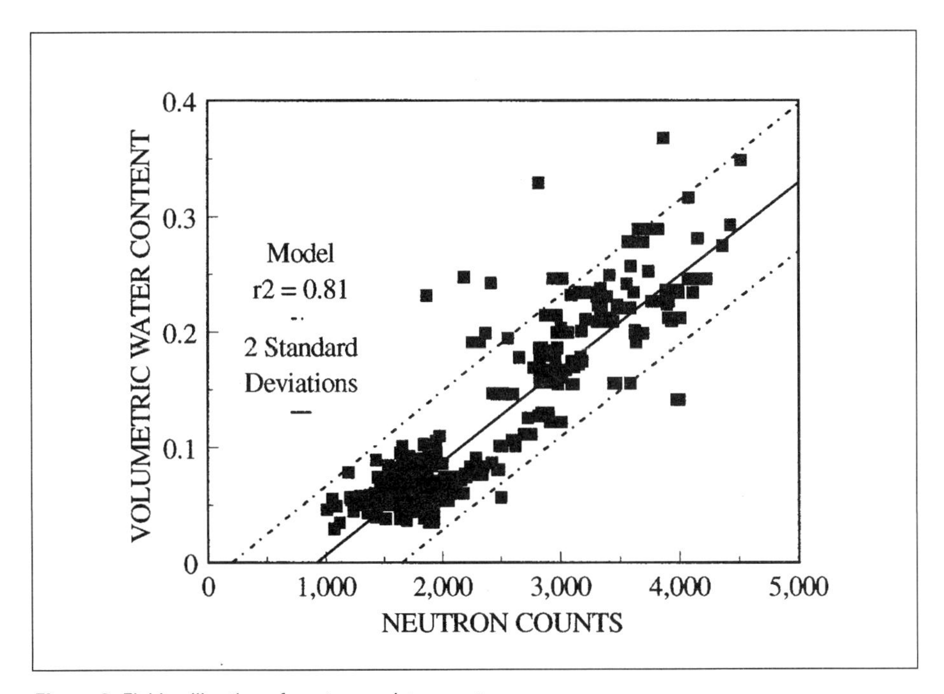

Figure 3. Field calibration of neutron moisture meters

the underlying tuff. The boreholes provided excellent cross-sections of many of the washes, stretching from sideslope to sideslope. An additional 24 boreholes, 18 to 78 m deep, were drilled from late 1991 through 1993. These boreholes improved the coverage of geomorphologic locations, especially the sideslope and ridgetop locations, and also provided deeper penetration, often extending through the nonwelded tuffs and into the densely welded tuffs. More detailed descriptions of the boreholes along with complete borehole designations and locations are given in Flint and Flint (1995).

With some exceptions, the data were collected on a monthly basis. Boreholes deeper than 30 m were only logged to 19.4 m monthly, and to total depth on a 6-month basis. The upper 5 m of each borehole were logged at 0.1-m-depth intervals, and the remaining depth was logged at 0.3-m-depth intervals.

RESULTS OF DATA ANALYSES

The data reported in this study were gathered from July 1984 to September 1995. Several runoff-producing storms were recorded in 1984 but the following 4 years, 1985 to 1988, experienced drought conditions that averaged 50 to 75 percent less rainfall than the normal 130 to 180 mm/year that ranges over the site. This resulted in a gradual decline in volumetric water content of the shallow subsurface. By 1990, volumetric water content in the boreholes had stabilized at between 2 and 5 percent. There was some relief from the drought in 1990, and precipitation during the winters of 1991-92, 1992-93, and 1994-95 was well above normal. Several storms produced runoff in 1990-95.

One- and Two-Dimensional Moisture Content Profiles

Figure 4 represents water content logs for three different geomorphologic locations. Figure 4A illustrates a typical pattern of the moisture content with depth at a channel location where there is little to no surface runoff into the channel and the thickness of soil is relatively deep (7 m). During the 3-year period, where measurement dates were chosen that represented the complete range of changes in volumetric water content, the moisture content changed from approximately 4 percent during a dry year to slightly more than 10 percent during a wet year in the upper 2 m. Figure 4B indicates that at a sideslope with a soil cover of 0.8 m, some water penetrates much deeper to over 10 m, possibly through the fractures. The change in water content in the bedrock is only about 2.5 percent at any given depth. Figure 4C illustrates the changes in the moisture content up to 5 percent to depths of 7 m at a ridgetop location with almost no soil cover and with open and filled fractures at the surface.

Evaluation of volumetric water content with depth and time is important for determining the condition of the borehole, and whether the profiles measured represent

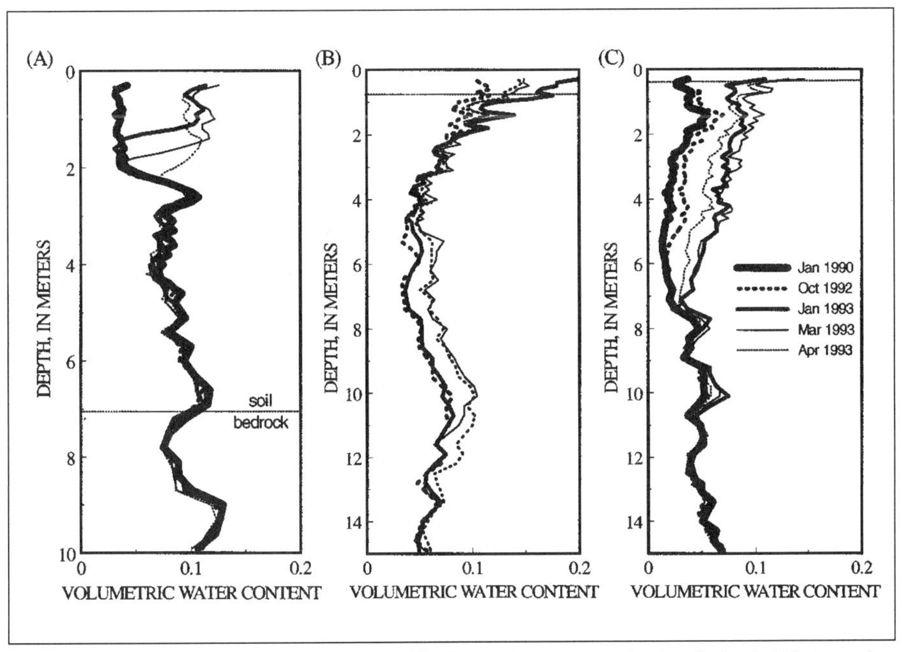

Figure 4. Volumetric water content measured for 5 months, representing boreholes in (A) channels, (B) sideslopes, and (C) ridgetops

changing water contents of the surrounding material or if the water is bypassing the soil and rock by flowing through the borehole annulus. Figure 5A portrays the changes in the volumetric water content with depth and time in borehole N1, which is located in an active channel with periodic runoff, and 8.3 m of alluvium overlying bedrock. The moisture content at the alluvium-bedrock contact is higher than the remaining soil profile and is shown in the figure as a thin dark line at 8.3 m. In this borehole, a transition of moderately welded to nonwelded tuff results in the highest porosity and water content at approximately 11 m. In 1984, the entire profile was relatively moist following a very wet season, and the upper 7 m zone gradually dried out over the next 10 years. Annual winter rains occurred occasionally, but the pulses did not extend beyond the depth at which roots are found, and therefore, all the water probably was lost to ET. The winter of 1994 to 1995 had greater than average precipitation with several runoff events in the channel, which resulted in wetting of the entire profile within several months. There is no evidence of borehole leakage (rapid drainage of water around the borehole casing) in this borehole.

Borehole N66 (Figure 5B) is located on a sideslope with no overlying alluvium in highly fractured, densely welded tuff. This represents an extreme example of bore-

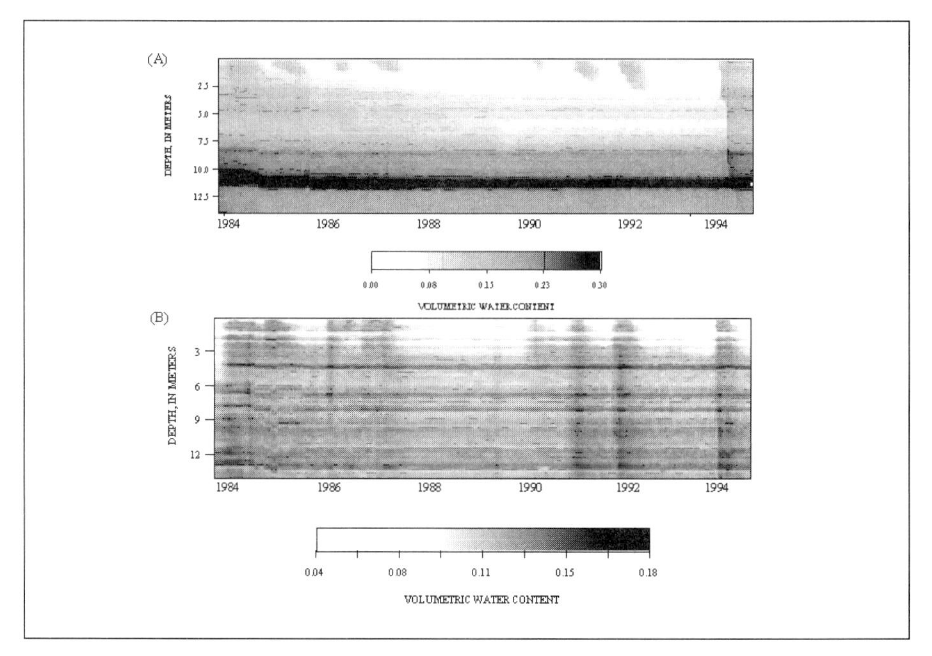

Figure 5. Volumetric water content with depth and time for boreholes (A) N1, in an active channel, and (B) N66, on a sideslope

hole leakage which shows immediate penetration to the bottom of the borehole following precipitation. A data set such as this would not be used for quantitative calculations of infiltration for this location.

The depth of water percolation can be elucidated on the basis of two-dimensional patterns of water content changes represented graphically in cross-section. Figure 6 shows an example of changes in the moisture content between August 17 and August 20, 1984. During that period, a runoff event occurred in the wash north of the potential repository location. There are eight boreholes that span a sideslope (N2), a small intermittent channel (N3), a small terrace (N4 and N5), an active channel (N6, N7, and N8), and a terrace (N9). Figure 6 shows that the small channel at N3 infiltrated water to depths over 6 m, and the active channel at N6, although it did not experience the largest change in water content, allowed water to move at least 15 m. The largest volume of water penetrated in the channel at borehole N8 and flowed laterally in the subsurface to borehole N5, where there was no surface infiltration. There was minimal surface infiltration in the terrace location N9, and some lateral distribution from N8, which is about 12 m away. This interpretation supports field evidence of stratification in the alluvial deposits in washes with shallow calcium carbonate deposition, and deeper textural layering.

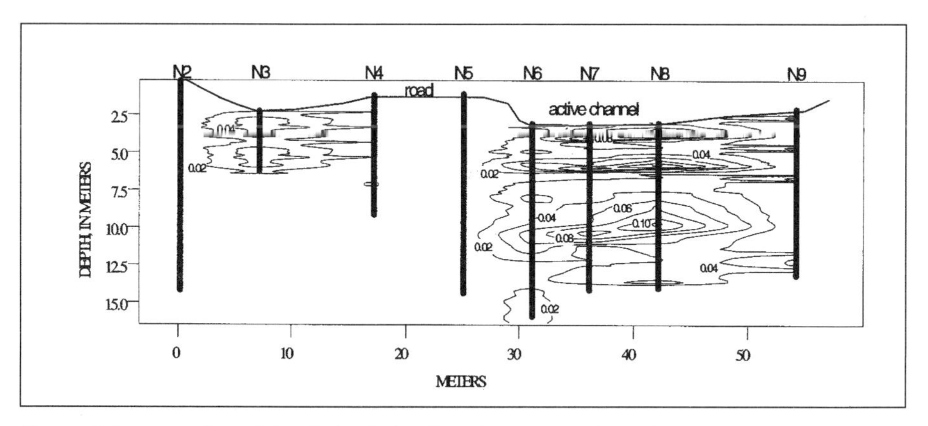

Figure 6. Cross-section of boreholes indicating change in volumetric water content between August 17 and August 20, 1984 (dotted line indicates estimated soil/bedrock contact)

Calculations of Depth of Effective Water Penetration

Water loss below the depth affected by ET processes can become net infiltration. This depth varies between locations with plant species of different rooting depths, as well as with radiation load, but typically this depth is no deeper than 5 m in alluvium and is approximately 1 to 2 m in bedrock. The degree of variation of long-term time series of water-content readings with depth in a given borehole can be used to determine the depth of effective water penetration. A simple method of standardizing readings with depth for each borehole and comparing the changes of the moisture content that occurred in different boreholes is to calculate the standard deviation of the water-content readings. The depth at which a standard deviation is within the range of a random measurement error is the depth of no detectable change in water content. In Figure 7A, watercontent changes for a borehole in a channel are shown as a series of lines representing different measurement dates that span the range of water contents. Below a depth of about 3 m there are no fluctuations. The standard deviation of all readings taken over the same time period at each depth is calculated and presented in Figure 7B. At 3 m, the variation is below the random error of the measurement, assumed less than 0.01, and this depth is considered the depth of penetration of water in this borehole. This depth is shallower than the deepest extent for the roots of the plants that grow in channel locations, and therefore, it is unlikely that net infiltration occurred at this borehole for this time period.

This method of calculating the depth of water penetration was used to illustrate the differences between geomorphologic locations. Figure 8A indicates little difference in depth of water penetration for several boreholes located in channels and terraces,

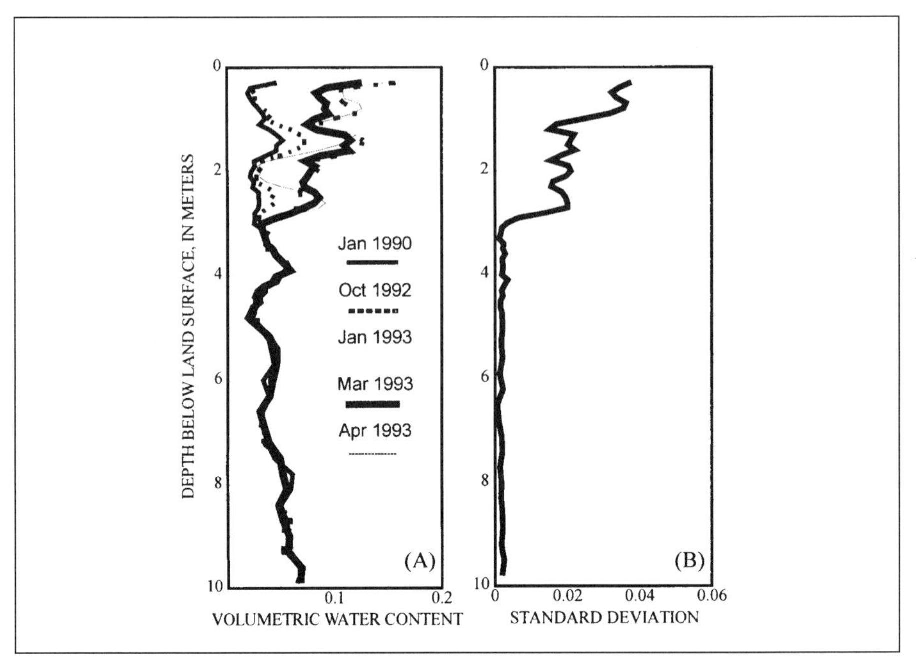

Figure 7. (A) Volumetric water content measured over 3 years, and (B) corresponding standard deviation of 3 years of data at each depth, indicating estimation of depth of penetration of water

with the exception of one channel that occasionally has runoff. This variation in water content to a greater depth probably indicates net infiltration. Figure 8B shows the comparison of the channel to two sideslope locations. The channel shows water penetration to about 2.5 m, while the south-facing sideslope, with high radiation load and thus greater ET, shows water penetration to about 5 m. The north-facing sideslope, which has lower radiation loads and lower ET, shows water penetration to over 13 m, illustrating the importance of incorporating slope and aspect (and thus the calculation of radiation load for estimates of ET), into any estimates of net infiltration. The greatest variation in water content profiles is found in the ridgetop boreholes (Figure 8C). In all boreholes, the depth of water penetration exceeds 1 m, which is usually estimated as the root zone in fractured, welded rock.

Figure 9 summarizes the estimation of the depth of water penetration using calculations of standard deviation of water content for a 4-year period for 34 boreholes representing two watersheds. This figure shows that for locations with the shallowest soil (black bars), generally ridgetops and sideslopes, the deepest water penetration (white bars) occurs; and for locations with deep soils, terraces and channels, shallow water penetration occurs. The exception is boreholes located in channels that receive occasional runoff.

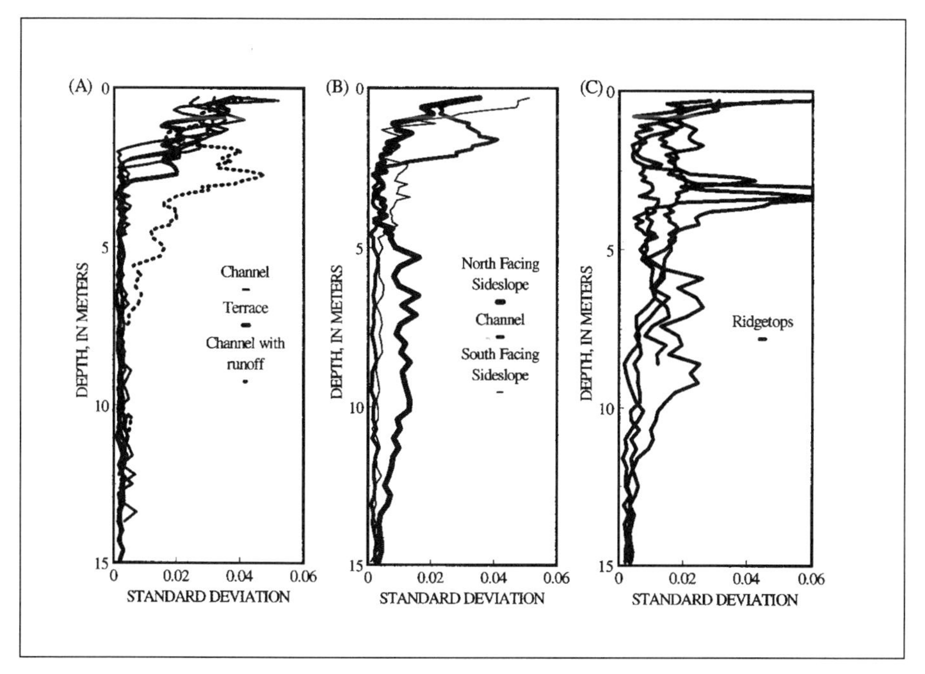

Figure 8. Standard deviation of volumetric water content readings for each depth, measured for 3 years, for various boreholes indicating differences in (A) channels, terraces, and channel with runoff, (B) north- and south-facing sideslopes, and (C) ridgetop boreholes

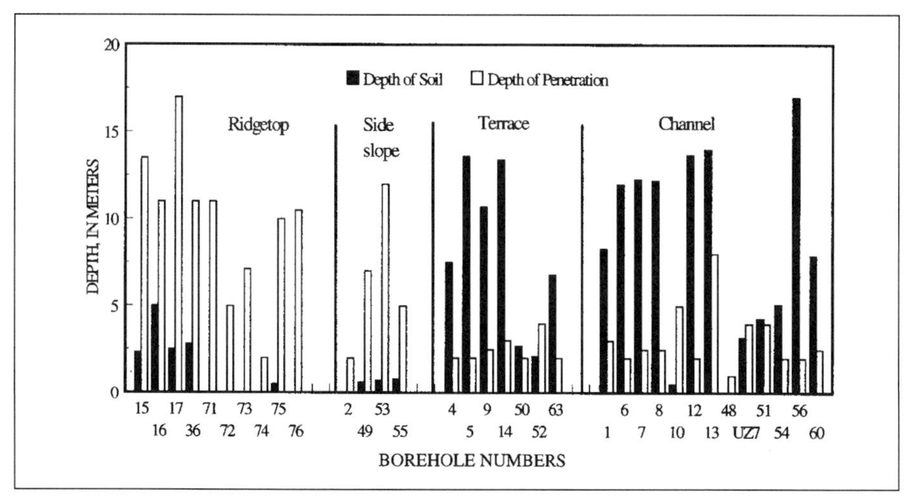

Figure 9. Average depth of penetration of the wetting front for years 1990 to 1993, and soil depth for 34 boreholes (Flint and Flint, 1995)

If the soil is shallow or missing, water can reach high saturations at the alluvium-bedrock interface providing conditions that initiate the flow of water into bedrock fractures. Once in the fractures, water can flow quickly downward below the zone of evapotranspiration. Figure 10 shows an example of the water content variation with depth in boreholes having no soil cover, when bedrock properties are different. The welded, fractured, low-porosity bedrock has the least storage capacity and the water is not held near the surface, but probably results in runoff or ET. The moderately welded, 20-percent porosity bedrock, which imbibes water, probably reduces the surface runoff, yet has some fracturing which creates the deepest water penetration. The nonwelded, high-porosity bedrock is relatively unfractured, and therefore, stores water near the surface resulting in minimal penetration.

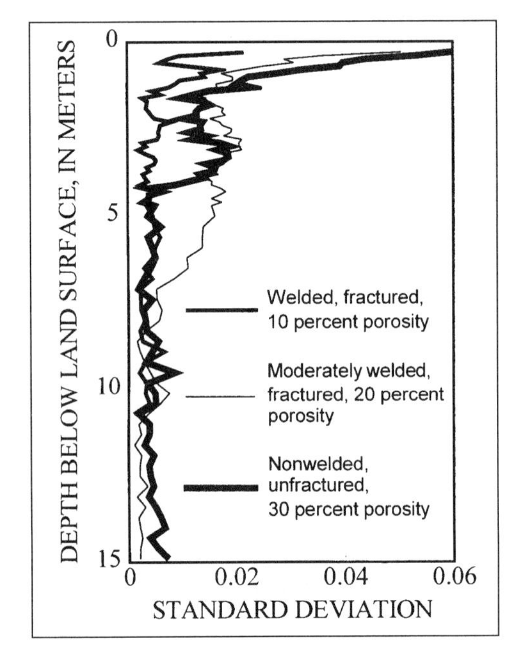

Figure 10. Standard deviation of volumetric water content measured over time, indicating differences in boreholes in various bedrock types

Estimates of Shallow Infiltration

Net infiltration was estimated by determining the increase in the water content below the ET zone, which is assumed to be 2 m below the soil-bedrock interface if the soil is thinner than 6 m, or 6 m into the soil for deeper soils. The sum of increases in volumetric water content for a depth interval multiplied by the length of this interval (McElroy, 1993) is a minimum (conservative estimate) flux, because fracture flow that bypasses the surrounding rock matrix is not included in the measurement. As shallow fractures at this site are often filled with carbonate materials, this omission is not considered significant, particularly on a volumetric basis. As an example, average water content is calculated for borehole N15 from a depth of 2.7 m

(0.7 m of alluvium and 2 m of bedrock) to the bottom of the hole (19 m) over a 3-year period (Figure 11). Four precipitation events, two in 1993 and two in 1995 (both El Niño years), increased water content for this interval. The decrease in water content measured over time is assumed to be an estimate of drainage. The slope of the line in Figure 12 reflects a decrease in the total water stored below 2.7 m and indicates a downward drainage flux of 23.2 mm/year. The actual water contents for this example are presented in Figure 12. The continual downward migration of water below 12 m can be seen between the initial event in 1993 and the subsequent events in 1995 (Figure 11). Because this water is still present in the measurement zone of the neutron probe, it is not calculated as part of the 23.2 mm/year, which is only calculated as water no longer detected by the neutron probe below 2.7 m.

Using calculations of flux for all boreholes and regression analysis, it was determined that five factors significantly affected net infiltration: geomorphologic location, soil thickness, precipitation, aspect and bedrock geology (Hudson *et al*. 1994). The values of flux affected by geomorphologic location and soil thickness are presented in Figure 13. This figure clearly shows that the soil thickness influences net infiltration. It also shows that for the suite of 90 boreholes for which flux (infiltration rate below the root zone) was calculated (8 holes were considered unsuitable for evaluation due to apparent leakage down the casing), only 4 channel holes with deeper than 2 m of soil cover had flux. These 4 holes were located in washes that received several runoff events. Channels located far up in the washes where there was no soil cover had significant fluxes. The average total volume of water, in m^3, contributed by the various geomorphologic locations per year can be calculated as the average flux for the boreholes from each location in Figure 13, in mm/year, mul-

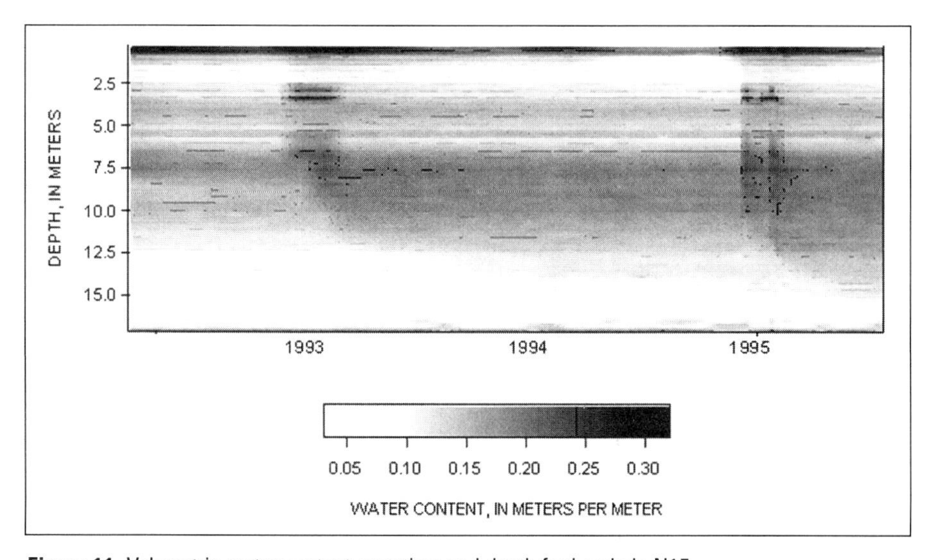

Figure 11. Volumetric water content over time and depth for borehole N15

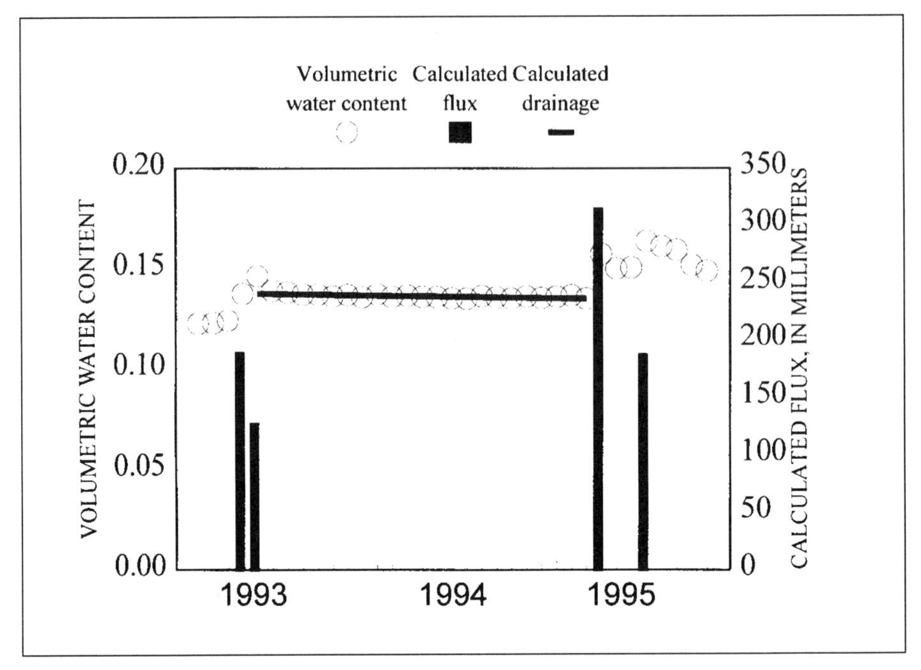

Figure 12. Volumetric water content of bedrock in borehole N15 and calculated flux (drainage, calculated as slope of points indicated, is 23.2 mm per year for the total depth of bedrock in the boreholes)

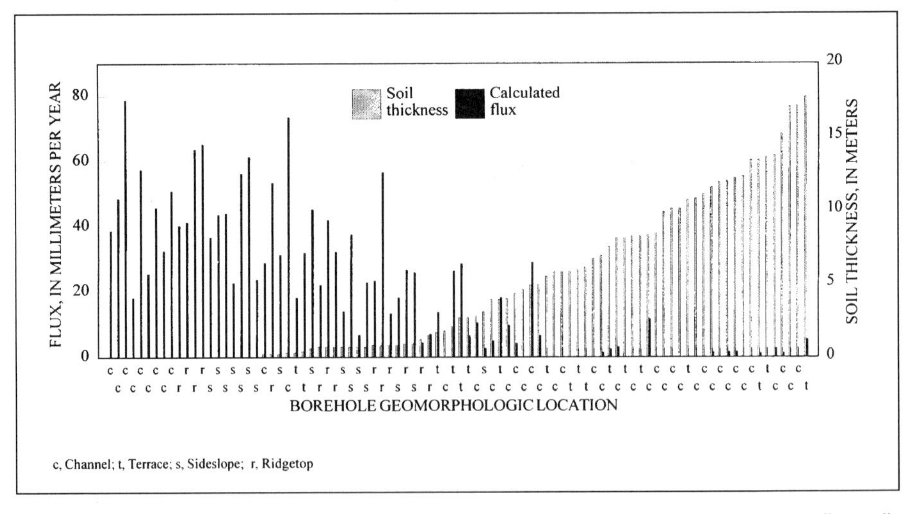

Figure 13. Flux, calculated from neutron borehole water content changes, and corresponding soil thickness for boreholes in four geomorphologic locations

tiplied by the area within the site occupied by those locations, as well as the percent of the total flux for the site per year. The results are: ridgetop, 1,109,170 m^3 (19 percent); sideslope, 4,310,280 m^3 (73 percent); terrace, 422,090 m^3 (7 percent); and channel, 65,006 m^3 (1 percent). This indicates that the sideslope contributes a far greater volume of net infiltration than the other geomorphologic locations. The channels, despite the large volumes of water available from concentrated runoff, contribute less.

OTHER LESSONS LEARNED

Borehole Rugosity

The accurate determination of volumetric water content is dependent upon the errors of neutron logging, which are caused by borehole rugosity (air gaps behind the casing). A neutron count rate is inversely proportional to the size of the air gap. A quantitative means of compensating the logs for the air gaps to provide a more accurate measurement of water content was developed by comparing the logs with four-direction, collimated, dual-spaced gamma-gamma density logs for the same borehole (Ellett *et al.* 1995). By comparing density measurements from the gamma-gamma logs with density measurements of core, void space behind the casing can be estimated. Empirical curves were developed for changes in count rates due to air gap size and predicted air gap sizes were compared favorably with three-arm caliper logs that measured borehole diameter when the casing was removed. An equation describing the change in water content as a function of air-gap size was developed and applied to the neutron logs to compensate for the air gaps. Compensated logs compared well to volumetric water content measured on core data (Figure 14).

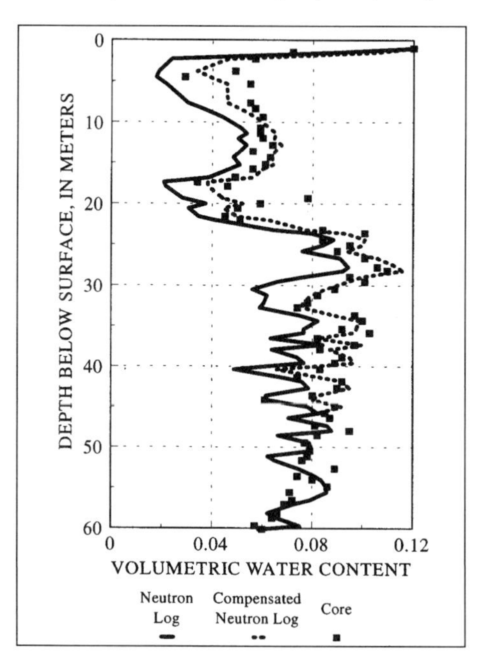

Figure 14. Volumetric water content measured on core samples and neutron logs, uncompensated and compensated for air gaps behind the casing

Another example of the significance of correcting for borehole rugosity is presented by Hudson *et al.* (1994). They conducted a ponded infiltration experiment at a neutron hole and monitored infiltration with time (Figure 15A). The corresponding model for this experiment could not reproduce the measured neutron hole data without using unrealistic soil hydraulic parameters. Analysis of the borehole using gamma-gamma logs provided an indication of washout zones behind the casing. The casing was carefully removed and the caliper readings were taken to verify the washout zones (Figure 15B). Using the appropriate hydraulic parameters for the soil allowed a realistic model to be constructed for most of the soil (Figure 15A). The results from the model closely matched the actual volume of water infiltrated in the experiment.

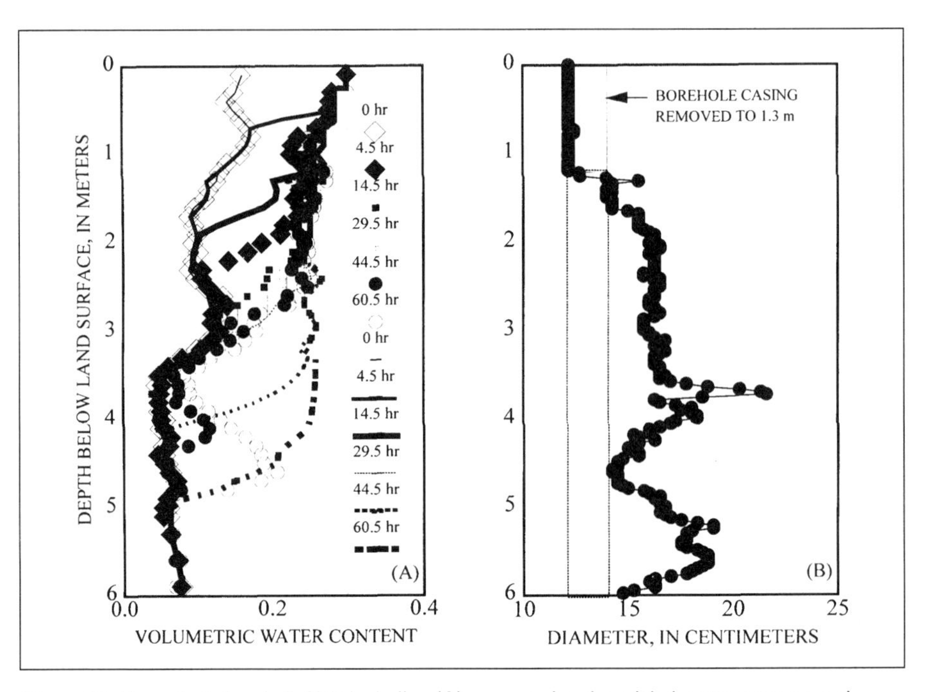

Figure 15. Rugosity in borehole N85 including (A) measured and modeled water content, and (B) caliper measurements of borehole diameter

SUMMARY AND CONCLUSIONS

Neutron moisture logging has been shown to be a useful tool in deep, large-diameter boreholes in arid environments for quantifying changes in water content. A comprehensive, large-scale data set of volumetric water content for 98 neutron boreholes at Yucca Mountain for the years 1984 to 1995 was used to develop a conceptual model of shallow infiltration processes in a mountainous, arid environment.

Using this data set, calculations of net infiltration were made, ranging between 0 and 80 mm/yr.

According to the conceptual model of infiltration developed using the data set and field observations, precipitation is the most important factor determining the amount and distribution of infiltration. Depth to the soil-bedrock interface is the second most important factor, followed by the hydraulic properties of the bedrock and the variation in ET due to differences in slope and aspect, and, thus, energy loads. Locations with deep alluvium, such as terraces and most channels, have little to no net infiltration because of large storage capacity and high ET. The exception are active channels, which receive periodic runoff, but these constitute a very small percentage of the total area of the site. Sideslopes have runoff, but little surface storage and water can penetrate the fractured bedrock to get below the zone affected by ET processes, and contribute the most to the total volume of water infiltrated over the site. Ridgetops are relatively flat, thus reducing runoff, but because soils are comparatively thin and underlying bedrock is fractured, these geomorphologic locations generally have deep infiltration.

REFERENCES

Chanasyk, D.S. and M.A. Naeth (1996), Field measurement of soil moisture using neutron probes, *Canadian J. Soil Science*, vol. 76, p. 317-323.

Ellett, K.M., J.W. Bullard, J.A. Hevesi and A.L. Flint (1995), *Compensating neutron moisture meter logs for borehole rugosity*, EOS, American Geophysical Union Fall meetings, vol. 76, no. 46, San Francisco, CA., December 1995.

Evett, S.R. and J.L. Steiner (1995), Precision of neutron scattering and capacitance type soil water content gauges from field calibration, *Soil Science Soc. Am. J.*, vol. 59, p. 961-968.

Flint, L.E. and A.L. Flint (1995), Shallow infiltration processes at Yucca Mountain— Neutron logging data, 1984-93, *U.S. Geological Survey Open-File Report 95-4035*, 46 p.

Hammermeister, D.P., D.O. Blout, and J.C. McDaniel (1985), *Drilling and coring methods that minimize the disturbance of cuttings, core and rock formation in the unsaturated zone, Yucca Mountain, Nevada*, Proceedings of the National Water Well Association Conference on Characterization and Monitoring of the Vadose (Unsaturated) Zone, National Water Well Association, p. 507-541.

Hevesi, J.A., Flint, A.L., and Istok, J.D., 1991, Precipitation estimation in mountainous terrain using multivariate geostatistics: II. Isohyetal maps, *Journal of Applied Meteorology*, vol. 31, no. 7, p. 677-688.

Hudson, D.B., A.L. Flint, and W.R. Guertal (1994), *Modeling a ponded infiltration experiment at Yucca Mountain, NV*, High Level Radioactive Waste Management, Proceedings of the Fifth Annual International Conference, Las Vegas, Nev., May 22-26, 1994, LaGrange Park, Ill., American Nuclear Society.

Klenke, J.M. and A.L. Flint (1991), A collimated neutron tool for the measurement of soil water content, *Soil Science Society of America Journal*, vol. 55, no. 4, p. 916-922.

McElroy, D.L. (1993), *Soil moisture monitoring results at the radioactive waste management complex of the Idaho National Engineering Laboratory, FY-1993*, EGG-WM-11066, EG&G Idaho, Inc., Idaho Falls, Idaho, 110 p.

Nixon, P.R., and G.P. Lawless (1960), Translocation of moisture with time in unsaturated soil profiles, *J. Geophys. Res.*, vol. 65, no. 2, p. 655-661.

O'Farrell, T.P. and E. Collins (1983), 1982 Biotic survey of Yucca Mountain, Nevada Test Site, Nye County, Nevada, EG&G Energy Measurements, Santa Barbara Operations, Goleta, Calif., 41 p.

Sharma, M.L., M. Bari, and J. Byrne (1991), Dynamics of seasonal recharge beneath a semi-arid vegetation on the Gnangara Mound, Western Australia, *Hydrological Processes*, vol. 5, p. 383-398.

Sternberg, P.D., M.A. Anderson, R.C. Graham, J.L. Beyers, and K.R. Tice (1996), Root distribution and seasonal water status in weathered granitic bedrock under chaparral, *Geoderma*, vol. 72, no. 1-2, p. 89-98.

Tyler, S.W. (1988), Neutron moisture meter calibration in large diameter boreholes, *Soil Science Society of America Journal*, vol. 52, no. 3, p. 890-892.

Winograd, I.J. and W. Thordarson, (1975), Hydrogeologic hydrochemical framework, south-central Great Basin, Nevada-California, with special reference to the Nevada Test Site, *U.S. Geological Survey Professional Paper 712-C*, 126 p.

CHARACTERIZATION AND MONITORING OF UNSATURATED FLOW AND TRANSPORT PROCESSES IN STRUCTURED SOILS

Philip M. Jardine, R.J. Luxmoore, and **J.P. Gwo,**
Oak Ridge National Laboratory

G.V. Wilson, *Desert Research Institute*

INTRODUCTION

The disposal of low-level radioactive waste generated at U.S. Department of Energy facilities within the Weapons Complex has historically involved shallow land burial in unconfined pits and trenches within the vadose zone. The lack of physical or chemical barriers to impede waste migration has resulted in the movement of radionuclides into the surrounding soil and bedrock and the formation of secondary contaminant sources. At the Oak Ridge National Laboratory (ORNL), located in eastern Tennessee, USA, the extent of the problem is massive, and thousands of underground disposal trenches have contributed to the spread of radioactive contaminants across tens of kilometers of landscape. Subsurface transport processes are driven by large annual rainfall inputs (~1400 mm/year) where as much as 50 percent of the infiltrating precipitation results in groundwater and surface water recharge (10 and 40 percent, respectively). Lateral storm flow and groundwater interception with the trenches enhances the migration of waste constituents into the surrounding subsurface environment, which promotes the formation of secondary contaminant sources. The subsurface media at ORNL are comprised of fractured saprolites and structured clays, which are conducive to rapid preferential flow, coupled with significant matrix storage. The macropores and fractures within the media surround highly porous matrix blocks that typically have water contents from 30 to 50 percent. Large hydraulic and geochemical gradients form between the various media regions, often resulting in nonequilibrium conditions for solute transport.

A significant limitation in defining the remediation needs of the secondary sources results from an insufficient understanding of the transport processes that control contaminant migration. Without this knowledge, it is impossible to assess the risk associated with the secondary source contribution to the total offsite migration of contaminants. The objective of our research was to improve the understanding of contaminant transport processes in highly structured, heterogeneous subsurface environments that are complicated by preferential flow and matrix diffusion. Our approach involves multiscale experimental research and numerical analysis of coupled hydrological and geochemical processes controlling the fate and transport of contaminants in structured subsurface soils. Through the use of novel tracer techniques and experimental manipulation strategies, in laboratory-, intermediate-, and field-scale experiments, we attempt to unravel the impact of coupled transport processes on the nature and extent of secondary contaminant sources.

EXPERIMENTAL STRATEGIES FOR ASSESSING FLOW AND TRANSPORT IN THE VADOSE ZONE

Two field installations on contrasting watersheds have been developed on the Oak Ridge Reservation, in Oak Ridge, Tennessee, (to investigate one-, two-, and three-dimensional flow and transport processes in unsaturated subsurface environments (Luxmoore and Abner 1987). The facilities were designed for quantifying nutrient and contaminant fluxes in soils with spatially variable physical and chemical characteristics. One site is located on the Walker Branch Watershed and is a deep, highly weathered Ultisol, with Chepultepec dolomite as the parent material. The soils are highly structured, acidic, cherty silt loams. The experimental subwatershed drains an area of 0.47 ha from an elevation of 334 to 322 m. The second site is located on the Melton Branch Watershed and is underlaid by Maryville limestone, which is a limy shale formation. The soils are weakly developed Inceptisols that have been weathered from interbedded shale-limestone sequences. The limestone has been weathered to massive clay lenses devoid of carbonate, and the more resistant shale has weathered to an extensively fractured saprolite. The experimental subwatershed drains an area of 0.63 ha from an elevation of 275 to 258 m (Figure 1a). Both soil types can accommodate large rainfall inputs since they are highly structured and conducive to rapid pore water fluxes in macropores and fractures.

Functional relationships between water content, pressure head, and hydraulic conductivity suggest that the soil behaves as a three-region medium consisting of macro-, meso-, and micropores (Wilson *et al.* 1992). Macro- and mesopores can be conceptualized as primary and secondary conduits or fractures, and micropores as the soil matrix. The three-region conceptualization of these structured soils is supported by field measurements of Wilson and Luxmoore (1988), who showed the convergent flow from mesopores into macropores and the subsequent bypass of the soil matrix. The tension infiltrometer developed by Perroux and White (1988) was used to assess multiregion flow at the field facilities. This device allows water to be infiltrated into the soil under tension, thereby excluding the larger pores from contributing to the infiltration process. By determining the infiltration rate under positive pressure (saturated) and selected negative pressures, successively, the hydrologic contribution of corresponding pore classes may be quantified. Wilson and Luxmoore (1988) found that macropores (pores $>$ 1.5 mm in diameter) accounted for as much as 85 percent of the saturated flow and that for unsaturated, conditions, mesopore fluxes remained large ($\sim 2 \times 10^{-5}$ m sec^{-1}), and were considered sufficient to infiltrate most precipitation without macropore filling. In order to improve a model for

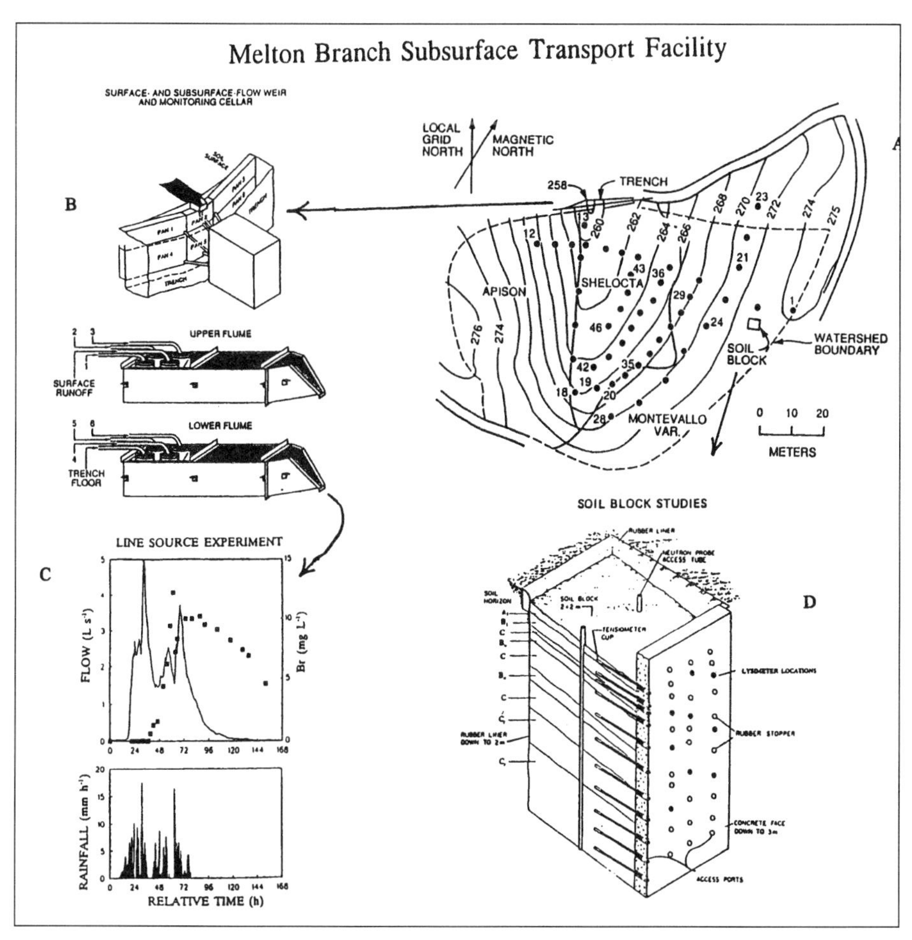

Figure 1. Subsurface transport facility for studying storm-driven tracer transport in the unsaturated zone of a fractured weathered shale (a-d). Multiple pore region sampling capabilities are available at both the pedon (d) and field scales. The unique subsurface weirs (b) allow for measurement of real-time subsurface tracer fluxes.

solute transport in unsaturated, structured soils, multiscale experimental and numerical investigations have been employed to quantify the rates and mechanisms of intrapore-region interactions.

LABORATORY-SCALE ASSESSMENT

Undisturbed columns (typically 15 cm in diameter by 40 cm in length) have been excavated from the field facilities and used in tracer transport experiments to assess the effect of hydrological and geochemical factors on the fate and transport of nonreactive and reactive solutes (Jardine *et al.* 1988; Jardine *et al.* 1993 a, b; Reedy *et al.* 1996). The primary purpose of the laboratory-scale research is to quantify transport mechanisms that are expected to affect unsaturated flow and transport at the field-scale. Several techniques have been employed to assess nonequilibrium processes that result from the large difference in hydraulic conductivity of macro- and mesopores vs. the soil matrix. These techniques include (1) controlling flow-path dynamics with manipulations of pore-water flux and soil-water tension, (2) isolating diffusion and slow geochemical processes with flow interruption, (3) using multiple tracers with different diffusion coefficients, and (4) using multiples tracers with grossly different sizes.

Variations in Pressure-Head

Controlling flow-path dynamics by manipulation of the soil water content with pressure-head variations is an excellent technique to assess nonequilibrium processes (Seyfried and Rao 1987; Jardine *et al.* 1993 a, b). The basic concept of the technique is to collect water and solutes from selected sets of pore classes to determine how each set contributes to the bulk flow and transport processes for the whole system. In heterogeneous soils, a decrease in pressure-head (that is, it becomes more negative) causes larger pores, such as fractures or macropores, to drain and become nonconductive for solute transport. Since advective flow processes tend to dominate in large pores, a decrease in pressure-head restricts flow and transport to smaller pores, and limits the disparity of solute concentrations among pore groups. By minimizing the concentration gradient in the system, the extent of physical nonequilibrium is decreased.

The experimental set-up is shown in Figure 2a, where pressure-heads are imposed on fritted glass plates at the inlet and outlet boundaries of a soil column using a constant head mariotte device and a vacuum source, respectively. Figure 2b shows the breakthrough curves of a nonreactive (Br^-) tracer at three different pressure-heads in the column of undisturbed weathered fractured saprolite from the Melton Branch field facility. The increasing asymmetry of the breakthrough curves with increasing saturation (less negative pressure-head) is indicative of enhanced preferential flow, coupled with mass loss, into the matrix. As the soil saturation decreases, breakthrough-curve tailings become less significant because the effect of larger pores (fractures) involved in the transport process decreases. The mass transfer between pores becomes increasingly negligible. These studies showed that the application of -10 and -15 cm pressure heads resulted in five- and 40-fold decreases in the mean pore water flux, respectively, with a relatively small change in soil water content ($0.55\ cm^3/cm^3$ at $h = 0$ to $0.51\ cm^3/cm^3$ at $h = -15$ cm). These results suggest that most of the water flux may be channeled through pores that hold water with tensions less than -10 cm (primary fractures, macropores) even though their surface area and contribution to the total soil system porosity is very small (Wilson and Luxmoore 1988; Wilson *et al.* 1989).

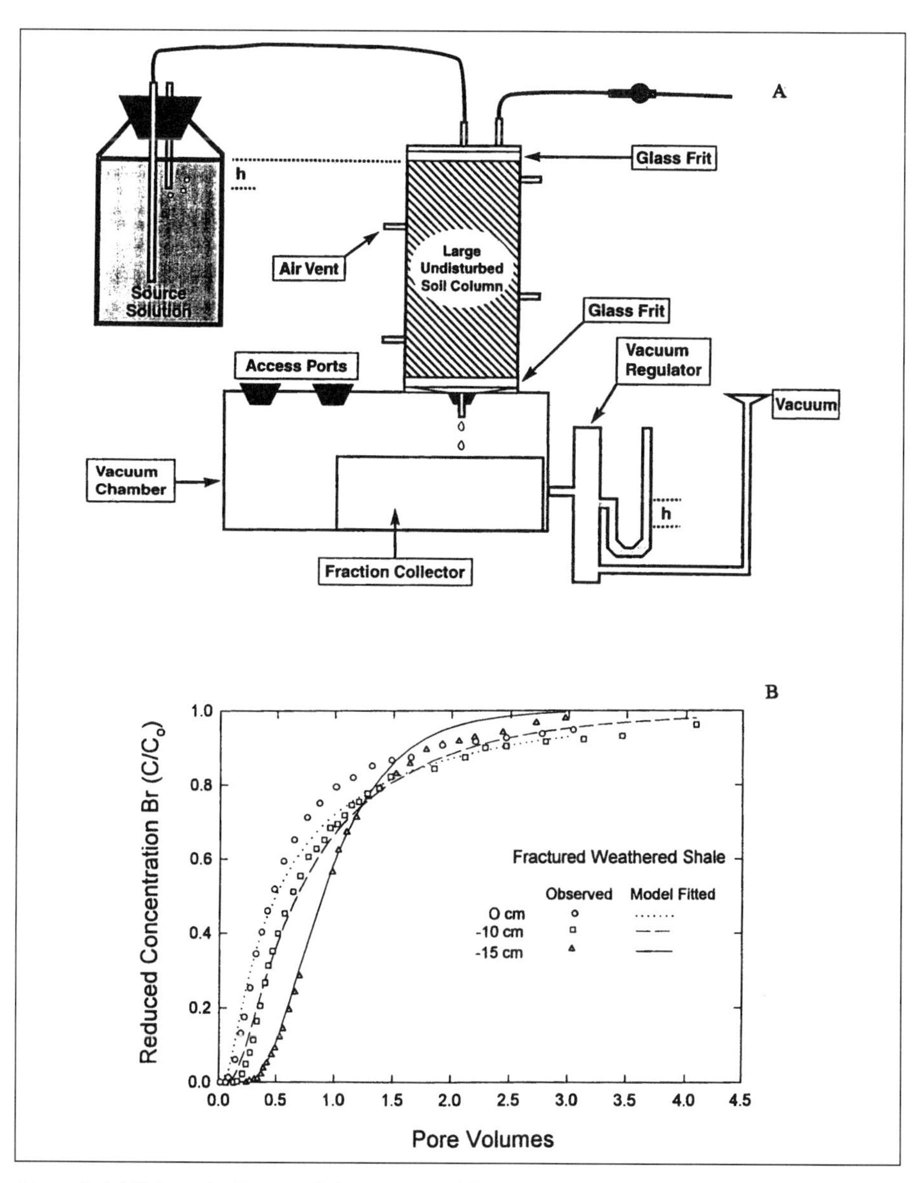

Figure 2. (a) Schematic diagram of the unsaturated flow apparatus illustrating a constant pressure head (h) design and (b) breakthrough curves for a nonreactive Br^- tracer as a function of h in an undisturbed column of fractured weathered shale. For conditions where h = 0 cm, transport occurred under saturated flow and the entire fracture network was conductive. When h = -10 cm the primary fracture network became nonconductive, and when h = -15 cm primary fractures and a portion of the secondary fractures became non-conductive. The model-fitted curves used the classical convective-dispersive model with optimization of the dispersion coefficient to the observed data (modified from Jardine, P.M., et al., *Soil Sci. Soc. Am. J. 57,* 945-953, 1993a. With permission).

Flow Interruption

Another useful technique for isolating diffusion or slow time-dependent geochemical reactions involves flow interruption during a portion of a tracer displacement experiment (Brusseau *et al.* 1989, Hu and Brusseau 1995, Reedy *et al.* 1996). The technique involves inhibiting the flow process during an experiment for a designated period of time, and allowing a new physical or chemical equilibrium state to approach. When physical nonequilibrium processes are significant in a soil, the flow-interruption method will cause concentration perturbation for a conservative tracer when the flow is resumed. Interrupting flow during tracer injection will decrease a tracer concentration when the flow is resumed, whereas interrupting flow during tracer displacement (washout) will increase the tracer concentration when the flow is resumed. The concentration perturbations that are observed after flow interruptions are indicative of solute diffusion between pore regions of heterogeneous soils. Conditions of preferential flow create concentration gradients between pore regions (physical nonequilibrium), resulting in diffusive mass transfer between the regions. Therefore, during injection, tracer concentrations within advection-dominated flow paths (that is, fractures and macropores) are higher than those within the matrix. Upon flow interruption, the relative concentration decrease indicates that solute diffusion occurs from larger, more conductive pores into the smaller pores. During tracer displacement, or washout, the concentrations within the preferred flow paths are lower than those within the matrix. Thus, solute diffusion is occurring from smaller pores into larger pores, and a concentration increases when a flow interruption has been imposed (Reedy *et al.* 1996).

Multiple Tracers with Different Diffusion Coefficients

A powerful technique for quantifying physical nonequilibrium in structured subsurface media is the simultaneous use of multiple tracers with different diffusion coefficients (Abelin *et al.* 1987; Maloszewski and Zuber 1990, 1993; Jardine *et al.* 1999). In general, when the technique is used to quantify physical nonequilibrium processes in soils and rock, two or more conservative tracers that have different diffusion coefficients are simultaneously displaced through the porous media. When physical nonequilibrium processes are significant in porous media, tracers with larger diffusion coefficients will be preferentially lost from advective flow paths (that is, fractures) due to more rapid diffusion into the surrounding solid-phase matrix. Likewise, tracers with smaller molecular diffusion coefficients (such as larger molecules) will remain in the advective flow paths for longer time periods due to slower diffusion into the matrix porosity. When advective processes are dominant in a system and matrix diffusion is negligible, multiple tracer breakthrough profiles will not differ considerably. (See Brusseau *et al.* 1993 for an example.)

The utility of multiple tracers for quantifying physical nonequilibrium processes in structured media can be observed in Figure 3. The simultaneous transport of two

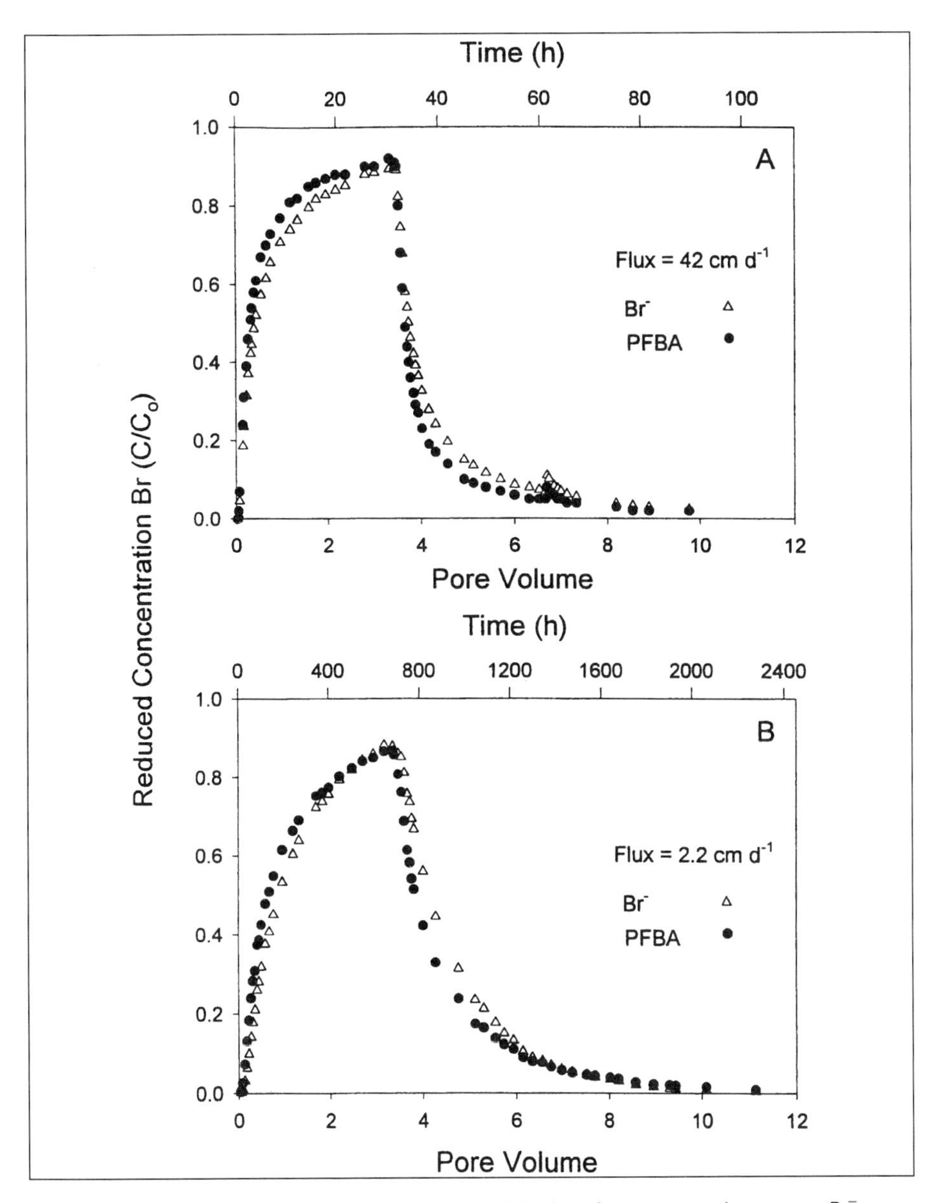

Figure 3. Breakthrough curves for the simultaneous injection of two nonreactive tracers, Br^- and PFBA, at a flux of (a) 42 cm d^{-1} and (b) 2.2 cm d^{-1} in an undisturbed column of fractured weathered shale. The free water diffusion coefficient for Br^- is 40 percent larger than that of PFBA. (O'Brien, R., 1994, ORNL, unpublished data).

nonreactive tracers, Br^- and (PFBA), was investigated in large, undisturbed columns of fractured, weathered shale at two different pore water fluxes. The molecular diffusion coefficient for PFBA is 40 percent smaller than the diffusion coefficient for Br^- (Bowman 1984). Differences in the breakthrough curves for these solutes can be attributed to differences in the rates of tracer diffusion into the soil matrix. Since the PFBA diffused more slowly into the weathered shale matrix, its breakthrough at the column exit was initially more rapid than Br^-, but required a greater amount of time to approach equilibrium (see Figure 3 a, b). Thus, Br^- had a larger mass loss to the matrix at any given time, and exhibited a more retarded breakthrough relative to PFBA. However, Br^- will reach equilibrium more rapidly than PFBA, and the tracer breakthrough curves will eventually cross at longer times. In contrast, the mobility of these two nonreactive tracers would be identical in a column of unstructured media since pore class heterogeneity would be minimal, thus limiting the significance of physical nonequilibrium during transport. (See Brusseau *et al.* 1993 for an example.)

Multiple Tracers with Grossly Different Sizes

Another sensitive technique for confirming and quantifying physical nonequilibrium in heterogeneous soil systems is the use of multiple tracers with distinctly different sizes. Specifically, this technique uses both dissolved solutes and colloidal tracers so that flow-path accessibility can be controlled. Viruses, bacteria, fluorescent microspheres, DNA-labeled microspheres, radiolabeled iron-oxide particles, and synthetic polymers have all been used as colloidal tracers in various subsurface media (Barraclough and Nye 1979; Gerba *et al.* 1981; Smith *et al.* 1985; Bales *et al.* 1989; Harvey *et al.* 1989, 1993, 1995; Toran and Palumbo 1991; McKay *et al.* 1993 a, b; Hinsby *et al.* 1996; Yang *et al.* 1996; Reimus 1996). Colloidal particles are typically large enough to be excluded from the matrix porosity of soils and geologic material. If they are not severely retarded by the porous media, colloidal particles serve as excellent tracers for quantifying advective flow velocities in systems conducive to preferential flow. When colloidal tracers are coupled with dissolved solutes that can interact with the matrix porosity, a unique technique emerges for assessing physical nonequilibrium processes in subsurface media.

Intermediate-Scale Assessment

A logical progression from laboratory-scale undisturbed columns is the use of intermediate-scale *in situ* pedons for assessing the interaction of coupled processes on the fate and transport of solutes in the fractured weathered shales (Figure 1d). This research scale, unlike the column scale, encompasses more macroscopic structural features common to the field (such as dip of bedding planes, more continuous fracture network, and convergent flow processes), yet allows for a certain degree of experimental control since the pedon can be hydrologically isolated from the surrounding environment. Each field facility contains a 2 m by 2 m by 3 m deep-isolated, undisturbed pedon, which is a soil block with three excavated sides refilled with compacted soil and a concrete wall with access ports placed in good contact with the front soil face. The pedon was instrumented with a variety of solution samplers designed to monitor water and solutes through various pore regimes as a function of depth. Fritted glass-plate lysimeters of varying porosity were held under different tensions to derive solutions from various pore regimes. Coarse-fritted glass samplers were held at zero tension and collected free-flowing advective pore water (macropores or primary fractures); medium porosity frits were held at 20 cm tension and collected pore water from mesopores or secondary fractures; and fine frits were held at 250 cm tension and collected pore water from the soil matrix.

Numerous tracer experiments have been conducted at these facilities using both nonreactive and reactive tracers with various steady-state infiltration rates or transient flow conditions driven by storm events (Jardine *et al.* 1989, 1990 a, b; Wilson, unpublished data). The purpose of the experiments was to determine transport properties and mass-transfer rates for the various pore regimes. An example of tracer mobility (Br^-) through a pedon of structured clay soil is shown in Figure 4. The tracer was applied slowly to the surface soil matrix in order to establish a concentration gradient between the large and small soil-pore regimes. The results show that during the vertical flux of storm water through the soil profile, solutes were transported by advection and diffusion from small-pore regions to large-pore regions via hydraulic and concentration gradients, respectively, with small pores being a source for solute transport in large pores. Solutes that were rapidly transported to greater depths through the large pores increased solute reserves in adjacent small pores at the same depth through a reversal of the mechanism described above. Thus, a cascading process of Br^- transfer between large and small pores was observed during sequences of drainage and redisturbution over the 375-day study period (Jardine *et al.* 1990a).

FIELD-SCALE ASSESSMENT

Waste migration issues are typically field-scale problems that are complicated by large-scale media heterogeneities that cannot be replicated at the laboratory or intermediate scale. In order to validate our conceptual understanding of vadose zone transport that was derived from laboratory- and pedon-scale observations, field-scale solute fate and transport experiments must be performed. The ORNL field facilities were constructed to assess storm-driven solute mobility in the vadose zone of the fractured saprolites (Figure 1). The facilities, which contain a buried line source for tracer release, to simulate leakage of trench waste from contaminated areas, are equipped with an elaborate array of water- and solute-monitoring devices. At each site, lateral subsurface flow is intercepted by a 2.5 m deep by 16 m long trench that

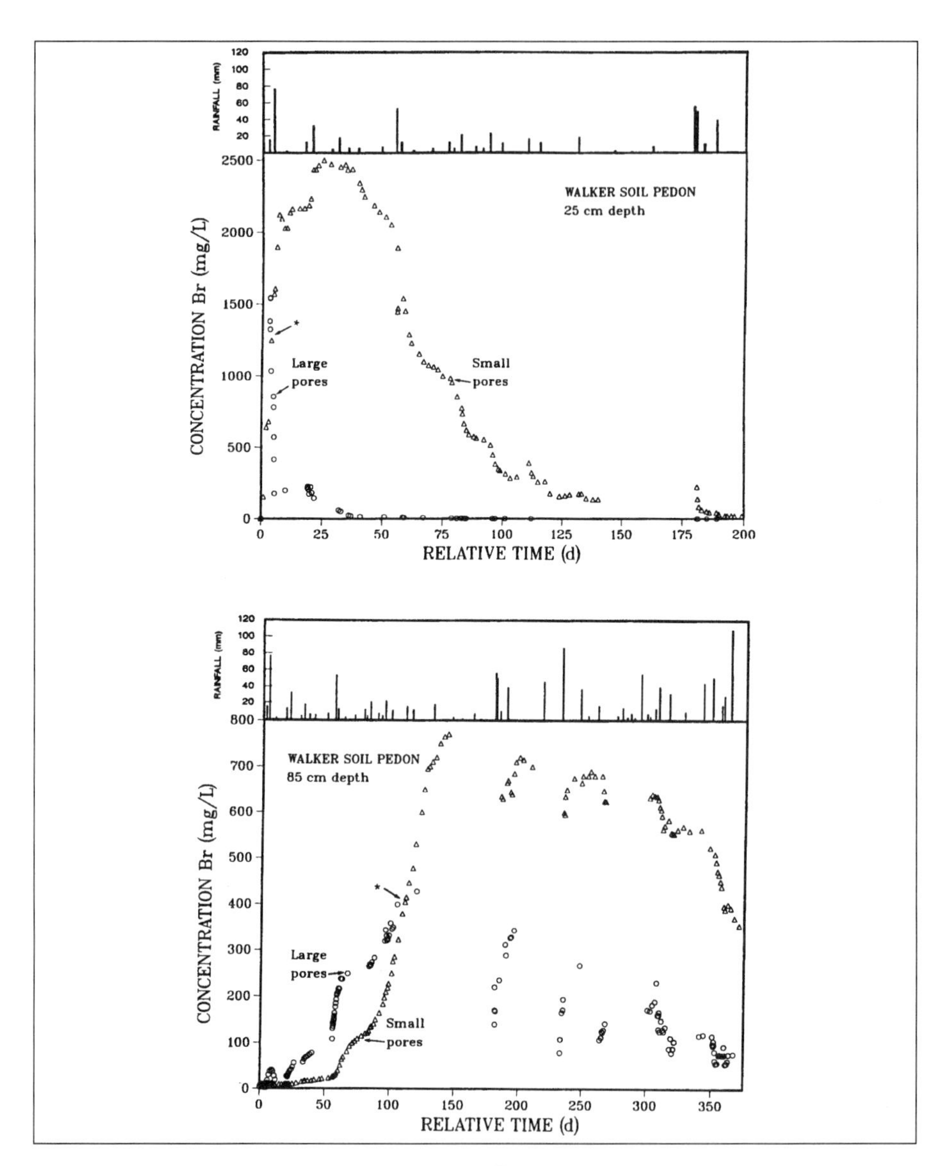

Figure 4. Storm-driven, unsaturated transport of Br^- at two depths through large and small pores of an *in situ* pedon using coarse and fine fritted glass solution samplers, respectively, as solute collection devices. The symbol (*) designates the point where a shift in the concentration gradient occurs between large and small pores. (Modified from Jardine, P.M., et al.(1990), *Geoderma, 46*,103-118. With permission).

has been excavated across the outflow region of the subwatershed (Figure 1b). Six massive stainless steel pans with 10 cm lips were pressed back into the soil face to capture subsurface drainage from different portions of the landscape. Subsurface drainage from the pans, as well as overland storm flow and drainage under the pans, is routed into tipping-bucket rain gauges that are situated in two H-flumes and are equipped with ultrasonic sensors for measuring water levels. The tipping buckets and ultrasonic sensors are equipped with computer-data-acquisition capability, which allows real-time monitoring of tracer fluxes during storm events. Besides the ability to capture subsurface drainage, the field facilities are equipped with numerous multilevel solution samplers, tensiometers, and piezometer wells, with the latter used to assess perched water table dynamics during storm events. The field facilities have been well characterized with respect to spatial variability of surface and subsurface hydraulic conductivities, basic geochemical properties, and mineralogical analyses (Watson and Luxmoore 1986; Wilson and Luxmoore 1988; Wilson *et al.* 1989, 1990, 1991 a, b, 1992, 1993, 1998; Jardine *et al.* 1988, 1989, 1990 a, b, 1993 a, b, 1998; Luxmoore *et al.* 1990; Kooner *et al.* 1995; Reedy *et al.* 1996).

In an effort to address field-scale fate and transport processes during transient flow in the fractured, weathered shale media at the Melton Branch facility, Wilson *et al.* (1993) released a Br^- tracer from the ridgetop buried-line source during a storm event and monitored its mobility through the subsurface for nearly 8 months. During the release they observed that a small portion of the total Br^- mass (approximately 5 percent) migrated very rapidly through the hillslope via lateral storm flow with subsequent export through the weirs, which were located 70 m from the line source (Figure 1c). The first arrival of Br^- at the weirs was 3 hours after the initiation of the release, which illustrates the incredibly large hydraulic conductivity of the primary fracture network during hillslope convergent flow. These results are consistent with the measurements of Wilson and Luxmoore (1988) that showed 85 percent of ponded water flux went through primary fractures that constitute only 0.1 percent of the total soil porosity. Subsequent storm events over the 8-month period resulted in the export of approximately 20 percent of the injected Br^- mass (as shown in Figure 5a). While the actual tracer release revealed a rapid transport through the fracture network of the soil, the mass transfer into the low-permeability matrix was significant, since more than 50 percent of the applied tracer was found to reside within matrix porosity of the soil, the tracer occurred primarily at a depth of 1 to 1.5 m, which is also where lateral storm flow occurs through the hillslope (Figure 5b). Large-scale structural heterogeneities of the subsurface media controlled solute mobility, since the tracer was preferentially transported along bedding planes that dipped in the direction opposing the hillslope topography.

Strong evidence indicating matrix-fracture interactions during transient storm flow can be inferred from tracer breakthrough patterns at the subsurface weirs. Storm events that followed the release of tracer resulted in tracer breakthrough pulses that were delayed relative to the subsurface flow hydrograph (Figure 5a). Stable isotope

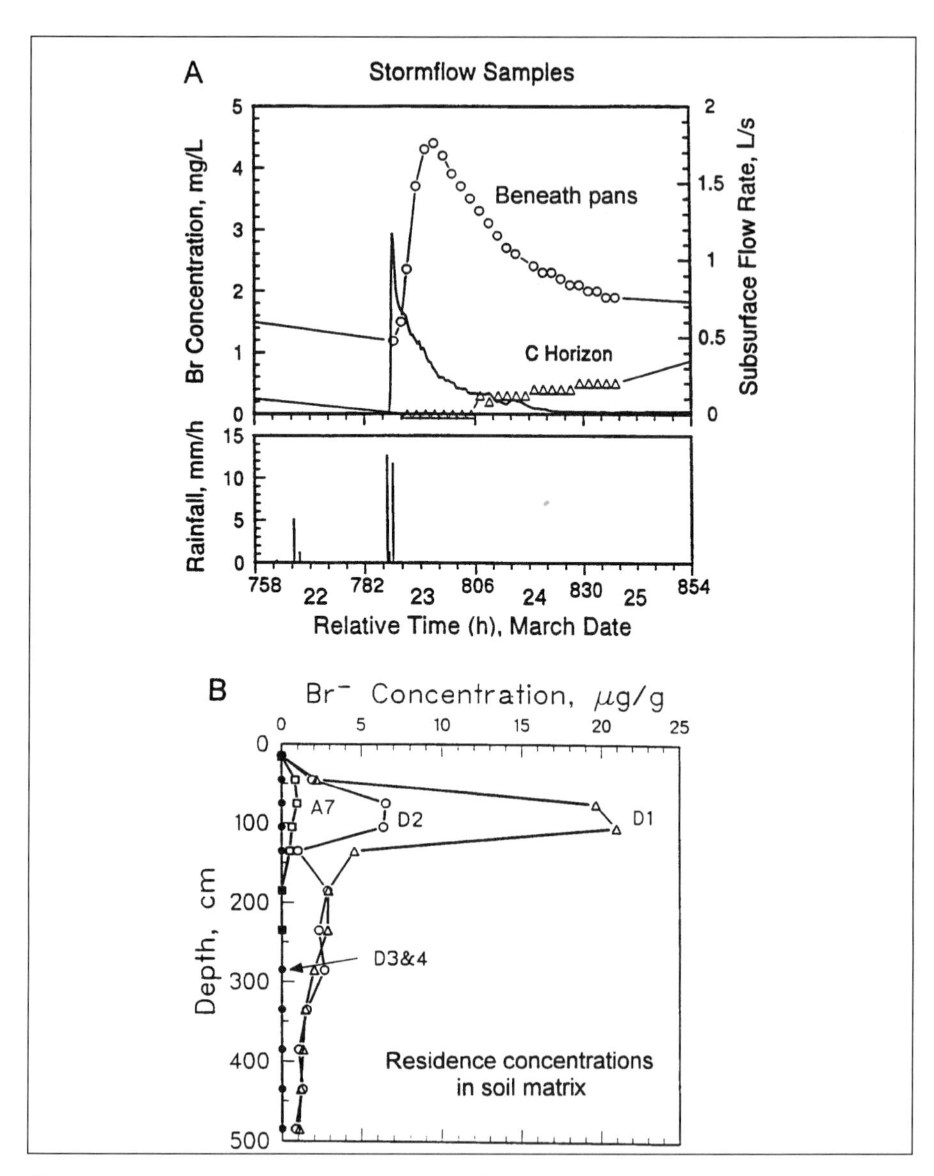

Figure 5. Storm flow and soil matrix tracer results following a release of Br^- at the field-scale tracer injection facility in the fractured weathered shale soil. The upper graph (a) shows an example of a subsurface flow hydrograph (solid line) for the lower flume that resulted from a storm event, with the corresponding Br^- concentrations from the C-horizon and from flow beneath the pans. The lower graph (b) shows Br^- residence concentrations from the soil matrix as a function of depth for six sampling locations downslope of the linesource where tracer was released (modified from Wilson, G.V., et al. (1993), *J. Hydrol., 145*, 83-109. With permission).

and solute chemistry analysis revealed that subsurface flow was predominately new water at peak flow, and almost exclusively old water during the descending limb of the subsurface hydrograph. These results suggested that the storm-driven export of Br^- through the weirs was the result of tracer mass transfer from the soil matrix into the fracture network, with subsequent mobility through the hillslope. The field-scale endeavors provided an improved conceptual understanding of how transient hydrodynamics and media structure controll the migration and storage of contaminants in the subsurface. The field-scale findings were consistent with the multiregion conceptual framework established with laboratory- and intermediate-scale observations of flow and transport through these heterogeneous systems.

MULTIREGION CONCEPTUAL TRANSPORT MODEL IN STRUCTURED VADOSE ZONE SOILS

The multiscale experimental endeavors have provided conceptual frameworks for multiregion flow and for transport mechanisms that control solute mobility in the fractured, weathered shales. Experimentally, we can distinguish three pore-size classes (primary fractures, secondary fractures, and soil matrix). Thus, a representative elemental volume (REV), at any point in the soil consists of three regions, each with its own flow and transport parameters (Figure 6). Intra-region mass transfer is described by flow from a physical point to a neighboring one, and inter-region mass transfer between the various pore regimes is controlled by both advective and diffusive processes, where hydraulic gradients (caused by differences in fluid velocity in different-sized pores) drive advective mass transfer, and concentration gradients drive diffusive processes. Another process inherent in multiregion flow and transport is the time-dependent nature of both the advective and diffusive mass transfer rates between the various pore domains. Time-dependent mass transfer rates take into account changes in concentration gradients as solute mass is transferred between pore regimes. Time-dependent rate coefficients also indirectly account for variabilities in matrix block sizes. The concept of multiregion flow and transport in structured media has been numerically implemented by Gwo *et al.* (1991), with the mathematical formulation of the code described in Gwo *et al.* (1994, 1995 a, b; 1996) and the simulation of experimental data illustrated in Gwo *et al.* (1998 and 1999).

Relevant web sites:

http://www.esd.ornl.gov/facilities/hydrology/WATERSHD/INDEX.HTM

http://www.esd.ornl.gov/programs/ETPI/

http://hbgc.www.esd.ornl/

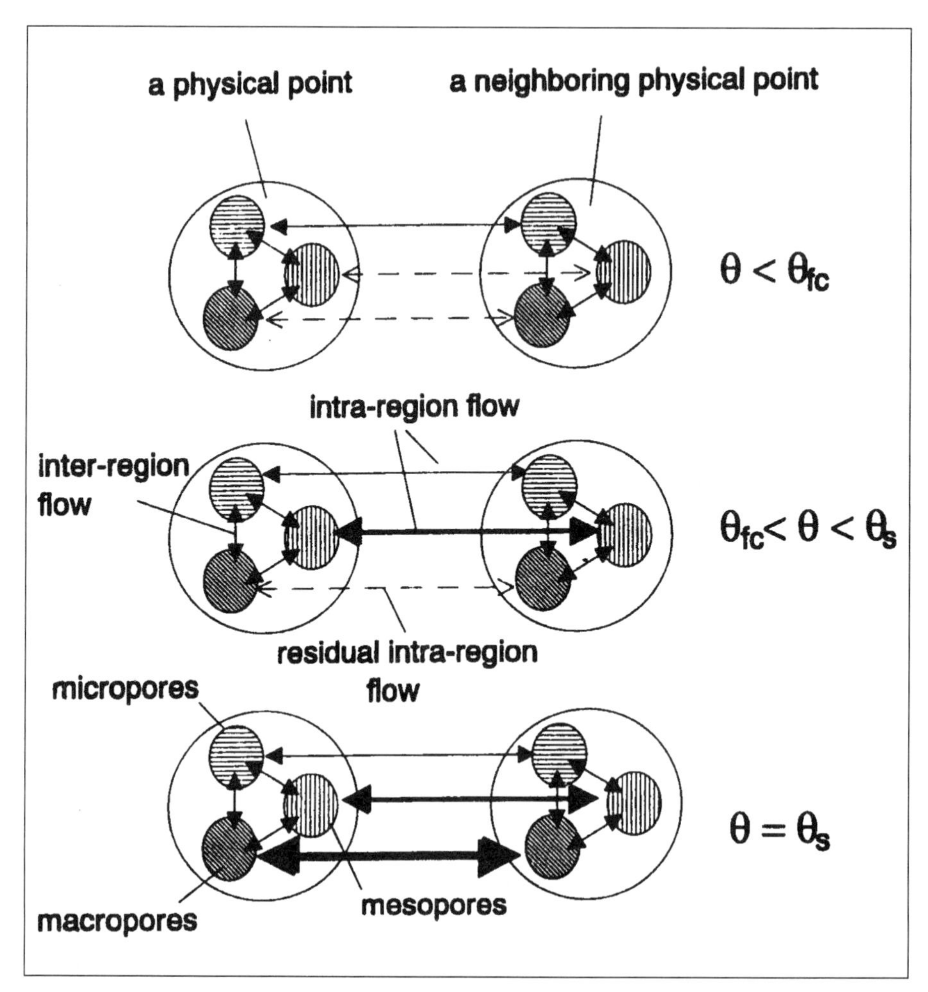

Figure 6. A three-region example of the multi-region concept, in which θ, θ_{fc}, θ_s stand for water content, water contents at field capacity and saturation, respectively. The pore regions can be either active or inactive depending on the overall water content of the medium as shown by the size of the connecting arrows. The mass exchange among pore regions, however, may be continuous when the water content is changed.

REFERENCES

Abelin, H., L. Birgersson, J. Gidlund, L. Moreno, I. Neretnieks, H. Wilden, and T. Agren (1987), 3-D migration experiment - Report 3, part I, Performed experiments, results and evaluation, *Tech. Rep., 87-21*, Stripa Proj., Stockholm, Nov.

Bales, R.C., C.P. Gerba, G.H. Grondin, and S.L. Jensen (1989), Bacteriophage transport in sandy soil and fractured tuff. *Appl. Environ. Microbiol., 55*, 2061-2067.

Barraclough, D., and P.H. Nye (1979), The effect of molecular size on diffusion characteristics in soil. *J. Soil Sci., 30*, 29-42.

Bowman, R. (1984), Evaluation of some new tracers for soil water studies. *Soil Sci. Soc. Am. J., 48*, 987-993.

Brusseau, M.L., P.S.C. Rao, R.E. Jessup, and J.M. Davidson (1989), Flow interruption: A method for investigating sorption nonequilibrium. *J. Contamin. Hydrol., 4*, 223-240.

Brusseau, M.L. (1993), The influence of solute size, pore water velocity, and intraparticle porosity on solute dispersion and transport in soil, *Water Resourc. Res., 29*, 1071-1080.

Gerba, C.P., S.M. Goyal, I. Cech, and G.F. Bogdan (1981), Quantitative assessment of the adsorptive behavior of viruses to soils, *Environ. Sci. Technol., 15*, 940-944.

Gwo, J.P., G.T. Yeh, and G.V. Wilson (1991), Proceedings of the International Conference on Transport and Mass Exchange Processes in Sand and Gravel Aquifers: Field and Modeling Studies, Ottawa, Canada, 2, 578-589.

Gwo, J.P., P.M. Jardine, G.T. Yeh, and G.V. Wilson (1994), MURF user's guide: A finite element model of multiple-pore-region flow through variably saturated subsurface media, Oak Ridge National Laboratory, ORNL/GWPO-011.

Gwo, J.P., P.M. Jardine, G.T. Yeh, and G.V. Wilson (1995a), MURT user's guide: A finite element model of multiple-pore-region transport through variably saturated subsurface media, Oak Ridge National Laboratory, ORNL/GWPO-015.

Gwo, J.P., P.M. Jardine, G.V. Wilson, and G.T. Yeh (1995b), A multiple-pore-region concept to modeling mass transfer in subsurface media, *J Hydrol. 164*, 217-237.

Gwo, J.P., P.M. Jardine, and G.T. Yeh (1996), Using a multiregion model to study the effects of advective and diffusive mass transfer on local physical nonequilibrium and solute mobility in a structured soil, *Water Resourc. Res. 32*, 561-570.

Gwo, J.P., R. O'Brien, and P.M. Jardine (1998), Mass transfer in structured porous media: embedding mesoscale structure and microscale hydrodynamics in a two-region model, *J Hydrol. 208*, 204-222.

Gwo, J.P., G.V. Wilson, P.M. Jardine, and E.F. D'Azevedo (1999), Modeling subsurface contaminant reactions and transport at the watershed scale, Proceedings of the 1997 Chapman/Outreach Conference, Riverside, CA, (in press).

Harvey, R.W., L.H. George, R.L. Smith, and D.L. LeBlanc (1989), Transport of microspheres and indigenous bacteria through a sandy aquifer: Results of natural- and forced-gradient tracer experiments, *Sci. Technol., 23*, 51-56.

Harvey, R.W., N.E. Kinner, D. MaDonald, D.W. Metge, and A. Bunn (1993), Role of physical heterogeneity in the interpretation of small-scale laboratory and field observations of bacteria, microbial-sized microsphere, and bromide transport through aquifer sediments, *Water Resourc. Res., 29*, 2713-2721.

Harvey, R.W., N.E. Kinner, D. MaDonald, D.W. Metge, and A. Bunn, (1995) Transport behavior of groundwater protozoa and protozoan-sized microspheres in sandy aquifer sediments, *Applied and Environmental Microbiology, Jan*, 209-217.

Hinsby, K., L.D. McKay, P. Jørgensen, M. Lenczewski, and C.P. Gerba (1996), Fracture aperture measurements and migration of solutes, viruses and immiscible creosote in a column of clay till, *Ground Water, 34*, 1065-1075.

Hu, Q., and M.L. Brusseau (1995), Effect of solute size on transport in structured porous media, *Water Resourc. Res., 31*, 1637-1646.

Jardine, P.M., G.V. Wilson, and R.J. Luxmoore (1988), Modeling the transport of inorganic ions through undisturbed soil columns from two contrasting watersheds, *Soil Sci. Soc. Am. J. 52*,1252-1259.

Jardine, P.M., G.V. Wilson, R.J. Luxmoore, and J.F. McCarthy (1989), Transport of inorganic and natural organic tracers through an isolated pedon in the field, *Soil Sci. Soc. Am. J. 53*, 317-323.

Jardine, P.M., G.V. Wilson, and R.J. Luxmoore (1990a), Unsaturated solute transport through a forest soil during rain events, *Geoderma. 46*, 103-118.

Jardine, P.M., G.V. Wilson, J.F. McCarthy, R.J. Luxmoore, and D.L. Taylor (1990b), Hydrogeochemical processes controlling the transport of dissolved organic carbon through a forested hillslope, *J. Contaminant Hydrology, 6*, 3-19.

Jardine, P.M., G.K. Jacobs, and G.V. Wilson (1993a), Unsaturated transport processes in undisturbed heterogeneous porous media, I. Inorganic Contaminants, *Soil Sci. Soc. Am. J. 57*, 945-953.

Jardine, P.M., G.K. Jacobs, and J.D. O'Dell (1993b) Unsaturated transport processes in undisturbed heterogeneous porous media, II. Co-Contaminants, *Soil Sci. Soc. Am. J. 57*, 954-962.

Jardine, P.M., R. O'Brien, Wilson, G.V., and J.P. Gwo (1998), Experimental techniques for confirming and quantifying physical nonequilibrium processes in soils, 243-271. In H.M. Selim and L. Ma. Physical Nonequilibrium in Soils: Modeling and Application. Ann Arbor Press, Inc. Chelsea, Michigan.

Jardine, P.M., W.E. Sanford, J.P. Gwo, O.C. Reedy, D.S. Hicks, J.S. Riggs, and W.B. Bailey (1999), Quantifying diffusive mass transfer in fractured shale bedrock, *Wat. Resourc. Res.* (in press).

Kooner, Z.S., P.M. Jardine, S. Feldman (1995), Competitive surface complexation reactions of SO_4^{2-} and natural organic carbon on soil, *J. Environ. Qual. 24*: 656-662.

Luxmoore, R.J. and C.H. Abner (1987), Field facilities for subsurface transport research DOE/ER-0329, U.S. Department of Energy, Washington, D.C., 32.

Luxmoore, R.J., P.M. Jardine, G.V. Wilson, J.R. Jones, and L.W. Zelazny (1990) Physical and chemical controls of preferred path flow through a forested hillslope, *Geoderma. 46*, 139-154.

Maloszewski, P., and A. Zuber (1990). Mathematical modeling of tracer behavior in short-term experiments in fissured rocks, *Water Resourc. Res., 26*, 1517-1528.

Maloszewski, P., and A. Zuber (1993), Tracer experiments in fractured rocks: Matrix diffusion and the validity of models, *Water Resourc. Res., 29*, 2723-2735.

McKay, L.D., J.A. Cherry, R.C. Bales, M.T. Yahya, and C.P. Gerba (1993a), A field example of bacteriophage as tracers of fracture flow, *Environ. Sci. Technol., 27*, 1075-1079.

McKay, L.D., R.W. Gillham, and J.A. Cherry (1993b), Field experiments in a fractured clay till, 2. Solute and colloid transport, *Water Resourc. Res., 29*, 3879-3890.

Perroux, K.M., and I. White (1988), Designs for disc permeameters, *Soil Sci. Soc. Am. J. 52*,1205-1215.

Reedy, O.C., P.M. Jardine, and H.M. Selim (1996), Quantifying diffusive mass transfer of non-reactive solutes in columns of fractured saprolite using flow interruption, *Soil Sci. Soc. Am. J. 60*, 1376-1384.

Reimus, P.W. (1996), The use of synthetic colloids in tracer transport experiments in saturated rock fractures, PhD thesis presented to the University of New Mexico, Albuquerque, New Mexico, LA-13004-T.

Seyfried, M.S., and P.S.C. Rao, (1987), Solute transport in undisturbed columns of an aggregated tropical soil: Preferential flow effects, *Soil Sci. Soc. Am. J., 51*, 1434-1444.

Smith, M.S., G.W. Thomas, R.E. White, and D. Ritonga (1985), Transport of *Escherichia coli* through intact and disturbed soil columns, *J. Environ. Qual., 14*, 87-91.

Toran, L., and A.V. Palumbo (1991), Colloid transport through fractured and unfractured laboratory sand columns, *J. Contam. Hydrol., 9*, 289-303.

Watson, K.W., and R.J. Luxmoore (1986), Estimating macroporosity in a forest watershed by use of a tension infiltrometer, *Soil Sci. Soc. Am. J., 50*, 578-582.

Wilson, G.V. and R.J. Luxmoore (1988). Infiltration, macroporosity, and mesoporosity distributions on two forested watersheds, *Soil Sci. Soc. Am. J. 52*, 329-335.

Wilson, G.V., J.M. Alfonsi, and P.M. Jardine (1989). Spatial variability of subsoil hydraulic properties of two forested watersheds, *Soil Sci. Soc. Am. J. 53*, 679-685.

Wilson, G.V., P.M. Jardine, R.J. Luxmoore, and J.R Jones (1990), Hydrology of a forested watershed during storm events, *Geoderma. 46*, 119-138.

Wilson, G.V., P.M. Jardine, R.J. Luxmoore, L.W. Zelazny, D.A. Lietzke, and D.E. Todd (1991a), Hydrogeochemical processes controlling subsurface transport from an upper subcatchment of Walker Branch watershed during storm events: 1. Hydrologic transport processes, *J. Hydrology., 123*: 297-316.

Wilson, G.V., P.M. Jardine, R.J. Luxmoore, L.W. Zelazny, D.E. Todd, and D.A. Lietzke (1991b), Hydrogeochemical processes controlling subsurface transport from an upper subcatchment of Walker Branch watershed during storm events: 2. Solute transport processes, *J. Hydrology. 123*, 317-336.

Wilson, G.V., P.M. Jardine, J.P. Gwo (1992), Modeling the hydraulic properties of a multiregion soil, *Soil Sci. Soc. Am. J. 56*, 1731-1737.

Wilson, G.V., P.M. Jardine, J.D. O'Dell, and M. Collineau (1993), Field-scale transport from a buried line source in unsaturated soil, *J. Hydrology 145*, 83-109.

Wilson, G.V., J.P. Gwo, P.M. Jardine, and R.J. Luxmoore (1998), Hydraulic and physical nonequilibrium effects on multi-region flow and transport, 37-61, In H.M. Selim and L. Ma. Physical Nonequilibrium in Soils: Modeling and Application, Ann Arbor Press, Inc., Chelsea, Michigan.

Yang, Z., R.S. Burlage, W.E. Sanford, and G.R. Moline (1996), DNA-labeled silica microspheres for groundwater tracing and colloid transport studies, Proc., 212th National Meeting, Am. Chem. Soc., Orlando, FL, Aug. 25-30.

DNAPL AND RESIDUAL WATER CHARACTERIZATION IN THE VADOSE ZONE USING THE PARTITIONING INTERWELL TRACER TEST (PITT)

James Studer and Paul E. Mariner, *Duke Engineering and Services*

INTRODUCTION

Approximately 362,872 kilograms (kg) of chlorinated solvents and other organic chemicals used in research, development, and facility operations were disposed in the Sandia National Laboratories/New Mexico (SNL/NM) Chemical Waste Landfill (CWL) from 1962 to 1982 (SNL/NM, 1992). These wastes, either containerized or as free liquids, were placed in shallow unlined pits or trenches. Some of the wastes migrated into the vadose zone as Dense Non-Aqueous Phase Liquid (DNAPL). Volatilization of the volatile organic compounds (VOCs) and subsequent migration through the 150 meter (m) thick vadose zone resulted in a relatively large soil gas plume and limited groundwater contamination. Trichloroethylene (TCE) was the chlorinated solvent disposed in the largest quantities, was the most prevalent VOC soil gas contaminant, and was the principal VOC detected in groundwater. Information on the in situ volume and extent of TCE DNAPL represented a critical data gap for the CWL Resource Conservation and Recovery Act (RCRA) corrective action since 1990, when TCE was first detected in the groundwater. No viable technology existed for in situ DNAPL characterization within the vadose zone, or saturated zone, until 1994. The Partitioning Interwell Tracer Test (PITT) was introduced to SNL/NM in 1994. In December 1995, the first PITT ever conducted to measure NAPL in the vadose zone was completed beneath two of the more important solvent disposal trenches at the CWL (Studer, et al., 1996; Mariner, et al., 1999).

THE PARTITIONING INTERWELL TRACER TEST (PITT)

The concept of using a PITT was originally developed for petroleum exploration. It was designed to measure the saturation of oil between wells in unsaturated formations where the use of enhanced oil recovery techniques were being considered. The use of a PITT to measure NAPL in aquifers was first presented in the spring of 1994 (Jin, et al., 1994). By 1998, more than 30 such PITTs had been conducted across the United States. Although most of these PITTs were conducted below the water table using aqueous tracers, several were conducted in the vadose zone using gaseous tracers (Studer, et al., 1996; Simon, et al., 1998; Deeds, et al., 1999). This paper presents a case study of the first PITT ever conducted in the vadose zone to measure either NAPL or water saturation in an environmental setting. Details of this PITT and its design can be found in Jin (1995), Studer, et al. (1996), Mariner, et al. (1999), and Whitley, et al. (1999).

The PITT technology is a new application of chromatography. Chromatography is the separation of compounds by injecting a mixture of the compounds into a gas or liquid carrier stream that flows past a stationary phase. The compounds partition into the stationary phase to different degrees, resulting in greater retardation of those compounds that have higher affinities for the stationary phase. In a vadose zone PITT, the mobile carrier is air and the stationary phase is typically water and/or NAPL residual. In a saturated zone PITT, the mobile carrier is water and the stationary phase is NAPL residual. A non-partitioning tracer is included in a PITT to measure the mean residence time of the mobile phase.

The relative chromatographic separation of partitioning and non-partitioning tracers in a PITT is measured directly from tracer breakthrough curves at each sample location. These breakthrough curves are used to compute the retardation factor, R_i, of each tracer within the swept region. R_i is equal to the ratio of the average rate at which tracer *i* travels to the sample location relative to the average rate at which a non-partitioning, unretarded tracer travels to the sample location. In a vadose zone PITT, R_i is related to tracer *i*'s partition coefficients and fluid saturation by the following equation (Mariner, et al., 1999):

$$R_i = 1 + \left(\frac{K_{w,i} S_w + K_{n,i} S_n}{1 - S_w - S_n} \right)$$

S_w and S_n are the average saturation of the water and NAPL stationary phases, and $K_{w,i}$ and $K_{n,i}$ are the respective water-air and NAPL-air partition coefficients. S_w and S_n are equal to the ratio of the average volume of water or NAPL to the volume of the pore space. Estimation of average vadose zone NAPL saturation, S_n, requires not only use of an NAPL-partitioning tracer, but also use of a water-partitioning tracer to estimate a value for S_w.

The retardation factor for each breakthrough curve is determined from first moment analysis. Because the non-partitioning tracer by definition will have a retardation factor (first temporal moment) of 1, comparison of the integrated mean residence times of the partitioning and non-partitioning tracers will determine retardation factors for the partitioning tracers. These retardation factors, combined with the known partition coefficients, allow calculation of the average water and NAPL saturation within the swept volume.

METHODS

A plan view of the test zone, targeted trenches, and design flow field at the CWL site is shown in Figure 1. One injection well and one extraction well were installed 16.8 m apart on opposite sides of the buried trenches. Each well was screened and packed off at intervals of 3.0-10.7, 12.2-18.3, and 19.8-24.4 m below ground surface

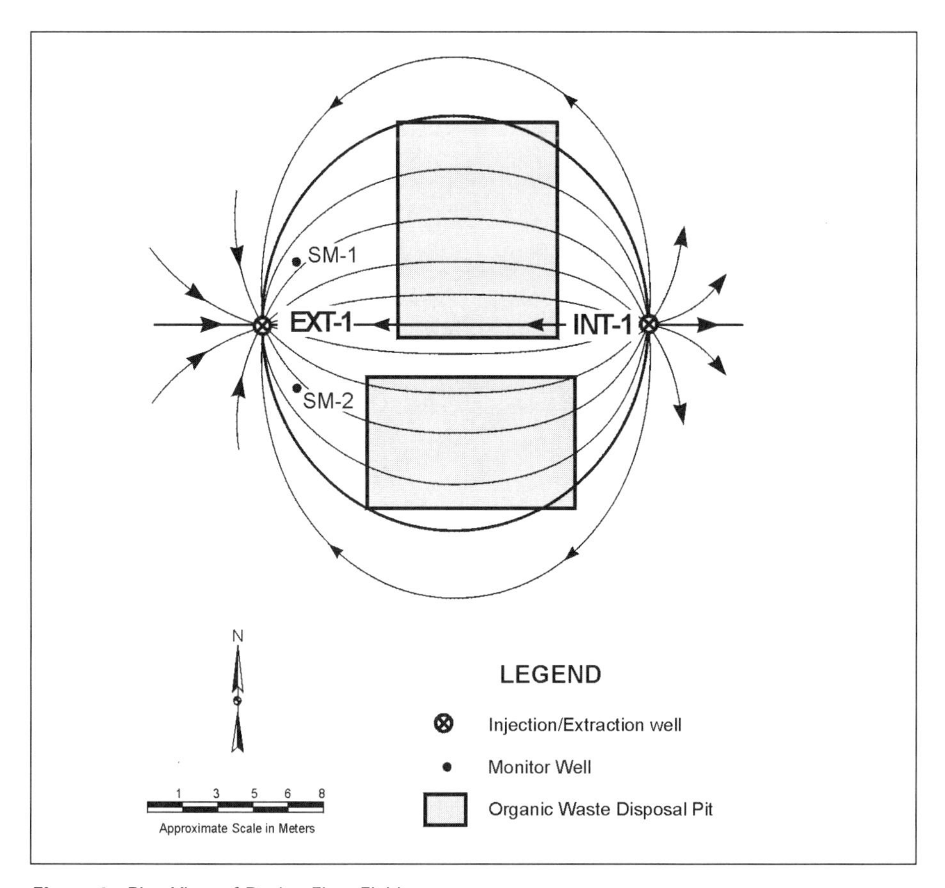

Figure 1. Plan View of Design Flow Field

(bgs) so that DNAPL and water residual could be measured at different depths. In addition, two soil gas monitoring wells, referred to as SM-1 and SM-2, were installed 3 m northeast and southeast of the extraction well to allow separate monitoring of flow paths to the north and south. These monitoring wells were constructed using SEAMIST™ membranes and contained sample points at 3 m intervals from 3 to 24.4 m bgs to allow relatively high vertical resolution of measured DNAPL and water residual saturation.

The PITT tracer slug was a mixture of TCE-partitioning, water-partitioning, and non-partitioning tracers. These tracers, their masses injected, and their partition coefficients are shown in Table 1. The non-partitioning tracers were methane (CH_4) and sulfur hexafluoride (SF_6), and the water-partitioning tracer was difluoromethane (CF_2H_2). Although the "non-partitioning" tracer SF_6 slightly partitions into TCE, the average saturation of TCE in the test zone was too low for perceptible SF_6 retarda-

TABLE 1 **Tracers used in the PITT.**

Tracer	Mass Injected (kg)	K_n (TCE)	K_w
methane (CH_4)	0.1	0	0
sulfur hexafluoride (SF_6)	1.0	1	0
difluoromethane (CF_2H_2)	1.0	2	1.7
octafluorocyclobutane (C_4F_8)	1.0	9	0
dodecafluorodimethylcyclobutane (C_6F_{12})	1.0	16	0
perfluoro-1,3-dimethylcyclohexane (C_8F_{16})	1.0	72	0
perfluoro-1,3,5-trimethylcyclohexane (C_9F_{18})	1.0	162	0

tion. The "water-partitioning" tracer also partitions into TCE; however, the effect of TCE on overall retardation is negligible when average TCE saturation is less than 5 percent of average water saturation, as it was in this case.

The PITT tracers were mixed with nitrogen and maintained in a tank under a slight positive pressure. After steady-state air injection and soil gas extraction rates were established, air injection was diverted through the tracer tank for 2 hours to inject the tracers. Injection and extraction rates at the time of tracer slug introduction were 212.4 liters per minute (LPM) to satisfy equilibrium partitioning requirements and to allow sufficient sampling and analysis at the monitoring locations. After 2.0, 4.0, and 6.7 days, the rates were increased to 339.8, 679.7, and 1,472.6 LPM, respectively. Concentrations of the tracers were measured at 19 sample locations using two automated stream selectors connected to two on-line gas chromatographs (GCs). Each GC could analyze approximately 30 samples per day.

RESULTS AND DISCUSSION

The PITT required 15 consecutive days to complete in the field. Cold weather on several nights caused water condensation and temporary malfunction of one of the on-line GCs. Nevertheless, a sufficient amount of data were collected over the course of the test to provide useful breakthrough curves. SF_6 and C_9F_{18} provided the most reliable non-partitioning and TCE DNAPL-partitioning tracer data, respectively. The results for each monitoring location are presented in Table 2.

TABLE 2 **Average TCE DNAPL saturation (percent) measured from the breakthrough curves at the extraction well (EXT-1) and monitoring wells (SM-1 and SM-2).**

Screened Interval (m)	EXT-1	Depth (m)	SM-1	SM-2
3–10.6	0.11 ± 0.02	3	0.18 ± 0.02	0.21 ± 0.05
		6.1	nd	0.10 ± 0.02
		9.1	nd	nd
12.2–18.3	nd	12.2	nd	nd
		15.2	nd	nd
		18.3	nd	nd
19.9–24.4	nd	21.3	nd	nd
		24.4	nd	nd

nd = not detected (less than 0.05 percent)

First moment analysis indicated that TCE DNAPL had penetrated non-uniformly to depths of 6.1 to 9.1 m bgs. The average TCE DNAPL saturation in the shallow zone between 3 m and 10.7 m bgs was measured to be 0.11 ± 0.02 percent, based on a calculated C_9F_{18} retardation factor of 1.2. One monitoring point at 3 m bgs indicated an approximate TCE-DNAPL saturation of 0.21 ± 0.05 percent for the stream tubes it captured, corresponding to a retardation factor of approximately 1.4 (Figure 2). No TCE DNAPL was detected in the intermediate or deep zones, as demonstrated by a lack of chromatographic separation of the tracers.

The pore volume swept by the tracers in the shallow zone was estimated to be between 110,000 L and 510,000 L. The lower value is based on unexplained low total tracer recoveries while the upper value is estimated based on the design flow field and observations at the monitoring locations. Applying the 0.11 ± 0.02 percent average TCE DNAPL saturation measured for the shallow zone to these estimates implies a total measured TCE DNAPL volume of approximately 150 to 680 L in the swept volume.

Average water saturation was determined from first moment analysis of SF_6 and CF_2H_2 breakthrough curves. Results indicated that average water saturation for the shallow, intermediate, and deep zones were 23, 13, and 10 percent, respectively.

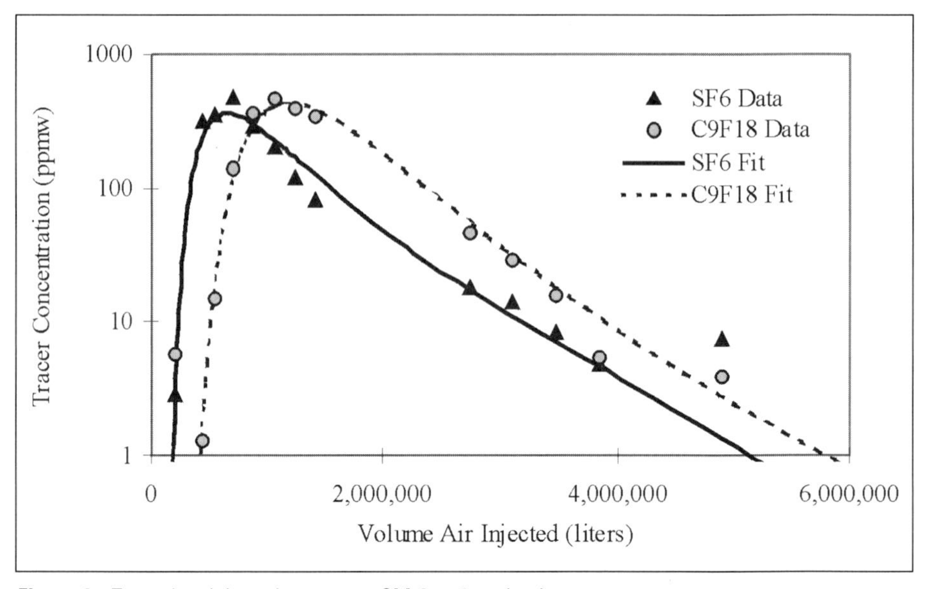

Figure 2. Tracer breakthrough curves at SM-2 at 3 m depth.

CONCLUSIONS

Prior to conducting the PITT, DNAPL was estimated to have penetrated downward in excess of 16 m beneath the two trenches, the average DNAPL saturation immediately below the trenches was estimated to be approximately 4 percent, and the estimated DNAPL volume was on the order of 10,000 to 50,000 L. These estimates were based on waste inventory records, soil gas sampling results, and the fact that DNAPL was encountered during exploratory drilling beneath the two trenches (coating the rods and dripping from the drill bit). A corresponding cleanup time using vapor extraction technology was speculated to be on the order of four years. The PITT showed that these estimates were too high. DNAPL penetration was shown to be less than 11 m and the volume of DNAPL present at the start of remediation was shown to be significantly less (150 – 680 L) than previously thought.

The results of this PITT were used to calm fears of deep DNAPL penetration and to successfully design an accelerated source removal program consisting of two voluntary corrective measures (VCMs) - a vapor extraction VCM and a landfill excavation VCM. Determining the volume and average saturation of DNAPL beneath the two trenches allowed a more accurate prediction of the time that would be necessary to remove the VOCs from beneath not only the two trenches, but also the entire site.

After a short-term vapor extraction pilot test and approximately nine months of active vapor extraction during the VCM proper, an estimated 350 L of VOCs (as TCE) were extracted using well EXT-1. Results from soil gas sampling of SM-1 and SM-2 during the active extraction period, and for approximately 5 months thereafter, provided strong evidence of dramatic declines in VOC concentration and volume, and concentration rebound of less than 20 percent. As can be seen on Figure 3, VOC concentrations indicated by photoionization detector (PID) at SM-2 ports during

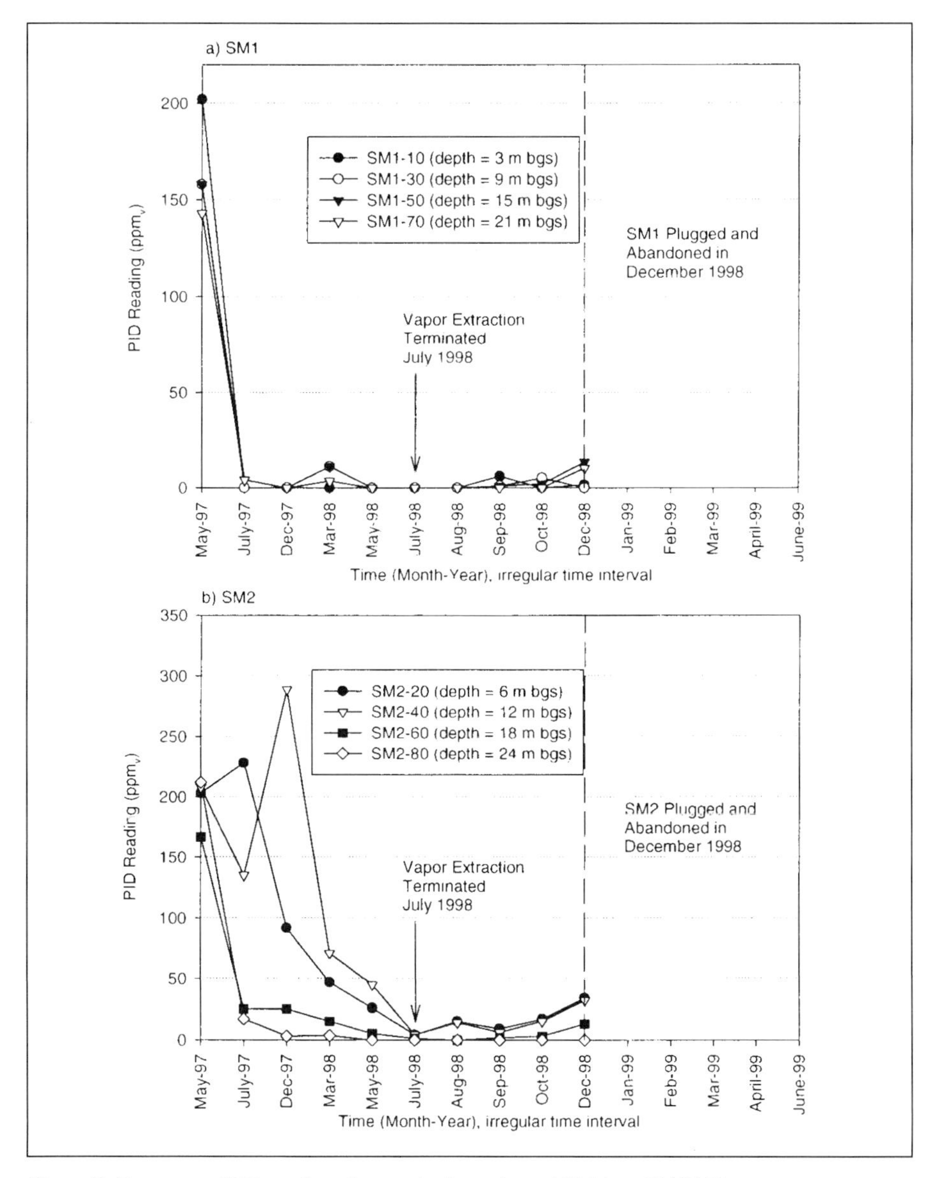

Figure 3. Summary of PID readings for monitoring points a) SM 1 and b) SM2.

active extraction and the rebound observation period were higher than at SM-1. This is consistent with the PITT conclusion that the presence of DNAPL was more extensive within the stream tubes intersected by SM-2.

A comparison of the estimated volume of VOCs extracted using well EXT-1 (350 L) against the PITT estimate of the DNAPL volume present prior to extraction (150–680 L) suggests that the pre-PITT estimate of DNAPL volume, i.e., 10,000 to 50,000 L, was approximately two orders of magnitude higher than the true volume.

ACKNOWLEDGMENTS

The authors gratefully acknowledge the following individuals for their continuing contributions to the development of vadose zone PITT technology: Gary Pope of the University of Texas at Austin and the following Duke Engineering & Services colleagues: Richard Jackson, Minquan Jin, Neil Deeds, Varadarajan Dwarakanath, and Jeff Silva. PITT technology is patented by the University of Texas and Duke Engineering & Services. We also wish to thank Sandia National Laboratories/New Mexico, and in particular David Miller, Richard Fate, James Phelan, Wu Ching Cheng, and Randy Roberts, for their support.

This work was supported by the United States Department of Energy under contract DE-AC04-94AL8500.

REFERENCES

Deeds, N.E., Pope, G.A., and McKinney, D.C., 1999. Vadose zone characterization at a contaminated field site using partitioning interwell tracer technology. Environmental Science & Technology, 33(16): 2745-2751.

Jin, M., 1995. A Study of Nonaqueous Phase Liquid Characterization and Surfactant Remediation. Ph.D. Dissertation, University of Texas, Austin, TX.

Jin, M., Delshad, M., McKinney, D.C., Pope, G.A., Sepehrnoori, K., Tilburg, C., and Jackson, R.E., 1994. Subsurface NAPL contamination: partitioning tracer test for detection, estimation and remediation performance assessment. In Toxic Substances and the Hydrologic Sciences. American Institute of Hydrology, Minneapolis, MN.

Mariner, P.E., Jin, M., Studer, J.E., and Pope, G.A., 1999. The first vadose zone partititioning interwell tracer test for nonaqueous phase liquid and water residual. Environmental Science & Technology, 33(16): 2825-2828.

Sandia National Laboratories/Mew Mexico, 1992. "Chemical Waste Landfill Final Closure Plan and Postclosure Permit Application," December 1992, Sandia National Laboratories, Albuquerque, New Mexico.

Simon, M.A., Brusseau, M.L., Golding, R., and Cagnetta, P.J., 1998. Organic and aqueous phase partitioning gas tracer field experiment. In Platform Abstracts of the First International Conference on Remediation of Chlorinated and Recalcitrant Compounds, Monterey, CA.

Studer, J.E., Mariner, P.E., Jin, M., Pope, G.A., McKinney, D.C., and Fate, R. 1996. Application of a NAPL partitioning interwell tracer test (PITT) to support DNAPL remediation at the Sandia National Laboratories/New Mexico Chemical Waste Landfill. In Proceedings of Superfund/Hazwaste West Conference 1996, Las Vegas, NV.

Whitley, G.A., McKinney, D.C., Pope, G.A., Rouse, B.A., and Deeds, N.E., 1999. Contaminated vadose zone characterization using partitioning gas tracers. Journal of Environmental Engineering, 125(6): 574-582.

ESTIMATION OF THE SOIL HYDRAULIC PROPERTIES

Marcel G. Schaap, Feike J. Leij, and **Martinus Th. van Genuchten,**
U.S. Salinity Laboratory, USDA-ARS

Many vadose zone studies use numerical models to simulate the movement of water and solutes in the subsurface. Knowledge about the soil hydraulic properties (for example, the water retention curve, and the saturated and unsaturated hydraulic conductivities) is essential for running most or all of these models. A broad array of methods currently exists to determine soil hydraulic properties in the field or in the laboratory (see Klute 1986 and van Genuchten *et al.* 1999). Field methods allow for *in-situ* determination of the hydraulic properties but have uncertainties about the actual sample volume. Laboratory measurements require more sample preparation but do allow a larger number of measurements and a better control of the experimental conditions. Most laboratory and field techniques, however, have specific ranges of applicability with respect to soil type and saturation (Klute 1986). Another limitation of direct measurements is they generally require a substantial investment in both time and money. Also, many vadose zone studies are concerned with large areas of land that may exhibit significant lateral and vertical spatial variability in the soil hydraulic properties. Performing measurements in these cases is virtually impossible, thus requiring alternative methods for estimating soil hydraulic properties.

A large number of indirect methods to generate soil hydraulic properties are now also available. Although these techniques vary widely in terms of methodology and complexity, all use some form of surrogate data to estimate soil hydraulic properties. In broad terms, three methods can be distinguished: pore-size distribution models, inverse methods and pedotransfer functions.

Pore-size distribution models are very often used to estimate the unsaturated hydraulic conductivity from the distribution, connectivity and tortuosity of pores. The pore-size distribution can be inferred from the water retention curve, which is normally much easier to measure than the unsaturated hydraulic conductivity function. One of the most popular models was developed by Mualem (1976). The model may be simplified into closed-form expressions when the water retention is described with the functions of Brooks-Corey (1964) or van Genuchten (1980). The latter function is given by:

$$S_e = \frac{\theta(h) - \theta_r}{\theta_s - \theta_r} = [1 + (\alpha h)^n]^{1/n-1} \quad (1)$$

where $\theta(h)$ represents the water retention curve, defining the water content, θ (e.g., cm^3/cm^3), as a function of the soil water pressure head h (cm), S_e is the effective saturation, θ_r and θ_s (cm^3/cm^3) are residual and saturated water contents, respectively, while α and n are curve shape parameters. The corresponding closed-form expression for the unsaturated hydraulic conductivity, K (e.g., cm/day) is:

$$K(S_e) = K_s S_e^L \{1-[1-S_e^{n/(n-1)}]^{1-1/n}\}^2 \quad (2)$$

in which K_s is the saturated hydraulic conductivity and L (-) is an empirical pore tortuosity/connectivity parameter that is normally assumed to be 0.5 (Mualem 1976 and van Genuchten, 1980). In spite of some theoretical complications, we found that using L=-1.0 gave better predictions, on average, for a large soil hydraulic database.

Closed-form expressions such as Eq. (1) and (2) provide a consistent description of water retention and unsaturated hydraulic conductivity curves and are convenient for use in numerical models simulating variably saturated flow. The same set of equations can also be used in inverse problems by combining a numerical model of the Richards equation with an optimization algorithm to estimate several or all of the unknown parameters K_s, L, θ_r, θ_s, α and n from observed time series of infiltration, water content and/or pressure head (see, for example, Kool *et al.* 1987, Simunek and van Genuchten 1996, and Abbaspour 1997). The resulting hydraulic properties may be considered effective properties in that they are obtained from data pertaining to real flow conditions. Unfortunately, inverse methods are often vulnerable due to non-uniqueness of the results, that is, two or more sets of optimized parameters may be applicable to the problem being studied (see, for example, Hopmans and Simunek 1999).

Pedotransfer functions (PTFs), offer a third method for estimating hydraulic properties (Bouma and van Lanen 1987) by using the fact that hydraulic properties are very dependent upon soil texture and other readily available taxonomic information (such as the particle size distribution, bulk density and/or organic matter content). For example, fine-textured soils are known to have very different water retention characteristics and much lower saturated hydraulic conductivities than coarse textured soils. PTFs take advantage of such information. Quasi-physical methods by Arya and Paris (1981), Haverkamp and Parlange (1986), and Tyler and Wheatcraft (1989) use the concept of shape similarity between pore- and particle-size distributions. However, the vast majority of PTFs are completely empirical and do not use any physical concepts. Although considerable differences exist among PTFs in terms of the required input data (Rawls *et al.* 1991), all of them use at least some information about the particle-size distribution. Three main distinctions can be made: class, continuous, and neural network PTFs.

Class PTFs assume that similar soils have similar hydraulic properties, thus permitting one to use lookup tables or databases to find the appropriate values for a particular soil textural class. Examples of class PTFs are given by Carsel and Parrish (1988) and Wösten *et al.* (1995) for water retention parameters of soil textural classes. Leij *et al.* (1996) developed a comprehensive database that permits discrimination among additional soil attributes, and allows relevant information about the soil water retention and unsaturated hydraulic conductivity measurements to be retrieved, among other features.

Continuous PTFs are simple linear or nonlinear regression equations that provide continuously varying soil hydraulic properties across the textural triangle. The predictions may be improved by extending the input data through addition of basic soil properties like bulk density, porosity or organic matter content (Rawls and Brakensiek 1985, Vereecken *et al.* 1989). Additional improvements are possible by including one or more water retention data points in the analysis (Rawls *et al.* 1992, Williams *et al.* 1992). Other authors have predicted soil hydraulic properties using more limited or extended sets of input variables (Rawls *et al.* 1992; Schaap and Bouten, 1996; Vereecken *et al.* 1989). Such hierarchical approaches are of great practical use since they permit more flexibility in using all available data.

Recently, neural network analysis was used to improve the predictions of empirical PTFs (Pachepsky *et al.* 1996; Schaap and Bouten 1996; Tamari *et al.* 1996). An advantage of neural networks, as compared to traditional PTFs, is that neural networks require no *a priori* model concept. The optimal, possibly nonlinear, relations that link input data (for example, particle size data, bulk density, etc.) to output data (for example, hydraulic parameters) are obtained and implemented in an iterative calibration procedure. As a result, neural network models typically extract the maximum information from the data, and thus provide the most powerful PTFs possible (Schaap *et al.* 1998a).

One problem that often hampers the practical application of all PTFs is their very specific data requirements. When developing PTFs, most authors tend to use mathematical expressions that provide the best results for their data. This approach sometimes produces models that require many input variables (Rawls *et al.* 1991).

However, users are frequently confronted with situations where several input variables needed for a PTF are not available, or are of a relatively poor quality. Using two or more PTFs that have different data requirements is not a solution for such situations since the PTFs most likely were calibrated on different data sets, and will thus likely provide inconsitent predictions (Schaap and Leij 1998). Another problem is that PTFs provide, at best, yield predictions with a modest level of accuracy (see, for example, Tietje and Tapkenhinrichs 1995). For this reason it would be very useful if predictions of the soil hydraulic properties could be accompanied with confidence intervals.

To address these issues, Schaap *et al.* (1998a) used bootstrap-neural network analyses to develop a hierarchical approach to predict water retention parameters in equation (1), as well as K_S. Five PTFs were developed with data requirements that should suit most practical situations. The combination with the bootstrap method (Efron and Tibshirani 1993) provided the confidence intervals for the predictions. Table 1 gives an overview of the required input data for each model, with the root-mean-square residuals showing the performance of the PTFs on independent data (Schaap *et al.* 1998a). The models were developed using a database containing 2,085 samples; Figure 1 gives the textural distribution of this database. Because the same data were used for all models, the predictions should be consistent among each other.

Model 1 is a class PTF; values of the hydraulic parameters for each textural class are shown in Table 2. The parameters in this table are based on measured data, and are, as such, probably more realistic than those given by Carsel and Parrish (1988) who obtained van Genuchten (1980) parameters by transforming predictions of Brooks-Corey (1964) parameters calculated with a PTF developed by Rawls and

TABLE 1 **Input requirements for the five models in the hierarchical approach. Root-mean-square residuals (RMSR) provide the performance of each model for water retention and saturated hydraulic conductivity.**

Model Input		Prediction Error (RMSR): Retention $[cm^3cm^{-3}]$	K_s $[\log(cm\ day^{-1})]$
Model 1	Textural class	0.108	0.740
Model 2	Sand, silt,clay	0.106	0.735
Model 3	+ bulk density	0.083	0.684
Model 4	+ θ at 33 kPa	0.066	0.611
Model 5	+ θ at 1500 kPa	0.063	0.610

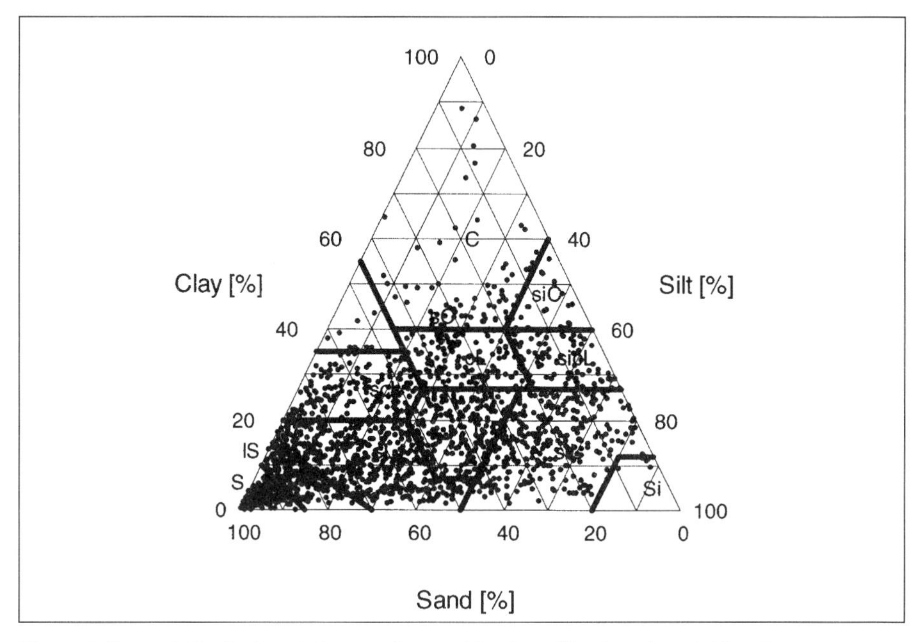

Figure 1. Textural distribution of the samples used for the calibration of model 1 to 5.

TABLE 2 **Class average values of the *van Genuchten* [1980] water retention parameters and the saturated hydraulic conductivity (Ks). N is the number of samples for each textural class.**

	N	θ_r	θ_s	α	n	K_s
		cm³/cm³		1/cm	–	cm/day
Sand	308	0.053	0.375	0.035	3.18	646
Loamy Sand	201	0.049	0.391	.032	1.76	105
Loam	242	0.062	0.400	0.0098	1.48	12.0
Sandy Loam	476	0.039	0.388	0.026	1.45	38.0
Silt Loam	330	0.065	0.439	0.0049	1.67	18.2
Sandy Clay Loam	172	0.065	0.384	0.017	1.34	13.2
Silty Clay Loam	87	0.090	0.484	0.0076	1.53	11.2
Clay Loam	140	0.083	0.444	0.012	1.44	8.1
Silt*	6	0.050	0.489	0.0066	1.68	43.7
Clay	84	0.100	0.457	0.011	1.27	14.8
Sandy Clay*	11	0.128	0.380	0.025	1.22	11.5
Silty Clay	28	0.115	0.476	0.014	1.33	9.6

* Parameter values may not be reliable because of small number of samples.

Brakensiek (1985). Note that in Table 2 the values for the silt and sandy clay classes may not be very reliable since they are based on only a few samples.

Models 2 through 5 in Table 1 use more input data to make predictions of soil hydraulic parameters, and thus become progressively more accurate as indicated by lower RMSR values. Model 2 uses the sand, silt and clay fractions, while model 3 also requires a measured bulk density value. Models 4 and 5 additionally require one or two retention points, respectively. Adding retention points may at first seem a bit counterintuitive the PTFs eventually will be used to predict water retention. However, the two water retention points referred to in Table 2 are routinely available in the large NRCS database (Soil Survey Staff 1995). Using this information leads to much smaller RMSR values than for model 3 (Table 1). As shown by Schaap *et al.* (1998a), neural network predictions can also be easily derived for parameter combinations and input values other than those used for the five models summarized here.

The combination of the neural network calibration with the bootstrap method further enables us to quantify model uncertainty on a per sample basis. Figure 2 (a-c) shows the confidence intervals as predicted by the model 3 (Table 1) for water retention, K_s and K(h), for a loamy sand and a clay sample. For water retention and the unsaturated hydraulic conductivity, 10 and 90% percentiles of the variability are shown; for K_s we show the entire probability distribution. Directly apparent from the figures is the larger uncertainty for the clay as compared to the loamy sand. This is caused by the lower number of fine-textured soils in the data set relative to coarse-textured soils. In general, the uncertainty estimates increased when predictions were made for samples that were less common in the calibration data sets.

We implemented the five hierarchical PTFs using a user-friendly Windows 98 program called Rosetta. This program allows one to select one of the models of Table 1 using available site-specific data, the user can easily generate the parameters K_s, L, θ_r, θ_s, α, and n of equations (1) and (2), as well as estimates of their uncertainties. The hydraulic parameters can then be immediately interfaced with simulation models; moreover, uncertainty estimates are produced for possible use in risk-based analyses of water and solute transport, among other applications. Version 1.0 of Rosetta is available from the U.S. Salinity Laboratory website (http://www.ussl.ars.usda.gov/MODELS/ROSETTA.HTM). We plan to regularly release updates of Rosetta when significant amounts of new calibration data become available.

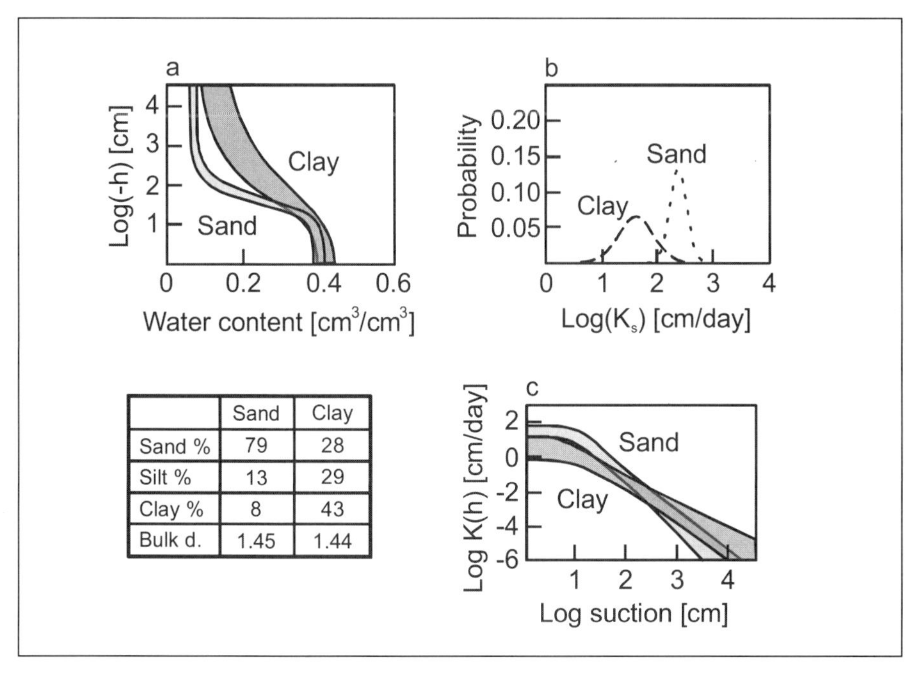

	Sand	Clay
Sand %	79	28
Silt %	13	29
Clay %	8	43
Bulk d.	1.45	1.44

Figure 2. Predictions with uncertainty intervals for water retention (a), saturated hydraulic conductivity (b), and unsaturated hydraulic conductivity (c). The table gives the input values that were using the neural network model (Schaap et al., 1999).

REFERENCES

Abbaspour, K.C., M.T. van Genuchten, R. Schulin, and W. Schllppi. 1997. A sequential uncertainty domain inverse procedure for estimating subsurface flow and transport parameters. *Water resour. Res*. 33(8):1879-1892.

Arya, L.M., and J.F. Paris. 1981. A physico-empirical model to predict the soil moisture characteristic from particle-size distribution and bulk density data. *Soil Sci. Soc. Am. J.* 45:218-1227.

Bouma, J. and J.A.J. van Lanen. 1987. Transfer functions and threshold values: From soil characteristics to land qualities. p 106-110. *In* K.J. Beek et al. (ed.). Quantified land evaluation. International Institute Aerospace Surv. Earth Sci. ITC publ. 6. Enschede, the Netherlands.

Brooks, R.H., and A.T. Corey. 1964. Hydraulic properties of porous media. Hydrol. paper 3, Colorado State Univ., Fort Collins, CO.

Carsel, R.F., and R.S. Parrish. 1988. Developing joint probability distributions of soil water retention characteristics. Water Resour. Res. 24:755-769.

Efron, B., and R.J. Tibshirani. 1993. An Introduction to the bootstrap. Monographs on statistics and applied probability. Chapman and Hall, New York, NY.

Haverkamp, R., and J.Y. Parlange. 1986. Predicting the water-retention curve from particle size distribution: 1. Sandy soils without organic matter. Soil Sci. 142:325-339.

Hopmans, J.W., and J. Simunek. 1999. Review of inverse estimation of soil hydraulic properties In: van Genuchten, M. Th., F.J. Leij, and L. Wu (eds.), 1999. Proc. Int. Workshop, Characterization and Measurement of the Hydraulic Properties of Unsaturated Porous Media, pp. 643-659, University of California, Riverside, CA (in press).

Klute, A. 1986. ed. Methods of soil analysis, Part 1: Physical and mineralogical methods, second Edition. *Monogr. 9.* ASA and SSSA, Madison, WI.

Kool, J.B., J.C. Parker, and M.Th. van Genuchten. 1987. Parameter estimation for unsaturated flow and transport models - a review, *J. Hydrol.*, 91:255-293.

Leij, F.J., W.J. Alves, M. Th van Genuchten, and J.R. Williams, 1996, The UNSODA unsaturated soil hydraulic database, version 1.0, *EPA report EPA/600/R-96/095*, EPA National Risk Management Laboratory, G-72, Cincinnati, OH, USA, (http://www.ussl.ars.usda.gov/MODELS/UNSODA.HTM).

Mualem, Y. 1976. A new model for predicting the hydraulic conductivity of unsaturated porous media, *Water Resour. Res.*, 12;513-522.

Pachepsky, Ya.A., D. Timlin, and G. Varallyay. 1996. Artificial neural networks to estimate soil water retention from easily measurable data. Soil Sci. Soc. Am. J. 60:727-733.

Rawls, W.J., T.J. Gish, and D.L. Brakensiek. 1991. Estimating soil water retention from soil physical properties and characteristics. In: Stewart, B.A. *Advances in Soil Science*, Springer-Verlag, New York.

Rawls, W.J. and D.L. Brakensiek. 1985. Prediction of soil water properties for hydrologic modeling. p 293-299. *In* Jones, E.B. and T.J. Ward (ed.) Watershed management in the eighties. Proc. Irrig. Drain. div., ASCE, Denver, CO. April 30 - May 1. 1985.

Rawls, W.J., L.R. Ahuja and D.L. Brakensiek. 1992. Estimating soil hydraulic properties from soils data. p 329-340. *In* M. Th. van Genuchten, M.Th., F.J. Leij, and L.J. Lund (ed.). Indirect methods for estimating the hydraulic properties of unsaturated soils. Proc. Int Worksh. Riverside, CA. Oct 11-13. 1989. University of California, Riverside, CA.

Schaap, M.G. and W. Bouten. 1996. Modeling water retention curves of sandy soils using neural networks. Water Resour. Res. 32:3033-3040.

Schaap, M.G., Leij F.J. and van Genuchten M.Th. 1998a. Neural network analysis for hierarchical prediction of soil water retention and saturated hydraulic conductivity. *Soil Sci. Soc. Am. J.* 62:847-855.

Schaap, M.G., and F.J. Leij, 1998. Database Related Accuracy and Uncertainty of Pedotransfer Functions, *Soil Science* 163:765-779.

Schaap, M.G., F.J. Leij and H. Hoffmann-Riem, 1998b, Prediction of Unsaturated Soil Hydraulic Conductivity with a Pore-Size Distribution Model, Submitted to *Water Resour. Res.*.

Simunek, J., and M.Th. van Genuchten, 1996. Estimating unsaturated soil hydraulic properties from tension disc infiltrometer data by numerical inversion, *Water. Resour. Res*, 32:2683-2696.

Soil Survey Staff, 1995. Soil survey laboratory information manual. Soil Surv. Invest. Rep. no 45. Natl. Soil Surv. Center, Lincoln, NE.

Stolte, J., J.I. Freyer, W. Bouten, C. Dirksen, J.M. Halbertsma, J.C. van Dam, J.A. van den Berg, G.J. Veerman, and J.H.M. Wösten. 1994. Comparison of six methods to determine unsaturated soil hydraulic conductivity. *Soil Sci Soc. Am. J.* 58:1596-1603.

Tamari, S., J.H.M. Wösten, and J.C. Ruiz-Suárez. 1996. Testing an artificial neural network for predicting soil hydraulic conductivity. Soil Sci. Soc. Am. J. 60:1732-1741.

Tietje, O., and M. Tapkenhinrichs. 1993. Evaluation of pedotransfer functions. Soil Sci. Soc. Am. J. 57:1088-1095.

Tyler, S.W., and S.W. Wheatcraft. 1989. Application of fractal mathematics to soil water retention estimation. Soil Sci. Soc. Am. J. 53:987-996.

van Genuchten, M.Th. 1980. A closed-form equation for predicting the hydraulic conductivity of unsaturated soils, *Soil Sci. Am. J.* 44: 892-898.

van Genuchten, M. Th., F.J. Leij, and L. Wu (eds.), 1999. Proc. Int. Workshop, Characterization and Measurement of the Hydraulic Properties of Unsaturated Porous Media, University of California, Riverside, CA.

Vereecken, H., J. Maes, J. Feyen, and P. Darius. 1989. Estimating the soil moisture retention characteristic from texture, bulk density, and carbon content. Soil Sci. 148:389-403.

Williams, R.D., L.R. Ahuja, and J.W. Naney. 1992. Comparison of methods to estimate soil water characteristics from limited texture, bulk density, and limited data. Soil Sci. 153:172-184.

Wösten J.H.M., P.A. Finke and M.J.W. Jansen. 1995. Comparison of class and continuous pedotransfer functions to generate soil hydraulic characteristics. Geoderma 66:227-237.

CHAPTER 4 CONTENTS

4

Performance Monitoring

Eric Lindgren

BACKGROUND/SCOPE

Monitoring can mean a number of different things in the life cycle of an environmental restoration project. For many sites, monitoring usually begins with government-regulated groundwater sampling and analyses after a site has been determined to be "contaminated" and groundwater has been found to be affected. Once the nature of the contamination has been characterized, it becomes necessary to initiate clean-up and verify that remediation objectives are being met through performance monitoring. This type of monitoring is usually used to optimize the remediation process in use and will be referred to as "process optimization monitoring." After the contamination area has been remediated, stabilized, or contained, some type of long-term performance monitoring should be expected to continue for years, if not decades.

Because process optimization monitoring is linked to a remediation effort, the time period over which it is conducted is limited. The cost for process optimization is usually considered as part of the cost of remediation. Due to the long time frames, long-term monitoring is projected to be very costly when using traditional methods, which rely on manual collection of groundwater samples for off-site analysis. As a landlord of many contaminated sites, the DOE is directly impacted by this problem. For example, the Savannah River Site currently has 1,400 monitoring wells that require analyses of 40,000 groundwater samples per year. Off-site analysis costs are typically in the range of $100 to $1,000 per sample. Additional significant costs include the manpower to collect and manage the samples, as well as the cost for disposal of the significant amounts of purge water generated as part of sampling activities. At some Sandia National Laboratories' operable units, the projected total-life-cycle cost of long-term monitoring exceeds the initial characterization and restoration costs.

The principle focus of this chapter is performance monitoring in the vadose zone, with an emphasis on significantly reducing long-term monitoring costs. Long-term vadose zone monitoring can provide an early warning system for impending groundwater contamination and allow timely, lower-cost corrective actions. An effective, highly automated vadose zone monitoring system coupled with a highly automated groundwater monitoring system can provide the justification for drastically reducing the frequency of labor intensive and, therefore, expensive traditional groundwater sampling and analysis.

Considerable attention is given in this chapter to emerging low-cost sensor technologies and to the synergistic integration of the point measurements that they produce with spatially distributed geophysical measurements. Taken together, these new sensors and new ways of integrating their data suggest an alternative path in developing cost-efficient, long-term performance monitoring systems.

CHARACTERIZING CONTAMINATION THROUGH MONITORING

Government-regulated groundwater monitoring traditionally consists of installing at least three monitoring wells (background, plume, and down gradient), but often many more are added. Typically, the wells are sampled periodically (quarterly, for example) for expensive offsite analysis of a large suite of contaminants. The purpose of this monitoring is to delineate the groundwater plume, the rate of its movement and expansion, and the concentration and identity of its contaminants. Government regulators use this information to quantify the impact on human health and the environment and to negotiate the further characterization and eventual remediation of the site.

DETERMINING THAT REMEDIATION OBJECTIVES ARE BEING MET

Performance monitoring takes on meaning once the characterization has been completed and a remedial action has been initiated. Process optimization monitoring is conducted to optimize the process operation and to verify that remediation objectives are being met. Successful process optimization monitoring, especially from the performance verification perspective, is vital for general acceptance of cleanup technologies, particularly for innovative technologies undergoing initial field demonstration (Gierke and Powers 1997). Depending on the situation, process optimization monitoring may encompass complex process control monitoring or detailed soil or gas analyses to track and verify an *in situ* process, or it may simply require visual confirmation that the backhoe has dug deeply enough. This activity also encompasses the monitoring required to demonstrate that *in situ* containment barriers are constructed continuously and have no initial leaks. As such, process optimization monitoring is generally highly specific to the remedial methods used and is limited in time to the period of active remediation. In the next section of this chapter, process optimization monitoring is discussed first in general terms, followed by a case study. Because process optimization monitoring is tightly linked to the remedial method used, details are provided in the appropriate chapters and sections on specific remedial options.

KEEPING AN EYE ON TREATED SITES

Monitoring usually continues after active remediation, stabilization, or containment has been completed. Only in the case of clean closure, where all contamination has been completely removed, is continued monitoring not expected. It is becoming increasingly evident that clean closure is not usually economically or technically possible at many sites. More typically, the situation will be that the contamination has not been completely removed, and residual levels are left to naturally attenuate; or that the contamination has been physically or chemically stabilized in place; or that an engineered barrier or cap has been constructed to mitigate contaminant migration. In these contexts, performance monitoring is conducted over years — and likely decades — to provide continuing evaluation of risk to human health and the environment. This activity, referred to as "long-term monitoring," is the focus of the remainder of this chapter. Because of the long time frame, the cost of long-term monitoring, using currently accepted technologies, can easily equal or exceed the cost of site characterization and remediation. For this reason, this chapter focuses on emerging low-cost sensor technologies and the synergistic integration of point measurements with spatially distributed geophysical measurements as means of reducing the costs of achieving spatial and temporal monitoring requirements.

PROCESS OPTIMIZATION MONITORING OF THE VADOSE ZONE

Process optimization monitoring, which includes performance verification, depends on site-specific conditions and the particular remedial options chosen. The objectives of this type of monitoring are to ensure that remedial objectives are being met by tracking the progress of contaminant removal, destruction, or stabilization and making adjustments to optimize the process; or to verify the physical integrity and continuity of a containment system and correct as needed. For example, consider the aerobic biological destruction of an organic contaminant. It is sometimes possible to track the destruction progress by measuring degradation products, such as carbon dioxide levels, in the vadose zone. However, if the targeted organic is not the only carbon source for biological degradation, more detailed (and complex) analyses will be required. In this case, it may be possible to use isotopic ratios of the carbon in the target contaminant to determine how much of the carbon dioxide can be attributed to the contaminant destruction. In the anaerobic biological dehalogenation of tetrachloroethylene, for some sites, the process may terminate at vinyl chloride, the most toxic member of the chlorinated ethylene series. If this is the case (process monitoring, such as soil gas analysis, must be performed to detect this problem), some action will be required to either facilitate the dehalogenation of vinyl chloride under anaerobic reducing conditions or to reverse the conditions, if possible, from anaerobic to aerobic to promote the biological oxidation to carbon dioxide. The monitoring methods used to verify the biological destruction of a given contaminant depend on the metabolic pathway for destruction and on the specific nature of the contaminant.

Site characterization techniques can be performed sequentially as a way to monitor the remediation performance and to verify that remedial objectives have been met. As discussed in Chapter 3, partitioning interwell tracer tests (PITT) can be performed in the vadose zone to estimate the mass of nonaqueous phase liquid (NAPL) present. Consider the remediation of NAPL contamination in vadose soils by soil vapor extraction. If the PITT is performed before remediation begins, this test will give an estimate of the mass of NAPL to be removed. Monitoring the concentration of NAPL in the extracted vapors will provide a measure of the mass of NAPL that has been removed. After the remediation is considered to be complete, another PITT, it is hoped, will show that most of the NAPL has been removed from the subsurface (Jin *et al.* 1995).

For *in situ* remediation methods using active processes, it is important to monitor process-specific variables to estimate mass balance calculations and contaminant removal/destruction. In electrokinetic extraction of heavy metals, the remediation can be tracked by monitoring the flux of contaminant being removed relative to the current applied, which establishes a measure of current efficiency (Lindgren *et al.* 1998). The current efficiency typically remains constant until the remediation is 60 to 70 percent complete, then tapers off.

In processes that both remove and destroy the targeted contamination, for example, steam stripping with hydrous pyrolysis oxidation (HPO), it is difficult to quantify how much contaminant has been destroyed (by comparison, quantification of removal is easy). The injection of steam and air can strip contaminants from the soil and, at the same time, stimulate the desired oxidative destruction of the targeted contaminants. However, the injection of steam and air can potentially relocate contaminants elsewhere in the subsurface. It is vital to distinguish between

The case study, "An Integrated Approach to Monitoring a Field Test of *In Situ* Contaminant Destruction," by Robin L. Newmark *et al.*, illustrates the use of noble gas tracers with varying degrees of solubility in water to determine how much injected water and gas is recovered as well as to evaluate the degree of *in situ* mixing, an important variable for the success of HPO. *See page 564.*

contaminant destruction and unintended contaminant relocation, as well as to distinguish between destruction by the intended mechanism (for example, HPO) and other possible natural processes (for example, biodegradation).

LONG-TERM MONITORING OF THE VADOSE ZONE

CONTAMINANT LOCATION IN VADOSE SOILS

Contaminants in vadose zone soils are comprised of three or four phases: a solid soil phase, an aqueous phase, a gaseous vapor phase and, possibly, a nonaqueous phase liquid (NAPL) phase. The phase of a contaminant is determined by its concentration and physical and chemical characteristics. Generally, classifying contaminants by volatility and solubility helps to illustrate where the problem areas are in monitoring, especially in the vadose zone. Table 4-1 indicates (via preferred sample type) the phase(s) that various contaminants favor in the vadose zone and in shallow groundwater. The table is divided into contaminant groups roughly based on solubility and volatility. The criteria for the various categories are similar to those presented in Thomas (1982). Groupings range from "soluble and volatile" to "nonsoluble and nonvolalile." Keep in mind that there are no truly nonvolatile or nonsoluble compounds, these operational definitions are only intended to qualitatively define general behaviors. Notably, "nonsoluble and nonvolatile" compounds have lower mobility and (for constituents of average toxicity) less significant impacts on underlying groundwater and the downgradient risk receptors, resulting in less overall concern and risk. For contaminants of extreme toxicity (e.g., plutonium), penetration of even a small amount of contaminant via complex processes (e.g., colloidal transport) can pose a significant risk and should be accommodated by developing long term monitoring plans for sites with such contaminants. For each grouping, the table indicates the most appropriate sampling phases, with the most easily and economically accessed phases shown in *italic type*. In general, soil gas (SG), groundwater (GW) and groundwater monitoring well headspace (HS) tend to be easier phases to sample compared to pore water (PW) or bulk soil or sediments (S). For each grouping, the degree of environmental mobility, and typical or dominant transport phases are noted along with example contaminants.

TABLE 4-1 Classification of contaminant characteristics.

Type of Contaminant	Soluble	Intermediate\ 1000>solubility>10 mg/l	Nonsoluble
Volatile (vp > 1 mmHg)	Vadose sampling: *SG*, PW	Vadose sampling: *SG*, PW, S	Vadose sampling: *SG*, S
	Aquifer sampling: *GW, HS*	Aquifer sampling: *GW, HS*, S	Aquifer sampling: *HS*, S
	Mobility: high	Mobility: varies	Mobility: medium/low
	Transport phase(s): gas, dissolved, nonaqueous liq.[a]	Transport phase(s): gas, dissolved, nonaqueous liq.[a]	Transport phase(s): gas, nonaqueous liq.[a], complex processes[b]
	Typical examples: benzene, trichloroethlene, chloroform, methylene chloride, phenol, methanol	Typical examples: chlorobenzene, terachloroethlene, ethylbenzene, toluene, xylenes, carbon tetrachloride	Typical examples: hexane, octane, decane
Nonvolatile (vp > 1 mmHg)	Vadose sampling: PW, S	Vadose sampling: PW, S	Vadose sampling: S
	Aquifer sampling: *GW*, S	Aquifer sampling: *GW*, S	Aquifer sampling: S
	Mobility: moderate–high	Mobility: moderate–low	Mobility: low/ very low tends to sorb to sediment
	Transport phase(s): Dissolved, nonaqueous liq.[a]	Transport phase(s): Dissolved, nonaqueous liq.[a]	Transport phase(s): Nonaqueous liq.[a], complex processes[b]
	Typical examples: High solubility anionic Inorganics ($CrO_4^{=}$, TcO_4^-)	Typical examples: pentachlorophenpl, napthalene, intermediate solubility anionic or cationic inorganic contaminants	Typical examples: heavy organics (DDT, dieldrin, PCBs, phenanthrene), low solubility cationic inorganics (Pu, Hg, Pb)

LEGEND:

SG = soil gas
PW = pore water
GW = groundwater
HS = head space
S = bulk soil or sediment

a = nonaqueous phase liquid (NAPL) if sufficient concentration present
b = complex processes, including enhanced mobility due to colloidal transport, chemical association and fracture flow

Note: Italic type indicates the ease of sampling

When considering the volatile constituents of many organic contaminants in the vadose zone, the soil gas phase is favored and is easy to sample; therefore, it is most practical to focus on soil gas for these compounds. For nonvolatile contaminants (primarily inorganic contaminants), the story is different. In the vadose zone, soluble, nonvolatile, primarily inorganic, contaminants reside in the pore water, which is very difficult to sample and monitor. Pore water sampling is usually performed using a suction lysimeter (see Chapter 3). Because of the difficulty encountered in directly sampling nonvolatile constituents in the vadose zone, the monitoring of indicator parameters, such as soil moisture and electrical properties, is of great utility (as discussed in detail in the following subsections).

Thus, in general, the contaminant type determines the phase to be monitored in the vadose zone. Volatile organic contaminants are most conveniently monitored in the soil gas, while inorganic contaminants require monitoring in the soil pore water. Organic and inorganic contaminants of intermediate solubility and/or volatility may provide more, or less, sampling options, depending on specific conditions and available technologies. These compounds require a site specific decision to determine the most appropriate strategy.

Functional Requirements and Consequences of Failure

Possible remedial actions can be grouped into four major categories: containment, *in situ* stabilization, *in situ* removal or destruction, and natural attenuation. For each of these remedial options, a different functional requirement and consequence of failure can be identified (Table 4-2). These functional requirements and consequences of failure help to define long-term monitoring objectives and temporal and spatial requirements for each remedial option. A distinguishing feature among the remedial options is the possibility of catastrophic failure of a containment structure or process that leads to fast changes at later times and at discrete points in space. The other remedial options are characterized by slow changes from more diffuse sources.

Temporal, Spatial, and Other Data Requirements

The determination of the spatial and temporal requirements for monitoring at a specific site requires a detailed understanding of the

TABLE 4-2 Remedial actions related to functional requirements and failure consequences.

Remedial Action	Functional Requirements	Consequence of Failure	Temporal Requirements	Spatial Requirements
Containment	No leaks	Catastrophic release from break(s)	Possible fast changes at any time; more likely at end of structural design life	Possible point source(s)
In situ stabilization	No contaminant movement	Slow leaching	Slow changes Possible increase with time	Diffuse source
In situ removal or destruction	Low residual No movement	Significant contamination migration	Slow changes	Diffuse source
Natural attenuation	No impact	Significant contaminant migration	Slow changes	Diffuse source

geohydrologic setting. Contaminant transport modeling and risk assessment will likely be required to optimize the long-term monitoring design. The focus here is not on the specifics, but on the general nature of the spatial and temporal requirements for long-term monitoring, and on using them as a guide for suggesting optimization strategies.

One way to see the difference between process optimization monitoring and long-term monitoring is to consider the temporal, spatial, and data quality requirements of each. Process optimization monitoring is conducted during the active remediation phase, during which much should be happening and the contaminant concentration in the soil should be dropping; during this phase, more frequent sampling is required over a relatively short period of time. Since active remediation or containment construction target a specific region of contaminated soil, the spatial requirements for *where* to monitor are better defined. Furthermore, since the desired information will be used to control the process, more detailed analyses will be required, with accuracy and precision being of great importance. In many cases, the soil, water, or gas samples are collected manually and samples are sent away for analysis. Because of the short time period over which process optimization monitoring is conducted, labor intensive, manual operations are acceptable.

Long-term monitoring (LTM), on the other hand, is expected to continue for years, if not for decades. The data quality requirements for long-term monitoring are more relaxed than for process optimization monitoring. With long-term monitoring, the concern is not so much with the precision and accuracy of the measurement, but with the certainty that a change has been detected. The idea is to monitor for changes that signal existing or developing contaminant migration. For the most part (with the exception of containment failure), these changes develop slowly, so that less frequent (but repetitive) sampling over a much longer period of time is justified or required. The long time frame and repetitive nature of sampling for LTM strongly suggests the need for a highly automated system.

The contaminant type also affects spatial sampling requirements. Pore water migration paths in the vadose zone are difficult to predict with certainty, and fingering can result in large fluxes of soluble contaminants in small spaces. Therefore, the spatial requirement for monitoring soluble inorganic contaminants with low volatility in the soil pore water, can only be achieved with a dense data grid over a large area. Without a dense grid, a pore water finger, and the contaminants contained within it, may escape detection. It is not likely that achieving the required data density using point measurements is practical, so the use of other methods for obtaining the needed data density, such as geophysical methods (as discussed below), is required.

The situation is different for volatile organic contaminants. When soil moisture is low, the larger size pores are air-filled, which permits relatively fast vapor phase transport by advection and diffusion. Soil gas migration precedes pore water migration and tends to be more evenly dispersed and predictable in the subsurface. Therefore, the spatial requirement for monitoring organic contaminants in the soil gas can be achieved with a less dense grid. So, even though volatile organics also move with the pore water, their presence will not escape detection because vapor phase advection and diffusion will transport them to the soil gas monitoring location in a short period of time. For this reason, soil gas surveys have successfully been used to track plumes of volatile organic contamination in groundwater. If sufficient organic contaminant is present to form a non-aqueous phase liquid (NAPL), the migration can be a hybrid of the above two situations: a NAPL can flow in fingers which are hard to detect, but significant volatility allows easy gas phase detection.

While gas phase detection of a NAPL finger flow by a long term monitoring system will work well to sound the alarm of a potential problem, it will not pinpoint where that problem is located. However, the purpose of the long-term monitoring system, is to sound the alarm, not to pinpoint where the NAPL finger is located. If the long term monitoring data suggests that there is a problem that warrants further investigation, one moves out of long term monitoring and back to characterization and, perhaps, remediation. Characterization and remediation have different temporal and spatial requirements, and labor intensive, manual sampling will likely need to be conducted.

VADOSE ZONE MONITORING

The advantage of monitoring in the vadose zone is that if a problem is detected, more time is available to take a careful look and possibly to conduct additional characterization and active remediation before an aquifer is affected. However, long-term groundwater monitoring will likely continue to be an important aspect of long-term monitoring of contaminated vadose sites (Cullen 1995).

At sites with large vadose zones, a vadose zone monitoring system may be advisable and could be required. If a vadose zone monitoring system effectively monitors beneath a facility, the Environmental Protection Agency (EPA), or other relevant regulatory agency, may decide it is appropriate to reduce the scope of groundwater monitoring and offset some of the cost of the vadose monitoring (Durant *et al.* 1993). This is not to say that groundwater monitoring will be eliminated entirely, but it might be effectively automated. Vadose zone monitoring can be used to detect contamination close to the source before it reaches the underlying aquifer. Early detection is, of course, advantageous because it provides an early warning for corrective action. Current vadose zone monitoring technologies are essentially similar to the characterization technologies described in Chapter 3. These technologies include: direct methods, such as soil sampling, suction lysimeter pore water sampling (Everett and Wilson 1986); soil gas sampling and indirect moisture detecting methods, such as neutron moderation logging (Cullen *et al.* 1995); tensiometers; and time domain reflectrometry (TDR) probes (described later).

The case study, "The Vadose Zone Monitoring System for the CAMU Project at Sandia National Laboratories," by Jim Studer, describes a state-of-the-art vadose zone monitoring system installed at Sandia National Laboratories in Albuquerque, New Mexico, as part of the first Corrective Action Management Unit (CAMU), permitted in New Mexico. *See page 576.*

Vadose zone monitoring systems are typically required for CAMU disposal cells, but usually in conjunction with a standard Resource Conservation and Recovery Act (RCRA) groundwater monitoring system. Usually, a reduction in the groundwater sampling frequency can be negotiated when the vadose zone monitoring system performs well. The design of the vadose zone monitoring system at the Sandia National Laboratories CAMU was considered to be so effective that a complete waiver from installing or operating a groundwater monitoring system was obtained. This is one of the first vadose zone monitoring systems in the nation to earn such a waiver.

The direct techniques that generate samples requiring analysis, such as soil gas or suction lysimetry, are also quite expensive. The indirect methods for moisture detection are cheaper because no samples are generated, and, therefore, no analysis required. However, all these techniques are labor intensive because they require a crew to go into the field and either make measurements or collect samples. The active nature of sampling using currently accepted methods, and the high cost of analysis, will keep the life-cycle cost of long-term monitoring unacceptably high for both the vadose zone and for traditional groundwater.

Reducing Long-Term Monitoring Costs

The key to reducing long-term monitoring costs is to significantly reduce active onsite sampling activities and related analytical costs. When the same information is gathered by passive means, crews needn't be deployed to collect samples or data, and there is no need for offsite analytical work. These passive systems must be highly automated. They must collect data, screen for alarm conditions and transmit the information to a central location for final analysis and archiving. A high degree of automation can be achieved through the synergistic combination of sensors for continuous point measurements and periodic geophysical measurements to provide information between the points. It is expected that highly automated long-term vadose zone and groundwater monitoring systems will result in life-cycle cost reductions of one to two orders of magnitude over present traditional methods based on manual groundwater sampling.

Numerous sensor systems have recently been developed that can supply the required information. Of particular importance are advanced electrochemical, fiber-optic sensors; radiation microsensors; and mass-sensitive piezoelectric or acoustic wave transducers. Each of these sensor types is discussed in more detail in the next section of this chapter. Furthermore, the incorporation of integrated microelectronic/micromachine components into these sensor systems has the potential to revolutionize many future activities through significant component miniaturization, automated data analysis, and integration. Many of these sensors can be deployed in a groundwater well to allow continuous unattended analysis of the water for contaminants of concern. Only when the sensors indicate an increase in contaminant level would traditional groundwater sampling be conducted.

However, all sensors only provide point measurements, which makes it difficult to achieve the required spatial data density required to detect some potential problems, especially in the vadose zone. Vadose zone moisture sensors, such as tensiometers and time domain reflectrometry (TDR) probes, are discussed below. The syncrgistic usc of the geophysical methods (to provide the required information between the point sensor measurements), including ground penetrating radar (GPR) and electric resistivity tomography (ERT), are also discussed in detail below.

SENSOR TECHNOLOGY

In the last chapter of the *Handbook of Vadose Zone Characterization & Monitoring* (Wilson *et al.* 1995), titled "Emerging Technologies for Detecting and Measuring Contaminants in the Vadose Zone" (Koglin *et al.* 1995), a number of emerging chemical sensor technologies are described. These chemical sensors can be broadly cast into categories of optical sensors, radionuclide sensors, electrochemical sensors, and mass sensors. The following subsections summarize new developments in optical sensor technology and explain in basic terms how the sensors work.

THE EMERGENCE OF MICROSENSOR TECHNOLOGY

During the late 1990s, great strides were made in sensor technology with particular focus on ultraminiaturization through the use of microlithography. Many new sensors are being developed through silicon chip fabrication technologies. The sensing element is essentially fabricated directly on a silicon chip to allow direct integration with any required signal and data processing circuitry, producing remarkably small sensors. Not only are these sensors small, but since they can be mass produced using standard silicon chip fabrication methods, they are expected to be inexpensive as well. New manufacturing processes should allow the economical deployment of a large number of sensors for many environmental monitoring purposes. These inexpensive sensors can be deployed in redundant numbers—as one sensor fails, another can simply be brought online in its place without the expense of retrieval.

Ultramicro-Electrochemical Sensors

Anodic stripping voltammetry (ASV) is one of the most useful electroanalytical techniques because of its low limits of detection for a number of high priority metal pollutants and its potential to be implemented at very low cost. This technique involves a preconcentration step in which the metal ion of interest is electrochemically deposited by applying a cathodic, or negative, potential into a thin film (usually of mercury or a polymer) that has already been placed on the electrode. In a typical system, a preconcentration time of a few minutes will yield sensitivities at the low parts-per-billion (ppb) level. With longer preconcentration times, parts-per-trillion can be achieved. After the metal is preconcentrated, the potential is scanned from the applied negative potential toward a more positive potential. The metals are then anodically stripped (reoxidized) from the electrode, which yields a current signal in the form of a peak at a redox potential characteristic of the metal. The magnitude of this peak is proportional to its concentration in solution. Thus, the voltage at which reoxidation occurs can be used to identify the metal ion. Multiple metals can be detected as long as the redox potentials of the different metals do not overlap (Herdan *et al.* 1998).

One example of such advanced microsensors is a type of mercury-plated iridium-based ultramicroelectrode array (UMEA) for ASV determination of many metal ions in water (Figure 4-1) (Kounaves *et al.* 1994). Mercury-plated iridium UMEAs are fabricated on standard silicon wafer substrate, and a thin silicon dioxide dielectric layer is grown on the substrate. Iridium and silicon nitride are successively evaporated onto the surface, then patterned by a photolithographic procedure. Each individual electrode is 5 μm to 10 μm in diameter (Silva *et al.* 1999) and spaced 50 to 300 μm (Belmont *et al.* 1996; Kounaves *et al.* 1994), center-to-center, from its nearest neighbor. This spacing is chosen to avoid any overlap of diffusion layers associated with individual electrodes during the time period of the electroanalytical technique selected. The iridium UMEA are plated with mercury before use. Having arrays of ultramicroelectrodes in a single sensing element increases the signal proportionally with the number of electrodes in the array. The number of electrodes in an array used by researchers has increased in recent years from 19 (Kounaves *et al.* 1994) to 100 (Belmont *et al.* 1996) to 900 (Silva *et al.* 1999).

Analysis with UMEAs typically includes a special form of ASV, where a square waveform is used to generate the signal, called "square wave ASV" (SWASV). The analysis can be performed faster and with more sensitivity (for example, detection limits of 10^{-12} M have been attained [Kounaves *et al.* 1994]). The use of UMEAs also requires the use of a special reference electrode. The use of the standard saturated calomel reference electrode can leak significant amounts of chloride ions which, if located in close proximity to the UMEA, causes local high concentrations leading to the accumulation of mercurous chloride and sensor failure (Kounaves *et al.* 1994; Nolan and Kounaves 1998). To prevent this, a solid–state reference electrode has been developed for *in situ* SWASV. This reference electrode is Ag/AgCl with an immobilized electrolyte protected by a polyurathane coating (Nolan *et al.* 1997).

Mercury-plated iridium UMEAs have been used to measure trace levels of cadmium, lead, copper, and zinc in natural waters without the

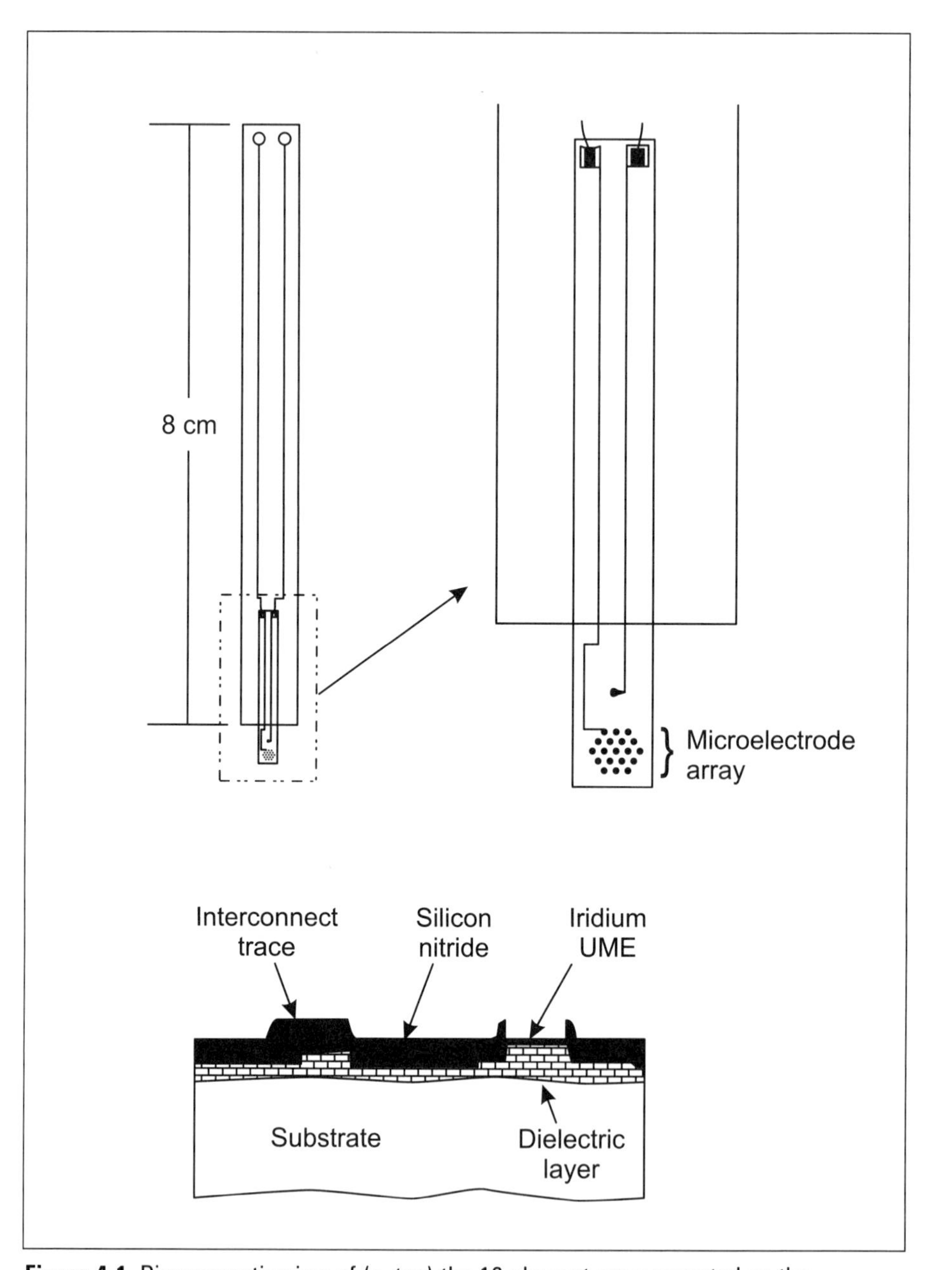

Figure 4-1. Diagrammatic view of (a, top) the 19-element array mounted on the PC board and (b, bottom) the structure of the iridium UME and traces on the silicon wafer (Kounaves *et al.* 1994).

need for deoxygenation or stirring. Over a concentration range of 0.5 to 15 ppb, Silva (1999) found the signal for both cadmium and lead to exhibit a strong linear correlation with not only the ion concentration, but also with the preconcentration time. Greater sensitivity can be achieved by increasing the preconcentration time. A proof-of-concept demonstration of *in situ* performance was conducted in groundwater wells at a metals-contaminated landfill site located at Hanscom Air Force Base in Bedford, Maryland (Herdan *et al.* 1998). Comparison was made to inductively-coupled plasma emissions spectroscopy (ICP) analysis by EPA method 200.7. The *in situ* measurements appeared to be systematically low, but trended with the ICP analysis. The UMEA analysis for lead revealed a range from 32 to 302 ppb, which compares with ICP analytical results of 174 to 898 ppb. Copper UMEA results showed a range of 4 to 12 ppb, and in ICP analysis, a range of 14 to 34 ppb was found. The UMEA did not detect any zinc, while the ICP method detected 20 to 65 ppb of zinc. The researchers mentioned working on an improved system to detect silver (III), chromium (VI), mercury (II), and selenium (IV).

One issue with mercury-plated iridium UMEAs is that of long-term stability, as it appears that the mercury plating needs to be replaced after 7 days in water (Kounaves *et al.* 1994). UMEAs can be fabricated in a similar fashion using gold or platinum rather than mercury-plated iridium (Uhlig *et al.* 1997). Gold UMEAs have been used to detect mercury with limits in the sub-ppb range (Uhlig *et al.* 1997) and to identify selenium (IV) with a detection limit of 0.42 ppb (Tan and Kounaves 1998). Since the gold electrode is not plated, long-term stability should not be as much of an issue. In general, the need to study the long-term stability and performance of these microlithographic electrode systems is grcat.

Wang *et al.* describe a micromachines stripping analytical system (1999). This analyzer integrates fluid-handling silicon microstructures and a three-electrode system on a chip. The detector is a mercury-plated iridium UMEA along with reference and counter electrodes all prepared by the same photolithographic process.

Mass Sensors

Mass sensors are microacoustic sensors well-suited to detecting changes in mass when analytes interact with the sensor surface; these devices can be used to monitor fluid properties such as density and viscosity directly. Microacoustic sensors consist of: (1) a physical layer which produces an acoustic wave when energized (that is, a piezoelectric material such as quartz or gallium arsenic) (Casalnuovo *et al.* 1998) and (2) a chemically selective layer, which is a special coating placed on the sensor surface to limit which analytes will interact with the surface and therefore be "sensed." Changes in the mass of the chemical coating result in changes in the resonant frequency of the physical layer, and are easily measured. Temperature has a pronounced effect on the performance of all microacoustic sensors; it must be either carefully controlled or a correction for the effect must be made. Generally, the higher the frequency of the sensor, the greater the temperature effect.

The piezoelectric effect is the phenomenon exhibited by certain crystals when subjected to an electric field. The crystal expands along one axis and contracts along the other axis. If the electric field is alternating, the expansion and contraction will cause vibrations in the crystal. The amplitude of the vibrations in the crystal will reach a maximum when the input signal matches the resonant frequency of the crystal, which depends on the physical dimension of and wave velocity in the crystal. Conversely, when a piezoelectric crystal is vibrated, an analogous electric signal will be generated. Microacoustic sensors have two main variants: that is, thickness shear mode (TSM) devices (also known as quartz crystal microbalances, or QCM), and surface acoustic wave (SAW) devices. The manner and location of the acoustic wave propagation is different for each class of device, which influences the sensing performance.

Quartz Crystal Microbalance Devices

In QCM devices, the acoustic shear wave is propagated perpendicular to the sensing surfaces through the bulk of the crystal. The fundamental acoustic wavelength is equal to twice the substrate thickness. A fundamental frequency of 10 MHz (thickness of quartz plate, 167 μm) is typically used (Dickert *et al.* 1999; Yan and Zhou 1998; Chang *et al.* 1995). These devices are commercially available. Thinner, higher frequency QCM are also available (30 MHz, 55.6 μm thickness) and have demonstrated a greater sensitivity (Bodenhofer *et al.* 1996). The sensitivity of QCM is proportional to the square of the resonant frequency and inversely proportional to the QCM surface area.

While QCMs can be used for liquid analysis, they are most commonly used for analysis of gases and vapors. QCMs have been considered for *in situ* vadose zone soil gas sensors (Koglin *et al.* 1995). The selectivity of the QCM is determined by the coating applied to the surface. Typically, both sides of the QCM are coated. The film thickness should be limited to 0.1 to 0.3 percent of the acoustic wavelength (Bodenhofer *et al.* 1996). For a 10 MHz QCM, the film thickness should be between 0.33 μm and 1.0 μm, and for a 30 MHz QCM, the film thickness should be between 0.11μm and 0.33 μm.

Bodenhofer *et al.* (1996) used a number of coatings on both QCM and SAW devices of different frequencies in side-by-side comparison of n-octane and tetrachloroethylene detection. As expected, 80 MHz and 433 MHz SAW devices were found to produce larger signals and to demonstrate enhanced surface sensitivity compared to the QCM devices. However, the SAW devices are more sensitive to temperature fluctuations. For monitoring volatile organic compounds, the QCM devices are easy to handle, and interpretation of the results is straightforward. The 10 MHz QCM was the most stable device, but the sensor response was relatively low. The 30 MHz QCM device was considered a good compromise between response and thermal stability. Yan and Zhou (1998) used poly (*o*-anisidine) (POA) and poly(allylamine) hydrochloride (PAH) coatings on 10 MHz QCMs to monitor phenols in the vapor phase. The limit of detection for phenol was 0.022 mg/L on the POA coating and 0.041 mg/L on the PAH coating. The coatings, however, also respond to variation in humidity, a significant problem for vadose zone soil gas monitoring.

QCM arrays and multivariate data analysis can be used to analyze multiple components of complex mixtures and correct for humidity effects. Dickert *et al.* (1999) used an array of four 10 MHz QMBs to independently monitor the concentration of the meta- and para-isomers of xylene, a difficult separation even for gas chromatography. The coating material for the four QCMs was chosen from among beta-cyclodextrins and calix[4] resorcinearenes in such a way that each coating responded to each xylene isomer and humidity, but to different degrees. After multivariate calibration, the sensor array obtained the accurate separation of m- and p-xylene mixtures (zero to 200 ppm) in variable humidity (zero to 60 percent relative humidity) environments.

QCM arrays are inexpensive and effective gas phase monitors suitable for vadose zone soil gas analysis. They can be designed to respond to specific compounds from complex mixtures or to general classes of compounds. However, there are limitations. The target compounds must be chosen as part of the design, but designers should be aware that unknown contaminants could also interact. Fortunately, there are methods for estimating polymer-coated acoustic wave vapor sensor responses using extensive tabulated parameters (Grate *et al.* 1995). The temperature stability of the subsurface environment should alleviate most problems with temperature drift, but the calibration of each individual device for thermal behavior is advised (Bodenhofer *et al.* 1996).

Surface Acoustic Wave Devices

In SAW devices, metallic inter-digital transducers (IDTs) are formed on the surface of the piezoelectric substrate by a microlithographic procedure. Upon application of a high-frequency signal to the input IDT, the piezoelectric crystal produces a Rayleigh surface acoustic wave. A second IDT, placed on the same surface a known distance from the first, will intercept the surface wave and transform it back into an electrical signal. Comparison of the input and output signals allows accurate measurement of the SAW velocity and attenuation. Because the SAW is concentrated on the surface, the velocity and attenuation of the SAW depend highly on the properties (that is, mass and viscoelasticity) of any surface material placed on the crystal. The placement of thin films on the surface that selectively adsorb target compounds will change the SAW velocity and allow detection of the compound. The high sensitivity of the SAW device allows effective detection with a small surface area (Krylov 1995).

The use of SAW devices for monitoring gases is very similar to the approach used with QCMs in that the selectivity is determined by a coating applied to the surface. For SAW devices, the coating is applied only to the side of the piezoelectric crystal having the IDT circuitry. Coatings on SAW devices are generally thinner than those on QCM devices. SAW devices operate at higher frequencies—typically 80 to 500 MHz—and the coating thickness should be on the order of 1 percent of the acoustic wavelength. For quartz substrates, this would be about 0.40 μm for 80 MHz and 0.06 μm for 500 MHz SAW devices. As mentioned above, in the comparison of QCM and SAW devices with the same coating material, Bodenhofer (1996) found that SAW devices produced larger signals and demonstrated enhanced surface sensitivity, but were also more sensitive to temperature fluctuations.

A field demonstration of a SAW-based vadose zone monitoring system was conducted at Hanford, Washington. A down hole probe was tested in two vadose zone wells, one containing high levels (20,000 to 25,000 ppm) of carbon tetrachloride (CCl_4) and the other containing low levels (40 to 50 ppm).

Data from the down hole probe compared very well with both an above-ground SAW-based analysis and a commercial infrared system for CCl4, as described in detail in the case study, "*In Situ* Field Screening of VOCs Using a Portable Acoustic Wave Sensor System," by Gregory C. Frye-Mason. *See page 580.*

Integrated SAW Devices Using Gallium Arsenide

Currently available SAW devices are standalone microdevices that require external circuitry for excitation and evaluation. Using quartz crystal for the required piezoelectric properties complicates integration of other semiconductor circuitry with SAW devices because quartz is not semiconductive and silicone is not piezoelectric. Gallium arsenide (GaAs) is a unique material for such integration because it is both piezoelectric (similar to quartz) and semiconductive. Manufacturing procedures for fabricating semiconductor devices (such as integrated high-frequency radio frequency ([RF]) microelectronics) from GaAs are well-established. This fact allows SAW devices to be fabricated directly onto the GaAs substrate, along with all the required microelectronics to energize and interpret the sensor (Casalnuovo *et al.* 1998).

Monolithic integration of SAW sensors and the supporting microelectronics is advantageous because of smaller size, simpler packaging, and lower power requirements which all lead to more economical fabrication. Furthermore, this integration improves the sensitivity of the device substantially. By removing the need for high-frequency interconnections, devices that operate at higher frequencies are possible. Higher frequency means greater sensitivity in a smaller sensor, as illustrated in Figure 4-2. This illustration shows that tripling the frequency results in a ten-fold response increase in a sensor one-tenth the size. It is anticipated that integrated GaAs SAW devices can be designed for operation at frequencies up to 1 GHz. Dual reference/sensor delay line devices have been fabricated that automatically compensate phase shifts caused by temperature changes. By fabricating the sensor and reference delay line on the same substrate, both devices experience the same temperature change, and the integrated phase comparator cancels the common drift (Casalnuovo *et al.* 1998).

The technology of SAW devices is improving rapidly with the advent of smaller, more stable and sensitive integrated devices, that are simpler to operate. SAW sensors are likely to be widely used for monitoring volatile organic compounds in vadose zone soil when used in arrays like those described for the QCM devices. The advantages of a fully integrated SAW array are its much greater sensitivity in a much smaller package (which will allow for more elements in the array), automatic temperature compensation, and the possibility of integrating electronics

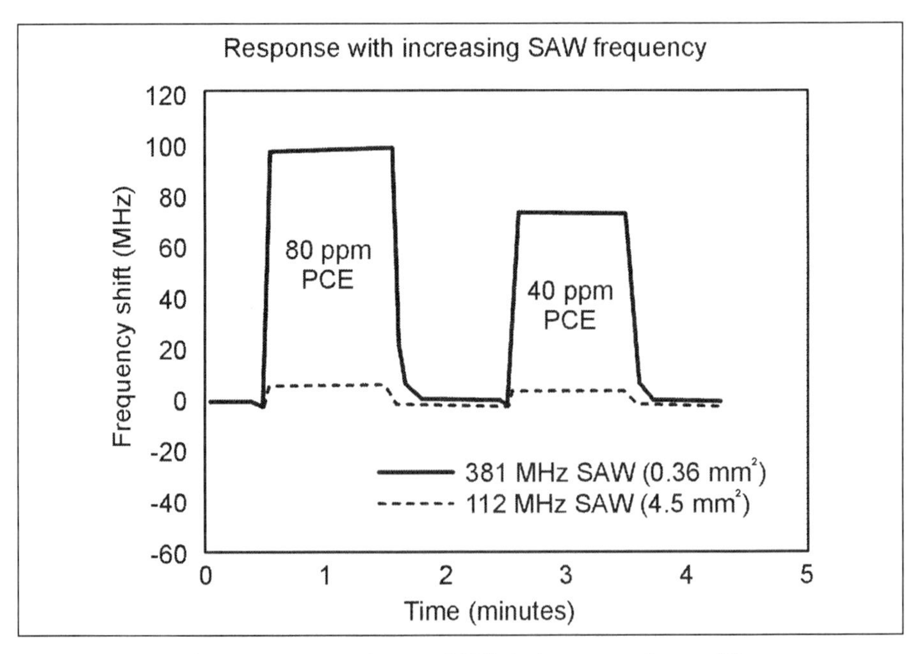

Figure 4-2. Plot of signal response for two SAW devices operating at different frequencies.

for performing the pattern recognition required for interpreting the results. As useful as these capabilities would be, they do not fully exploit the potential of these tiny sensors. The great improvement in size and sensitivity of integrated SAW arrays has allowed further integration, for example, as the sensor in a microgas chromatographic system under development at Sandia National Laboratories.

μChemLab™

Complex mixtures of organic compounds are best quantified after high-resolution separation by gas chromatography (GC). Sensor arrays, as already discussed, will always have some uncertainty regarding the possible presence of an unexpected constituent. The additional information of retention time on a separation column is usually sufficient to at least provide notice of the presence of an unexpected constituent and often leads to identification. SAW sensors have been utilized in a recently developed and commercialized gas chromatography instrument (SAW/GC). This instrument is a small, powerful man-portable field instrument (DOE 1998). Figure 4-3 shows a schematic diagram of a SAW/GC system under development, called "μChemLab™," which takes miniaturization much farther. The major components of the system are modules for sample collection, separation, detection, and gas flow control. The detection system is based on a four SAW arrays. Microscale GC capillary columns, fabricated by microlithographic methods, have been tested (Hudson *et al.* 1998). Fabrication of the first prototype will be completed by the fall of 1999. The initial application is for chemical weapon detection, so the analytical system is designed more for semi-volatile organic analysis, which is more challenging than volatile organic analysis. This initial prototype will consist of interconnected modules, but eventually, monolithic integration is anticipated. *In situ* vadose soil gas monitoring with such a powerful device at strategic locations is expected to provide a great deal of certainty to a monitoring system otherwise based primarily on simple sensor arrays.

Fiber-Optic Sensors

Fiber-optic technology was first commercially developed for use in the telecommunications industry but now has widespread applications. One such application is as a chemical sensor. Total internal reflection

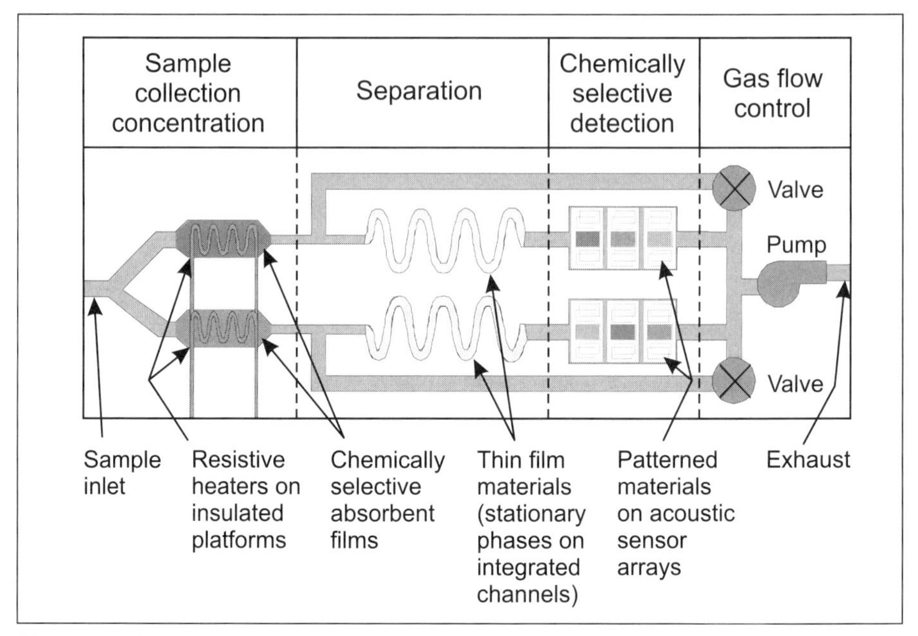

Figure 4-3. Schematic of the gas-phase ChemLab™ system. The system incorporates Sandia-designed and fabricated concentration, separation, and detection components and commercially available diaphragm pumps and valves.

allows the efficient transmission of light waves (infrared, visible, or ultraviolet) through optical fibers, and the interaction of the light with an analyte is the basis for sensing. The interaction may be luminescence, fluorescence, scattering, absorption, reflection, or any interaction that causes changes in the light properties, for example, intensity, wavelength, phase, or spectral distribution. Measurement of these changes provides information on the analyte that caused the change. Fiber-optic sensors offer distinct advantages over other sensors because a conductive wire is not required to transmit the data to the surface and therefore is immune to electromagnetic interference (Choudhury 1998).

Fiber-optic sensors consist of an optical fiber with an indicating surface at one end and an optical detector at the other end. There are many fiber-optic sensor configurations, but they tend to fall into two main categories: end-of-fiber and length-of-fiber (porous fiber, or evanescent wave). For either type of sensor, the optical detection can be as simple as the measurement of light intensity changes or as complex as using fiber-optic spectrometers to measure changes in spectral distribution (Rodgers and Poziomek 1996).

The end-of-fiber sensor, sometimes called an optrode, uses the optical fiber as a conduit to bring light to and from the sample located at the end of the fiber and provide a point measurement capability with a very small sampling volume. The small sample volume can have great advantages when a discrete, minimally intrusive point measurement is required, but sensitivity can be an issue.

With length-of-fiber sensors, all or part of the fiber's length is used for sensing, which allows either distributed measurement capability, if the fiber is stretched out, or point measurement of a larger volume, if the fiber is coiled. The longer the sensing length, the greater the sensitivity. Porous fiber sensors are specially made with the indicator chemistry incorporated directly into the structure of the fiber. Sol-gel-glass techniques can be used to produce a range of porosity and chemistry inside the fiber. Because of the large surface area of the porous core, this method is well-suited for absorbency measurements. Other length-of-fiber sensors rely on evanescent wave interactions. When light is totally reflected at an interface between the optical fiber wall and the surrounding medium, part of the light penetrates a certain distance (usually less than 100 nm) into the evanescence zone of the medium, as illustrated in Figure 4-4. This penetration of light is called the "evanescent wave" and can be used to detect the presence of optical indicators or

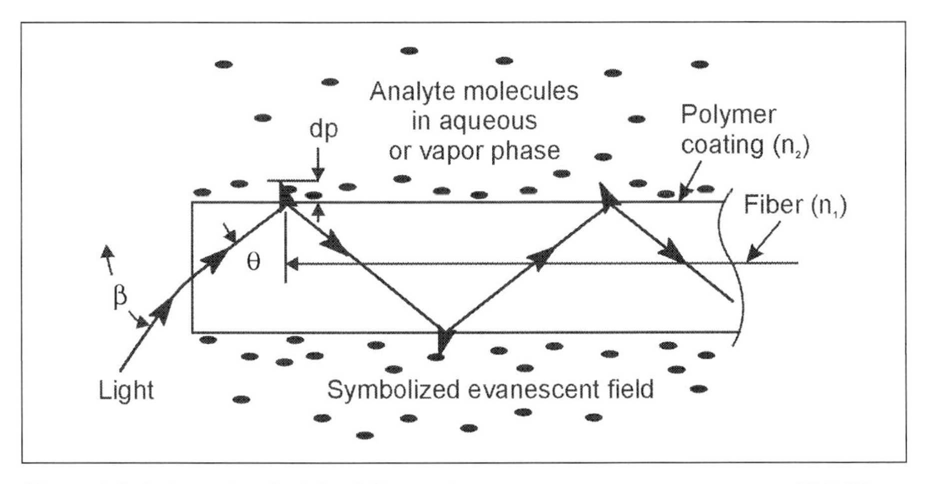

Figure 4-4. Schematic principle of fiber-optic evanescent wave spectroscopy (FEWS). When light is totally reflected at an interface between an optical waveguide and the surrounding medium, part of the light penetrates a certain distance into the adjacent phase (Mizaikoff *et al.* 1999).

changes in refractive index at the surface of the fiber. Evanescent wave sensors are easier and cheaper to manufacture because it is simpler to treat the fiber surface than to modify the porosity of the fiber core.

As with mass detectors, the selectivity of fiber-optic detection can be controlled with special coatings placed on the fiber end (end-of-fiber sensors) or along the surface (evanescent wave sensors). The coating can act preferentially to adsorb targeted analytes or to serve as an indicator. Changes in the intensity of the returning light for a given range of wavelengths depends on the absorbency or fluorescence of the analytes, or indicator layer located near the end of the fiber. The indicator is immobilized on the end of the fiber either covalently or by incorporation into a polymer or membrane. For monitoring purposes, reversible indication is clearly preferable to irreversible indication that requires regeneration.

Therefore, with simple fiber-optic sensors that monitor changes in return light intensity, the coating imparts selectivity. As with mass sensors, arrays of sensors with different coatings and multivariant analysis can be used to monitor the concentration of complex mixes of targeted contaminants. Aqueous phase inorganic analytes and both aqueous and vapor phase organic analytes can be detected with fiber-optic sensors; however, organic applications are more common. Methods measuring the absorbency of reversible dye indicators have been reported for the monitoring of pH; benzene, toluene and xylene; and the heavy metal ions Pb^{2+}, Cd^{2+} and Zn^{2+}. Methods measuring the refractive index change due to the selective and reversible sorption of volatile organic compounds and dense nonaqueous phase liquid (DNAPL) have also been reported (Rodgers and Poziomek 1996). These types of sensors offer continuous monitoring using inexpensive detectors. However, as with mass sensor arrays, the selectivity is provided by the particular indicator or coating used, and the identification of an unknown or unexpected contaminant is generally not possible. Depending on the particular system, response from nontarget compounds can lead to false positive readings.

The use of more sophisticated fiber-optic spectrometric methods is less prone to false positive readings because compound identification is possible. When fiber-optic sensors are coupled with spectrometric detectors, the coating concentrates the targeted analytes but does not solely impart selectivity. The coating can selectively adsorb and concentrate a broad class of compounds, such as hydrophobic organics, and the spectrometric detection method is used to identify and quantify individual analytes present in the coating. Mid-infrared fiber-optic evanescent wave sensors (MIR-FEWS) can probe the fingerprint region of the infrared spectrum when coupled with Fourier transform infrared (FTIR) detection, and can identify many organic analytes in air or water. MIR-FEWS methods, however, require special and expensive silver halide optical fiber and a more expensive FTIR detector (Mizaikoff 1999). These factors can limit the number of such sensors deployed.

Ultraviolet (UV) fiber-optic evanescent wave sensors (UV-FEWS) can detect and identify aromatic and polyaromatic hydrocarbons (PAHs) in air and water using standard, less expensive fibers and UV detectors (Schwotzer *et al.* 1997). Some heavy metals (chromate, for example) may also be detected by UV absorbency methods (Rodgers and Poziomek 1996).

Radionuclide Monitoring

The U. S. Department of Energy (DOE) is faced with unique environmental problems with respect to radionuclide contamination and related contaminant migration issues. A necessary and significant component of the DOE's site characterization and monitoring programs is to identify and track the nature and extent of radionuclide contamination. Monitoring and identifying radionuclide transport in the vadose zone is a particular need, as exemplified by the much-publicized leaking radioactive waste storage tanks at Hanford. Commercially available radiation detectors (for alpha and beta particles, X-rays and gamma-rays) rely on lithium-drifted silicon and high-purity germanium sensors. These sensors offer high resolution spectroscopic capabilities but are limited to operation at cryogenic temperatures (77°K). The cooling requirement of these sensors precludes the use of them for vadose zone monitoring.

Fortunately, material and electronic advances have been made that have resulted in the production of radiation detection systems which operate at room temperature. Performance of the new systems rivals that of cryogenically cooled systems in terms of energy resolution, signal-to-noise ratio, collection efficiency, and sensitivity. These advances have made it possible to develop a number of semiconductor materials with the properties necessary for high-performance spectrometers. Cadmium zinc telluride (CZT) is especially suited because its high atomic number allows it to stop high-power energetic charged particles, and its high-bulk resistivity (~10^{11} W/cm) ensures a low dark current during detector operation. CZT can directly measure the number and energy of X-ray and gamma-ray photons and charged alpha and beta particles that interact with the detector. Thus, not only can it detect radiation, but it can also identify the atom emitting the radiation.

The detection and identification attribute, coupled with the small physical size of these sensors (Figure 4-5), make them extremely valuable in vadose zone and groundwater monitoring systems. These sensors provide essentially continuous *in situ* scintillation counting of soil or water. The zone of detection of the sensor in soil or water depends on the type and energy of the radiation. For example, because energetic charged particles (alpha and beta) have poor penetration ability, the contaminant would need to contact the surface of the detector. For all practical purposes, this fact essentially eliminates *in situ* detection of particle-emitting radionuclides, such as plutonium (an alpha emitter)

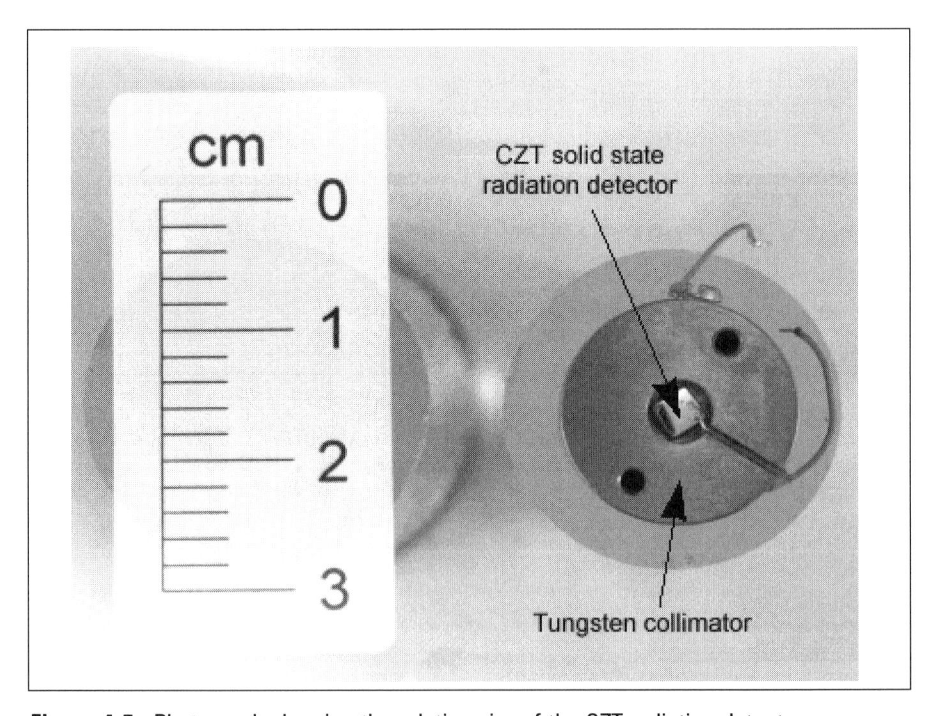

Figure 4-5. Photograph showing the relative size of the CZT radiation detector.

and technetium (a beta emitter) in vadose soil and groundwater. Energetic photons have greater penetrating power, which increases with energy. Detection distances of a few centimeters for X-rays and many centimeters for gamma-rays can be expected. Table 4-3 lists the gamma emitting radionuclides of most concern in low-level waste, the energy of the gamma radiation, and the approximate sampling volume of the CZT radiation detector in water.

TABLE 4-3 Approximate detection diameter for selected gamma-emitting radionuclides.

Radionuclide	Half Life (Year)	Gamma (MeV)	Water Linear Attenuation Coefficient (1/cm)	Approx Detection Diameter (90% Atten.) (cm)
I^{129}	1600000	0.040	0.2400	19.2
U^{234}	240000	0.117	0.1592	28.9
Ra^{226}	1602	0.186	0.1394	33.0
Am^{241}	4558	0.060	0.1920	24.0
Cs^{137}	30	0.662	0.0862	53.4
Co^{60}	5	1.332	0.0619	74.3

Microsensor Summary

Electrochemical sensors are useful in detecting contaminants (primarily metals) in water, although some gas- and vapor-phase applications exist. Ultramicro-electrochemical sensors offer exceptionally low detection limits for metals in water and are extremely small in size. Modern photolithographic mass production techniques offer the potential for very low unit costs. However, the long-term stability of the electrode surfaces has not been studied. Determination of the effects of long submersion times and fouling on performance are needed before these sensors can be considered for long-term monitoring.

QCM mass sensors find application in both vapor and liquid organic analysis, but are best-suited for vadose soil gas or well-head-space monitoring systems. Groundwater monitoring is problematic because any fouling that forms on the surface will be detected as a change in mass. For volatile organics, well-head-space monitoring is a viable alternative to direct groundwater monitoring. SAW mass sensors find application primarily for vapor analysis, so they would also be useful for both vadose zone soil gas monitoring and well-head-space monitoring.

Optical sensors can be used for either groundwater or vadose zone soil gas analysis. Optical sensors do not require wires to bring the sensor signal to the surface, which may be important when considering electromagnetic geophysical conditions. Optical sensors have been used to measure both organic and inorganic contaminants in both aqueous and vapor phases, and have demonstrated a great deal of versatility for this class of sensor. Simple, inexpensive optical sensors can be designed to continuously monitor for either selected or for a general range of contaminants. More expensive optical sensor systems, coupled with spectrophotometers, can positively identify and quantify contaminants if the simpler systems detect a change. As with the other classes of sensors, study of the long-term stability of optical sensors is required, especially with respect to the various coatings used.

The CZT radiation sensor can be deployed in both the groundwater and vadose zone soils. This ambient temperature sensor can quantify and identify the source of radiation that interacts with it. While the detector is sensitive to alpha and beta particles, the penetration depth of these particles in water or soil is so low that minimal detection can be expected. Gamma emitting radionuclides, however, are easily detected and identified in both soil and groundwater.

Vadose Zone Moisture Sensors

As discussed earlier, monitoring the pore water for inorganic contaminants is a difficult task and requires the use of indirect methods. There are, however, a number of sensing techniques for detecting moisture or the electrical changes due to moisture change that look promising for long-term monitoring. These techniques include tensiometers and TDR probes.

Tensiometers

Soil moisture alone does not provide information on the driving force for unsaturated fluid flow in the vadose zone or allow the direction of water movement to be determined. The driving force for water movement in vadose zone soils is the matrix (or suction) potential, which is usually expressed in terms of a vacuum (using negative pressure units). The matrix potential of a given soil is determined by the texture and moisture content of the soil. If two soils of two different textures (one fine, the other coarse) are placed in contact with each other and allowed to reach equilibrium, water will move between them until they come to the same matrix potential. The moisture content of the fine-textured soil, having more small pores, will be appreciably higher than that of the coarse-textured soil. With knowledge of the soil texture, the matrix potential can be estimated from the measured soil moisture content; however, it is much better to measure the matrix potential directly with a tensiometer.

As described in Chapter 3, a tensiometer typically consists of a porous ceramic cup or plate attached to a nonporous tube, with some means to measure a vacuum inside the tube. When the porous cup is placed in hydraulic contact with soil, and the cup and tube filled with water and sealed, a vacuum will develop in the tube as water flows from the ceramic into the soil. At equilibrium, the measured vacuum inside the tensiometer is equal to the matrix potential of the soil. The maximum vacuum attainable in a tensiometer is limited by the vapor pressure of water so, at normal ambient temperatures, it is limited to about 1 atmosphere. At sea level, this amounts to a 10 m hanging water column. However, the effective range of a typical tensiometer is reduced by the height of the water column above the ceramic cup, which is a serious limitation for any tensiometer installed deeper than a few meters.

Deep Tensiometers

Hubbell and Sisson (1996; 1998 a, b; Sisson and Hubbell, 1999) have developed and field-tested the "advanced" tensiometer specifically for extending tensiometer measurements to deep vadose zone applications. The advanced tensiometer has a permanently installed porous ceramic cup attached to an outer guide tube that extends to the surface. A removable inner tube, which includes a specially designed pressure transducer, is installed down the outer tube guide and seals onto the top of the porous ceramic cup. The instrument can be installed from 0.15 m to any depth greater than 30 m. The transducer can be either calibrated in place or removed for replacement. In field testing, the tensiometer provided continuous measurements with no operator intervention for periods of about 3 months. Typically, all that is required after that time is for the tensiometer to be de-aired and refilled with water from the surface, which is a minimal effort. It appears that the advanced tensiometer is well-suited to long-term monitoring applications even at great depths, with little instrument maintenance.

An air-filled tensiometer initially described and demonstrated by Faybishenko (1986) and further tested by Tokunaga and Slave (1994), also circumvents problems of deep operation by partially filling the tensiometer tube with water and measuring the vacuum generated in the headspace at the land surface. These instruments are inexpensive, but have demonstrated slow response and substantial loss of sensitivity.

Time Domain Reflectrometry

TDR is an established water measurement technique that uses guided high-frequency electromagnetic wave reflections to measure, independently, the electric permitivity and conductivity of the soil between the wave guide probes. The wave guide probes are metal rods (at least two), ranging from a few centimeters to half a meter long. Design details are given in Chapter 3 and in a number of other standard texts on vadose monitoring (for example, Wilson *et al.* 1995; Stephens 1996). The frequency shift of the reflected wave relates to the electric permitivity or dielectric constant of the soil. For most soils (with low clay content), the dielectric constant can be accurately and universally correlated with moisture content (Topp *et al.* 1980). The universal correlation is a desirable feature because soil-specific calibration is not required. The amplitude change of the reflected wave relates independently to the electric conductivity of the soil. It is the inflection point of this amplitude change that is used as a timing mark for determining the frequency shift. As the electric conductivity of the soil increases, the amplitude shift decreases, which is why for highly conductive soils (typical of high-content clay soils or highly contaminated soils), the frequency shift cannot be determined (and thus the TDR moisture measurement fails). The electrical conductivity of the soil is a function of its moisture content and the concentration of each ion in the pore water (Heimovaara *et al.* 1995).

By knowing the moisture content, independent from the measured phase shift, the ionic strength of the pore water can be estimated with a single TDR measurement, which makes TDR uniquely suited for vadose monitoring in that it can be used to measure moisture content and, indirectly, to probe the concentration of ions in the pore water. Moisture content monitoring alone can miss a solute transport problem. While it is likely that a solute transport event in the vadose zone will involve an increase in moisture, it may not always be the case. If unsaturated hydraulic flow is at steady state and some event causes a previously stabilized waste to dissolve, moisture monitoring alone will not detect a problem, but electrical conductivity monitoring may. Therefore, TDR fills an important monitoring niche for both moisture and soluble nonvolatile contaminant concentrations in the pore water.

Sensor Limitation

Because all sensors essentially make point measurements, they suffer from the same limitation of sensing volume. Although there is a distinct advantage to making point measurements continuously in time, the inherent spatial discontinuity suggests that a dense grid of sensors may be required. While it is likely that the cost of sensor's will drop, the cost of placement will likely remain high; therefore, a dense grid of sensors is technically and economically unattractive. The synergistic coupling of point sensor measurements with spatially distributed geophysical measurements is suggested as a way to achieve the spatial and temporal monitoring requirements with a sparser grid of sensors.

GEOPHYSICAL MEASUREMENTS

Geophysical use of electromagnetic waves for the characterization of the subsurface has been developed to aid in the exploration of oil and mineral deposits. The measurement of wave propagation velocity and amplitude attenuation can be correlated with subsurface attributes of interest such as density or dielectric constant. Subsurface structures or heterogeneity can be resolved by using both the refraction and reflection of waves. Surface-to-surface reflection techniques are the easiest to perform because they are nonintrusive (Figure 4-6). If the data sampling is very dense, and the location of the transmitter and receiver are known, high-resolution three-dimensional velocity estimates can be made (Eppstein and Dougherty 1998; Hole 1992). However, since estimating velocity depends on reflector depth, and determining reflector depth depends on wave propagation, the simultaneous determination of sub-

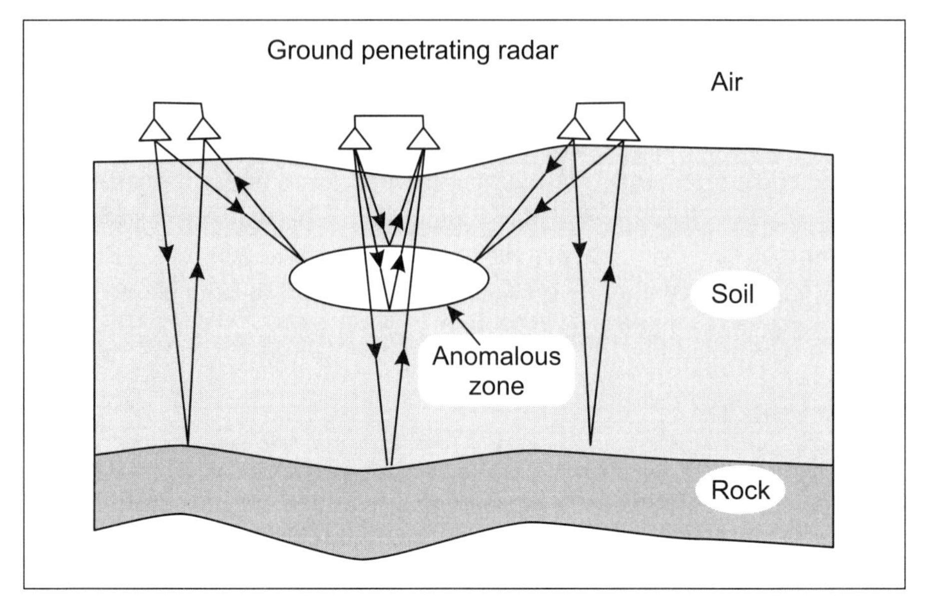

Figure 4-6. Conceptual illustration of surface-to-surface geophysical measurement technique; the radar is being used in the reflection profiling mode on soil over bedrock (*Geophysical Prospecting* 1989).

surface structure and velocity is mathematically ill-posed. A synergistic approach of integrating surface geophysical measurements with core descriptions, geophysical well logs, or point sensor measurements can be used to define the subsurface structure and to facilitate a unique solution. This enables the accurate interpolation of core and well log information, provides continuity in subsurface data, and reduces the number of wells necessary for hydrogeologic characterization (Eddy-Dilek *et al.* 1997). Sheets and Hendricks (1995) demonstrated that surface environmental monitoring (EM) methods combined with borehole neutron attenuation soil moisture data, provided accurate soil water estimates once site-specific calibration between electrical conductivity and soil water content was developed.

Cross borehole techniques, in which the transmitter and receiver are placed in different boreholes (Figure 4-7), simplifies the mathematics because the transmitter/receiver separation distance is precisely known, and the only unknown is the wave propagation velocity or attenuation (Eppstein and Dougherty 1998). These techniques can achieve much

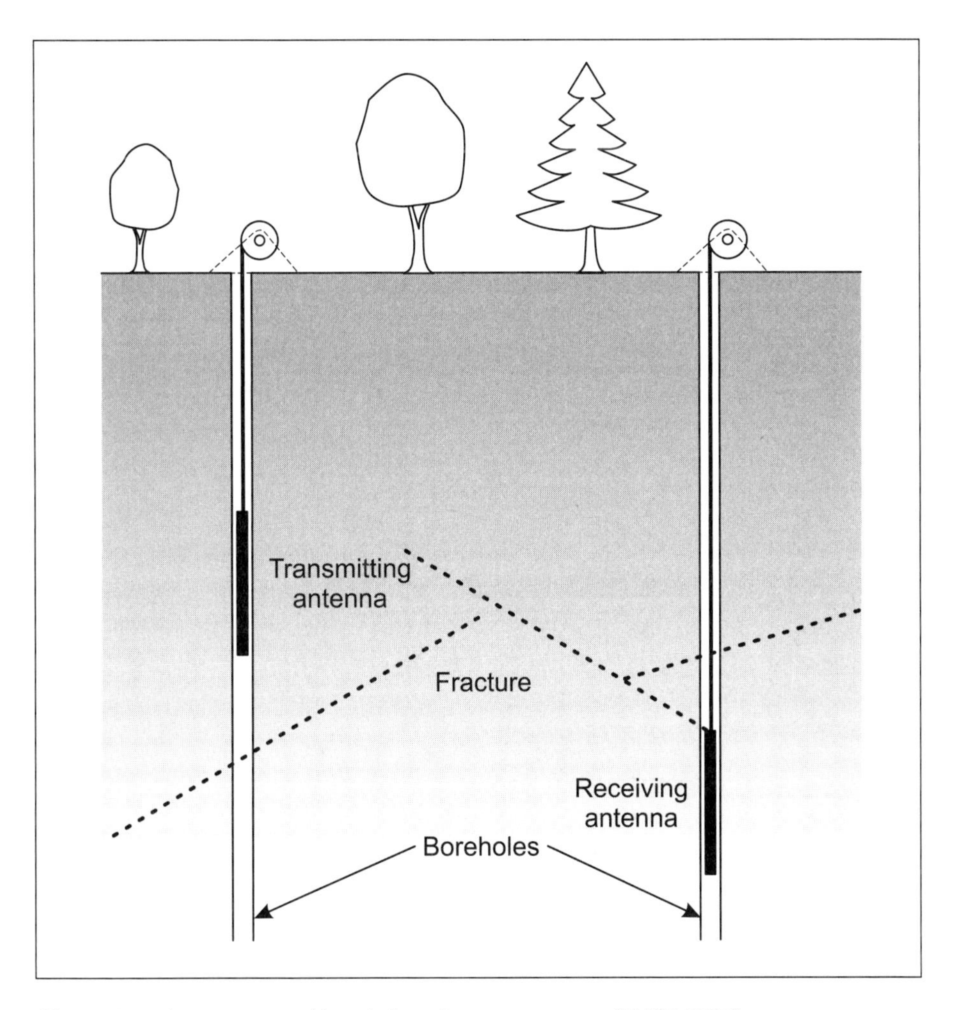

Figure 4-7. Arrangement of borehole radar measurement (IE ICE 1993).

greater resolution of the subsurface structure than surface-to-surface reflection techniques, and a number of three-dimensional cross borehole inversion algorithms have been developed (Alumbaugh and Newman 1997; Eppstein and Dougherty 1998). The use of boreholes is more expensive and intrusive, so the benefits of the inherent higher resolution must be weighed against the cost and logistical problems of providing access and collecting cross borehole data.

Ground-Penetrating Radar

GPR is an electromagnetic geophysical technique most often used as a tool for detecting buried objects. Recently, it has been used to detect contrasts in subsurface moisture content (Weiler *et al.* 1998). The measured velocity of unguided wave propagation can be used to estimate the dielectric constant of the media as well as its spatial variability. The estimated dielectric constant, which is related to the hydraulic properties of the material, has been used to probe soil moisture (Chanzy *et al.* 1996; Eppstein and Dougherty 1998; Weiler *et al.* 1998; van Overmeeren *et al.* 1997) and hydraulic conductivity (Hubbard *et al.* 1997). GPR techniques are best used in unsaturated, coarse-grained soils. GPR performance deteriorates in electrically conductive environments, such as in saturated systems, or in systems containing significant amounts of expanding clays (that is, smectite or vermiculite). GPR techniques, however, have been successfully used in saturated soils and soils with substantial nonexpanding clay. The suitability of GPR for a particular application depends on site-specific soil conditions.

Ground penetrating radar uses short pulses of 10 to 1000 MHz electromagnetic energy (50 to 200 MHz typical) to probe the subsurface. The frequency used affects the depth and resolution of the measurement. Higher-frequency signals sample shallower, dryer layers with higher resolution, while lower-frequency signals sample deeper and more conductive layers with lower resolution. The use of multiple frequency measurements enables sampling over a greater depth range with optimal resolution for each depth (Hubbard *et al.* 1997). The physics of the measurement are very similar to that used in TDR, although the frequencies used are somewhat different. As with TDR, GPR measures the electric permitivity or dielectric constant of the soil. Weiler *et al.* compares use of TDR and GPR as *in situ* soil moisture probes (1998). The calibration curves for the soil dielectric constant were very similar for the two methods, and close agreement was achieved. Since the same correlation between moisture content and dielectric constant is used, comparison of moisture content with field samples was also very good.

With cross borehole GPR, the data is easier to interpret. Using optical theory, an electromagnetic wave traveling between the transmitter in one borehole and the receiver in another can be analyzed as a ray. When the wave encounters a change in dielectric permitivity, it bends because of refraction, thus changing the travel time to the receiver. By making many measurements with the source and receiver at different locations in the boreholes, the wave velocity structure between the boreholes can be imaged using tomographic or inversion methods. The wave velocity is simply related to the dielectric permitivity of the soil. The boreholes should be deeper than the region to be imaged, and the depth should be at least three times greater than the borehole spacing. The spacing of the boreholes and the spatial interval of transmitter and receiver movement between measurements determine resolution quality (Paprocki and Alumbaugh 1999).

Like TDR, the GPR attenuation is related to soil electric conductivity. GPR attenuation can be calculated by inverting the waveform amplitudes using the same algorithm and inputs used to obtain the wave velocity tomogram. Therefore, cross borehole GPR provides information on the spatial distribution of both moisture and pore water ionic strength, and the TDR point measurements can be used as independent calibration checks (Peterson and Majer 1999).

Electric Resistivity Tomography

Electrical resistivity (or resistance) tomography (ERT) is a geophysical technique for imaging the electrical resistivity (the inverse of conductivity) distribution between two boreholes. To implement ERT, several small electrodes are permanently placed in electrical contact with the soil at known locations in the boreholes, as shown in Figure 4-8. A known direct current is applied between two electrodes, and the resulting voltage drop is measured between all other electrode pairs. To avoid polarization of the excitation electrodes, the applied current can be pulsed as a square wave with alternating polarity. Then, two different electrodes are driven by a known current, and the voltage drops are again measured using all electrode pairs not being used as source electrodes. This process is repeated, using an automatic switching and data collection system, until all linearly independent combinations have been measured. Each ratio of measured voltage to applied current is called a "transfer resistance."

ERT lends itself to the collection of large data sets using automated, relatively simple and inexpensive equipment. The excitation source can be simple direct current voltage supplies or battery banks, and the

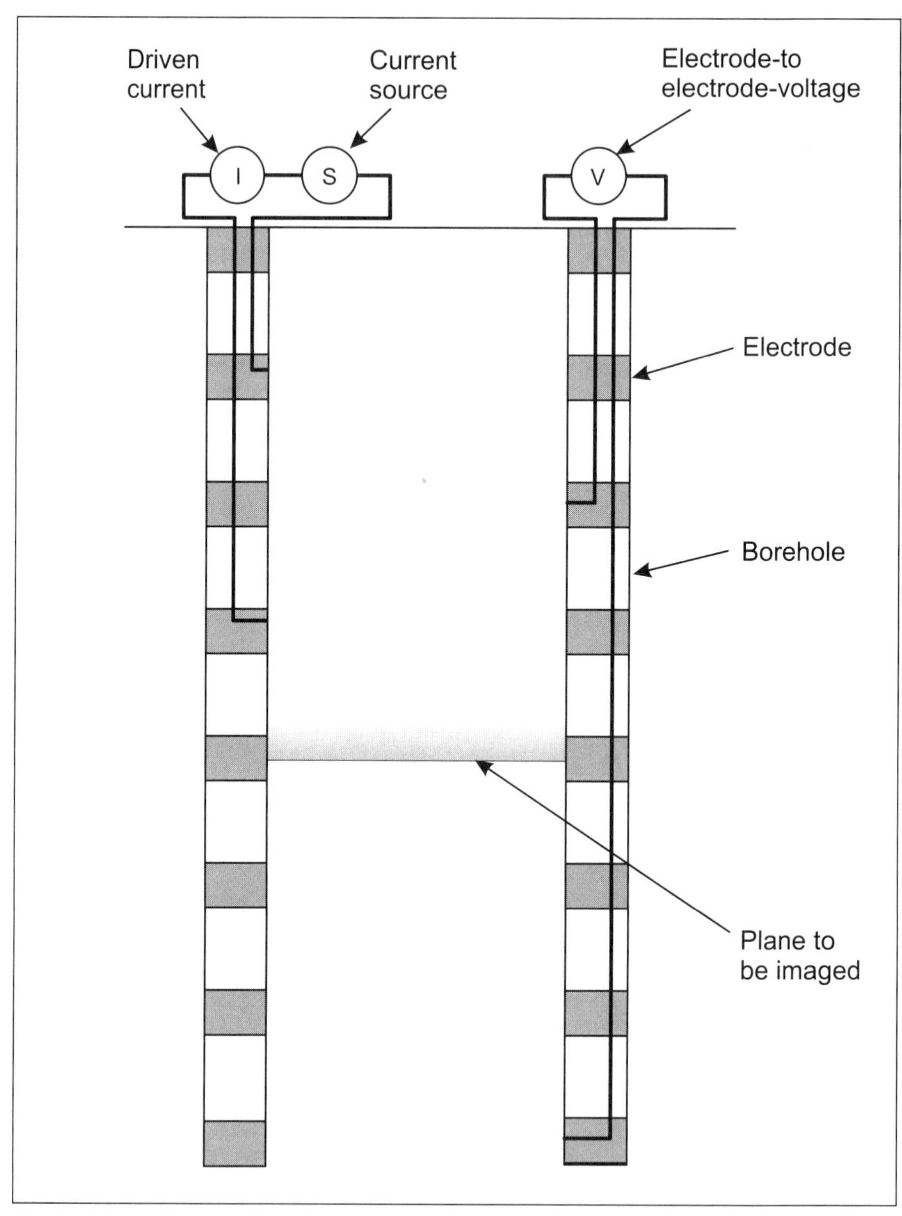

Figure 4-8. Schematic of resistivity measurements made between two boreholes: electrodes in each borehole make electrical contact with the formation (Daily and Ramirez 1992).

voltage and current measurements are made with a standard high impedance voltmeters. Automation is achieved with programmable relays and data acquisition systems that determine the sequence of excitation and voltage measurement to be made and stored.

Electrode arrays are made from metal screen electrodes; each electrode is fastened to a plastic pipe and connected to a wire that runs to the surface. The array is placed down a borehole slightly larger in diameter than the array and grouted in place. Installations have been made through small diameter (1 inch) holes with a Geoprobe™ pneumatic hammer. The layout and spacing between electrodes on a given array, and the spacing between arrays, determine the resolution quality of the images. The electrode arrays extend below the region to be imaged—ideally to a depth twice the electrode array spacing. Electrode array spacing of 0.5 m to 60 m has been used. Having more electrodes generally increases resolution, but the data collection time increases also, scaling with the square of the number of electrodes. Permanently installed electrode arrays facilitate a high level of repeatability and make the method well-suited to long-term monitoring (LaBrecque *et al.* 1996).

The goal is to use these measured transfer resistances to calculate the distribution of soil resistivity between the boreholes. This problem is highly nonlinear because the current paths depend on the resistivity distribution (Daily and Ramirez 1992). The complexity of the data set requires interpretation using sophisticated inversion algorithms based on forward and reverse modeling (Maillol *et al.* 1999). Recently, rapid 3D resistivity forward modeling and inversion algorithms have been described that allow the use of commonly available computers (Zhang, Mackie, and Madden 1995; Zhang, Rodi, and Mackie 1996; Morelli and LaBrecque 1996).

ERT measures soil electrical resistivity (or its inverse, the soil electrical conductivity) directly. As discussed above, soil electrical resistivity is related to both the volumetric soil moisture content and pore water conductivity, or ionic strength. Therefore, ERT alone cannot be used to distinguish between changes in volumetric moisture content and pore water ionic strength without additional information from other measurements. This is not a great drawback, since any unexpected change in soil conductivity, whether due to increased moisture or increased pore water salinity, deserves additional investigation. TDR point measurements of moisture content can provide this additional information, and TDR soil conductivity measurements can be used to verify the ERT images.

Compatibility Problems With Geophysical Methods And Sensors

Electrically conductive materials, especially when oriented in a linear manner (for example, metallic pipe, metallic well casings, and wires), can cause problems with electrical and electromagnetic geophysical measurements. Nonconductive pipe and well casings must be used if electric or electromagnetic geophysical methods are to be used for the transmission of data. The combination of sensors and geophysical methods requires careful consideration in design. Fiber optics is an alternative system that can be used with electrically conductive materials.

For GPR, both grounded and nongrounded wires cause problems (although grounded wires are worse). For GPR surveys, the removal of as many wires as possible may be required. This is quite a problem with TDR probes since the probe is permanently installed, and the disconnection and eventual reconnection would have to be made at depth. The problem of wires interfering with GPR has not been subjected to systematic study (Alumbaugh, personal communication, 1999). The orientation of the wires (vertical vs. horizontal) may make a difference, and perhaps the vertical wires can remain if the horizontal wires are removed.

For ERT, any grounded wire in electrical connection with the soil will cause large perturbations in the result and could easily damage the sensor monitoring equipment. Fortunately, the problem is eliminated when the connection to ground is broken. Either the sensors and associated monitoring equipment must be isolated from ground by design, or all wires must be physically disconnected. When ERT is considered for routine use at a site, it will pay to carefully design all other sensor systems to be isolated from ground to at least the maximum voltage expected to be used for ERT. TDR probes, because they are in intimate contact with the vadose soils, must be isolated or disconnected. Because simply disconnecting wires or otherwise isolating them from ground is easier than removal, ERT is potentially more compatible with long-term monitoring of sensor-instrumented vadose zone soil than is GPR.

Placing, Replacing, and Calibrating of Sensors

Sensor emplacement is a vital aspect of any long-term monitoring program, especially in the vadose zone. Placement, replacement, and periodic calibration of sensors in groundwater is greatly facilitated by using existing groundwater monitoring wells, which are typically large enough to accommodate many different sensor systems. There are obvious advantages in having sensors easily retrievable for replacement, repair, testing, and calibration. Sensor placement in the vadose zone is not as easy. The most difficult installations are for those sensors and samplers that require intimate hydraulic connection with the soil, such as the ceramic cups of suction lysimeters and tensiometers. These instruments require installation in dedicated boreholes, and replacement is likely to be quite difficult, if not impossible. Furthermore, these devices are not compatible with one another, or with any other moisture monitoring sensor, so boreholes must be adequately separated. However, it may be possible to use the ceramic cup of an advanced tensiometer system as a suction lysimeter.

The next most difficult type of sensor to install is the TDR probe, which requires intimate physical contact with the soil. TDR probes are usually installed vertically at the bottom of a borehole and typically are not considered retrievable. The borehole is typically backfilled and additional TDR probes installed into the backfill. Installation of the probes into backfill is not as desirable as installation into the undisturbed soil because the hydraulic nature of the backfill will not necessarily match the adjacent soil and may skew the moisture measurement. Methods for installing sensors in the side walls of boreholes are discussed in Chapter 3.

The easiest vadose zone sensors to install are those that monitor soil gas. These sensors are installed by lowering them to the screened interval of a vadose zone well. Some sort of packing device may be used to seal inside the well above the screened interval, especially if multiple screened intervals are used in the same well. Care must be taken in the design and installation of such wells to adequately seal the well casing to the formation so that movement of soil gas in this annular region is prevented. The open space created inside the screened section is exposed to the vadose soils and comes to equilibrium with the soil gas. Mass sensor arrays and fiber-optic sensor arrays can be used to continuously monitor the volatile organic compound content of the soil gas. This continuous monitoring will not perturb the distribution of contaminants in the vadose zone soil gas as would happen if a continuous gas sample were to be extracted for surface analysis.

Larger, permanent vadose zone wells can be justified if some or all the sensors in the well are designed to be retrievable. These wells are expensive to install, but the cost savings from sensor replacement may eventually offset the cost of the well. An alternative is to use a lower cost placement technology and install all sensors permanently. If a sensor fails, it must then be abandoned and a new sensor placed. Or, as mentioned previously, low-cost sensors can be deployed in redundant numbers, and as one sensor fails, another can be brought online. Push technologies, such as those offered by cone penetrometers or small pneumatic hammer equipment (the Geoprobe™, for example) offer lower cost placement capabilities with the added benefit of not generating soil cuttings, which require disposal. The diameter of well that can be installed is limited to 1 to 2 inches, so sensors must be sized accordingly. Fortunately, many of the fiber-optic and SAW-based sensors are small enough to be deployed inside the smallest of these wells.

The idea of technology insertion is an important one for reducing the lifecycle cost of long-term monitoring. Because long-term monitoring will be conducted over many decades, it is quite reasonable to assume that many technological advancements will be made in sensing and monitoring over this time. As these advancements occur, these higher performing, more cost-effective approaches can be incorporated into new and existing installations. Future retrofitting of existing installations can be facilitated by designing with anticipated changes in mind. These designs relate mostly to placement technology and the ease with which new technologies can be incorporated.

A LOOK AT LONG-TERM MONITORING SYSTEM DESIGN

Proper design of a long-term monitoring system requires a detailed understanding of the geohydrologic setting at the specific site. Contaminant transport modeling and risk assessment are required to determine the specific spatial and temporal requirements for monitoring at the site. A systems engineering approach is required to achieve synergies in the available monitoring methods and to optimize the system so that it meet's the spatial and temporal measurement requirements in the most cost-effective manner. As measurements are made, data must be collected and transmitted to a central location for final analysis and archiving.

Measurement System Design

Through the synergistic combination of point sensor measurements and spatially distributed geophysical measurements, the design of a more cost-effective long-term monitoring system is possible. Geophysical data, collected periodically (approximately annually), can be used to meet the spatial requirements of monitoring in the vadose zone so that the density of point measurements can be reduced. Data from vadose zone point sensor systems can be collected automatically at a frequency that meets the temporal requirement. If groundwater monitoring is required, data from groundwater sensor systems can be collected automatically and used to justify drastically reducing groundwater sampling frequency (to about once every 5 years). The important attributes of the various monitoring techniques discussed in this chapter are summarized in Table 4-4.

Monitoring of the vadose zone pore water for nonvolatile, non-gamma emitting inorganic contaminants is complicated by the difficulty in sampling the pore water for analysis. Monitoring of soil moisture content and electrical conductivity can provide indications of contaminant transport. The standard technique is to use neutron attenuation measurements to directly probe the moisture (θ) in the soil. This method, however, does not lend itself to automation, and therefore, measurements must be made manually and only periodically. The advanced tensiometer can provide continuous measurements of the soil matrix potential (ψ), which provides the information required to evaluate unsaturated flow gradients. TDR probes can continuously monitor the soil electrical permitivity (ε) which correlates to the moisture content. TDR probes also can continuously monitor the soil electrical conductivity (σ), which is directly related to the total concentration of ions in the pore water. Soil electrical conductivity measurement is the best method available today for probing, *in situ*, the inorganic contaminant content of the vadose pore water. Both GPR and ERT can provide spatially distributed measurement of soil properties. GPR can provide independent information on both the soil moisture and electrical conductivity, which is a favorable attribute, but is seriously affected by the presence of wires. ERT can provide distributed information on the soil's electrical conductivity, but is adversely affected only by wires connected to ground. Disconnecting or isolating all other sensor wire from

TABLE 4-4 **Monitoring technique attribute summary.**

Monitoring Technique	Media Sampled vadose	aquifer	Analyte Detected	Measurement Frequency	Identify?	Comments
Electrochem. sensors						
UMEA		GW	Inorganic	Continuous	Yes	
Mass sensors						
QCM	SG	HS	Organic	Continuous	No	
SAWs	SG	HS	Organic	Continuous	No	
μChemLab™	SG	HS	Organic	Near continuous	Yes	
Fiber optic sensors						No wires
Refrac. index	SG	GW/HS	Organic	Continuous	No	
Absorbance	SG	GW/HS	Organic, inorganic	Continuous	No	
Spectroscopic						
UV-FEWS	SG	GW/HS	Organic, inorganic	Near continuous	Yes	
MIR-FEWS	SG	GW/HS	Organic	Near continuous	Yes	
Radiation sensors						
CZT	S/PW	GW	γ-Rad	Continuous	Yes	
Point moisture, θ						
Tensiometer	S/PW		$\psi(\theta)$	Continuous	No	
TDR	S/PW		$\varepsilon(\theta)$, $\sigma(\theta)$	Continuous	No	
Neutron atten.	S/PW		θ	Periodic	No	Not automated
Distributed moisture, θ						
GPR	S/PW		$\varepsilon(\theta)$, $\sigma(\theta)$	Periodic	No	Wires may interfere
ERT	S/PW		$\sigma(\theta)$	Periodic (can be automated)	No	Grounded wires interfere

Legend:
ψ = soil matrix potential
θ = volumetric moisture content
ε = soil electrical permitivity
σ = soil electrical conductivity
γ-Rad = gamma radiation

ground is feasible (and, perhaps can be automated), which would make ERT more compatible in an overall monitoring system.

Continuous soil gas analysis for organic contaminants or tracers emplaced in the waste, can be performed using SAW sensor arrays, QCM sensor arrays, and/or fiber-optic sensor arrays connected to data logging/analysis/transmission systems. These sensors can be located in the vadose zone to monitor soil gas or in a groundwater monitoring well to monitor the headspace above the water table. Selective adsorptive or indicating coatings allow continuous, low maintenance and economical monitoring of targeted contaminants. Continuous groundwater monitoring may include electrochemical microsensors for inorganics, primarily metal ions, and fiber-optic sensor arrays for heavier organics. If any sensors indicate the increasing presence of a targeted organic contaminant, the use of more sophisticated sensors such as the µChemLab™ or MIR/UV-FEWS can be added to confirm identification of the targeted contaminant and to identify any unknown or unexpected compounds. Monitoring and identification of gamma-emitting radionuclides may be performed continuously in either vadose zone soil or in groundwater using a CZT radiation detector.

System Assembly

A long-term monitoring system is more than just a bunch of sensors connected to data acquisition systems and collecting gigabytes of data every day. In order to be useful, rather than a burden, raw data must be analyzed and transformed into information from which decisions can be made. As much as possible, preliminary data analysis and reduction should be performed automatically. Many computerized data acquisition systems are capable of performing simple data reduction while continuously monitoring for preset alarm conditions. As long as parameters remain within acceptable bounds, all that is needed are average values (hourly, daily or even weekly depending on the monitoring situation) along with some basic statistics like minimum, maximum and standard deviation. Data can be automatically transmitted to a central location by a number of means. The communications medium may be a hard-wired Internet link, a standard phone line, or if these are not available, a cellular phone or satellite link (McKinnon and Hubbard, 1997). Only when a parameter triggers an alarm would a more detailed data set be required.

It is possible to set up a data storage buffer that contains the most recent detailed data set for a fixed interval of time. In the event of an alarm condition, the data in this buffer could be transmitted and analyzed to determine the details of the situation leading to the alarm. In this manner, the amount of data transmitted is minimized and the process is automated to the greatest extent possible. Automation is the key to minimizing life-cycle costs of long-term monitoring.

Figure 4-9 shows a schematic view of an idealized, integrated sensor system that collects data related to the environmental status of a closed environmental restoration site. Measurements from the surface, the vadose zone, and the underlying aquifer, are collected and merged with meteorological data and transmitted to a central location. The data can be used to test conceptual models and calibrate numerical models, and thereby to develop a detailed picture of the subsurface processes at work. This understanding builds confidence in the monitoring system

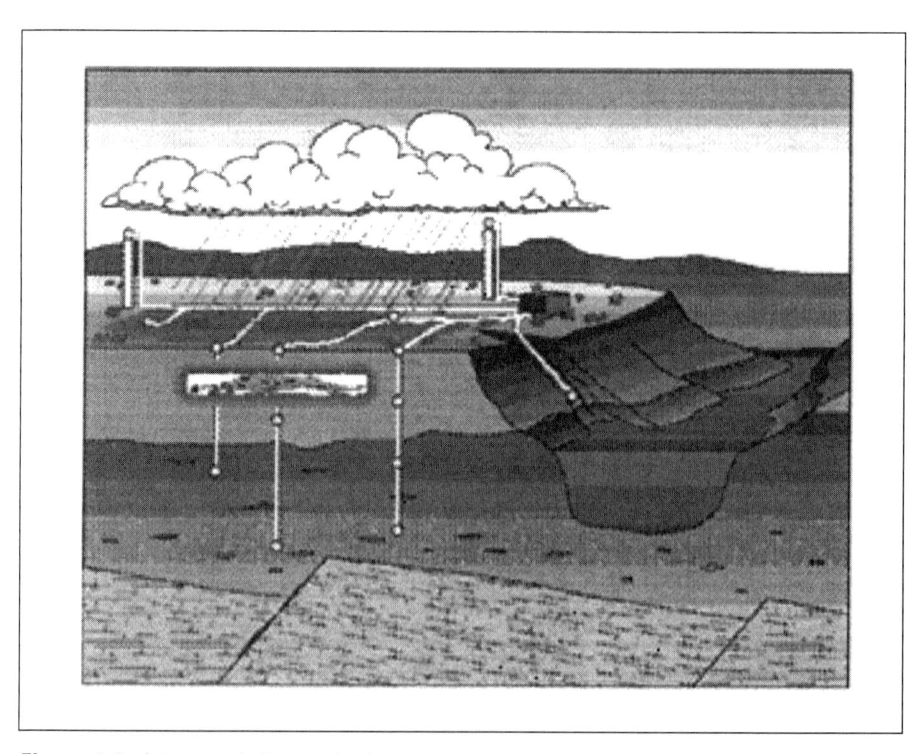

Figure 4-9. Integrated site monitoring system.

and provides a basis for optimization of its operation in order to achieve further cost reduction.

SUMMARY OF NEEDS AND GAPS

In general, all the sensor systems discussed in this chapter require further study of the their long-term stability for long-term monitoring. Many specific questions must be addressed: How long will they last in use? How long between calibrations? Is *in situ* calibration possible?

There is also a great need to study integrated systems and to optimize monitoring combinations that meet all monitoring requirements rather than to optimize individual methods independent of the whole. For example, TDR and GPR produce highly compatible data sets, but TDR cables can interfere with GPR measurements. Methods of installing TDR probes in vadose zone wells with retrievable cables can alleviate this problem. ERT also provides compatible data, but with only half the data that TDR provides. However, ERT is less hampered by wires, so in some situations using ERT may be a better approach overall.

Once the spatial and temporal monitoring requirements and data quality objective have been determined, a systems engineering approach is needed to consider all aspects of designing the most cost-effective long-term monitoring system. It's most efficient to plant all point sensor monitors, such as soil gas or pore water monitors, in a single borehole. How can the borehole spacing be optimized for both soil gas monitoring and pore water point monitoring when combined with a periodic geophysical imaging program? Optimization of the geophysical tomography measurements will allow the greatest spacing between boreholes for a given image resolution. Will the data be comparable to the point sensor measurements, considering the volume averaging involved in producing the images? Will this affect the design? A systems engineering approach must also consider the possible synergies and incompatibilities of various integrated systems and optimize logistics of collecting the data. What are the tradeoffs between using fiber-optic sensors versus mass or electrochemical sensors for data transmission? How do the available placement technologies affect the choice of sensor system? Are the data sets all compatible? Can the data be used in an automatic decision-making processes, triggering more detailed analyses when warranted?

In validating cleanup and site closure decisions, site owners all over the world need significantly lower costs and more effective methods to monitor dynamic contaminant migration over tens to hundreds of years. The magnitude of the cost for environmental monitoring at both public and private sites makes this a major national and global issue. Given the long time horizon of monitoring activities, the probability of significant technology leaps in some of these areas (for example, sensor technology) is very high. With the appropriate application of innovative monitoring systems and technology, the potential for savings is significant.

REFERENCES

Alumbaugh, D.L. and G.A. Newman. "3-Dimensional Massively-Parallel Electromagnetic Inversion .2. Analysis of a Crosswell Electromagnetic Experiment," *Geophysical Journal International*, 128(2) (1997): 355-363.

Belmont, C., M.L. Tercier, J. Buffle, G.C. Fiaccabrino, and M. Koudelkahep. "Mercury-Plated Iridium-Based Microelectrode Arrays for Trace-Metals Detection By Voltammetry: Optimum Conditions and Reliability," *Analytica Chimica Acta*, 329(3) (1996): 203-214.

Bodenhofer, K., A. Hierlemann, G. Noetzel, U. Weimar, U., and W. Gopel. "Performances of Mass-Sensitive Devices for Gas-Sensing: Thickness-Shear Mode and Surface-Acoustic-Wave Transducers," *Analytical Chemistry*, 68 (1996): 2210-2218.

Casalnuovo, S.A., E. J. Heller, V. M. Hietala, A. G. Baca, R. Kottenstette, S. L. Hietala, J. L. Reno, and G. C. Frye-Mason. "Acoustic Wave Chemical Microsensors in GaAs," SAND 98-1925C, Sandia National Laboratories, Albuquerque, NM (1998).

Chang S.M., Y.H. Kim, J.M. Kim, Y.K. Chang, and J.D. Kim. "Development of Environmental Monitoring Sensor Using Quartz-Crystal Micro-Balance," *Molecular Crystals and Liquid Crystals Science and Technology Section A-Molecular Crystals and Liquid Crystals*, 267 (1995): 405-410

Chanzy A., A. Tarussov, A. Judge, and F. Bonn. "Soil-Water Content Determination Using a Digital Ground-Penetrating Radar," *Soil Science Society of America Journal*, 60(5) (1996): 1318-1326.

Choudhury, P.K., "Optical Sensors in Environmental Monitoring," *Current Science*, 74(9) (1998): 723-725.

Cullen S.J., J.H. Kramer, and L.R. Luellen. "A Systematic Approach to Designing a Multiphase Unsaturated Zone Monitoring Network," *Ground Water Monitoring and Remediation*, 15(3), (1995): 124-135.

Cullen. S.J., "Vadose Zone Monitoring: Experiences and Trends in the United States," *Ground Water Monitoring and Remediation*, 15(3) (1995): 136-144.

Daily, W. and A. Ramirez. "Electrical Resistivity Tomography of Vadose Water Movement," *Water Resources Research*, 28(5), (1992): 1429-1442.

Davis, J.L., and A.P. Annan. "Ground-Penetrating Radar for High-Resolution Mapping of Soil and Rock Stratigraphy," *Geophysical Prospecting,* 37 (1989): 531-551.

Dickert, F.L., O. Hayden, and M.E. Zenkel. "Detection of Volatile Compounds With Mass Sensitive Sensor Arrays in the Presence of Variable Ambient Humidity," *Analytical Chemistry* 71, (1999): 1338-1341.

Durant, N.D., V.B. Myers, and L.A. Eccles. "EPA's Approach to Vadose Zone Monitoring at RCRA Facilities," *Ground Water Monitoring and Remediation*," 13(1) (1993): 151-158.

Eddy-Dilek, C.A., Looney, B.B. Hoekstra, P. Harthill, N. Blohn, and M. Phillips. "Definition of a Critical Confining Zone Using Surface Geophysical Methods," *Ground Water*, 35(3) (1997): 451-462.

Eppstein, M.J., and D.E. Doughterty. "Efficient 3-Dimensional Data Inversion: Soil Characterization and Moisture Monitoring from Cross-Well Ground-Penetrating Radar at a Vermont Test-Site." *Water Resources Research*, 34(8) (1998): 1889-1900.

Everett, L.G., and L.G. Wilson. "Permit Guidance Manual on Unsaturated Zone Monitoring for Hazardous Waste Land Treatment Units," Office of Solid Waste Report No. EPA/530-SW-86-040 (1986): 87-106.

Faybishenko, B.A. "Water-Salt Region of Soils Under Irrigation," (Russian), *Agroproizgat,* Moscow, Russia (1986).

Grate, J.W., S.J. Patrash, and M.H. Abraham. "Method for Estimating Polymer-Coated Acoustic-Wave Vapor Sensor Responses," *Analytical Chemistry*, 67(13) (1995): 2162-2169.

Geophysical Prospecting 37 (1989): 531-551.

Gierke, J.S., and S.E. Powers. "Increasing Implementation of *In Situ* Treatment Technologies Through Field-Scale Performance Assessments," *Water Environment Research*, 69(2), (1997): 196-205

Heimovaara, T.J., A.G. Focke, W. Bouten, and J.M. Verstraten. "Assessing Temporal Variations in Soil-Water Composition With Time-Domain Reflectometry," *Soil Science Society of America Journal,* 59(1995): 689-698.

Herdan, J., R. Feene, S.P. Kounaves, A.F. Flannery, C.W. Storment, G.T.A. Kovacs, and R.B. Darling. "Field-Evaluation of An Electrochemical Probe for In-Situ Screening of Heavy-Metals in Groundwater," *Environmental Science & Technology,* 32(1) (1998): 131-136

Hubbard, S.S., Y. Rubin, and E. Majer. "Ground-Penetrating-Radar-Assisted Saturation and Permeability Estimation in Bimodal Systems," *Water Resources Research,* 33(5) (1997): 971-990.

Hubbell, J.M., and J.B. Sisson. "Portable Tensiometer Use in Deep Boreholes," *Soil Science*, 161(6) (1996): 376-382.

Hubbell, J.M., and J.B. Sisson. "Advanced Tensiometer for Shallow or Deep Soil Water Potential Measurements," *Soil Science*, 163(4) (1998a): 271-277.

Hubbell, J.M., and J.B. Sisson. "Advanced Tensiometer for Water Potential Measurements," *Measurements & Control,* 191(10), (1998b): 81-85.

Hudson, M.L., R.J. Kottenstette, C.M. Matzke, G.C. Frye-Mason, K.A.Shollenberger, D.R. Adkins, and C.C. Wong. "Design, testing, and simulation of microscale gas chromatography columns," SAND 98-1551C, Sandia National Laboratories, Albuquerque, NM (1998).

IE ICE *Trans. Commun.,* E76-B(10) (1993).

Jin, M., M. Delshad, V. Dwarakanath, D.C. McKinney, G.A. Pope, K. Sepehrnoori, C.E. Tilburg, and R.E. Jackson. "Partitioning Tracer Test for Detection, Estimation and Remediation Performance of Subsurface Nonaqueous Phase Liquids," *Water Resources Research,* 31(5), (1995): 1201-1211.

Jin, M., G.W. Butler, R.E. Jackson, P.E. Mariner, J.F. Pickens, G.A. Pope, C.L. Brown, and D.C. McKinney. "Sensitivity Models and Design Protocol for Partitioning Tracer Tests for in Alluvial Aquifers," *Ground Water*, 35(6) (1997): 964-972.

Koglin, E.N., E.J. Poziomek, and M.L. Kram. "Emerging Technologies for Detecting and Measuring Contaminants in the Vadose Zone," in *Handbook of Vadose Zone Characterization & Monitoring*, L.G. Wilson, L.G. Everett and S.J. Cullen, (Eds.), Lewis Publishers, Boca Raton, FL (1995).

Kounaves, S.P., W. Deng, P.R. Hallock, G.T.A. Kovacs, and C.W. Storment. "Iridium-Based Ultramicroelectrode Array Fabricated By Microlithography," *Analytical Chemistry* v 66(3) (1994): 418-423

Krylov, V.V., "Surface-Properties of Solids, and Surface Acoustic-Waves : Application to Chemical Sensors and Layer Characterization," *Applied Physics A-Materials Science & Processing,* 61 (1995): 229-236.

Labrecque D.J., A.L. Ramirez, W. Daily, A.M. Binley, and S.A. Schima. "ERT Monitoring of Environmental Remediation Processes," *Measurement Science and Technology,* 7 (1996): 375-383.

Lindgren, E.R., M.G. Hankins, E.D. Mattson, and P.M. Duda. "Electrokinetic Demonstration at the Unlined Chromic Acid Pit," SAND 97-2592, Sandia National Laboratories, Albuquerque, NM (1998).

Maillol, J.M., M.K. Seguin, O.P. Gupta, H.M. Akhauri, and N. Sen. "Electrical Resistivity Tomography Survey for Delineating Uncharted Mine Galleries in West Bengal, India," *Geophysical Prospecting*, 47 (1999): 103-116.

McKinnon, A.D., and C.W. Hubbard. "Automating Communications with and Developing User Interfaces for Remote Data Acquisition and Analysis," *IEEE Transactions on Nuclear Science*, 44(3) 1062 (1997).

Mizaikoff, B., M. Jakusch, and M. Kraft. "Infrared Fiber-Optic Sensors – Versatile Tool for Water Monitoring," *Sea Technology* (1999): 25.

Morelli, G., and D.J. LaBrecque. "Robust Scheme for ERT Inverse Modeling," *Proceedings of the SAGEEP*, Denver, CO (1996): 629-638.

Nolan, M.A., and Kounaves S.P. "Effects of Mercury Electrodeposition On the Surface Degradation of Microlithographically Fabricated Iridium Ultramicroelectrodes," *Journal of Electroanalytical Chemistry,* 453(1-2) (1998): 39-48.

Nolan, M.A., S.H. Tan, and S.P. Kounaves "Fabrication and Characterization of A Solid-State Reference Electrode for Electroanalysis of Natural-Waters With Ultramicroelectrodes," *Analytical Chemistry,* 69(6) (1997): 1244-1247.

Paprocki, L., and D. Alumbaugh. "An Investigation of Cross-Borehole Ground Penetrating Radar Measurements for Characterizing the 2D Moisture Content Distribution in the Vadose Zone," *Proceedings of the Symposium on the Application of Geophysics to Engineering and Environmental Problems*, M.H. Powers, L. Cramer, and R. S. Bell (Eds.), Oakland, CA (1999): 583-592.

Peterson, J.E. Jr, and E.L.Majer. "Hydrogeological Property Estimation Using Tomographic Data at the Boise Hydrogeophysical Research Site," *Proceedings of the Symposium on the Application of Geophysics to Engineering and Environmental Problems*, M.H. Powers, L. Cramer, and R. S. Bell (Eds.), Oakland, CA (1999): 629-638.

Ramirez, A., and W. Daily. "Monitoring an Underground Steam Injection Process Using Electrical Resistance Tomography," *Water Resources Research*, 29(1) (1993): 73-87.

Rodgers, K.R., and E.J. Poziomek. "Fiber Optic Sensors for Environmental Monitoring," *Chemosphere*, 33(6) (1996): 1151-1174.

Sato, M., and T. Suzuki, "Recent Progress in Borehole Radars and Ground Penetrating Radars in Japan," IE ICE *Trans. Commun.,* E76-B(10) 1993.

Schwotzer, G., I. Latka, H. Lehmann, and R. Willsch. "Optical Sensing of Hydrocarbons in Air Or in Water Using UV Absorption in the Evanescent Field of Fibers," *Sensors and Actuators B,,* 38-39 (1997): 150 –153.

Sheets, K.R., and J.M.H. Hendricks. "Noninvasive Soil-Water Content Measurement Using Electromagnetic Induction," *Water Resources Research,* 31(10) (1995): 2401.

Silva, P.R.M.and M.A. El Khakani, M. Chaker, G.Y. Champagne, J. Chevalet, L. Gastonguay, R.Lacasse, and M. Ladouceur. "Development of Hg-electroplated-iridium based microelectrode arrays for heavy metal traces analysis" *Analytica Chimica Acta*, 385(1-3) (1999): 249-255.

Sisson, J.B., and J.M. Hubbell. "Water Potential to Depths of 30 Meters in Fractured Basalt and Sedimentary Interbeds," in *Characterization and Measurement of Hydraulic Properties of Unsaturated Porous Media*, M. Th. van Genuchten and F.J. Leij, (Eds.), U.S. Salinity Laboratory, Riverside, CA., (1999): (in review)

Tan, S.H., and S.P. Kounaves. "Determination of Selenium(Iv) At A Microfabricated Gold Ultramicroelectrode Array Using Square-Wave Anodic-Stripping Voltammetry," *Electroanalysis*, 10(6) (1998): 364-368.

Thomas, R. "Volatilization from Water," in *Handbook of Chemical Property Estimation Methods*, W.J. Lyman, W.F. Reehl, and D.H. Rosenblatt (Eds.), McGraw Hill Book Company, New York, NY (1982).

Tokunaga, T., and R. Slave, "Gauge Sensitivity Optimization in Air Pocket Tensiometry: Implications for Deep Vadose Zone Monitoring," *Soil Science*, 158(6) (1994).

Topp, G.C., J.L. Davis, and A.P. Annan. "Electomagnetic Determination of Soil Water Content: Measurement in Coaxial Transmission Lines," *Water Resources Research*, 16(3), (1980): 574-582.

Uhlig, A., U. Schnakenberg, and R. Hintsche. "Highly Sensitive Heavy-Metal Analysis On Platinum-Ultramicroelectrode and Gold-Ultramicroelectrode Arrays," *Electroanalysis*, 9(2) (1997): 125-129.

U.S. Department of Energy (DOE). "Surface Acoustic Wave/Gas Chromatography for Trace Vapor Analysis," *Innovative Technology Summary Report*, OST Reference #282, (1998).

van Overmeeren, R.A., S.V. Sariowan, and J.C. Gehrels. "Ground-Penetrating Radar for Determining Volumetric Soil-Water Content: Results of Comparative Measurements at 2 Test Sites," *Journal of Hydrology*, 197(1-4) (1997): 316-338.

Ward, A.L., R.G. Kachanoski, and D.E. Elrick. "Analysis of Water and Solute Transport Away From A Surface Point-Source," *Soil Science Society of America Journal*, 59, (1995): 699-706.

Wang, J., B.M. Tian, J.Y. Wang, J.M. Lu, C. Olsen, C. Yarnitzky, K. Olsen, D. Hammerstrom, and W. Bennett. "Stripping analysis into the 21st century: faster, smaller, cheaper, simpler and better," *Analytica Chimica Acta*, 385, (1999): 429-435.

Weiler, K.W., T.S. Steenhuis, J. Boll, and K.J.S. Kung. "Comparison of Ground-Penetrating Radar and Time-Domain Reflectometry As Soil-Water Sensors," *Soil Science Society of America Journal*, 62 (1998): 1237-1239.

Yan, L., and X.C. Zhou. "Quartz crystal microbalance sensor deposited with LB films of functional polymers for the detection of phenols in vapour phase," *Chemical Research in Chinese Universities,* 14(4) (1998): 433-437.

Zhang, J., R.L. Mackie, and T.R. Madden. "3-D Resistivity Forward Modeling Using Conjugate Gradients," *Geophysics*, 60 (1995): 1313-1325.

Zhang, J., W. Rodi, and R.L. Mackie. "Regularization in 3-D DC Resistivity Tomography," Proceedings of the SAGEEP, Denver, CO (1996): 687-694.

CASE STUDIES

AN INTEGRATED APPROACH TO MONITORING A FIELD TEST OF *IN SITU* CONTAMINANT DESTRUCTION

Robin L. Newmark, Roger D. Aines, G. Bryant Hudson, Roald Leif, Marina Chiarappa, Charles Carrigan,John Nitao, and **Allen Elsholz,** *Lawrence Livermore National Laboratory, Livermore, CA*

Craig Eaker, *Southern California Edison Company, Rosemead, CA*

Presented at the 1999 Symposium on the Application of Geophysics to Engineering and Environmental Problems (SAGEEP), Oakland, California, March 14–17, 1999

ABSTRACT

The development of *in situ* thermal remediation techniques requires parallel development of techniques for monitoring physical and chemical changes for purposes of process control. Recent research indicates that many common contaminants can be destroyed *in situ* by hydrous pyrolysis/oxidation (HPO), which eliminates the need for costly surface treatment and disposal. Steam injection, combined with supplemental air, can create the conditions in which HPO occurs. Field testing of this process, conducted in the summer of 1997, indicated rapid destruction of polycyclic aromatic hydrocarbons (PAHs). Previous work established a suite of underground geophysical imaging techniques that provide sufficient knowledge of the physical changes in the subsurface during thermal treatment at sufficient frequencies to be used to monitor and guide the heating and extraction processes. In this field test, electrical resistance tomography (ERT) and temperature measurements provided the primary information regarding the temporal and spatial distribution of the heated zones.

Verifying the *in situ* chemical destruction posed new challenges. We developed field methods for sampling and analyzing hot water for contaminants, oxygen, intermediates, and products of reaction. Since the addition of air or oxygen to the contaminated region is a critical aspect of HPO, noble gas tracers were used to identify fluids from different sources. The combination of physical monitoring with noble gas identification of the native and injected fluids, and accurate fluid sampling, resulted in an excellent temporal and spatial evaluation of the subsurface processes. In turn, this evaluation made it possible to quantify the amount of *in situ* destruction occurring in the treated region. The experimental field results constrain the destruction rates throughout the site and enable site managers to make accurate estimates of total *in situ* destruction based on the recovered carbon. As of October 1998, more than 400,000 kg (900,000 lb) of contaminant have been removed from the site; about 18 percent has been destroyed *in situ*.

INTRODUCTION

A key issue in the application of *in situ* contaminant destruction methods is verifying that destruction has taken place. Methods involving the injection of fluids into contaminated soils have the inherent problem of proving that the process has actually reduced the contaminant, and that contaminant reduction did not occur merely by dilution or displacement of contaminated ground water. In a recent field test of combined thermal methods at the Visalia Pole Yard (Visalia, CA), we combined a suite of techniques to monitor the physical changes and progress of chemical destruction *in situ*. At Visalia, Southern California Edison Co. (Edison) is applying the Dynamic Underground Stripping (DUS) thermal remediation method to clean up a large (4.3 acre) site contaminated with pole-treating compounds. This is a full-scale cleanup, during which extraction of contaminants is augmented by combined steam/air injection to enhance the destruction of residual contaminants by hydrous pyrolysis/oxidation (HPO) (Cummings 1997). The site currently contains dense non-aqueous phase liquid (DNAPL) product composed of pole-treating chemicals (primarily creosote and pentachlorophenol) and an oil-based carrier. The chemicals are present in the alluvial soil and groundwater from near the surface, through the shallow aquifers (from 11–23 m [35–70 ft]), to the base of the intermediate aquifer at 32 m (105 ft) below ground surface (Geraghty and Miller 1992). The most important groundwater resource is found in the deep aquifer, at about 37 m (120 ft) below ground surface. The thermal remediation system was designed to remove contaminant from the intermediate and shallow aquifers without disturbing the deep aquifer.

DUS combines soil heating techniques (steam injection and/or electrical heating) and vacuum extraction to heat the soil, mobilize, and remove organic contaminants (Newmark *et al.* 1997, 1998). Laboratory results indicated that HPO can play a significant role in the final removal, or polishing, of residual contaminant from the soil (Knauss *et al.* 1997, 1998a-c; Leif *et al.* 1997), generally viewed to be the most difficult part of a cleanup. Where groundwater is oxygenated (that is, at the edges of the contaminated area, the up-gradient side in particular), hydrous pyrolysis will occur during the initial steaming of the site. After most waste product has been removed from the site, this process continues. Oxygen can be added to the injected steam to promote the hydrous pyrolysis reaction after collapse of the steam zone. The injected oxygen dissolves in water held as residual saturation during the steam injection pulse, and can readily diffuse from that water into the rest of the aqueous phase after collapse of the steam zone. Therefore, the supplemental oxygen can be readily delivered to the heated "reaction zone." Because of the difficulty of extracting pentachlorophenol by vapor extraction, its destruction by HPO was anticipated to be a significant factor in the final cleanup of the Visalia site.

Previous work (Newmark 1994; Ramirez 1995) established a suite of underground geophysical imaging techniques that provide sufficient knowledge of the physical changes in the subsurface during thermal treatment at sufficient frequencies to be used to monitor and guide the heating and extraction processes. Electrical resistance tomography (ERT) and temperature measurements provided the primary information about the temporal and spatial distribution of the heated zones. Verifying the chemical destruction *in situ*, along with integrating the various field measurements to determine how the process progresses *in situ*, posed new challenges.

Laboratory evaluation of HPO is easy in a closed system. Field application, however, constitutes an open environment in which native flowing and pumped groundwater are free to move through the test area. In addition, vacuum extraction enhances the supply of atmospheric gases through the vadose zone to interact with the ground water. The field tests were designed to track the physical and chemical changes, identify individual fluids, and to enable us to quantify the oxygen depletion, carbon dioxide (CO_2) production, and contaminant destruction.

FIELD TESTING *IN SITU* CHEMICAL DESTRUCTION

To evaluate the progress of the HPO chemical destruction *in situ*, we conducted two field experiments at the Visalia site. In both experiments, steam was injected into well S4, and physical and chemical monitoring was performed at wells MW36 (a plastic monitoring well) and S13 (a steel-cased extraction well), located 24 m and 28 m away, respectively (Figure 1). Steam movement was tracked by ERT and by monitoring thermocouples in nearby cone penetrometer-emplaced vertical electrode arrays. Fluids were regularly analyzed at well MW36 (using the system described below), and occasionally at well S13, where the down-hole pump delivered an unpressurized high-volume sample.

HPO is an aqueous-phase reaction, and it is essential to capture the fluid chemistry for evaluation. At elevated temperatures, many of the key constituents are sufficiently volatile that traditional sampling techniques are not suitable. We developed high-temperature packer and pump systems capable of delivering a pressurized, isolated fluid stream to the surface, where in-line analysis can be performed (Newmark *et al.* 1998). To measure key chemical constituents, an in-line analytical system was developed for both dissolved oxygen and organic analyses. In this system, a manifold maintains pressures above hydrostatic on the fluid stream until analyses are performed. For the gas chromatograph, this involved keeping pressure on the sample stream until a fixed volume sample was collected for input to a purge and trap collection sample handling system. An additional challenge was to protect the existing plastic monitoring wells from temperature-induced collapse. The packer systems were modified to circulate cooling fluids to both protect these wells and to permit sampling during steam injection. These systems have performed successfully to date, both in protecting the wells and in providing valuable fluid samples.

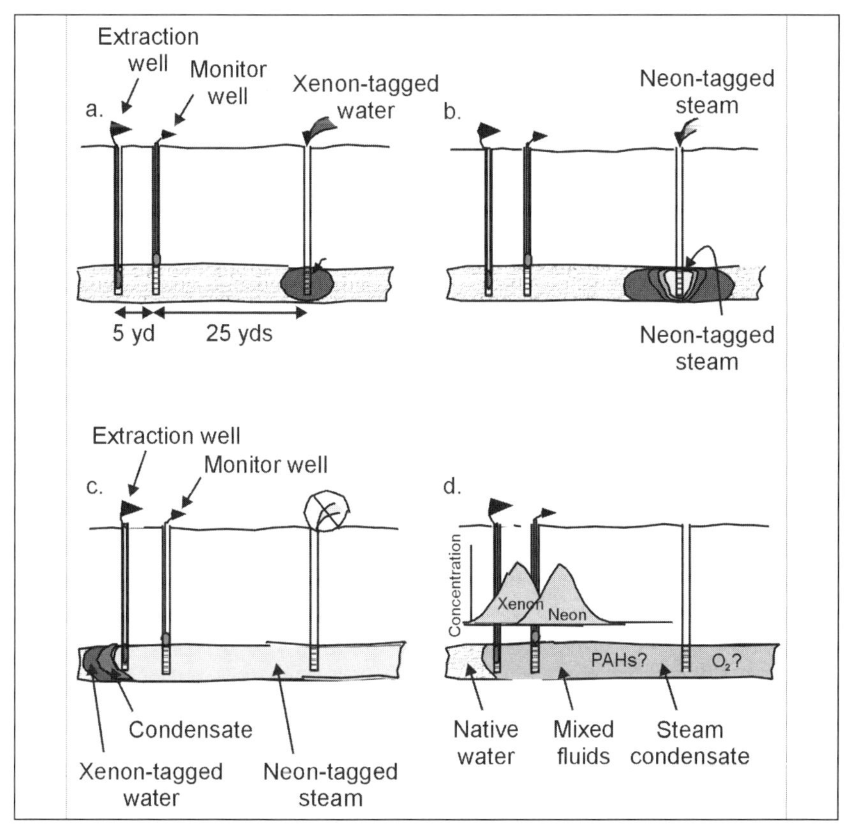

Figure 1. Conceptual design for the first tracer experiment. a) Xenon-tagged water is pumped into the injection well. b) Neon-tagged steam is injected. c) Steam passes the extraction well, injection ceases, and collapse is monitored. d) A reaction zone remains, in which heat and oxygen are available to degrade contaminants.

Since the addition of air or oxygen to the contaminated region is a critical aspect of HPO, noble gas tracers (Xe, Kr, Ne, and He) were used to track the movement of the steam (and subsequent condensation to liquid water) and other gases initially present in the steam (Hudson *et al.* 1998; Newmark *et al.* 1998). In most experiments in which the chemical contamination is observed to decrease, after some additive (such as nutrients or oxidative chemicals) is introduced into the subsurface, it is difficult to definitively attribute the decrease in contaminant concentration to a particular process. Introducing fluid into the subsurface can dilute and/or displace the native ground water, and contaminant destruction cannot be quantified without some means of identifying the fluids present. The tracers are utilized to identify those fluids directly derived from the injectate of interest, as opposed to native fluids or condensate from injection wells elsewhere in the field. Tracers permit us to follow the injected steam/water/oxygen pattern from a single injection well, to measure how well mixing occurs, how oxygen is consumed, how CO_2 is produced/transported, and how the intermediate HPO destruction products correlate with temperature and oxygen. Isotopic measurements made on samples of the extracted carbon forms were used to determine the mass of carbon derived from contaminant oxidation as opposed to (modern) atmospheric gas origins. The combination of accurate fluid sampling and tracking of the injected fluids allowed us to determine the amount of *in situ* destruction occurring in the treated region.

The subsurface conditions are complex and involve multiple phases and phase changes combined with mass and heat transport. Numerical simulations were run using a non-isothermal, unsaturated zone transport (NUFT) code developed at Lawrence Livermore National Lab (LLNL) (Nitao 1995), which includes individual gas properties in the multi-phase system. These simulations were used as design tools for the field experiments and as aids in data interpretation. Model calculations give tracer concentrations in time and space that are directly comparable to measurements. The experimental results showed remarkably close agreement with the simulations (Newmark *et al.*1998). The modeling predicted steam and tracer movement to within an hour or two in most instances. Modeling also effectively predicted the times of "thermal breakthrough," which occurs when sufficient heat has built up in the subsurface for vaporization of contaminants to begin, and "steam collapse," which is the opposite phenomenon, reflecting the condensation of the steam zone (Figure 2).

FIRST EXPERIMENT: MIXING AND DISPERSION

A key question in the design of *in situ* HPO systems is the degree to which the heat, oxygen, and contaminant can be mixed. Strong dispersive mixing of oxygen, contaminant, and hot water is a critical aspect in promoting HPO. Initial concerns focused on whether or not piston-like flow conditions would dominate, which would reduce the mixing zones crucial to the success of HPO as an *in situ* remediation technique. The first tracer experiment was designed to address hydrology issues of mixing and dispersion (Figure 1).

About 80,000 liters of xenon-tagged water was introduced to injection well S4, screened from 24 to 30 m. Water injection was followed by neon-tagged steam injection. The tracers were chosen based on their solubilities; high-solubility xenon will mimic the aqueous phase, where low-solubility neon will mimic the vapor phase. Sampling was conducted in wells MW36 and S13 until the steam front had passed beyond both monitoring wells. Thermocouple measurements made in the S13 extraction well clearly demonstrate the temperature rise as the steam front approaches and passes the monitoring point (Figure 2). Steam injection ceased, and the front was permitted to collapse back to the injection well. The timing of these events is given in Table 1. Fluid samples were evaluated for the presence of tracers, intermediates of reaction, contaminant concentration, and dissolved oxygen.

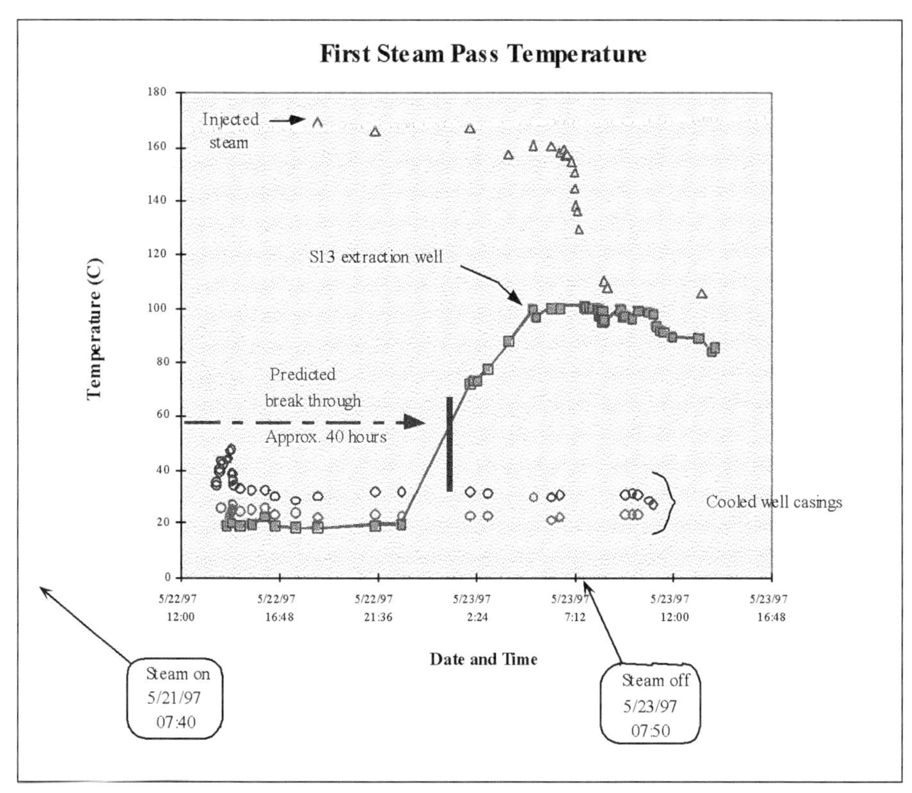

Figure 2. Temperature measurements made in the S13 extraction well show rapid rise on steam breakthrough, which occurred about 40 hours after injection began, as predicted by the NUFT models. Temperatures in the protected plastic monitoring well casings remained low.

TABLE 1 **Time history of the two tracer experiments.**

Date and Time	Event
First Tracer Experiment: Mixing and Dispersion	
5/19/97 15:55	Xenon tracer injection begun
5/20/97 13:30	Xenon tracer injection complete
5/21/97 8:00	Steam and neon injected into well S4
5/23/97 6:15	Well S13 exceeds 100° C (steam passes well)
5/23/97 6:40	Injection is halted and steam zone collapses
Second Tracer Experiment: Hydrous Pyrolysis/Oxidation	
6/17/97 8:00	Steam, krypton, and helium tracer injection begins in well S4
6/19/97 13:33	Well S13 exceeds 100° C (steam passes well)
6/20/97 10:37	Injection is halted and steam zone collapses

Large dispersive mixing of the tracers was observed (Figure 3A), and this observation is in agreement with calculations in the subsurface model when realistic random permeability field realizations are used. Calculations using uniform media underestimate dispersive mixing by more than a factor of ten, in clear contrast to field measurements. The ratio of tracer gas to natural air mixed into water was much greater than that predicted by the initial assumption in this model of no mixing between injected steam and native ground water. These results showed the process to be more efficient than envisioned and favored the success of HPO in the field (Figure 4). In short, the results of this first experiment yielded excellent results that favor the success of HPO in the field.

SECOND EXPERIMENT: HYDROUS PYROLYSIS/OXIDATION

The second tracer experiment focused on HPO. Compressed air was injected along with the steam to enhance HPO. Helium and krypton tracers were injected along with the steam and compressed air; the tracer solubilities mimic the aqueous and vapor phases in a similar manner to the xenon and neon used in the first experiment (Figure 3B). Once again, the gradual increase in tracer concentrations support

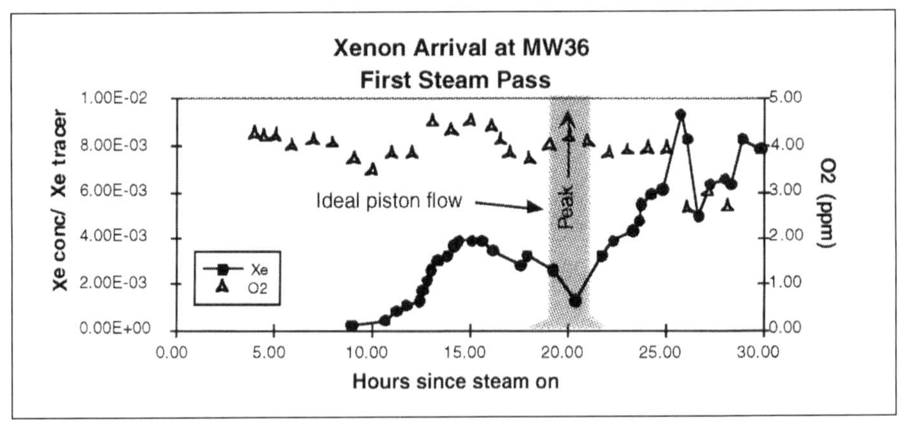

Figure 3A. Xenon arrival at MW36 during the first tracer experiment indicates extensive mixing and dispersion within the aquifer. Xenon concentrations rose gradually over a period of approximately 30 hours. The drop in concentration from 15 to 20 hours resulted from a drop in injection pressure; a mini-collapse of the steam zone caused the approaching steam front to temporarily recede. Upon restored injection pressure, the steam front approached, then passed the monitoring well. If pure piston-flow displacement occurred, xenon would be expected to arrive during an approximately 2-hour window, with peak concentrations many times those observed (shaded zone). Oxygen concentration begins to decrease after the first 24 hours.

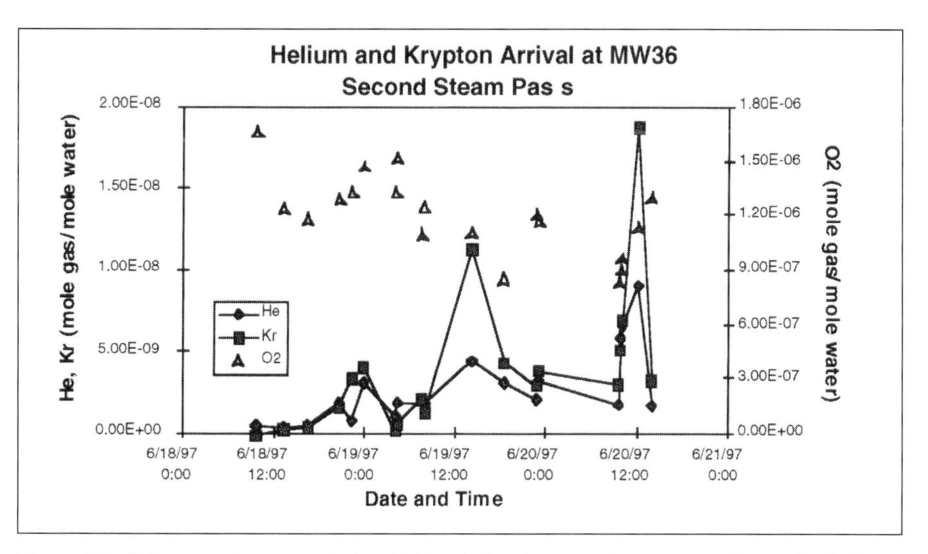

Figure 3B. Helium and krypton arrival at MW36 during the second tracer experiment. Gradual increase in concentrations support continued mixing and dispersion. Decreasing dissolved oxygen reflects the oxidation reactions consuming oxygen as it contacts contaminated fluids.

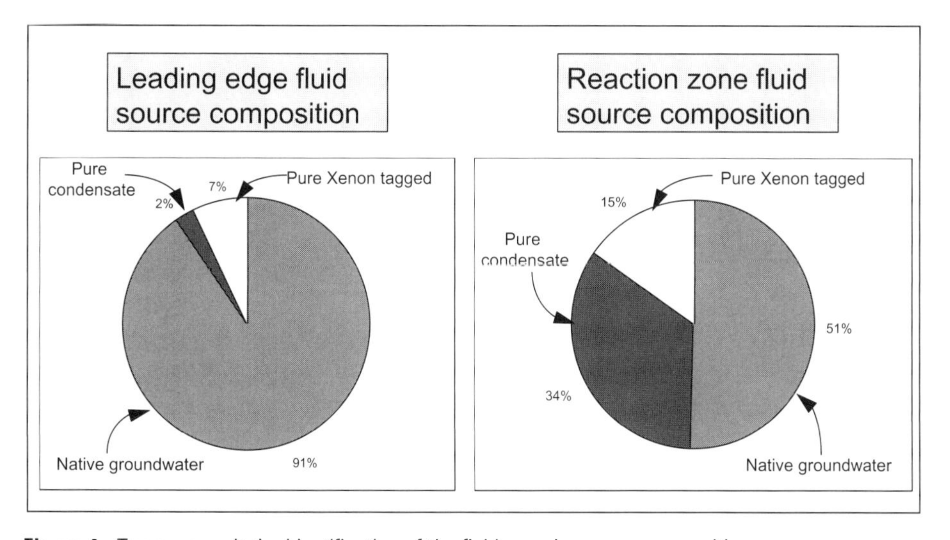

Figure 4. Tracers permit the identification of the fluid sample source composition.

significant mixing and dispersion, which is accompanied by a decrease in dissolved oxygen, all of which indicate that HPO is occurring along the leading edge of the steam front.

Evidence of the progress of HPO comes from a number of sources, including the disappearance of dissolved oxygen (consumed through the HPO reactions), the appearance of oxidized intermediates, the production of CO_2 (a final product of HPO), and the isotopic provenance of the carbon it contains (indicating the destruction of contaminant rather than other carbon sources). In the second experiment, steam injection was supplemented with air injection, which increased the dissolved oxygen available for HPO. Positive evidence of HPO was found on all counts (Newmark *et al.* 1998). A surprising finding was evidence of HPO occurring during and after the first hydrology-focused experiment, which did not include the supplemental air (and thus oxygen).

The chemical results are described in detail in Aines *et al.* (1998) and Leif *et al.* (1998). Dissolved oxygen decreased in fluids after steam collapse in both experiments, which indicated the consumption of oxygen during HPO. Oxidized intermediates were measured in fluid samples within a few days of steam collapse (oxidized intermediate concentrations increased dramatically in the extraction wells when steam injection ceased in the western wells in mid-July as well). Carbon dioxide, the final product of the HPO reaction, was measured in the vapor phase present in both the injection well and in S13. This steam-rich vapor consists of a steam-CO_2 mixture with small amounts of air. The extraction well also contained large amounts of oxidized intermediate products, consistent with laboratory results.

The combined results from the chemical and isotopic measurements of oxidation at Visalia indicate that HPO of creosote compounds in groundwater occurs readily at steam-derived temperatures of 100 to 110° C. This result is consistent with the observed laboratory rates of degradation. The heated Visalia groundwater system was observed to be saturated with both calcium carbonate (calcite) and dissolved air. Newly created inorganic carbon (CO_2 or bicarbonate) was determined by the change in isotopic composition of the groundwater, with measurements of the vadose zone gas and precipitated calcite supporting the groundwater analyses. During and directly after injection of steam into an oxygenated subsurface aquifer, ^{14}C and ^{13}C ratios were observed to change by approximately 20 percent toward the composition of the hydrocarbon contaminant at the site, indicating significant carbon oxidation. This change, along with the tracer determination of fluid source composition, requires the occurrence of an active exchange of fresh air in the aquifer to supply the added oxygen required to oxidize this amount of PAH-carbon (Aines *et al.* 1998).

RELATIONSHIP TO REMEDIATION RESULTS FOR THE ENTIRE SITE

The results of the field experiments were applied directly to the overall recovery in the Visalia field. After the initial heating phase (over 380,000 m^3 [500,000 yd^3] at a temperature of 100° C or above), Southern California Edison (Edison) adopted a "huff and puff" mode of operation, where the steam is injected for about a week, then injection ceases for about a week while extraction continues. Maximum contaminant removal is obtained during this steam-off period as the formation fluids flash to steam under the applied vacuum of about 1/4 atmosphere (at 2500 scfm [approximately 1 m^3/s]). In September, 1997, Edison began to add supplemental oxygen as compressed air along with steam injection; *in situ* HPO destruction rates increased from about 140 kg/day (300 lb/day) to about 360 kg/day (800 lb/day). Oxygen concentrations in the central extraction well were still well below 1 ppm, which indicated that HPO was still limited by oxygen availability.

By evaluating the carbon balance throughout the system, we quantified the extent of *in situ* oxidation of PAHs at Visalia (Aines *et al.* 1998). Two nearly independent methods were used. Isotopic analyses of Visalia effluent samples determined that a substantial amount of the extracted CO_2 derived from contaminant oxidation. By balancing the depletion of "modern" carbon, we were able to estimate that 10,100 kg of the extracted CO_2 was generated by oxidation of organic contaminants. By evaluating the carbon sources and sinks in the Visalia field, a geochemical carbon balance could be calculated, based on carbonate solubility alone, that indicated that 9,300 kg of the extracted CO_2 was generated by oxidation of organic contaminants. The two estimates agree to within 8 percent.

Using these constraints, we are able to evaluate the overall recovery achieved throughout the Visalia field during the first year of operations. As of October 1998, more than 400,000 kg (900,000 lb) of contaminant had been removed or destroyed. About 18 percent of the total amount has been destroyed *in situ* by using the HPO method. Free product recovery is continuing slowly. Contaminant concentrations are dropping in the extraction wells; the site is being cleaned from the periphery inward.

SUMMARY

The Visalia field tests confirmed *in situ* HPO destruction in soil and groundwater at rates similar to those observed in the laboratory under realistic field remediation conditions. HPO appears to work as fast as oxygen can be supplied, at rates similar to those measured in the laboratory. The predictive models used to design HPO steam injection systems have been validated by using conservative tracers to confirm mixing rates, oxygen consumption, CO_2 release, and effects of real-world heterogeneity. Accurate field measurements of the critical fluid parameters (destruction chemistry, oxygen content, steam front location) were demonstrated, using existing monitoring wells and portable data systems, with minimal capital cost.

The combination of physical monitoring with noble gas identification of the native and injected fluids and accurate fluid sampling resulted in an excellent temporal and spatial evaluation of the subsurface processes, from which the amount of *in situ* destruction occurring in the treated region could be quantified. The experimental field results constrain the destruction rates throughout the site and enable site management to make accurate estimates of total *in situ* destruction based on the recovered carbon. As of October 1998, more than 400,000 kg (900,000 lb) of contaminant have been removed from the site; about 18 percent of this amount has been destroyed *in situ*.

ACKNOWLEDGMENTS

We gratefully acknowledge the support of the U.S. Department of Energy's Office of Environmental Restoration and Office of Science and Technology, and LLNL's Laboratory-Directed Research and Development program. This work could not have been performed without the support of the Southern California Edison Co. and its Visalia Pole Yard employees, who cheerfully worked a 24-hour schedule in support of this project. SteamTech Environmental Services of Bakersfield, California, provided field support with steaming, temperature, and ERT data. We are indebted to Allen Elsholz for his tireless operation and maintenance of all the field measuring systems and for his work with Ben Johnson and George Metzger in designing and constructing the down-hole monitoring equipment. This work was performed under the auspices of the U.S. Department of Energy by the Lawrence Livermore National Laboratory under Contract W-7405-Eng-48.

REFERENCES

Aines, R.D., Leif, F., Knauss, K., Newmark, R.L., Chiarappa, M., Davison, M.L., Hudson, G.B., Weidner, R., and Eaker, C. "Tracking inorganic carbon compounds to quantify *in situ* oxidation of polycyclic aromatic hydrocarbons" during the *Visalia Pole Yard hydrous pyrolysis/oxidation field test*, (1998 [in prep]).

Cummings, M.A. "Visalia Steam Remediation Project: Case Study of an Integrated Approach to DNAPL Remediation." *Los Alamos National Laboratory Report* LA-UR-9704999 (1997): 9.

Geraghty and Miller, Inc. "Remedial Investigation/Feasibility Study," Southern California Edison, Visalia Pole Yard, Visalia, CA, *Southern California Edison Co.*, (1992).

Hudson, G.B., Davisson, M. L., Carrigan, C. R., Nitao, J., Aines, R. D., Newmark, R. L. and Eaker, C. "Tracing and modeling of subsurface thermal treatment," *Proceedings, First International Conference on Remediation of Chlorinated and Recalcitrant Compounds,* Monterey, CA (1998).

Knauss, K.G., Aines, R.D., Dibley, M.J., Leif, R.N., and Mew, D.A. "Hydrous Pyrolysis/Oxidation: In-Ground Thermal Destruction of Organic Contaminants." *Lawrence Livermore National Laboratory Report* UCRL-JC 126636 (1997).

Knauss, K.G., Dibley, M.J., Leif, R.N., Mew, D.A., and Aines, R.D. "Aqueous Oxidation of Trichloroethene (TCE): A Kinetic and Thermodynamic Analysis," in *Physical, Chemical and Thermal Technologies, Remediation of Chlorinated and Recalcitrant Compunds, Proceeding of the First International Conference on Remediation of Chlorinated and Recalcitrant Compounds*; Wickramanayake, G.B., and Hinchee, R.E. (Eds.), Battelle Press, Columbus, OH, (1998a):359-364. Also available as *Lawrence Livermore National Laboratory Report* UCRL-JC-129932 (1998a).

Knauss, K.G., Dibley, M.J., Leif, R.N., Mew, D.A., and Aines, R.D. "Aqueous Oxidation of Trichloroethene (TCE): A Kinetic analysis." Accepted for Publication, *Applied Geochemistry* (1998b).

Knauss, K.G., Dibley, M.J., Leif, R.N., Mew, D.A., and Aines, R.D. "Aqueous Oxidation of Trichloroethene (TCE) and Tetrachloroethene (PCE) as a Function of Temperature and Calculated Thermodynanmic Quantities," submitted to *Applied Geochemistry* (1998c).

Leif, R.N., Aines, R.D., and Knauss, K.G. "Hydrous Pyrolysis of Pole Treating Chemicals: A) Initial Measurment of Hydrous Pyrolysis Rates for Napthalene and Pentachlorophenol; B) Solubility of Flourene at Temperatures Up To 150°C," *Lawrence Livermore National Laboratory Report,* UCRL-CR-129938 (1997).

Leif, R.N., Chiarrappa, M., Aines, R.D., Newmark, R.L., and Knauss, K.G. "*In situ* Hydrothermal Oxidative Destruction of DNAPLS in a Creosote Contaminated Site," in P*hysical, Chemical and Thermal Technologies, Remediation of Chlorinated and Recalcitrant Compunds, Proceeding of the First International Conference on Remediation of Chlorinated and Recalcitrant Compounds*; Wickramanayake, G.B., and Hinchee, R.E. (Eds.), Battelle Press, Columbus, OH, (1998): 133-138. Also available as *Lawrence Livermore National Laboratory Report,* UCRL-JC-129933, (1998).

Newmark, R.L., and Aines, R.D. "Dumping Pump and Treat: Rapid Cleanups Using Thermal Technology," *Lawrence Livermore National Laboratory Report,* UCRL-JC 126637 (1997).

Newmark, R.L., Aines, R.D, Knauss, K.G., Leif, R., Chiarappa, M., Hudson, B., Carrigan, C., Tompson, A., Richards, J., Eaker, C., Weidner, R., and Sciarotta T. "*In situ* Destruction of Contaminants via Hydrous Pyrolysis/Oxidation: Visalia Field Test," Lawrence Livermore National Laboratory Report (1998 [in prep]).

Newmark, R.L., Boyd, S., Daily, W., Goldman, R., Hunter, R., Kayes, D., Kenneally, K., Ramirez, A., Udell, K., and Wilt, M. "Using geophysical techniques to control *in situ* thermal remediation," *Proceedings, Symposium on the Application of Geophysics to Engineering and Environmental Problems* (SAGEEP), Boston, MA (1994): 195-211.

Nitao, J.J. "Reference Manual for the NUFT Flow and Transport Code, Version 1.0," *Lawrence Livermore National Laboratoy Report*, UCRL-JC-113520 (1995).

Ramirez, A.L., Daily, W.D., and Newmark, R.L. "Electrical resistance tomography for steam injection monitoring and process control," *Journal of Environmental and Engineering Geophysics*, 0(1) (1995): 39-52.

THE VADOSE ZONE MONITORING SYSTEM FOR THE CAMU PROJECT AT SANDIA NATIONAL LABORATORIES

Jim Studer, P.E., P.G., *Duke Engineering and Services, Inc. Albuquerque, New Mexico*

Sandia National Laboratories (SNL) in Albuquerque, New Mexico, is a federal facility undergoing environmental restoration of multiple past waste disposal sites. Present plans call for using a temporary unit (TU) and corrective action management unit (CAMU) as part of a strategy to cost-effectively manage the hazardous wastes expected to be generated by numerous voluntary and final corrective measures. The TU has been granted a permit for 2 years of temporary storage, and construction has been completed. In September 1997, the United States Environmental Protection Agency (U. S. EPA), Region VI, approved a permit modification request for addition of the CAMU to SNL operations. This permit is the first granted for a CAMU in the Department of Energy and in New Mexico. The CAMU includes staging, storage, and treatment areas and a final containment cell. A vadose zone monitoring system is associated with the containment cell. Given the arid nature of the site, a groundwater monitoring waiver request was approved as part of the permit modification. The first phase of CAMU construction has been completed, and the facility is ready to accept remediation wastes.

The dimensions of the containment cell are 200 by 300 feet; groundwater is located 500 feet below the surface. The concept behind the CAMU vadose zone monitoring system was based on the premise that more accurate and more timely monitoring of the containment cell for leakage would be possible through vadose zone monitoring and that groundwater monitoring would not be necessary. The system was designed to operate over the CAMU operational period and the subsequent 30-year post-closure monitoring period. As shown in Figure 1, the vadose zone monitoring system consists of three primary subsystems: primary sub-liner, vertical sensor array, and chemical waste landfill and sewer.

The primary sub-liner subsystem was designed to detect primary leaks in the near field directly beneath the geosynthetic clay liner. There are several functional aspects of this system: the principal feature is a network of five horizontal access tubes for neutron probe moisture logging. The access tubes run the length of the disposal cell and are nominally 25 feet apart. As much as possible, the access tubes coincide with vulnerable leakage areas such as drainage trenches and the leachate collection sump. In lieu of suction lysimeters for sampling soil pore water, the horizontal section of the horizontal access tubes are constructed of moisture permeable, extra-strength vitrified clay pipe. If the neutron probe detects excess moisture, the access system is designed to allow the deployment of a SEAMIST™ membrane equipped with absorbent pads. The absorbent pads collect a small moisture sample that can be analyzed for contaminant identification.

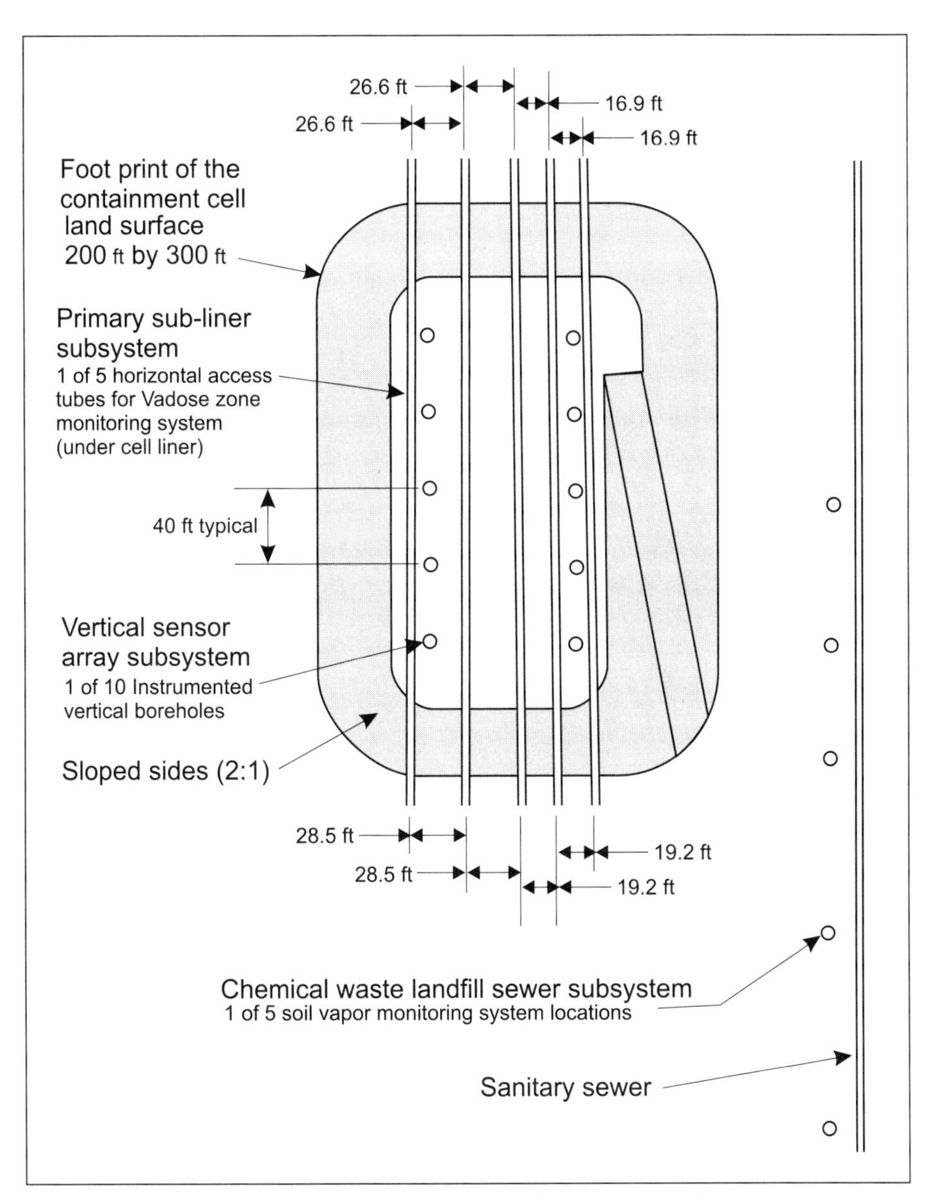

Figure 1. The vadose zone monitoring system consists of three primary subsystems: 1) primary sub-liner, 2) vertical sensor array, and 3) chemical waste landfill and sewer. The primary sub-liner subsystem detects primary leaks in the field directly beneath the geosynthetic clay liner. The vertical sensor array and chemical waste landfill and sewer sub-systems detect secondary leaks.

The vertical sensor array subsystem detects secondary leaks and involves integrated moisture measurement (time domain reflectrometry (TDR) system), temperature measurement (thermistors), and vapor sampling (active soil gas sampling tubes) equipment. Two TDR sensors are installed in each of 10 bore holes. The bottom TDR sensor is located 15 feet below the floor of the disposal cell and was pushed into undisturbed soil. The upper TDR sensor was installed 5 feet below the cell floor into a bore hole backfill material consisting of 90 to 95 percent native soil and 5 to 10 percent bentonite clay. Each TDR probe was part of an assembly that included a thermistor temperature probe and a soil gas sampling tube. Five vertical sensor arrays were spaced 40 feet apart along each of the two outer primary subliner horizontal access tubes.

The chemical waste landfill and sewer sub-system also detects secondary leaks and consists of five 20-foot-long vertical access tubes for neutron moisture logging that are screened for 2 feet at the bottom to allow for soil gas sampling. These vadose zone wells are located in a line between the CAMU and a nearby sewer line and chemical waste landfill. The integrated design allows for assessment of other factors that could lead to false-positive detection, and this capability significantly enhances the value of the system's overall leak detection capability. Examples of factors that could lead to false-positives include condensation buildup beneath the cell liner, moisture migration due to environmental gradients, the nearby sanitary sewer, and the nearby chemical waste landfill that has generated a large VOC vapor plume.

The monitoring frequency of the three sub-systems depends on the phase of operation at the CAMU. All systems will be monitored monthly for 12 months before waste emplacement to establish a base line. Monthly monitoring will continue through the active operation of the CAMU, during which waste is being placed, and will continue until closure is complete. Closure will be considered complete 1 year after the cover has been constructed, or until stable readings have been obtained, whichever is longer. Then, during post-closure, which will last a maximum of 30 years, monitoring of all three systems will be conducted quarterly for the first 3 years; after the first 3 years, monitoring frequency will be open to negotiation.

The groundwater monitoring waiver request was based on the projected effectiveness of the vadose zone monitoring system in detecting future possible leaks and in responding to these leaks. The U. S. EPA agreed with the technical superiority of the approach put forth in the permit modification request and accepted the waiver request. Therefore, no groundwater monitoring wells will be required for the CAMU. This will have significant implications for not only the CAMU itself, but also for other waste management facilities within U. S. EPA Region VI and, perhaps, elsewhere. Of interest is the fact that the waiver was granted despite the fact that the final cover system for the containment cell is an innovative, but, as yet, unproven, alternative Resource Conservation and Recovery Act (RCRA) design.

Design of this innovative final cover was based on alternative landfill capping technology concepts for arid land environments. This arid-cap design places more emphasis on use of natural earth materials for native vegetation support, capillary pressure enhancement, and evapo-transpiration, as opposed to the more traditional RCRA-style caps that utilize multiple layers of geosynthetic materials and compacted clay. The arid-cap design is more robust in terms of long-term effectiveness than the RCRA-style cap technology.

The CAMU cell vadose zone monitoring system and final cover designs, led by Duke Engineering and Services Inc. (formally INTERA Inc.), are considered to represent the leading edge for the waste management industry. As a result of these designs, Sandia National Laboratories will realize: lower construction costs; lower operation, maintenance, and monitoring costs; and, enhanced performance and reliability compared to the more traditional alternatives of a groundwater monitoring system and a RCRA-style cap.

REFERENCE

U.S. Department of Energy (DOE). "Class III Permit Modification Request for the management of Hazardous Remediation Waste in the Corrective Action Management Unit, Technical Area III, Sandia National Laboratories/New Mexico Environmental Restoration Project," Final Class III Permit Modification Request to Module IV of the RCRA Hazardous Waste Facility Operating Permit Number NM5890110518-1, September 1997, DOE, Albuquerque Operations Office, Albuquerque, NM (1997).

IN SITU FIELD SCREENING OF VOLATILE ORGANIC COMPOUNDS USING A PORTABLE ACOUSTIC WAVE SENSOR SYSTEM

Gregory C. Frye-Mason, *Sandia National Laboratories, Albuquerque, New Mexico*

ABSTRACT

Portable acoustic wave sensor (PAWS) systems have been developed for real-time, online, and *in situ* monitoring of volatile organic compounds (VOCs). These systems utilize the high sensitivity of surface acoustic wave (SAW) devices to changes in the mass or other physical properties of a film cast onto the device surface. Using thin polymer films that rapidly (within a few seconds) and reversibly absorb the chemical of interest, these sensors can be used to detect and monitor a wide range of VOCs. Current minimum detection levels for typical VOCs in a real-time mode range from about 1 to 10 ppm. Sensor responses are reproducible and yield accurate measurements, and the devices operate over a wide concentration range. Aboveground and down-hole systems have been demonstrated at environmental restoration sites for *in situ* monitoring of contaminants in vadose zone monitoring wells and have been confirmed by real-time analysis of gas samples pulled to the surface from a cone penetrometer probe.

INTRODUCTION

A significant number of government and industrial sites have been previously contaminated with large amounts of volatile organic compounds, including chlorinated hydrocarbons (CHCs) (Riley 1993). One key need for these sites is an effective characterization of the distribution of contamination, along with continual monitoring of down-hole concentration changes during the remediation. This characterization and monitoring can best be accomplished by *in situ* analysis of contaminant concentrations using sensitive, accurate, and robust sensors that will not be overloaded by the high concentrations (for example, tens of thousands of ppm) that may be present. Since a common and effective technique for removing volatile organics from soils is to extract them using large vacuum pumps and wells in the contaminated site, another need is to be able to provide real-time, online monitoring of contaminant concentrations in off-gas streams from these soil vapor extraction systems. In some cases, onsite treatment systems are used to destroy the VOCs prior to emission to the environment. For remediating sites contaminated with CHCs, these treatment systems generally produce corrosive hydrochloric acid vapors, which in turn require monitoring systems to provide online analysis of low residual VOC concentrations in corrosive streams.

In addition to site restoration, monitoring, and characterization, new environmental regulations imposed over the past decade, including the new Clean Air Act and the Montreal Protocol, have had a significant impact on industrial processes involving volatile organics. Many of these regulations require chemical monitoring to document compliance in numerous industrial settings.

Currently, most chemical analyses at industrial and restoration sites are performed by taking grab samples and sending them off for laboratory analysis. This process provides only periodic chemical information and often requires highly trained operators using sophisticated and expensive equipment. Online monitoring systems tailored to the chemical environment of interest are an attractive alternative because they offer real-time and continuous information using a relatively inexpensive system with automatic operation.

There are several requirements for useful real-time, online monitors. The monitor's detection method must be sufficiently sensitive to detect the chemical species of interest at relevant concentrations, and have a rapid enough response time for timely chemical information. The system must be insensitive to changes in temperature, pressure, or humidity, and must be easy to integrate into the industrial or site remediation process with simple and automatic operation. Some discrimination among chemical species is desired, but not at the cost, complexity, or time required for spectroscopic or chromatographic systems. Ideally, the monitor of choice will be inexpensive and small, with high reliability and durability.

Monitoring systems developed around chemical sensors have the potential to meet these requirements. A chemical sensor generally consists of a coating material that either absorbs or adsorbs the chemical species of interest, and a transduction mechanism for detecting this absorption/adsorption (Wohltjen 1984; Grate, *et al.* 1993). This paper focuses on describing a chemical sensor system based on a surface acoustic wave device coated with a viscoelastic polymer that absorbs a wide variety of volatile organics. The discussion covers the design, characterization, and field testing of prototype sensor systems which use a single active SAW sensor for real-time, online, and *in situ* monitoring of VOCs.

SURFACE ACOUSTIC WAVE SENSORS

A schematic diagram of a SAW sensor is shown in Figure 1. A quartz substrate containing two interdigitated transducers is coated with an absorptive polymer film. Application of an alternating voltage to the input transducer generates a strain field in the underlying quartz because of its piezoelectric properties. This strain field launches a surface acoustic wave that travels along the substrate and interacts with the polymer overlayer before being converted back into an electrical signal by the output transducer. The wave/film interaction results in a perturbation of the wave propagation properties, specifically the wave velocity, v, and the wave attenuation,

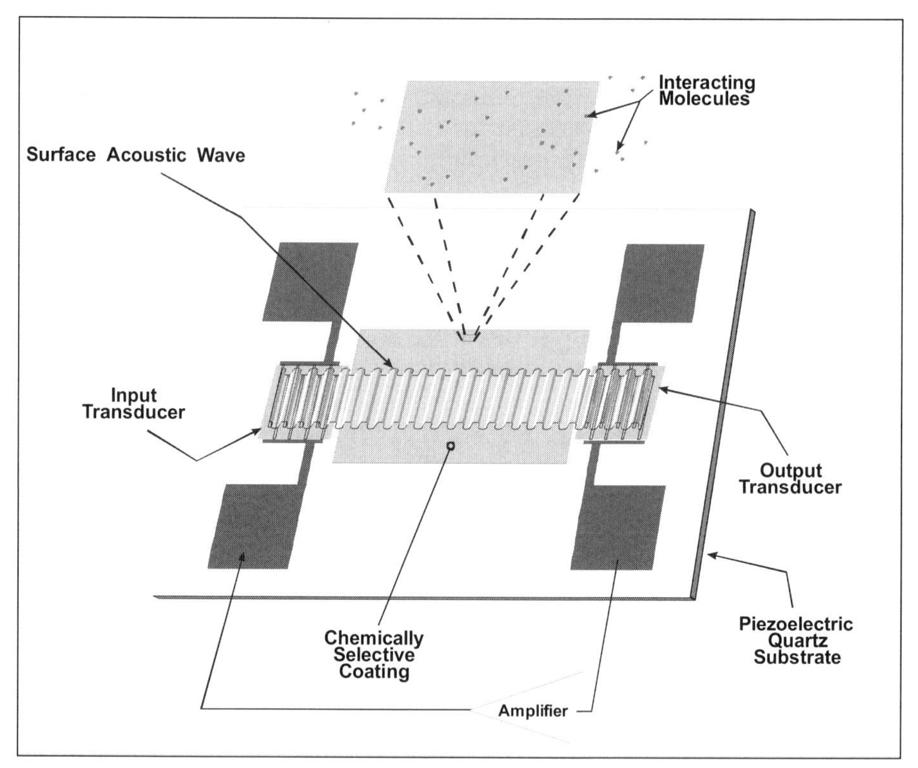

Figure 1. Schematic diagram of SAW sensor showing uptake of VOC molecules in the polymer coating (Wohltjen 1984, Grate *et al.* 1993).

α (that is, the rate of wave diminution with distance). This sensitivity to thin film properties makes SAW devices ideally suited for chemical sensors to monitor gas and vapor species. Using viscoelastic polymer films, studies show that by simultaneously monitoring both velocity and attenuation from a single device, it is possible to identity an isolated molecular species and to determine its concentration as well (Frye and Martin 1991, 1993).

In practice, these two responses are measured by using the SAW device as the feedback element of an oscillator circuit. In such a configuration, relative frequency changes ($\Delta f/f$ where f is frequency) are proportional to relative velocity changes ($\Delta v/v$), which means that a frequency counter can be used to monitor velocity (Wohltjen 1984). Wave velocity can be resolved to a precision of approximately 0.01 ppm (1 Hz noise out of 97 MHz operating frequency). Insertion loss is evaluated by measuring the power level after the SAW device using a radio frequency (RF) detector to convert RF power to a DC voltage. Noise levels as low as 0.0004 dB have been measured using this technique (Frye *et al.* 1992). When using elastomeric polymers (such as polyisobutylene) as the absorptive sensor layers, the rapid diffusional properties result in rapid (few seconds) and reversible sensor responses, enabling online continuous monitoring of concentration. The reversibility of the absorption makes it possible to evaluate the sensor baseline online by periodically removing the chemical from the gas stream using a stream of chemical-free ambient air.

PORTABLE ACOUSTIC WAVE SENSOR SYSTEMS

Basic PAWS System Design

The basic PAWS system consists of two SAW devices contained in a common test case and configured with appropriate oscillator circuitry, data acquisition and control electronics, and environmental sampling. One sensor is the active sensor and the other is a reference. Each SAW device is part of a separate oscillator with the sensing device exposed to the chemical while the reference is isolated. The resultant frequencies from the two oscillators are mixed to obtain a difference frequency. This signal is easier to measure and is partially compensated for temperature-induced baseline drift. During calibration, the sensor device is challenged with known concentrations of the chemical of interest while both wave velocity and attenuation are measured. Either response can then be used to determine chemical concentration, while the ratio of responses can be used to verify the identity of an isolated chemical species (Frye and Martin 1991, 1992).

Elements of the RF oscillator electronics are high-gain, low-noise amplifiers, a low-pass filter, impedance matching for the SAW transducers, and couplers to remove RF signals for frequency and power measurements. Since the viscoelastic properties used to obtain the two independent device responses change with temperature, some thermal control is required to overcome the resulting temperature sensitivity. Overall thermal stability is provided by an electronic feedback system to control absolute and relative temperature fluctuations in the test case. In addition, frequency mixing is used to remove (to first order) temperature-dependent response in the two similar devices. All components of the oscillator electronics, the thermal system, and the test case itself are integrated onto two PC-boards.

The standard PAWS environmental sampling system consists of a pump to move the gas across the sensor, a Nafion membrane tube, and a three-way valve. The Nafion membrane tube is configured so that sample gas goes through the inside while ambient air is passed in countercurrent flow on the outside (the Nafion membrane tube is contained in a larger diameter Teflon tube). The selective permeation of water over VOCs through the Nafion membrane allows this system to bring the sample gas to ambient humidity without altering contaminant concentration. This prevents condensation at high humidity and minimizes errors introduced by the small but significant humidity sensitivity of SAW sensors (Frye and Pepper 1999). Since SAW devices are sensitive to the mass of chemical adsorbed on the device surface, they will detect adsorption of water vapor—a phenomenon that occurs on even the most hydrophobic surfaces. The solenoid-controlled three-way valve is used to periodically direct ambient air to the sensor to purge the sensor of the contaminant and to reestablish sensor baseline (that is, sensor response at zero concentration). System stability (drift rate) dictates how often sensor baseline must be reevaluated.

Above-Ground and Down-Hole PAWS Prototypes

For above-ground monitoring, the SAW sensors, oscillator circuitry, data acquisition and control electronics, sampling hardware, and power conditioners are housed in a single modular package with a total volume of approximately 0.4 ft^3. Communication to a notebook computer uses an RS-232 serial interface since data requirements are small. The power conditioning enables any 12 V power source, including a battery for sites where AC power is not available, to be used to power the PAWS module. Gas inlet/outlet ports with 1/8-inch Swagelok® connections are provided on the front of the module.

Vadose zone monitoring can be implemented in one of two ways: (1) packaging the sensor, electronics, and gas handling hardware and inserting them into the monitoring well at a point above the groundwater for *in situ* analysis or (2) extracting a gas sample from the well through a long tube and delivering it to an above-ground sensor package. The down-hole PAWS probe was designed for the former; however, this *in situ* system is configured to also push or pull a sample to the surface for comparison with above-ground analysis. The PAWS down-hole system is illustrated in Figure 2. The probe, constructed to fit into 4-inch diameter and larger wells, houses the same components as the above-ground PAWS module. RF oscillator electronics and data acquisition/control electronics are placed in separate, shielded compartments within the probe to reduce interference during measurements. Environmental sampling and power conditioning hardware are contained in a third isolated section. The sampling system sends a portion of the sample gas to the surface using a 1/8-inch Teflon or stainless steel tube. The probe superstructure consists of a 2.5-foot-long anodized aluminum cylinder. An anodized aluminum cover protects the electronic components from the surrounding down-hole environment. The well environment is accessed through a port at the top of the probe that contains a frit to prevent particulate contamination inside the chamber.

Packers are attached above and below the probe and sealed against the well by inflating to 25 to 50 psi through a gas line from a surface 12 V pump. The bottom packer can be separated from the main section using a long gas line and tether to enable longer screened sections to be isolated. This surface gas line is also used with an in-line carbon-based scrubber to provide chemical-free air for the Nafion countercurrent flow and for the gas to establish sensor baseline. The gas handling equipment also contains a purge system consisting of a fritted port above the

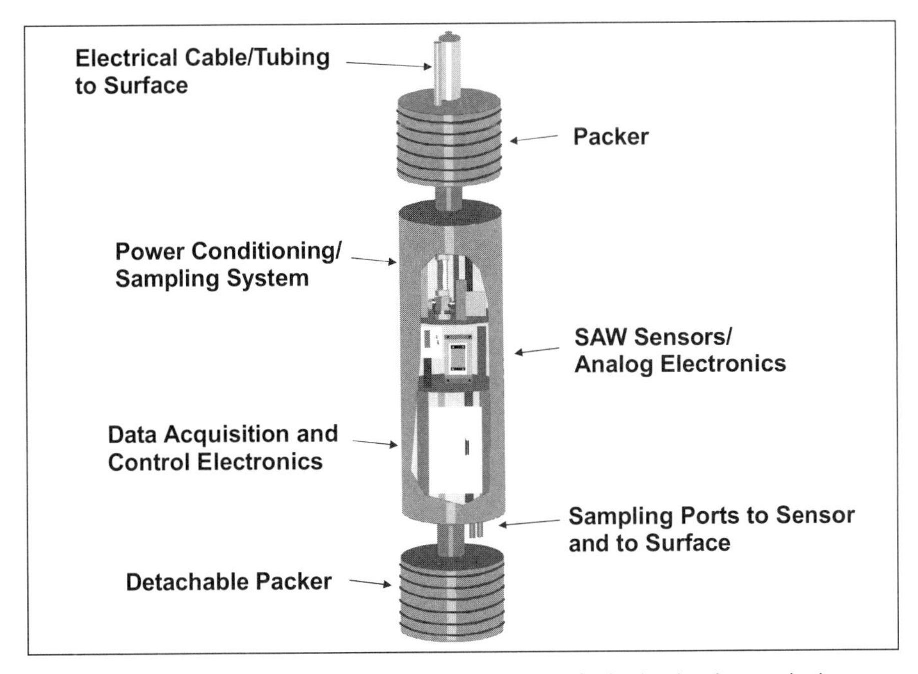

Figure 2. Schematic diagram of the PAWS down-hole probe for *in situ* site characterization.

bottom packer, a one-way valve, and a port that exits the probe above the top packer. A pump can be placed down hole above the top packer, or a 3/8-inch Teflon line to the surface can be used with a pump at the surface. Because of condensation problems with the down-hole pump, the surface configuration was found to be more reliable. Purge pumping rates of 20 l/min or more can be achieved. The probe is lowered into the well using a portable tripod, a winch, and a braided cable. Connections to the computing system and power supply at the surface use a cable containing four twisted shielded pairs. The serial communication between the probe and the up-hole computer requires use of the RS-422 interface as long line lengths may be necessary.

Down-Hole Probe Evaluations for Site Characterization

Three demonstrations of PAWS down-hole probes have been performed to date. The first used a fully integrated probe that only worked in 4-inch-diameter wells (Cernosek *et al.* 1993), while the second two used a smaller, lighter-weight probe that had changeable packers for use in 4-, 6-, and 8-inch-diameter wells. In the latter two tests, gas samples were pumped to the surface using a long sample line, and comparison analysis was performed using an above-ground PAWS system and a Brüel & Kjaer Model 1302 Photoacoustic IR system (B&K). Results from these two demonstrations are shown in Figures 3 and 4. In Figure 3, analysis from a very high concentration well is shown. Agreement between the above-ground PAWS and the commercial B&K instrument was within 2 percent from 100 ppm to over 20,000 ppm, a fact that proved the accuracy of the PAWS system.

However, differences were observed between the above-ground and down-hole systems. At high concentrations, the down-hole readings were higher (by about 20 percent), while at low concentrations in subsequent wells, the above-ground measurements were higher. These differences result from perturbations in the sample due to the Teflon sampling line adsorbing and desorbing VOCs. The data in the inset figure were taken prior to turning on the sample pump to the well; both the above-ground PAWS system and the B&K were seeing about 100 ppm of CCl_4 added to the ambient air being pulled in through the sampling line, which had previously been exposed to high CCl_4 concentrations. These additions and losses of VOC were not seen in the following test where stainless steel tubing was used for the sample line. This tubing, however, is expensive and difficult to deploy into the well.

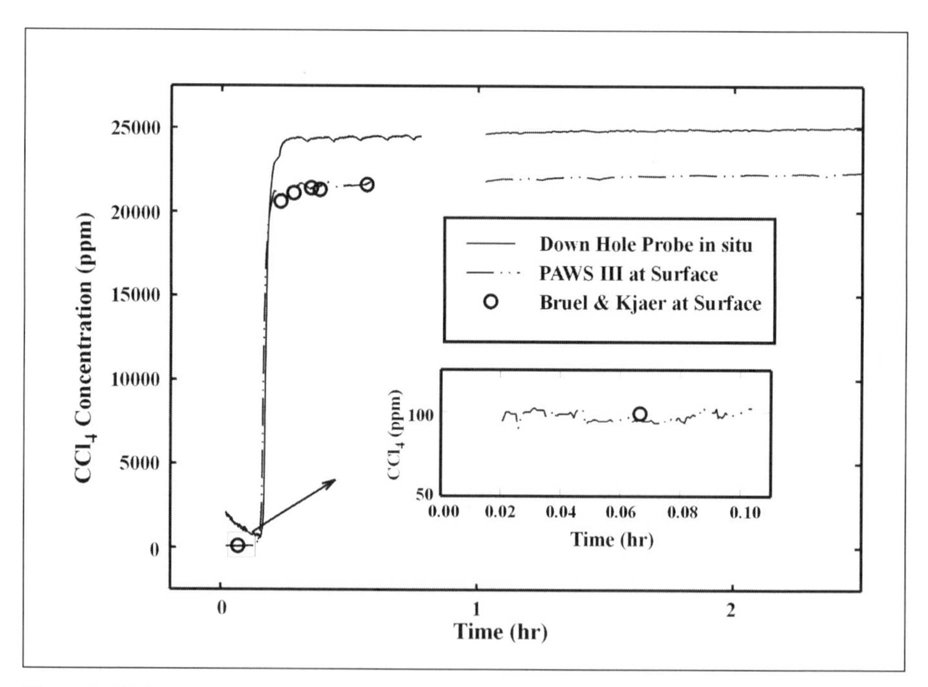

Figure 3. High concentration vadose zone well at the DOE's Hanford site: *In situ* PAWS analysis with above-ground comparison data from a separate PAWS system and a commercial IR system for CCl_4.

Figure 4 shows data from a lower concentration well. These measured trends are useful for showing the ability of the PAWS sensors to monitor low concentrations in a real-time mode. No detectable CCl_4 was observed by any of the instruments prior to turning on the purge pump (pump turned on 8 minutes before the zero point in Figure 4). With the purge pump on, a slow rise in concentration was seen during the test. This rise was easily tracked by the PAWS systems, even though it is small compared with the system minimum detection level for CCl_4, approximately 10 ppm. Again, agreement was high among the down-hole probe, the surface PAWS system, and the Brüel & Kjaer instrument.

The ability to take real-time data during purging of a well can be useful for determining when a steady-state concentration has been reached. Tests of this type demonstrated that large purge volumes may be required for the relatively porous soils typical at the Hanford carbon tetrachloride site. For example, during a period of high atmospheric pressure, no CCl_4 was observed even after purging 150 well volumes from a well where previously more than 1,000 ppm had been observed during a low atmospheric pressure period. The ability to place the PAWS probe in a well and leave it for long periods should allow the unperturbed concentration (that is,

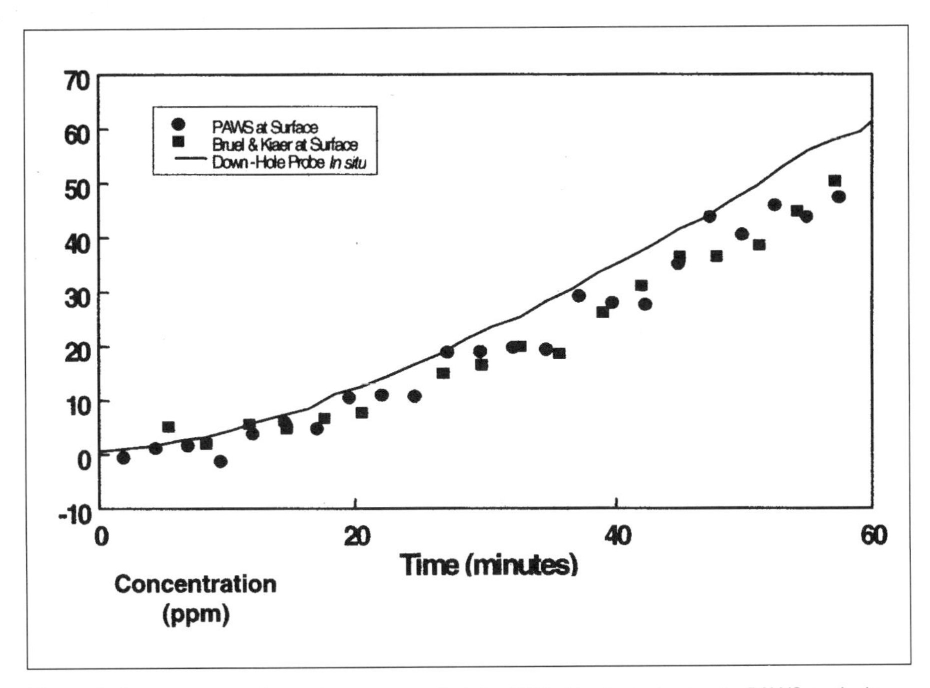

Figure 4. Low concentration vadose zone well at the DOE's Hanford site: *In situ* PAWS analysis with above-ground comparison data from a separate PAWS system and a commercial IR system for CCl_4.

where the base sampling rate is low: <100 ml/min) to be determined and tracked as a function of time.

CONCLUSIONS

PAWS systems have been designed, developed, and field-demonstrated for real-time, online, and *in situ* monitoring of VOCs at remediation sites. Systems have also been demonstrated for industrial process and emissions monitoring (Frye *et al.* 1992; Cernosek *et al.* 1993). PAWS systems are portable and easy to implement. For example, system set up for above-ground use takes about 10 minutes. The system has been successfully used by an untrained operator with fewer than three pages of instructions. Demonstrations indicate sensitive and accurate analysis of isolated VOCs and agreement with baseline instruments typically within 10 percent. Minimum detection levels for typical VOCs are about 1 to10 ppm. The systems have a very wide dynamic range and can handle concentrations to near saturation (25,000 ppm observed in the field and over 100,000 ppm in laboratory tests). The calibrations are stable; for example, in the last field demonstration, calibrations before and after the test were within 1.5 percent, even following the large amount of handling experienced during the field demonstration. The systems are designed to enable online baseline analysis so they can be used continuously for long-term monitoring. Response is rapid and reversible, so concentration fluctuations can be monitored in real time on a few-second time scale. The systems can be configured for either above-ground monitoring or *in situ* analysis in vadose zone monitoring wells. Future systems are being designed to have built-in preconcentration for trace detection, accessory modules for water and soil analysis, and an ability to provide molecular identification and quantification of multiple VOCs in simple mixtures using arrays of SAW sensors.

ACKNOWLEDGMENTS

This work was funded in part by the U.S. Department of Energy (DOE) Office of Technology Development and was performed at Sandia National Laboratories, a multi-program laboratory operated by Sandia Corporation, a Lockheed Martin Company, for the DOE under contract number DE-AC04-94AL85000. The author thanks the Sandia PAWS team (Don Gilbert, Terry Steinfort, Chris Colburn, Richard Cernasek, and Steve Silva) and the following for their support in the PAWS field demonstrations: Joe Rossabi of the Savannah River Technology Center; John Fisler, Jon Fancher, Kim Koegler, Dave Blumenkranz, Bruce Cassem, Randy Coffman, Jeff Gale, Torry Webb, and Cary Martin of Westinghouse Hanford, and John Evans of Pacific Northwest Laboratory.

REFERENCES

Cernosek, R.W., Frye, G.C., and Gilbert, D.W. "Portable Acoustic Wave Sensor Systems for Real-Time Monitoring of Volatile Organics," in *Proc. Ideas in Science and Electronics,* ISE: Albuquerque, NM (1993): 44-51.

Frye, G.C., and Martin, S. J. U.S. patent 5,076,095, to U.S. Department of Energy (1991).

Frye, G.C., and Martin, S. J., "Utilization of Polymer Viscoelastic Properties to Enhance Chemical Sensor Performance," in *Proceed of Symposium on Chem. Sensors II*, Electrochemical Society: Pennington, NJ (1993): 51-58.

Frye, G.C., Martin, S.J., Cernosek, R.W., and Pfeifer, K.B. *International Journal of Environmentally Conscious Manufacturing,* 1 (1992): 37.

Frye, G.C., and Pepper, S. H. *AT-On Site, The Journal of On Site/Real-Time Analysis*, Info-Science Services, Inc.: Northbrook, IL (1994): 62-70.

Grate, J.W., Martin, S. J., and White, R. M. *Anal. Chem.,* 65 (1993): 987A-996A.

Riley, R.G. "Arid Site Characterization and Technology Assessment: Volatile Organic Compounds—Arid Integrated Demonstration," Report #PNL-8662 ,*Off. Sci. Tech. Info*., Oak Ridge, TN, (1993).

Wohltjen, H. *Sensors and Actuators,* 5 (1984): 307.